Figure		*Volume*
Rectangular solid		$V = lwh$
Cube		$V = s^3$
Right circular cylinder		$V = \pi r^2 h$
Right circular cone		$V = \frac{1}{3}\pi r^2 h$
Sphere		$V = \frac{4}{3}\pi r^3$

Beginning Algebra

with Applications

SECOND EDITION

Linda L. Exley
Vincent K Smith

DeKalb College
Decatur, Georgia

PRENTICE HALL, Englewood Cliffs, New Jersey 07632

Library of Congress Cataloging-in-Publication Data

Exley, Linda L.
    Beginning algebra: with applications / Linda L. Exley, Vincent K
Smith. -- Student ed., 2nd ed.
      p.   cm.
    Includes index.
    ISBN 0-13-067257-2
    1. Algebra.   I. Smith, Vincent K.   II. Title.
QA152.2.E95   1994
512′ .9--dc20   93-36451
                                        CIP

Dedicated to Our Children
Amanda and Chuck Exley
Martin and Rowland Smith

Editor-in-Chief: Tim Bozik
Development Editor: Bobbie Lewis
Editorial Assistant: Audra Walsh
Production Editor: Tom Aloisi
Copy Editor: Bill Thomas
Supplements Editor: Mary Hornby
Design Director: Florence Dara Silverman
Designer/Page Layout: Lee Goldstein
Cover Designer: William McCloskey
Photo Editor: Lori Morris-Nantz
Photo Research: Teri Stratford
Production Coordinator: Trudy Pisciotti

© 1994 by Prentice-Hall, Inc.
A Paramount Communications Company
Englewood Cliffs, New Jersey 07632

Printed in the United States of America
10 9 8 7 6 5 4 3 2 1

ISBN: 0-13-067257-2

Prentice-Hall International (UK) Limited, *London*
Prentice-Hall of Australia Pty. Limited, *Sydney*
Prentice-Hall Canada Inc., *Toronto*
Prentice-Hall Hispanoamericana, S. A., *Mexico*
Prentice-Hall of India Private Limited, *New Delhi*
Prentice-Hall of Japan, Inc., *Tokyo*
Simon & Schuster Asia Pts. Ltd., *Singapore*
Editora Prentice-Hall do Brazil, Ltda., *Rio de Janeiro*

Contents

Preface

This book is a beginning algebra textbook with a review of arithmetic and geometry. It can be used as a developmental text to prepare students for intermediate algebra or liberal arts mathematics.

GETTING STARTED

Chapter 0 reviews topics from arithmetic and geometry. The sections in this chapter may be covered in detail, treated as a review, or omitted entirely. If the chapter is omitted, the individual topics can be covered as required with the material of the later chapters. A short review of geometry is included in Chapter 0, and geometric problems are integrated throughout the book.

PEDAGOGY

This book is written in a language that students understand. Wordy statements and mathematical jargon are kept to a minimum, while methods and techniques are illustrated with worked examples whenever possible. Although simple language is used and rigor is not stressed, careful mathematics is.

The heart of the book is found in over 5500 carefully selected exercises that develop and illustrate the ideas of algebra in a modern setting. Extensive problem sets occur at the end of each section. Each set starts with a section titled "Warm-Ups." Warm-Ups are a collection of problems graded in difficulty from easy to medium, which are keyed to worked examples in the text. Warm-Ups are followed by a set of exercises called "Practice Exercises." This longer set of problems contains the drill and practice exercises so necessary for students to develop their manipulative algebra skills. The Practice Exercises are also graded in

difficulty, but are mixed and not keyed to worked examples. The problem sets also have a few "Challenge Problems" that allow the student to probe into the natural extensions of the ideas presented in the text. The problem sets contain sufficient numbers of problems to provide the instructor considerable flexibility. The pedagogy is never dictated by a limited selection of problems.

The new material and all the teaching points of each section are presented with worked and annotated examples. Problem-solving steps are identified and explained. The students see what is to be done and how to do it. Most examples and all of the teaching points are reinforced with Warm-Up problems keyed to the examples. The topics are arranged in such a manner as to allow the student to progress through the book, developing skills as they are needed. The student is not put into a position of using a concept that has not already been developed.

The topics are developed so that instructors can tailor a course to fit their needs.

FEATURES OF THE TEXT

A Problem-Solving Approach A problem-solving approach is used throughout the book. Worked examples with explanatory text provide the key to an extensive collection of applied problems.

Connections Each chapter opens with an introductory paragraph putting the upcoming material in context, and showing its connections with other chapters, other courses, and historical development of mathematical ideas.

Key Problems The Warm-Ups and Practice Exercises contain key problems, marked in color. These problems illustrate the necessary teaching points of the section. The instructor can choose to work some or all of them in class as blackboard examples. They can be assigned as hand-in homework to be graded, or they can be used for any purpose where the instructor wishes to be sure of a limited but representative cross-section of the material.

Let's Not Forget   The review problems in every chapter, except Chapter 0, end with a novel feature called "Let's Not Forget." This segment contains a few carefully selected problems that repeat earlier themes, particularly certain sticky ideas and problems that historically give students trouble. The Let's Not Forget problems expand with each chapter.

Check Ups   Each chapter summary ends with a section called "Check Ups." These are worked examples from the text and are stated as problems. Check Ups can be used as a self-test for students to verify their understanding of the main points in the chapter. Each problem is keyed to an example number so that students can check their work by referring back to the text.

Be Careful!   Common student errors are prominently flagged in the margin with the admonition, "Be Careful!" The adjoining text explains the caution.

In Your Own Words   The problem sets in each section end with a few questions marked, "In Your Own Words." These may be used to encourage good writing in mathematics and to test comprehension of certain ideas.

Calculator Boxes   Calculators are introduced where appropriate, and instructional material on the proper use of calculators is included. The calculator boxes are not mere window dressing, but provide detailed instructions on meaningful problems for new calculator users. However, the calculator material, including exercises, is segregated and clearly marked for those who wish to omit it. In the calculator boxes and elsewhere in the text, the distinction between exact, approximated, and estimated answers is emphasized.

Chapter Summary   Each chapter contains a summary of the material and glossary of new terms.

Chapter Review   Each chapter ends with a set of review problems and a chapter test to reinforce the material.

NEW 2ND EDITION FEATURES

Problem Sets   All problem sets have been carefully edited and enhanced with more real-world applications added throughout. Problems involving geome-

try have been integrated throughout the text. Answers to *all* Warm-Up problems are included in the back-of-book answers.

Calculator Boxes   The calculator boxes have been edited and modernized to contain information about graphing calculators.

You Decide   A *critical-thinking* feature has been added to every chapter except Chapter 0. These "You Decide" questions are open-ended problems requiring some algebra skills but also forcing an analysis of a situation. Each requires that a decision be made and a written answer provided.

In General   The major theme of progressing from *numbers* to *expressions* to *equations* to *modeling* has been enhanced and further emphasized in this edition. The ac-method of factoring trinomials has been added to this edition. A section on graphing quadratic functions has been added. More photographs have been added throughout the text.

SUPPLEMENTS

Instructors using this text may obtain the following supplements from the publisher:

Instructor's Solutions Manual   Prepared by Irene Doo, Austin Community College. Contains worked-out solutions to all even-numbered problems. (0-13-075607-5)

Instructor's Solutions Manual with Tests and Syllabus   Prepared by Linda Kyle, Tarrant County Community College. Contains chapter tests, suggested teaching assignments, and teaching outlines. (0-13-075607-5)

Test Item File   Prepared by Nancy Carpenter, Johnson County Community College. Contains hardcopy of test questions from PH Test Manager and Make Test for the Macintosh. (0-13-075599-0)

PH Test Manager 2.0 (3.5″ or 5.25″)   Contains a bank of test questions which can be edited and manipulated to create tests: 3.5″ (0-13-075730-6); 5.25″ (0-13-075748-9); demo (0-13-075763-2).

Make Test for Macintosh   is a state-of-the-art Macintosh test generator that makes full use of the superior MAC graphics capabilities as well as the user-friendly interface. (0-13-075755-1); demo (0-13-075771-3)

PH Algo Test (3.5″ or 5.25″)   is a state-of-the-art IBM compatible program that allows instructors to produce exams by generating test questions from algorithms.

Students using this text may obtain the following supplements from the publisher:

Student Solutions Manual Prepared by Odene Forsythe, Westart Community College. Contains worked-out solutions to all odd-numbered problems. (0-13-075623-7)

Study Guide contains chapter summary material, practice exercises with additional examples and self-tests. (0-13-075649-0)

Math Master Tutor IBM (3.5" or 5.25"); Math Master Tutor Macintosh carefully keyed to the text with page references, Math Master Tutor contains four modes of instruction for each section in the text: Exploration (a text-on-disk feature), Summary (a concise review), Exercises (open-ended, algorithmically generated drills with step-by-step solutions), and Quiz (timed tests which teach students test-taking strategies). Available upon adoption (with cite license) for IBM and Macintosh.

- IBM 3.5" (0-13-075656-3)
- IBM 5.25" (0-13-075664-4)
- IBM demo (0-13-075680-6)
- MAC (0-13-075672-5)
- MAC demo (0-13-075776-3)

Lecture Videos contain important lecture topics and solving solutions presented by the authors and keyed to the book. Free upon adoption of 100 copies. (0-13-075714-4); demo lecture video (0-13-075722-5)

REVIEWERS AND CHECKERS

We wish to thank the following reviewers:
Wayne Barber, Chemeketa Community College (Salem, OR); Robert Davidson, East Tennessee State University (Johnson City, TN); Kendall Griggs, Hutchinson Community College (Hutchinson, KS); Peter Herron, Suffolk County Community College (Selden, NY); Maureen Kelley, Northern Essex Community College (Haverhill, MA); Jerome Krakowiak, Jackson Community College (Jackson, MI); Anita Lessmeister, The University of North Dakota (Grand Forks, ND); Joe Mathasius, Spokane Community College (Spokane, WA); Robert Prior, Riverside Community College (Norco, CA); Bessie Tucker, Jackson State University (Jackson, MS); Brenda Wood, Florida Community College at Jacksonville (Jacksonville, FL); Richard Semmler, Northern Virginia Community College (Annadale, VA); Kathy V. Rogers, The University of Southern Indiana (Evansville, IN)

Finally, we would like to thank the staff and faculty of DeKalb College for their continued support; Peggy Estes who insisted we do this work; Tom Aloisi, our diligent production editor; and development editor Bobbie Lewis.

Linda L. Exley
Vincent K Smith

0 Reviewing Arithmetic and Geometry

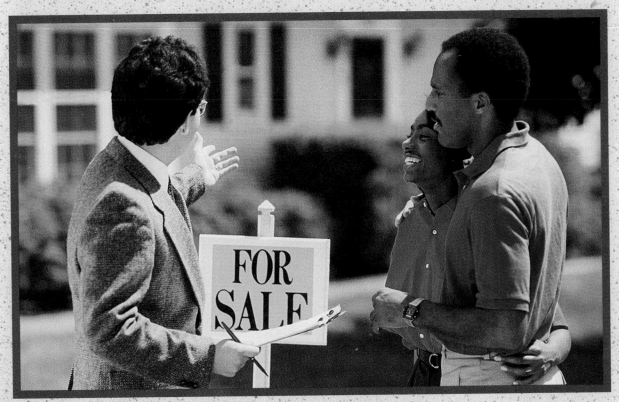

See Problem Set 0.3, Exercise 114.

1

Connections

Arithmetic books dating back to the Middle Ages have attempted to teach students how to calculate with counting numbers and fractions. For thousands of years, people have sought ways to calculate more rapidly and accurately. Various devices have been invented to carry out computations automatically. The oldest such device is the abacus, which often consisted of a series of beads strung on rows of wire, each representing a power of 10.

Today calculators and computers are used to make calculations in fields ranging from engineering to coaching basketball. Computers make thousands of arithmetic calculations for meteorologists in preparing a weather forecast for the National Weather Service.

So, why should we study arithmetic? Understanding how to add, subtract, multiply, and divide with natural numbers, fractions, and decimals is the foundation upon which algebra is developed. Algebra can be described as arithmetic with letters used to represent numbers. To understand the connection between arithmetic and algebra, we must first understand the concepts of arithmetic. This chapter contains a review of arithmetic.

0.1 Order of Operations

TAPE 1

Arithmetic is the study of numbers. The symbols that we use to name numbers are called **numerals.** For example, 2 and II are names for the number two. The symbol 2 is an Arabic numeral, whereas II is a Roman numeral. We use Arabic numerals almost exclusively in mathematics.

There are many kinds of numbers. In this chapter we will use the counting numbers, or natural numbers. They are 1, 2, 3, ... The three dots (called an *ellipsis*) indicate that the numbers continue.

In arithmetic, we perform four operations (addition, subtraction, multiplication, and division) with numbers. We use symbols to indicate these operations. The following chart indicates the symbols that we use for each operation with the numbers 7 and 5.

Operation	Symbol(s)	Read
Addition	$7 + 5$	7 plus 5
Subtraction	$7 - 5$	7 minus 5
Multiplication	7×5; $7(5)$; $(7)5$; $(7)(5)$	7 times 5
Division	$7 \div 5$; $\dfrac{7}{5}$; $7/5$; $5\overline{)7}$	7 divided by 5

The **sum** of two numbers is the result of adding them. The **difference** in two numbers is the result of subtracting one from the other. The **product** of two numbers is the result of multiplying them. The **quotient** is the result of dividing one number (dividend) by another (divisor).

► **EXAMPLE 1**

Express each phrase in symbols.

(a) The sum of 8 and 3.

(b) The difference when 6 is subtracted from 9.

(c) 6 less than 9 squared.

(d) The product of 2 and 7.

(e) The quotient when 12 is divided by 19.

(f) One more than twice 7.

Solutions

(a) $8 + 3$ (b) $9 - 6$ (c) $9^2 - 6$

(d) $2 \cdot 7$ (e) $12 \div 19$ (f) $2(7) + 1$ ◄

Mathematicians have developed a shorthand notation for writing products with the same number. Instead of writing $2 \cdot 2 \cdot 2$, we write 2^3. When we write 3^5, we mean $3 \cdot 3 \cdot 3 \cdot 3 \cdot 3$.

In an expression such as 3^5, we call 3 the **base** and 5 the **exponent.** The exponent tells us how many times to use the base in the product. We read 3^5 as 3 to the fifth **power.** We use the second and third powers so often that we read 7^2 as 7 *squared* and 4^3 as 4 *cubed.*

► **EXAMPLE 2**

Find the value of each expression.

(a) 3^2 (b) 2^3

Solutions

(a) $3^2 = 3 \cdot 3$ (b) $2^3 = 2 \cdot 2 \cdot 2$
       $= 9$ $= 8$ ◄

Often, more than one operation is involved in simplifying an expression. For example, $2 + 3 \cdot 4$ contains multiplication and addition. We must decide which operation to perform first. **Grouping symbols** such as parentheses ( ), brackets [ ], or braces { } are used to make the order of operations clear. We perform operations inside grouping symbols first.

If we write $(2 + 3) \cdot 4$, we perform the addition first and then the multiplication. So

$$(2 + 3) \cdot 4 = 5 \cdot 4$$
$$= 20$$

However, if we write $2 + (3 \cdot 4)$, we perform the multiplication first, followed by the addition. So

$$2 + (3 \cdot 4) = 2 + 12$$
$$= 14$$

If there are no grouping symbols, we do multiplications and divisions from left to right before we do additions and subtractions. So

$$2 + 3 \cdot 4 = 2 + 12$$
$$= 14$$

Fraction bars are also grouping symbols. For example,

$$\frac{2 + 10}{2 + 4} = \frac{12}{6}$$
$$= 2$$

This expression is the same as $(2 + 10) \div (2 + 4)$. The fraction allows us to omit the parentheses.

The following rules determine which operation to perform first.

Order of Operations

If grouping symbols are present, perform operations inside them, starting with the innermost grouping symbol, in the following order.

1. Perform any exponentiations.
2. Perform all multiplications and divisions in order from left to right.
3. Perform all additions and subtractions in order from left to right.

If grouping symbols are *not* present, perform operations in the above order.

A commonly used memory aid is the sentence, "Please excuse my dear Aunt Sally." The first letters of each word give the order of operations.

Parentheses
Exponents
Multiplication and Division from left to right
Addition and Subtraction from left to right

Be Careful!   Note that multiplication does not have to be done before division, nor addition before subtraction. Multiplications and divisions are done in order as they occur working from left to right. Additions and subtractions are done in order as they occur working from left to right.

► EXAMPLE 3

Perform the operations indicated.

(a) $6 \cdot 3 - \dfrac{4}{2}$ (b) $10 - 3 + 2^2$

(c) $14/2 \cdot 3$ (d) $12 - 2(6 - 3)$

(e) $(3^2 - 4)^2$ (f) $\dfrac{2 \cdot 3 + 8}{10 - 3}$

Solutions

(a) We perform the multiplication and division first.

$$6 \cdot 3 - \frac{4}{2} = 18 - 2$$

$$= 16$$

(b) We work with the exponent first.

$$10 - 3 + 2^2 = 10 - 3 + 4$$

$$= 11$$

(c) We perform multiplications and divisions from left to right in order as they occur. So we divide first and then multiply.

$$14/2 \cdot 3 = 7 \cdot 3$$

$$= 21$$

(d) We work inside the parentheses first.

$$12 - 2 (6 - 3) = 12 - 2 \cdot 3$$

$$= 12 - 6$$

$$= 6$$

(e) $(3^2 - 4)^2 = (9 - 4)^2$

$$= 5^2$$

$$= 25$$

(f) We simplify the numerator and denominator first.

$$\frac{2 \cdot 3 + 8}{10 - 3} = \frac{6 + 8}{10 - 3}$$

$$= \frac{14}{7}$$

$$= 2 \quad \blacktriangleleft$$

A number that is the square of a natural number is called a **perfect square.** For example, 9 and 16 are perfect squares because $3^2 = 9$ and $4^2 = 16$. The reverse of squaring is taking a square root. We write $\sqrt{9}$ to indicate the square root of 9. The square root of 9 is 3 because $3^2 = 9$.

► EXAMPLE 4

Find the square roots indicated.

(a) $\sqrt{4}$ (b) $\sqrt{16}$ (c) $\sqrt{25}$ (d) $\sqrt{18^2}$

Solutions

(a) $\sqrt{4} = 2$ because $2^2 = 4$

(b) $\sqrt{16} = 4$ because $4^2 = 16$

(c) $\sqrt{25} = 5$ because $5^2 = 25$

(d) $\sqrt{18^2} = 18$ because $18^2 = 18^2$ $\blacktriangleleft$

The $\sqrt{\ }$ symbol is another grouping symbol. That is, we perform operations *inside* the $\sqrt{\ }$ *before* we evaluate it. The next example illustrates this.

► **EXAMPLE 5**

Evaluate each expression.

(a) $\sqrt{144 + 25}$ (b) $\sqrt{144} + \sqrt{25}$

Solutions

(a) Since $\sqrt{\ }$ is a grouping symbol, we *must* perform the operations inside first.
$$\sqrt{144 + 25} = \sqrt{169}$$
$$= 13 \qquad \text{because } 13^2 = 169$$
(b) $\sqrt{144} + \sqrt{25} = 12 + 5 \qquad 12^2 = 144 \quad \text{and} \quad 5^2 = 25$
$$= 17 \qquad\qquad\qquad ◄$$

In Example 4, we found square roots of perfect squares. If we want to find square roots of natural numbers that are not perfect squares, we use a calculator or a square root table to *approximate* them. For example, $\sqrt{2}$ is approximately 1.414. We write $\sqrt{2} \approx 1.414$ to indicate that 1.414 is an approximation.

Calculator Box

Computations with a Calculator

Read *Using Your Calculator* in the Preface before reading this material. Most calculators have the order of operations built in. That is, they will calculate $3 + 5 - 2$ correctly by pressing the keys in the order given. Press

$\boxed{3}\ \boxed{+}\ \boxed{5}\ \boxed{-}\ \boxed{2}\ \boxed{=}$ (or $\boxed{\text{ENTER}}$ or $\boxed{\text{EXE}}$)

and the correct result ▮▮▮▮▮▮ 6 appears on the screen.

However, parentheses $\boxed{(}$ and $\boxed{)}$ are needed to preserve the order of operations in a calculation like $\dfrac{9 - 5}{3 + 1}$. Since the fraction bar is a grouping symbol, we must hold the numerator and denominator together before division. Press

$\boxed{(}\ \boxed{9}\ \boxed{-}\ \boxed{5}\ \boxed{)}\ \boxed{\div}\ \boxed{(}\ \boxed{3}\ \boxed{+}\ \boxed{1}\ \boxed{)}\ \boxed{=}$ (or $\boxed{\text{ENTER}}$ or $\boxed{\text{EXE}}$)

and read ▮▮▮▮▮▮ 1 on the display.

Let's approximate $\sqrt{5}$.

Scientific Calculator:

Press $\boxed{5}$ $\boxed{\sqrt{}}$ and read 2.236067977 on the display

Graphing Calculator:

Press $\boxed{\sqrt{}}$ $\boxed{5}$ $\boxed{=}$ (or $\boxed{\text{ENTER}}$ or $\boxed{\text{EXE}}$) for the same result.

Rounding to the nearest thousandth, $\sqrt{5} \approx 2.236$. It is important to estimate calculator computations so that we can determine if our answer is reasonable. $\sqrt{5}$ should be a little larger than 2, since $\sqrt{4} = 2$. So our answer is reasonable.

Calculator Exercises

Perform the following computations on a calculator and check the result with pencil and paper.

1. $71 - (8 - 3)$ **2.** $\dfrac{45 - 9}{3 \cdot 4}$ **3.** $5(18 - 11) - 2(234 - 222)$

4. $1 \div (3 + 9) \cdot 24$ **5.** $4\sqrt{9} + 16$ **6.** $\dfrac{\sqrt{173 - 109}}{2(11 - 9)}$ $\frac{64 - 8}{4}$

Approximate the square roots indicated to three decimal places. Be sure to use the $\approx$ symbol to indicate an approximation. Estimate the answer first.

7. $\sqrt{3}$ **8.** $\sqrt{10}$ **9.** $\sqrt{47}$ **10.** $\sqrt{85}$

Answers

1. 66   **2.** 3   **3.** 11   **4.** 2   **5.** 20   **6.** 2

7. Estimate is between 1 and 2; $\sqrt{3} \approx 1.732$.

8. Estimate is about 3; $\sqrt{10} \approx 3.162$.

9. Estimate is a little less than 7; $\sqrt{47} \approx 6.856$.

10. Estimate is about 9; $\sqrt{85} \approx 9.220$.

PROBLEM SET 0.1

Warm-Ups

In Problems 1 through 10, express each phrase in symbols. See Example 1.

1. The sum of 3 and 4   **2.** 7 more than 5.

3. 3 times 9.   **4.** 6 subtracted from 11.

5. 21 divided by 8.   **6.** Twice 14.

7. One-half of 17. 8.5^0 **8.** 5 more than 1.

9. 4 less than 6 squared.   **10.** One less than twice 13. $\frac{26}{-\frac{1}{25}}$

$6^2 \; \frac{36}{-\frac{4}{32}}$

In Problems 11 through 14, find the value of each. See Example 2.

11. 2^3 **12.** 3^2 **13.** 4^2 **14.** 5^3

In Problems 15 through 26, perform the operations indicated. See Example 3.

15. $6 + 3 \cdot 2$

16. $12 - 2^2$

17. $(3 + 4)(6 - 4)$

18. $(3 + 4)6 - 4$

19. $3 + 4(6 - 4)$

20. $\dfrac{8^2 - 6 \cdot 7}{2 \cdot 3 + 5}$

21. $9 + 42/7 - 2$

22. $2 \cdot 3^2 - 2^2$

23. $45 - (2 \cdot 3)^2$

24. $18 \div 6 \cdot 2$

25. $28 \div 4 - 2 \cdot 3$

26. $2^2 \cdot (2 \cdot 3)^2$

In Problems 27 through 32, find the square roots indicated. See Example 4.

27. $\sqrt{81}$

28. $\sqrt{36}$

29. $\sqrt{400}$

30. $\sqrt{49}$

31. $\sqrt{13^2}$

32. $\sqrt{15^2}$

In Problems 33 through 38, evaluate each expression. See Example 5.

33. $\sqrt{9 + 16}$

34. $\sqrt{9} + \sqrt{16}$

35. $\sqrt{8^2} + \sqrt{6^2}$

36. $\sqrt{8^2 + 6^2}$

37. $\sqrt{289 - 64}$

38. $\sqrt{289} - \sqrt{64}$

Practice Exercises

In Problems 39 through 52, find each value.

39. The product of 3 and 4.

40. 5 less than 17.

41. 3 added to 53.

42. 1 subtracted from 6.

43. 77 divided by 11.

44. One-third of 12.

45. 12 squared.

46. 11 less than 15.

47. Eleven squared plus six.

48. The square of the sum of 5 and 3.

49. Twice the sum of 8 and 7.

50. 4 cubed.

51. 13 squared minus 6.

52. One more than twice 7.

In Problems 53 through 82, perform the operations indicated.

53. $6 \cdot 3 - 2$

54. $2^2 - 3$

55. $\dfrac{16 - 9}{7}$

56. $(6 - 4)(6 + 4)$

57. $(7 - 4)6 - 4$

58. $12 - 4(6 - 4)$

59. $12 + 2(8 - 7)$

60. $\dfrac{8 \cdot 7 - 6^2}{2 \cdot 5 - 6}$

61. $22 - (20 - 14)$

62. $6^2 - \dfrac{16 - 4}{4}$

63. $23 - 63 \div 7 - 3^2$
$23 - 9 - 9.3$

64. $2^2 \cdot 3^2 - 2^2$

65. $\dfrac{2\left(\dfrac{48}{16}\right) - 2}{2^2}$

66. $(6 + 3)^2$

67. $25 - 2 \cdot 3^2$

68. $3 \cdot 8 \div 6$

69. $36 \div 9 \cdot 2 - 3$

70. $2 \cdot (4 \cdot 3)^2$

71. $2(3^3 - 3^2)$

72. $4^2 + 5 \div 5 - 2 \cdot 3$

73. $2 + (3 \cdot 2^3 - 2 \cdot 3^2)$

74. $2 + 3[2 + 3(7 - 6)]$

75. $19 - 3[2 + 2(11 - 3^2)]$

76. $[13 - 2(3 \cdot 5 - 12)] - 2^2$

77. $3(4^2 - 2^3)$

78. $3^2 - 3 \div 3 + 3 \cdot 3$

79. $\sqrt{25 - 9}$

80. $\sqrt{25} - \sqrt{9}$

81. $\sqrt{13^2 - 100/4}$

82. $\sqrt{4^2 + 3^2}$

In Problems 83 through 94, answer each question.

83. Mae has 9 books and Sue has 5 books. How many books do they have together?

84. Jo has 4 times 3 skirts and Geisla has 2 times 8 skirts. How many skirts do they have together?

85. Tony has 24 quarters and Ellen has 10 fewer quarters than Tony. How many quarters does Ellen have?

86. Carl drank 9 sodas over the last 3 days. His wife Cathy drank 3 fewer sodas. How many sodas did Cathy drink?

Problem 86

87. John has 6 records and Mary has 10 records. How many records do they have together?

88. Susan has 3 times 9 dollars and Judy has 2 times 8 dollars. How many dollars do they have?

89. Wei has 26 nickels and Choi has 10 more nickels than Wei. How many nickels does Choi have?

90. Juan has 12 marbles and Maria has 7 fewer marbles than Juan. How many marbles does Maria have?

91. The hypotenuse of a right triangle is $\sqrt{5^2 + 12^2}$ units long. What is the length of the hypotenuse of this triangle?

92. One leg of a right triangle is $\sqrt{11^2 - 7^2}$ units long. What is its length to the nearest tenth?

93. What is the sum of the square root of 64 and the square of 14?

94. Approximate to the nearest hundredth the square root of 50 subtracted from the square root of 60.

Challenge Problems

95. Find $\sqrt{200}$ to two decimal places. Compare it to the value of $\sqrt{2}$.

96. Find $\sqrt{2000}$ to two decimal places. Compare it to the value of $\sqrt{20}$.

In Your Own Words

97. What do we mean by $\sqrt{21}$?

98. What do we mean by 21^2?

99. State the order of operations.

Write in words what each expression means.

100. $8 + 7$

101. $11 - 14$

102. $5 \cdot 23$

103. $6 \div 7$

104. Explain two ways to interpret the statement "The sum of 7 and 2 squared."

0.2 Fractions

TAPE 1

Fractions are commonly used in algebra. This section will review operations with fractions in arithmetic. When working with fractions, it is sometimes necessary to write a number as a product of numbers. This process is called **factoring.** The numbers multiplied in the product are called **factors.** Most often we deal with natural numbers.

A **prime** number is a natural number larger than 1 whose only natural number factors are itself and 1. For example, 5 is prime because $5 = 5 \cdot 1$. This is the only way to write 5 as a product of natural numbers. Likewise, 2, 3, 7, and 11 are prime. We don't call 1 a prime number. The smallest prime number is 2.

Sometimes we need to factor a natural number that is not prime. Often there are several ways to do this. For example, 12 can be factored as follows:

$$12 = 2 \cdot 2 \cdot 3 \qquad 12 = 2 \cdot 6 \qquad 12 = 3 \cdot 4 \qquad 12 = 12 \cdot 1$$

Frequently, the most useful of these factored forms is $2 \cdot 2 \cdot 3$ because it is a product of prime numbers. There is only one way to factor a natural number using only prime factors. When we write a number as a product of prime numbers, we say that we have **factored it completely** or that we have found the **prime factorization.**

In factoring large numbers, it is helpful to know when a number is divisible by a prime.

Some Tests for Divisibility

1. A natural number is divisible by 2 if it is even, that is, if its last digit is 0, 2, 4, 6, or 8.

2. A natural number is divisible by 3 if the sum of the digits of the number is divisible by 3.

3. A natural number is divisible by 5 if its last digit is 0 or 5.

To factor a number into prime factors, we start with the smallest prime that will divide into the number and continue until we have all prime factors.

► **EXAMPLE 1**

Write each number as a product of prime factors.

(a) 45   (b) 108   (c) 392

Solutions

(a) We notice that 45 is not even, so it is not divisible by 2. However, the sum of its digits is 9, so it is divisible by 3.

$$45 = 3 \cdot \boxed{15}$$
$$= 3 \cdot \boxed{3 \cdot 5}$$
$$= 3^2 \cdot 5$$

(b) Since 108 is even, it is divisible by 2.

$$108 = 2 \cdot \boxed{54}$$

Now 54 is also divisible by 2. So

$$108 = 2 \cdot 2 \cdot \boxed{27}$$
$$= 2 \cdot 2 \cdot 3 \cdot 3 \cdot 3$$
$$= 2^2 \cdot 3^3$$

(c) Since 392 is even, we divide by 2.

$$392 = 2 \cdot 196$$
$$= 2 \cdot 2 \cdot \boxed{98}$$
$$= 2 \cdot 2 \cdot \boxed{2 \cdot 49}$$
$$= 2 \cdot 2 \cdot 2 \cdot 7 \cdot 7$$
$$= 2^3 \cdot 7^2$$

◄

Short division is long division without all the arithmetic. It is a convenient way to factor numbers. Let's factor 588 using short division. We begin with the smallest prime that will divide into 588. It is 2:

$$2 \overline{)588} \atop \overline{294}$$

We do the same thing with the number 294. The smallest prime that will divide into 294 is 2:

$$
\begin{array}{r}
147 \\
2 \overline{)294} \\
2 \overline{)\ 588}
\end{array}
$$

Now 147 is not divisible by 2. However, since the sum of its digits is 12, it is divisible by 3:

$$
\begin{array}{r}
49 \\
3 \overline{)147} \\
2 \overline{)\ 294} \\
2 \overline{)\ 588}
\end{array}
$$

The smallest prime that will divide into 49 is 7:

$$
\begin{array}{r}
7 \\
7 \overline{)49} \\
3 \overline{)147} \\
2 \overline{)\ 294} \\
2 \overline{)\ 588}
\end{array}
$$

Since 7 is prime, we stop the process. Thus, $588 = 2 \cdot 2 \cdot 3 \cdot 7 \cdot 7$.

Throughout this section, we will use the letters p, q, r, and s to represent natural numbers. We do this in order to state the basic rules and principles of operations with fractions. If we divide 5 by 9, we can write the quotient as the fraction $\frac{5}{9}$. In general, we say if we divide a *number p* by a *number q*, we can write the quotient as the fraction, $\frac{p}{q}$. We call p the **numerator** and q the **denominator.**

When working problems with fractions, we sometimes see fractions in which the numerator is larger than or equal to the denominator. Such fractions are called **improper fractions.** In *arithmetic* it is common to change improper fractions into mixed numbers. For example,

$$\frac{22}{9} = 2\frac{4}{9}$$

Notice that the mixed number $2\frac{4}{9}$ means $2 + \frac{4}{9}$.

In algebra we prefer to do arithmetic with a fraction rather than a mixed number (because we indicate *multiplication* when we write numbers next to each other, not addition.) So in algebra, when improper fractions arise, we leave them. When we find mixed numbers, we write them as fractions.

If a problem is stated using mixed numbers, we convert the mixed numbers to fractions and solve the problem. We answer the question using mixed numbers, since we usually answer questions in the form in which they are stated.

The fractions $\frac{1}{2}$, $\frac{2}{4}$, and $\frac{3}{6}$ are different ways of writing the same number. This is the idea behind the Fundamental Principle of Fractions.

Fundamental Principle of Fractions

$$\frac{p}{q} = \frac{p \cdot r}{q \cdot r}$$

If we say $\frac{3}{6} = \frac{1}{2}$, we are using the Fundamental Principle to reduce the fraction. It tells us that we can remove **factors** that are common to both the numerator and the denominator. If no factors are common to both the numerator and the denominator, we say the fraction is **reduced to lowest terms.**

► **EXAMPLE 2**

Reduce $\frac{48}{60}$ to lowest terms.

Solution

First, we factor the numerator and denominator.

$$\frac{48}{60} = \frac{2 \cdot 2 \cdot 2 \cdot 2 \cdot 3}{2 \cdot 2 \cdot 3 \cdot 5}$$

The common factors in numerator and denominator are 3 and $2 \cdot 2$.

$$\frac{48}{60} = \frac{2 \cdot 2}{5}$$

$$= \frac{4}{5}$$ ◄

If we see that 12 is a factor of both 48 and 60, we can do Example 1 without factoring completely.

$$\frac{48}{60} = \frac{12 \cdot 4}{12 \cdot 5}$$

$$= \frac{4}{5}$$

If we say $\frac{1}{2} = \frac{3}{6}$, we are using the Fundamental Principle to build up a fraction. The Fundamental Principle tells us that we can multiply both numerator and denominator by the same number and not change the value of a fraction.

► **EXAMPLE 3**

Write $\frac{2}{15}$ with a denominator of 45.

Solution

Since $45 = 15 \cdot 3$, we must multiply both numerator and denominator by 3.

$$\frac{2}{15} = \frac{2 \cdot \boxed{3}}{15 \cdot \boxed{3}}$$

$$= \frac{6}{45}$$ ◀

To multiply two fractions, we multiply numerators and multiply denominators.

Multiplication of Fractions

$$\frac{p}{q} \cdot \frac{r}{s} = \frac{p \cdot r}{q \cdot s}$$

► **EXAMPLE 4**

Find the products indicated and reduce answers to lowest terms.

(a) $\left(\dfrac{2}{3}\right)^3$ (b) $\dfrac{12}{35} \cdot 1\dfrac{1}{14}$

Solutions

(a) $\left(\dfrac{2}{3}\right)^3 = \dfrac{2}{3} \cdot \dfrac{2}{3} \cdot \dfrac{2}{3}$

$$= \frac{2 \cdot 2 \cdot 2}{3 \cdot 3 \cdot 3}$$

$$= \frac{8}{27}$$

(b) $\dfrac{12}{35} \cdot 1\dfrac{1}{14} = \dfrac{12}{35} \cdot \dfrac{15}{14}$ Express as an improper fraction

$$= \frac{12 \cdot 15}{35 \cdot 14}$$

Instead of multiplying the numbers in the numerator and denominator, we factor so that we can look for common factors to reduce.

$$\frac{12}{35} \cdot 1\frac{1}{14} = \frac{2 \cdot 2 \cdot 3 \cdot 3 \cdot 5}{7 \cdot 5 \cdot 7 \cdot 2}$$

$$= \frac{2 \cdot 3 \cdot 3}{7 \cdot 7}$$ Common factors are 2 and 5

$$= \frac{18}{49}$$ ◀

To divide two fractions, multiply the first fraction by the reciprocal of the second.

Division of Fractions

$$\frac{p}{q} \div \frac{r}{s} = \frac{p}{q} \cdot \frac{s}{r} = \frac{p \cdot s}{q \cdot r}$$

▶ **EXAMPLE 5**

Perform the division indicated $\dfrac{27}{16} \div \dfrac{9}{8}$, and reduce the answer to lowest terms.

Solution

$$\frac{27}{16} \div \frac{9}{8} = \frac{27}{16} \cdot \frac{8}{9} \qquad \text{Change to multiplication}$$

$$= \frac{27 \cdot 8}{16 \cdot 9}$$

$$= \frac{9 \cdot 3 \cdot 8}{2 \cdot 8 \cdot 9} \qquad \text{Factor}$$

$$= \frac{3}{2} \qquad\qquad\qquad\qquad ◀$$

We add fractions that have the same denominator by adding the numerators and using the common denominator as the new denominator. Subtraction of fractions is done in a similar manner.

Addition and Subtraction of Fractions

$$\frac{p}{q} + \frac{r}{q} = \frac{p + r}{q}$$

$$\frac{p}{q} - \frac{r}{q} = \frac{p - r}{q}$$

▶ **EXAMPLE 6**

Perform the operations indicated.

(a) $\dfrac{7}{13} + \dfrac{5}{13}$ (b) $\dfrac{16}{23} - \dfrac{11}{23}$

Solutions

(a) $\dfrac{7}{13} + \dfrac{5}{13} = \dfrac{7 + 5}{13}$ (b) $\dfrac{16}{23} - \dfrac{11}{23} = \dfrac{16 - 11}{23}$

$\qquad\qquad\quad = \dfrac{12}{13}$ $\qquad\qquad\qquad = \dfrac{5}{23} \qquad ◀$

When the denominators are not the same, we must find an appropriate denominator and write each fraction with this denominator. The denominator should be the smallest number divisible by each denominator. We call this denominator the **least common denominator.**

Notice how we can change the denominator of a fraction using the Fundamental Principle. We are actually multiplying the fraction by 1. Suppose we wish to write the fraction $\frac{5}{8}$ with a denominator of 56. Since $56 = 8 \cdot 7$, we need to multiply the denominator by 7.

$$\frac{5}{8} = \frac{5}{8} \cdot 1$$

$$= \frac{5}{8} \cdot \frac{7}{7} \qquad 1 = \frac{7}{7}$$

$$= \frac{35}{56} \qquad \text{Multiplication of fractions}$$

Procedure for Finding the Least Common Denominator (LCD)

1. Factor each denominator completely, using exponents.
2. List all the different prime factors from all the denominators.
3. The LCD is the product of the factors in step 2 each raised to the largest power of that factor in any single denominator.

► **EXAMPLE 7**

Find the least common denominator for each pair of denominators.

(a) 12 and 15
(b) 8 and 24
(c) 72 and 108

Solution

(a) First we factor each number.

$$12 = 2^2 \cdot 3$$

$$15 = 3 \cdot 5$$

The different prime factors are 2, 3, and 5.

2 occurs twice in 12 and no times in 15. So the LCD must contain 2 as a factor twice.

3 occurs one time in each number. The LCD must contain one factor of 3.

5 occurs no times in 12 and 1 time in 15. The LCD must contain one factor of 5.

The LCD is $2^2 \cdot 3 \cdot 5$, or 60.

(b) $24 = 2^3 \cdot 3$
    $8 = 2^3$

The different prime factors are 2 and 3.

2 occurs three times in both numbers.
The LCD must contain three factors of 2.

3 occurs one time in 24 and no times in 8.
The LCD must contain one factor of 3.

The LCD is $2^3 \cdot 3$, or 24.

(c) $72 = 2^3 \cdot 3^2$
    $108 = 2^2 \cdot 3^3$

The different prime factors are 2 and 3.

2 occurs three times in 72 and two times in 108.
The LCD must contain 2^3.

3 occurs two times in 72 and three times in 108.
The LCD must contain 3^3.

The LCD is $2^3 \cdot 3^3$, or 216.   ◀

▶ **EXAMPLE 8**

Perform the operations indicated and reduce the answers to lowest terms.

(a) $\dfrac{5}{24} + 1\dfrac{5}{8}$ (b) $\dfrac{7}{12} - \dfrac{8}{15}$

Solution

(a) We must first find the LCD. From Example 7(b), it is 24.

$$\frac{5}{24} + 1\frac{5}{8} = \frac{5}{24} + \frac{13}{8} \qquad \text{Express as an improper fraction}$$

$$= \frac{5}{24} + \frac{13 \cdot 3}{8 \cdot 3}$$

$$= \frac{5}{24} + \frac{39}{24}$$

$$= \frac{5 + 39}{24}$$

$$= \frac{44}{24}$$

$$= \frac{11 \cdot 4}{6 \cdot 4}$$

$$= \frac{11}{6} \qquad \text{Reduce}$$

$$= 1\frac{5}{6} \qquad \text{Express as a mixed number}$$

(b) From Example 7(a), the LCD is 60.

$$\frac{7}{12} - \frac{8}{15} = \frac{7 \cdot 5}{12 \cdot 5} - \frac{8 \cdot 4}{15 \cdot 4}$$

$$= \frac{35}{60} - \frac{32}{60}$$

$$= \frac{3}{60}$$

$$= \frac{1}{20}$$ ◄

The order of operations that we learned in Section 0.1 applies to fractions.

► **EXAMPLE 9**

Perform the operations indicated and reduce to lowest terms.

(a) $\dfrac{3}{4} + \dfrac{2}{3} \cdot \dfrac{1}{2}$ (b) $\left(\dfrac{2}{3} - \dfrac{1}{5}\right) \div \dfrac{7}{3}$

Solutions

(a) We multiply before we add.

$$\frac{3}{4} + \left(\frac{2}{3} \cdot \frac{1}{2}\right) = \frac{3}{4} + \frac{1}{3}$$

$$= \frac{9}{12} + \frac{4}{12}$$

$$= \frac{13}{12}$$

(b) Subtract inside the parentheses first.

$$\left(\frac{2}{3} - \frac{1}{5}\right) \div \frac{7}{3} = \left(\frac{10}{15} - \frac{3}{15}\right) \div \frac{7}{3}$$

$$= \frac{7}{15} \div \frac{7}{3}$$

$$= \frac{7}{15} \cdot \frac{3}{7}$$

$$= \frac{7 \cdot 3}{15 \cdot 7}$$

$$= \frac{1}{5}$$ ◄

► **EXAMPLE 10**

Jean has $5\frac{1}{2}$ bushels of peaches, which she wants to divide evenly among her three children. How many bushels will each child receive?

Solution

We must divide $5\frac{1}{2}$ by 3.

$$5\frac{1}{2} \div 3 = \frac{11}{2} \div 3 \qquad \text{Express as a fraction}$$

$$= \frac{11}{2} \cdot \frac{1}{3} \qquad \text{Multiply by the reciprocal}$$

$$= \frac{11}{6}$$

$$= 1\frac{5}{6} \qquad \text{Express as a mixed number}$$

Each child will receive $1\frac{5}{6}$ bushels of peaches.   ◄

Problem Set 0.2

Warm-Ups

In Problems 1 through 10, write each number as a product of prime factors. See Example 1.

1. 24   **2.** 36   **3.** 25

4. 81   **5.** 80   **6.** 200

7. 108   **8.** 972   **9.** 1008

10. 3969

In Problems 11 and 12, use the Fundamental Principle to reduce each fraction to lowest terms. See Example 2.

11. $\dfrac{12}{18}$ **12.** $\dfrac{36}{72}$

In Problems 13 through 16, use the Fundamental Principle to write each fraction with the given denominator. See Example 3.

13. $\dfrac{1}{2}$; denominator of 10   **14.** $\dfrac{3}{4}$; denominator of 12   **15.** $\dfrac{4}{3}$; denominator of 18

16. $\dfrac{11}{16}$; denominator of 144

In Problems 17 through 48, perform the operations indicated. Reduce answers to lowest terms. See Example 4.

17. $\frac{3}{4} \cdot \frac{1}{3}$

18. $\left(\frac{1}{3}\right)^3$

19. $\frac{15}{33} \cdot \frac{44}{35}$

20. $\left(1\frac{5}{6}\right) \cdot \left(1\frac{7}{11}\right)$

21. $\left(1\frac{3}{4}\right)^2$

22. $\frac{2}{3} \cdot 6$

For Problems 23 through 26, perform the divisions. Reduce answers to lowest terms. See Example 5.

23. $\frac{3}{4} \div \frac{1}{3}$

24. $\frac{4}{7} \div 2$

25. $\left(3\frac{7}{9}\right) \div \left(3\frac{2}{3}\right)$

26. $\frac{25}{24} \div \frac{45}{42}$

In Problems 27 through 32, find the LCD for each pair of numbers. See Example 7.

27. 3 and 7

28. 8 and 16

29. 6 and 8

30. 20 and 24

31. 36 and 27

32. 135 and 54

For Problems 33 through 38, perform the indicated operation. Reduce answers to lowest terms. See Examples 6 and 8.

33. $\frac{1}{3} + \frac{4}{15}$

34. $\frac{3}{4} - \frac{1}{3}$

35. $\frac{2}{3} - \frac{3}{8}$

36. $2\frac{11}{12} - 1\frac{13}{18}$

37. $\frac{3}{7} + 1$

38. $2 - \frac{4}{5}$

In Problems 39 through 44, perform the indicated operations. Reduce answers to lowest terms. See Example 9.

39. $\frac{1}{18} + \frac{7}{12} - \frac{5}{9}$

40. $\frac{3}{8} \cdot \frac{2}{3} \div 4$

41. $\frac{3}{2} \cdot \frac{1}{3} \cdot \frac{4}{5}$

42. $\frac{3}{2} \cdot \frac{1}{3} \div \frac{4}{5}$

43. $\frac{3}{2} \div \frac{1}{3} + \frac{4}{5}$

44. $\left(\frac{3}{2} + \frac{1}{3}\right) \cdot \frac{4}{5}$

In Problems 45 through 48, answer each problem with a statement. See Example 10.

45. Subtract $3\frac{1}{3}$ from the product of $8\frac{1}{3}$ and $\frac{3}{5}$.

46. Add $\left(\frac{2}{3}\right)^2$ to $6\frac{1}{3}$.

47. Joan cuts a wire that is $2\frac{1}{3}$ yards (yd) long in half. How long will each half be?

48. Marge is in charge of making pancakes at her club's breakfast. She must have 15 pounds (lb) of flour. If Virginia donated $5\frac{1}{2}$ lb, Jane gave $8\frac{2}{3}$ lb, and Madelyn gave $5\frac{1}{6}$ lb of flour, does Marge have enough flour?

Practice Exercises

In Problems 49 through 87, perform the operations indicated. Reduce answers to lowest terms.

49. $\frac{3}{5} + \frac{1}{3}$

50. $\frac{3}{5} - \frac{1}{3}$

51. $\frac{3}{5} \cdot \frac{1}{3}$

52. $\frac{3}{5} \div \frac{1}{3}$

53. $\left(\frac{1}{3}\right)^2$

54. $\left(\frac{1}{2}\right)^3$

55. $\left(\frac{2}{5}\right)^3$

56. $\left(\frac{1}{10}\right)^2$

57. $\frac{1}{3} + \frac{5}{6}$

58. $\dfrac{3}{7} - \dfrac{1}{14}$ **59.** $\dfrac{2}{3} - \dfrac{3}{7}$ **60.** $\dfrac{5}{6} - \dfrac{5}{9}$

61. $\dfrac{7}{12} + \dfrac{5}{18}$ **62.** $\dfrac{24}{50} \cdot \dfrac{40}{27}$ **63.** $\dfrac{36}{35} \div \dfrac{42}{5}$

64. $2\dfrac{5}{6} + 1\dfrac{1}{6}$ **65.** $4\dfrac{1}{8} - 3\dfrac{3}{8}$ **66.** $\left(1\dfrac{3}{7}\right) \cdot \left(2\dfrac{4}{5}\right)$

67. $\left(3\dfrac{3}{8}\right) \div \left(2\dfrac{1}{4}\right)$ **68.** $5\dfrac{3}{4} - 3\dfrac{1}{2}$ **69.** $\left(1\dfrac{4}{5}\right)^2$

70. $\dfrac{3}{4} + 1$ **71.** $3 - \dfrac{3}{2}$ **72.** $\dfrac{4}{5} \cdot 10$

73. $\dfrac{6}{7} \div 3$ **74.** $\dfrac{2}{27} + \dfrac{35}{54} - \dfrac{5}{9}$ **75.** $\dfrac{5}{8} \cdot \dfrac{2}{5} \div 6$

76. $\dfrac{1}{4} \cdot \dfrac{4}{3} \cdot \dfrac{3}{4}$ **77.** $\dfrac{1}{4} \cdot \dfrac{4}{3} \div \dfrac{3}{4}$ **78.** $\dfrac{1}{4} \div \dfrac{4}{3} + \dfrac{3}{4}$

79. $\dfrac{1}{4} + \dfrac{4}{3} \cdot \dfrac{3}{4}$ **80.** $\dfrac{1}{4} \div \left(\dfrac{4}{3} + \dfrac{3}{4}\right)$ **81.** $\left(\dfrac{1}{4} + \dfrac{4}{3}\right) \cdot \dfrac{3}{4}$

82. $\left(\dfrac{3}{4} - \dfrac{1}{5}\right) \div \dfrac{11}{5}$ **83.** $5\left(\dfrac{2}{3}\right)^2 \div \dfrac{10}{3}$ **84.** $\left(\dfrac{1}{2}\right)^2 + \left(\dfrac{1}{3}\right)^2$

85. $\left(\dfrac{2}{3} - \dfrac{1}{6}\right)^2$ **86.** $\left(\dfrac{1}{3} + \dfrac{1}{4}\right)\left(\dfrac{1}{3} - \dfrac{1}{4}\right)$ **87.** $\left(\dfrac{3}{2} - 1\right)^2$

In Problems 88 through 95, answer each question with a statement.

88. Spencer bought a 25-lb bag of fertilizer. If Linda gave $12\dfrac{3}{4}$ lb of it to her friend, how much fertilizer does Spencer have left?

89. What is $7\dfrac{2}{3}$ squared?

90. Last month Bonnie's phone bill showed that she talked $2\dfrac{1}{2}$ minutes (min) to her mother, $6\dfrac{1}{2}$ min to her sister, 30 seconds (s) to her boss, and 12 min to her son. For how many minutes did she talk last month?

91. Shirley wishes to cut a cord that is $2\dfrac{1}{3}$ yd long into three equal parts. How long will each part be?

92. Martin bought 6 pictures and Rowland bought $\dfrac{3}{2}$ as many as Martin. How many pictures did Rowland buy?

93. Tim bought 12 pencils and Irene bought two-thirds as many as Tim. How many pencils did Irene buy?

Problem 94

94. Ky has 6 sandwiches and Nyugen has $\dfrac{7}{3}$ as many sandwiches as Ky. How many sandwiches does Nyugen have?

95. Pierre cuts a board into three pieces each of length 5 m. How long was the board originally? It was 15 m long.

In Your Own Words

96. What does it mean to factor a natural number?

97. Describe the steps in finding the prime factors of 294.

Write out in words what each expression means.

98. $\dfrac{1}{5} \cdot 6 \cdot 7$ **99.** $\dfrac{1}{5} \cdot 6 + 7$ **100.** $\dfrac{1}{5}(6 + 7)$ **101.** $8 + 9 \div 10$ **102.** $(8 + 9) \div 10$ **103.** $\dfrac{8 + 9}{10}$

0.3   Decimals and Percents

TAPE 1

Fractions are quotients. The fraction $\frac{2}{5}$ is another way of writing $2 \div 5$. If we perform long division, we obtain a decimal.

$$\begin{array}{r} 0.4 \\ 5\overline{)2.0} \\ \underline{2\,0} \end{array}$$

Every fraction can be represented as a decimal by dividing the denominator into the numerator.

► **EXAMPLE 1**

Express each fraction as a decimal.

(a) $\frac{1}{2}$ (b) $\frac{1}{3}$ (c) $\frac{1}{6}$

Solutions

(a)
$$\begin{array}{r} 0.5 \\ 2\overline{)1.0} \\ \underline{1\,0} \end{array}$$

$$\frac{1}{2} = 0.5$$

Notice that we write 0 before the decimal point. This is a good practice. We prefer to write a digit on each side of the decimal point.

(b)
$$\begin{array}{r} 0.3333 \\ 3\overline{)1.0000} \\ \underline{9} \\ 10 \\ \underline{9} \\ 10 \\ \underline{9} \\ 10 \\ \underline{9} \\ 1 \end{array}$$

To indicate that the 3 continues without end, we write $\frac{1}{3} = 0.\overline{3}$, or $\frac{1}{3} = 0.333 \ldots .$

(c)
$$\begin{array}{r} 0.1666 \\ 6\overline{)1.0000} \\ \underline{6} \\ 40 \\ \underline{36} \\ 40 \\ \underline{36} \\ 40 \\ \underline{36} \\ 4 \end{array}$$

$$\frac{1}{6} = 0.1\overline{6}, \text{ or } \frac{1}{6} = 0.1666 \ldots$$ ◄

In Example 1, notice that each problem involved the quotient of two natural numbers. In each case, we wrote each fraction as a decimal, which either terminated or repeated. Anytime that we divide a natural number by a natural number, the decimal number will either repeat or terminate.

We perform operations with decimals as we do with other numbers. The next example illustrates the operations of addition, subtraction, multiplication, and division with decimals.

► **EXAMPLE 2**

Perform the operations indicated.

(a) $127.32 + 24.1 + 7.003$

(b) $128.73 - 38.658$

(c) $(22.3)(12.25)$

(d) $32.4 \div 8.1$

Solutions

(a) We line up the decimal points and add.

$$
\begin{array}{r}
127.320 \\
24.100 \\
+\quad 7.003 \\
\hline
158.423
\end{array}
$$

(b) Again, we line up the decimal points.

$$
\begin{array}{r}
128.730 \\
-\quad 38.658 \\
\hline
90.072
\end{array}
$$

(c)

$$
\begin{array}{r}
12.25 \\
\times\ 22.3 \\
\hline
3675 \\
2450 \\
2450 \\
\hline
273.175
\end{array}
$$

2 decimal places
1 decimal place

3 decimal places

Notice that the answer has three decimal places because there are two decimal places in 12.25 and one in 22.3.

$$(22.3)(12.25) = 273.175$$

(d) First, we move the decimal place to the right in both the dividend and the divisor until the divisor is a natural number.

$$
\begin{array}{r}
4 \\
8.1\overline{)32.4} \\
\hline
32\ 4
\end{array}
$$

$$32.4 \div 8.1 = 4$$

◄

► **EXAMPLE 3**

Subtract 8.73 from the sum of 5.2 and 6.9.

Solution

Since we are to subtract a number for the *sum* of 5.2 and 6.9, we add them first; then we subtract.

$$(5.2 + 6.9) - 8.73 = 12.1 - 8.73$$
$$= 3.37 \qquad \blacktriangleleft$$

Fractions and decimals are often written as *percents*. Writing 15% is another way of writing $\dfrac{15}{100}$, or 0.15. Although percents are common in writing and talking, they are usually written as decimals or fractions in computation.

Percent

$$1\% = 0.01 = \frac{1}{100}$$

First, let's look at how to write a decimal as a percent. To write 0.25 as a percent, we notice that

$$0.25 = 25(0.01)$$
$$= 25\%$$

Notice that we write 0.25 with a factor of 0.01 and that we replace 0.01 with the % symbol. Another observation is that when we wrote 0.25 as 25% we *moved the decimal point* **two** *places to the right*.

► **EXAMPLE 4**

Write each decimal as a percent.

(a) 0.35   (b) 0.5   (c) 0.125
(d) $0.\overline{3}$ (e) 2.0

Solutions

(a) We move the decimal point two places to the right and write the % symbol.

$$0.35 = 0.3\,5\,\%$$
$$= 35\%$$

(b) $0.5 = 0.5\,0\,\%$
$$= 50\%$$

(c) $0.125 = 0.1\,2\,5\,\%$
$$= 12.5\%$$

(d) First, we write three digits of the repeating decimal so that we can move the decimal two places to the right.

$$0.\overline{3} = 0.33\overline{3}$$

Now we move the decimal point and write the %.

$$0.33\overline{3} = 0.3\,3\,\overline{3}\,\%$$

$$= 33.3\%$$

(e)
$$2.0 = 2.0\,0\,\%$$

$$= 200\% \qquad \blacktriangleleft$$

To write a percent as a decimal, we reverse the procedure. That is, we replace the % symbol with the factor 0.01 and multiply.

$$41\% = 41(0.01)$$

$$= 0.41$$

Notice that we *moved the decimal point **two** places to the left.*

► **EXAMPLE 5**

Write the following percents as decimals.

(a) 17%   (b) 88%   (c) 7%

(d) 41.6%   (e) $99\frac{1}{2}\%$ (f) $33\frac{1}{3}\%$

Solutions

(a) We move the decimal point two places to the left and remove the % symbol.

$$17\% = \underset{\smile}{1\,7.}$$

$$= 0.17$$

(b) $88\% = \underset{\smile}{8\,8.}$ (c) $7\% = \underset{\smile}{0\,7.}$

$\qquad = 0.88$ $\qquad = 0.07$

Notice that it was necessary to place a leading zero in front of the 7 in order to move the decimal point *two* places.

(d) $41.6\% = \underset{\smile}{4\,1.6}$

$\qquad = 0.416$

(e) If the percent includes a fraction, we convert it to a decimal.

$$99\frac{1}{2}\% = 99.5\%$$

$$= \underset{\smile}{9\,9.5}$$

$$= 0.995$$

(f) $33\frac{1}{3}\% = 33.\overline{3}\%$

$$= \underset{\smile}{3\,3.\overline{3}}$$

$$= 0.33\overline{3}$$

$\blacktriangleleft$

Percent–Decimal Conversion

Decimal to Percent:

1. Move the decimal point *two* places to the right.
2. Write the % symbol.

Percent to Decimal:

1. Replace any fraction with an equivalent decimal.
2. Move the decimal point *two* places to the left.
3. Remove the % symbol.

To write a fraction as a percent, we write the fraction as a decimal and then follow the procedure given.

▶ **EXAMPLE 6**

Write each fraction as a percent.

(a) $\dfrac{1}{4}$ (b) $\dfrac{1}{8}$ (c) $\dfrac{1}{3}$

Solutions

(a) $\dfrac{1}{4} = 0.25$ (b) $\dfrac{1}{8} = 0.125$ (c) $\dfrac{1}{3} = 0.33\overline{3}$

 $\quad\ = 25\%$ $\quad\ = 12.5\%$ $\quad\ = 33.\overline{3}\%$ ◀

Another useful way to write percents is as fractions. Since

$$0.01 = \frac{1}{100}$$

we can replace the % symbol with the fraction $\dfrac{1}{100}$ and multiply.

▶ **EXAMPLE 7**

Write the following percents as fractions.

(a) 17%   (b) 88%   (c) 7%

(d) 41.6%   (e) $99\dfrac{1}{2}\%$ (f) $33\dfrac{1}{3}\%$

Solutions

(a) $17\% = 17\left(\dfrac{1}{100}\right) = \dfrac{17}{100}$

(b) $88\% = 88\left(\dfrac{1}{100}\right) = \dfrac{88}{100}$

However, the fraction $\dfrac{88}{100}$ can be reduced to $\dfrac{22}{25}$.

$$88\% = \frac{22}{25}$$

(c) $7\% = 7\left(\dfrac{1}{100}\right) = \dfrac{7}{100}$

(d) $41.6\% = 41.6\left(\dfrac{1}{100}\right) = \dfrac{41.6}{100}$

We should not leave a decimal point in a fraction. So we multiply the numerator and denominator by 10.

$$\frac{41.6}{100} = \frac{416}{1000}$$

Since 8 is a common factor, this fraction can be reduced.

$$\frac{416}{1000} = \frac{52}{125}$$

(e) We can work with $99\frac{1}{2}\%$ using fractions or we can work with 99.5% using decimals. Using fractions,

$$99\frac{1}{2}\% = \left(99\frac{1}{2}\right)\left(\frac{1}{100}\right) = \frac{199}{2} \cdot \frac{1}{100} = \frac{199}{200}$$

Using decimals,

$$99\frac{1}{2}\% = (99.5)\left(\frac{1}{100}\right) = \frac{99.5}{100} = \frac{995}{1000} = \frac{199}{200}$$

(f) Again, we may use decimals or fractions. Using fractions,

$$33\frac{1}{3}\% = \left(33\frac{1}{3}\right)\left(\frac{1}{100}\right) = \frac{100}{3} \cdot \frac{1}{100} = \frac{1}{3} \qquad \blacktriangleleft$$

To say a solution is 32% alcohol is to say $\frac{32}{100}$ of the solution is alcohol. The amount of alcohol in 500 liters (L) of a 32% alcohol solution is given by

$$\frac{32}{100} \cdot 500 = 32 \cdot 5 = 160$$

There are 160 L of alcohol in the solution.

To find a specified percent of a number, we change the percent to a fraction or a decimal and then multiply.

▶ **EXAMPLE 8**

Find each of the following.

(a) 23% of 3300   (b) $16\frac{1}{2}\%$ of 98

(c) $66\frac{2}{3}\%$ of 291   (d) $9\frac{1}{4}\%$ of \$24,000

Solutions

(a) Since $23\% = \dfrac{23}{100}$, we have

$$\dfrac{23}{100} \cdot 3300 = 23 \cdot 33 = 759$$

23% of 3300 is 759.

(b) Because $16\frac{1}{2}\% = 0.165$, we have

$$(0.165)(98) = 16.17$$

$16\frac{1}{2}\%$ of 98 is 16.17.

(c) Since $66\frac{2}{3}\% = \dfrac{200}{3} \cdot \dfrac{1}{100} = \dfrac{2}{3}$, we have

$$\dfrac{2}{3} \cdot 291 = \dfrac{582}{3} = 194$$

$66\frac{2}{3}\%$ of 291 is 194.

(d) Since $9\frac{1}{4}\% = 9.25\% = 0.0925$, we have

$$(0.0925)(24,000) = 2220$$

$9\frac{1}{4}\%$ of \$24,000 is \$2220.   ◀

Problem Set 0.3

Warm-Ups

In Problems 1 through 5, write each fraction as a decimal. See Example 1.

1. $\dfrac{3}{5}$ **2.** $\dfrac{3}{8}$ **3.** $\dfrac{7}{6}$ **4.** $3\dfrac{4}{5}$ **5.** $\dfrac{8}{3}$

In Problems 6 through 9, perform the operations indicated. See Example 2.

6. $12.9 + 23.67$

7. $234.9 - 123.45$

8. $(23.45)(2.03)$

9. $100.4 \div 50.2$

For Problems 10 through 12, see Example 3.

10. Subtract 13.8752 from the sum of 24.6 and 8.2.

11. Subtract the sum of 22.09 and 5.345 from 50.

12. One-half of 48.6 added to one-third of 66.9.

In Problems 13 through 21, write the given numbers as percents. See Examples 4 and 6.

13. $\dfrac{91}{100}$

14. 0.77

15. 0.09

16. 0.8

17. $\dfrac{7}{50}$

18. 0.08

19. $0.\overline{6}$

20. 0.888

21. $\dfrac{3}{8}$

In Problems 22 through 30, write the given percents as fractions and as decimals. See Examples 5 and 7.

22. 87%   **23.** 11%   **24.** 50%

25. 8%   **26.** 98%   **27.** $22\frac{1}{2}$%

28. 0.3%   **29.** $3\frac{1}{3}$%   **30.** 125%

In Problems 31 through 39, find the quantity described. See Example 8.

31. 10% of 400   **32.** 38% of 10,000   **33.** 80% of 245
34. 5% of 720   **35.** 99% of 5280   **36.** 1% of 550

37. $8\frac{1}{2}$% of \$50,000   **38.** $33\frac{1}{3}$% of 963   **39.** 0.7% of 21,200

Practice Exercises

In Problems 40 through 44, write each fraction as a decimal.

40. $\frac{7}{8}$ **41.** $\frac{9}{4}$ **42.** $\frac{1}{7}$

43. $\frac{11}{6}$ **44.** $5\frac{2}{5}$

In Problems 45 through 56, write the given percents as fractions and *as decimals.*

45. 78%   **46.** 33%   **47.** 10%

48. 55%   **49.** 2%   **50.** 25%

51. 11%   **52.** 89%   **53.** $11\frac{1}{2}$%

54. 0.9%   **55.** $6\frac{2}{3}$%   **56.** 250%

In Problems 57 through 65, write the given numbers as percents.

57. $\frac{97}{100}$ **58.** 0.99   **59.** 0.03

60. $\frac{1}{25}$ **61.** $\frac{11}{50}$ **62.** 0.008

63. $0.\overline{2}$ **64.** 0.555   **65.** $\frac{5}{8}$

In Problems 66 through 74, find the quantity described.

66. 12 % of 500   **67.** 34% of 10,000   **68.** 80% of 555
69. 3% of 700   **70.** 77% of 5280   **71.** 1% of 470

72. 0.5 % of \$3000   **73.** $66\frac{2}{3}$% of 963   **74.** 0.4% of 21,200

In Problems 75 through 94, perform the operations indicated.

75. $28.91 - 20.07$

76. $34.9 + 123.45$

77. $(23.15)(2.03)$

78. $84.3572 \div 68.03$

79. $(12.5 - 6.2) + 3.8$

80. $4.8(2.1) - 3.2$

81. $12.4 \div 0.062 - 6.3$

82. $(34.25 - 24.5) - (23.01 - 16.7)$

83. $0.8/0.04 + 0.2/0.4$

84. $6.3 - (5.7 - 1.2)$

85. $2.1(2.5 + 6.7)$

86. $(8.8 - 2.5) \div 0.7$

87. $12 - (2 - 1.5) - 8.5$

88. $4(1.2) \div 1.6 - 2.1$

89. $(2.1)^2$

90. $(0.2)^3$

91. $(0.1)^2$

92. $4(1.2)^2 - (0.1)^3$

93. $(2.01)(10.2)^2$

94. $(1.01)^2(2.01)^2$

95. Subtract 3.2345 from the sum of 4.6 and 8.022.

96. Subtract the sum of 12.09 and 5.009 from 25.

97. Find one-half of 48.06 added to one-third of 66.009.

Answer each question with a sentence.

98. Greg has 50 nickels and six \$5 bills. How much money does he have?

99. Jamie has 8 dimes, 3 quarters, and five \$10 bills. How much money does he have?

100. Jennifer has 24 pennies, 33 nickels, 15 dimes, and 7 quarters. How much money does she have?

101. Kristen wanted to buy \$5.35 worth of chicken, \$0.89 worth of onions, \$1.89 worth of potatoes, and \$0.99 worth of broccoli at the grocery store. She had \$7.50 in her purse. Did she have enough money?

102. Sondra paid phone bills of \$34.24, \$45.89, \$29.00, \$57.90. Her budget allows for \$175. Did she stay within her budget?

103. Suzanne bought a diamond ring for \$5000. She discovered that its value has increased by 18%. What is the ring worth now?

104. A variety store advertised a 10% discount on drills. If the original price of a drill was \$59.95, what is the sale price?

105. Joseph receives an allowance of \$50 per week. If his father increases his allowance by 5%, what will his new allowance be?

106. Ray has 22 nickels and twelve \$5 bills. How much money does he have?

107. Frank has 2 dimes, 4 quarters, and seven \$20 dollar bills. How much money does he have?

108. Joan has 4 pennies, 3 nickels, 12 dimes, and 5 quarters. How much money does she have?

109. Deli ham sells for \$2.29 per $\frac{1}{4}$ lb. What is the price of deli ham per pound?

110. Jo Ann bought \$4.19 worth of chicken, \$.79 worth of onions, \$1.49 worth of potatoes, and \$1.19 worth of broccoli at the grocery store. She had \$7.50 in her purse. Did she have enough money?

111. Carolyn buys hamburger every week. For four weeks she has bought 2.12 lb, 4.65 lb, 3.15 lb, and 5.05 lb. Has she bought more than 15 lb?

112. Sara bought 2.01 lb of grapes, 6.9 lb of apples, 3 lb of pears, 4.33 lb of bananas, and 3.05 lb of nectarines. She cannot carry more than 20 lb because of a bad back. Will she be able to carry all her purchases?

113. A dress was marked down 20%. If the original price was \$59, what was the sale price?

114. Jack decided to sell his house for 7% more than he paid for it. If he paid \$62,000 for it, what will his selling price be?

115. John studied his math for 2 hours (hr) every day. He thinks he can make an A if he increases his daily study time by 25%. How much longer each day will he study?

In Your Own Words

116. Describe how to convert 24% into a fraction.

117. How do we add decimals?

118. How do we multiply decimals?

119. How do we divide decimals?

0.4 Angles, Lines, Plane Figures

TAPE 1

This section is a review of some basic ideas and terminology from geometry. Such fundamental ideas as point, line, angle, and length are left undefined. No attempt is made to use precise notation. Angles with the same measure are said to be equal. The symbol $\angle$ indicates the measure of an angle. Lowercase letters are used to designate the length of line segments.

Angles

An *acute angle* is an angle of less than 90°.

A *right angle* is an angle of exactly 90°. The symbol $\llcorner$ is used in figures to indicate a right angle.

An *obtuse angle* is an angle of greater than 90°, but less than 180°.

A *straight angle* is an angle of exactly 180°. Its sides form a line.

Complementary angles are two angles whose sum is 90°. Each is called the *complement* of the other.

Supplementary angles are two angles whose sum is 180°. Each is called the *supplement* of the other.

Acute Angle   Right Angle   Obtuse Angle   Straight Angle

Complementary Angle   Supplementary Angle

Two intersecting lines form the angles numbered 1, 2, 3, and 4.

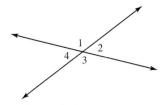

Angles 1 and 3 are *vertical* angles as are angles 2 and 4. Vertical angles are equal. Angles that have a common vertex and a common side are called *adjacent* angles. Angles 1 and 2 are adjacent angles as are angles 2 and 3, and angles 3 and 4, and angles 1 and 4.

► **EXAMPLE 1**

Are the following angles complementary or supplementary?

Solution

The sum of the measures of the two angles is 90°. The angles are complementary. ◄

Lines

Two lines that lie in the same plane and do not intersect are called *parallel lines.*

Two lines that intersect at right angles are called *perpendicular.*

Parallel Lines   Perpendicular Lines

A *transversal* is a line that intersects two or more lines at different points. A transversal of two lines forms some angles that are given special names.

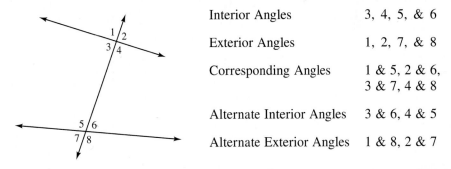

Interior Angles	3, 4, 5, & 6
Exterior Angles	1, 2, 7, & 8
Corresponding Angles	1 & 5, 2 & 6, 3 & 7, 4 & 8
Alternate Interior Angles	3 & 6, 4 & 5
Alternate Exterior Angles	1 & 8, 2 & 7

Parallel Lines Cut by a Transversal

If two lines are cut by a transversal, the lines are parallel if any of the following statements are true.

1. Corresponding angles are equal.
2. Alternate interior angles are equal.
3. Alternate exterior angles are equal.

If two lines are parallel, all the statements are true.

► EXAMPLE 2

Determine the measure of each angle in the figure. Lines l_1 and l_2 are parallel. $\angle 1 = 120°$.

Solution

Angles 1 and 2 are supplementary angles. So $\angle 2$ is $60°$.

Angles 1, 4, 5, and 8 are equal. So each is $120°$.

Angles 2, 3, 6, and 7 are equal. Thus, each is $60°$.   ◄

A *polygon* is a closed plane figure bounded by line segments. Triangles, quadrilaterals, pentagons, hexagons, and octagons are polygons with 3, 4, 5, 6, and 8 sides, respectively. If all the sides are of equal length, a polygon is called *equilateral*. If all the angles are equal, it is said to be *equiangular*. A *regular* polygon is one that is both equilateral and equiangular. Several polygons are shown next. The dotted lines are called *diagonals*.

Equilateral

Equiangular

Regular

A *triangle* is a polygon with three sides.

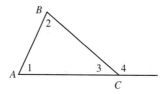

The three points *A*, *B*, and *C* are the *vertices* (singular, *vertex*) of the triangle. The three angles 1, 2, and 3 are the *interior angles* of the triangle, or simply the angles of the triangle. The three sides of a triangle and the three angles are sometimes called the *parts* of a triangle. Angle 4 is called an *exterior angle* of the triangle.

Two important relationships exist for these angles. The sum of the measures of the interior angles is $180°$. The measure of an exterior angle is equal to the sum of the measures of the opposite interior angles.

$$\angle 1 + \angle 2 + \angle 3 = 180°$$

$$\angle 4 = \angle 1 + \angle 2$$

► **EXAMPLE 3**

Determine $\angle A$ in each figure.

(a)

(b)

Solutions

(a) Since the figure shows $\angle B = 90°$, $\angle A$ must be $60°$.

(b) $\angle A + 20° = 115°$
     $\angle A = 95°$ ◄

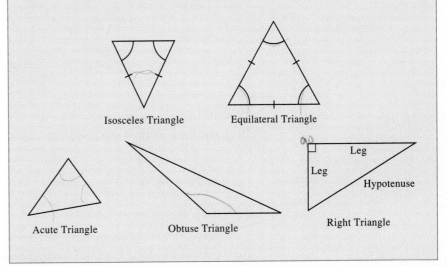

Triangles

An *isosceles triangle* has two sides of equal length. The angles opposite the equal sides are equal.

An *equilateral triangle* has all three sides equal. It is also equiangular.

An *acute triangle* has three acute angles.

An *obtuse triangle* has one obtuse angle.

A *right triangle* contains a right angle. The longest side of a right triangle (the side opposite the right angle) is called the *hypotenuse.* The other two sides are called the *legs* of a right triangle.

Isosceles Triangle   Equilateral Triangle

Acute Triangle   Obtuse Triangle   Right Triangle

The following result is one of the most useful relationships in all mathematics.

Pythagorean Theorem

In the right triangle above, the side of length c units is the hypotenuse and the sides of lengths a and b units are the legs. The square of the length of the hypotenuse equals the sum of the squares of the lengths of the legs. That is,

$$c^2 = a^2 + b^2$$

► **EXAMPLE 4**

Verify the Pythagorean Theorem for a right triangle with sides of 3, 4, and 5 inches.

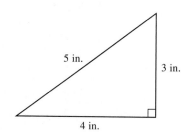

Solution

The hypotenuse is the longest side. So it must be 5.

$$5^2 = 25$$

The legs are 3 and 4.

$$3^2 + 4^2 = 9 + 16$$
$$= 25$$

Thus, the square of the length of the hypotenuse equals the sum of the squares of the lengths of the two legs.   ◄

A geometric property that is of great practical use involves triangles that have the same shape.

Similar Triangles

Two triangles are similar if any one of the following statements is true.

1. Two angles of one triangle equal two angles of the other.

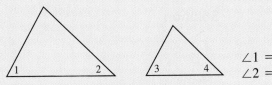

$$\angle 1 = \angle 3$$
$$\angle 2 = \angle 4$$

2. Corresponding sides are proportional.

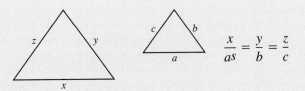

$$\frac{x}{as} = \frac{y}{b} = \frac{z}{c}$$

3. Two corresponding pairs of sides are proportional and the angle between them is equal.

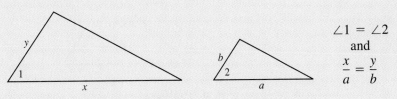

$$\angle 1 = \angle 2$$
and
$$\frac{x}{a} = \frac{y}{b}$$

If two triangles are similar, then all these statements are true.

► EXAMPLE 5

Are the triangles in the figure similar?

Solution

As we examine the figure, we notice a large triangle and a small triangle. Both are right triangles and they share the angle at the vertex A. This makes the angles at B and D equal. We see that they have two equal angles. Thus, by statement 1 above, they are similar triangles.   ◄

► **EXAMPLE 6**

The triangles are similar. Set up proportions for corresponding sides.

Solution

$$\frac{x}{4} = \frac{y}{5} = \frac{z}{2}$$ ◄

Another important situation arises when two triangles have the same shape and the same size.

Congruent Triangles

Two triangles are congruent if any one of the following statements is true.

1. Two angles and the included side of one triangle are equal to two angles and the included side of the other triangle (angle–side–angle).

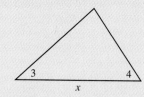

$$\angle 1 = \angle 3$$
$$\angle 2 = \angle 4$$
$$a = x$$

2. Corresponding sides are equal (side–side–side).

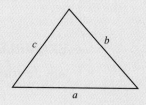

$$a = x$$
$$b = y$$
$$c = z$$

3. Two sides and the included angle of one triangle are equal to two sides and the included angle of the other triangle (side–angle–side).

$$a = x$$
$$b = y$$
$$\angle 1 = \angle 2$$

If two triangles are congruent, then all these statements are true.

▶ **EXAMPLE 7**

Are the triangles in the figure congruent?

Solution

We notice that the third angle in the left triangle can be calculated.

$$180° - 105° - 17° = 58°$$

So the left triangle has a side of length 10 between angles of 17° and 58°. Because the triangle on the right also has a side of length 10 between angles of measure 17° and 58°, we conclude that the triangles are congruent (angle–side–angle).   ◀

A *quadrilateral* is a four-sided plane figure.

Quadrilaterals

A *trapezoid* is a quadrilateral with exactly two sides parallel. The parallel sides are called *bases* and the other sides are called *legs*. A trapezoid with equal legs is called an *isosceles trapezoid*.

A *parallelogram* is a quadrilateral in which both pairs of opposite sides are parallel. The diagonals of a parallelogram bisect each other, and any two consecutive angles are supplements of each other.

A *rhombus* is a parallelogram with equal sides.

A *rectangle* is a parallelogram whose angles are right angles.
A *square* is a rectangle with equal sides.

Circles

A *circle* is the set of all points in a plane the same distance from a given point called the *center*. A *radius* of a circle is a line segment connecting the center with a point on the circle. A *chord* is a line segment connecting two points on the circle. A *diameter* is a chord that contains the center. A *tangent* to a circle is a line in the plane of the circle that intersects the circle in exactly one point.

Problem Set 0.4

Warm-Ups

1. Find the measure of each angle in the figure. Lines l_1 and l_2 are parallel. Lines l_3 and l_4 are perpendicular. $\angle b = 135°$. See Examples 1 and 2.

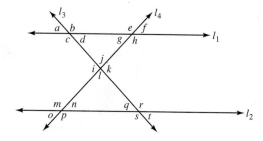

In Problems 2 through 5, find $\angle A$. *See Example 3.*

2.

3.

4.

5.

6. Verify the Pythagorean Theorem for a right triangle with sides of lengths 6, 8, and 10 units. See Example 4.

7. Why is each pair of triangles similar? See Example 5.
(a)

(b) $l_2 \| l_2$

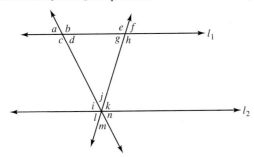

8. Why is each pair of triangles congruent? See Example 7.
(a) the parallelogram

(b) $l_1 \| l_2$

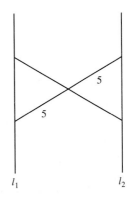

Practice Exercises

In Problems 9 through 20, determine if each statement is true or false. Use the following figure. Lines l_1 and l_2 are parallel. $\angle d = 60°$ and $\angle g = 75°$.

9. $\angle i = 75°$

10. $\angle j + \angle n = 180°$

11. $\angle g + \angle j = 105°$

12. $\angle g = 45°$

13. $\angle m + \angle n = 135°$

14. Angles h and k are supplementary.

15. $\angle l = 45°$

16. Angle b is an acute angle.

17. Angle f is obtuse.

18. Angles k and n are complementary.

19. $\angle b = 105°$

20. $\angle m + \angle n = 105°$

In Problems 21 through 26, find $\angle A$.

21.

22.

23.

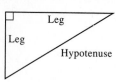

Leg

Leg

Hypotenuse

Right Triangle

24.

$40°$ A

25.

$88°$

A

$28°$

26.

A

$115°$ $133°$

27. Verify the Pythagorean Theorem for a right triangle with sides of lengths 5, 12, and 13 units.

In Problems 28 through 31, determine if each pair of triangles is congruent or similar but not congruent. Why?

28.

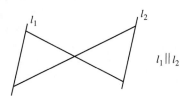

l_1 l_2

$l_1 \| l_2$

29.

3

4

6

8

30.

31.

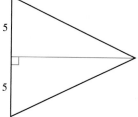

5

5

In Problems 32 through 38, determine whether each statement is true or false.

32. A rectangle is equiangular.

33. An isosceles triangle is an acute triangle.

34. A square is a rhombus.

35. A square is a regular polygon.

36. A triangle is a polygon.

37. The acute angles in a right triangle are complementary.

38. A rhombus is an equilateral polygon.

Challenge Problems

39. Figure out a formula for the number of diagonals that can be drawn from one vertex of a regular polygon.

In Your Own Words

40. What are congruent triangles?

41. What does the Pythagorean Theorem say?

▶ Chapter Summary

Glossary

Numeral   A symbol used to name a number.

Sum   The result of adding two or more numbers.

Difference   The result of subtracting one number from another number.

Product   The result of multiplying two or more numbers.

Quotient   The result of dividing one number by another number.

Exponent   The number of times the base is used as a factor.

Base   The number being multiplied the number of times given by the exponent.

Grouping symbols   Symbols such as (), [], {}, $\sqrt{}$, and fraction bars used to make the order of operations clear.

Factors   Numbers that are multiplied together.

Prime number   A natural number larger than 1 whose only factors are itself and 1.

Reduced to lowest terms   A fraction with no factors common to the numerator and denominator.

Square root   A positive number that when squared is a given number.

Order of Operations   If grouping symbols are present, perform operations inside them, starting with the innermost grouping symbol, in the following order.

1. Perform any exponentiations.

2. Perform all multiplications and divisions in order from left to right.

3. Perform all additions and subtractions in order from left to right.

If grouping symbols are not present, perform operations in the above order.

Operations with Fractions   Operations with fractions follow these rules ($p, q, r,$ and s are natural numbers).

Fundamental Principle of Fractions

$$\frac{p}{q} = \frac{p \cdot r}{q \cdot r}$$

Multiplication of Fractions

$$\frac{p}{q} \cdot \frac{r}{s} = \frac{p \cdot r}{q \cdot s}$$

Division of Fractions

$$\frac{p}{q} \div \frac{r}{s} = \frac{p}{q} \cdot \frac{s}{r} = \frac{p \cdot s}{q \cdot r}$$

Addition and Subtraction of Fractions

$$\frac{p}{q} + \frac{r}{q} = \frac{p + r}{q}, \qquad \frac{p}{q} - \frac{r}{q} = \frac{p - r}{q}$$

Procedure for Finding the Least Common Denominator (LCD)

1. Factor each denominator completely, using exponents.
2. List all the different prime factors from all the denominators.
3. The LCD is the product of the factors in step 2 each raised to the largest power of that factor in any single denominator.

Percent–Decimal Conversion

Decimal to Percent

1. Move the decimal point *two* places to the right.
2. Write the % symbol.

Percent to Decimal

1. Replace any fraction with an equivalent decimal.
2. Move the decimal point *two* places to the left.
3. Remove the % symbol.

Perimeter and Area Formulas

Figure	Sketch	Perimeter	Area
Triangle		$P = a + b + c$	$A = \dfrac{1}{2}bh$
Rectangle		$P = 2l + 2w$	$A = lw$
Parallelogram		$P = 2a + 2b$	$A = bh$
Square		$P = 4s$	$A = s^2$
Trapezoid		$P = b_1 + s + t + b_2$	$A = \dfrac{1}{2}h(b_1 + b_2)$
Circle		$C = 2\pi r$	$A = \pi r^2$

CHECK UPS

In Problems 1 and 2, express each phrase in symbols.

1. The product of 2 and 7. Section 0.1; Example 1d
2. The quotient when 12 is divided by 19. Section 0.1; Example 1e

In Problems 3 and 4, perform the operations indicated.

3. $10 - 3 + 2^2$ Section 0.1; Example 3b
4. $12 - 2(6 - 3)$ Section 0.1; Example 3d

In Problems 5 and 6, evaluate the expressions.

5. $\sqrt{25}$ Section 0.1; Example 4c

6. $\sqrt{144 + 25}$ Section 0.1; Example 5a

7. Factor 392 into prime factors.   Section 0.2; Example 1c

In Problems 8 and 9, perform the operations indicated.

8. $\dfrac{27}{16} \div \dfrac{9}{8}$ Section 0.2; Example 5

9. $\dfrac{3}{4} + \dfrac{2}{3} \cdot \dfrac{1}{2}$ Section 0.2; Example 9a

10. Write 41.6% as a decimal.   Section 0.3; Example 5d

11. Find 23% of 3300.   Section 0.3; Example 8a

▶ Review Problems

In Problems 1 through 10, perform the operations indicated.

1. $\dfrac{6}{7} - \dfrac{2}{3}$

2. $\dfrac{1}{2} \cdot \dfrac{4}{9} + \dfrac{5}{6}$

3. $18/6 \cdot 5$

4. $2^2 - (10 - 8)$

5. $\dfrac{6(3) - 3^2}{6 - 4}$

6. $12 - 2 \cdot 3 + 6 \div 3$

7. $\left(\dfrac{1}{2}\right)^2 - \dfrac{4}{9} \cdot \dfrac{9}{16}$

8. $6.3(2.1) + 8.4/0.04$

9. $7 - \dfrac{2}{3}$

10. $\left(1 + \dfrac{1}{2}\right)^2$

In Problems 11 through 15, express each phrase in symbols.

11. Twice 5 increased by 3.

12. One less than 9.

13. The sum of one-half of 7 and twice 4.

14. The difference when twice 2 is subtracted from twice 8.

15. The sum of 4 and 6 subtracted from 15.

In Problems 16 through 20, answer each question.

16. Jack has 7 baseball cards. Roberto has 2 fewer baseball cards than Jack. How many cards does Roberto have?

17. Sue bought 4.32 lb of chicken, 3.72 lb of hamburger, 8.18 lb of steak, and 1 lb of bacon. How many pounds of meat did she buy?

18. Wendell budgeted 17% of his money for entertainment. If he has $212, how much did he budget for entertainment?

19. Find 50% of 300.

20. Bill, Joe, and Sam need 50 ft of rope. Bill has $12\frac{1}{2}$ ft, Joe has $20\frac{1}{4}$ ft, and Sam has $15\frac{7}{8}$ ft. Do they have enough rope?

In Problems 21 through 25, express each percent as a fraction and a decimal.

21. 15%

22. 3%

23. 13.5%

24. 12 1/2%

25. 83 1/3%

In Problems 26 through 30, express each number as a percent.

26. $\dfrac{13}{100}$

27. 0.18

28. 0.08

29. $\dfrac{5}{6}$

30. 2.5

In Problems 31 through 35, factor each number completely.

31. 16

32. 84

33. 100

34. 1944

35. 400

In Problems 36 through 43, evaluate the given expressions.

36. $\sqrt{4}$

37. $\sqrt{4900}$

38. $\sqrt{169 - 25}$

39. $\sqrt{6400}$

40. $\sqrt{400}$

41. $\sqrt{169} - \sqrt{25}$

42. $\sqrt{144 + 81}$

43. $\sqrt{144} + \sqrt{81}$

44. Find the area and perimeter of a rectangle with dimensions $12\frac{1}{2}$ cm by $8\frac{1}{3}$ cm.

45. Find the area and circumference of a circle with radius $6\frac{3}{4}$ m.

► Chapter 0 Test

In Problems 1 through 5, choose the correct answer.

1. When reduced, $\dfrac{15}{60} = (?)$.

(a) $\dfrac{1}{4}$ (b) $\dfrac{1}{5}$

(c) $\dfrac{1}{2}$ (d) $\dfrac{1}{3}$

2. When factored completely, $96 = (?)$.

(a) $16 \cdot 6$ (b) $24 \cdot 4$

(c) $32 \cdot 3$ (d) $2^5 \cdot 3$

3. $12\frac{1}{2}\% = (?)$.

(a) 0.125   (b) 1.25

(c) 12.5   (d) 125

4. $1.2/0.06 - (1.3)(0.05) = (?)$.

(a) 1.935   (b) 19.935

(c) 199.935   (d) 374

5. $\sqrt{25 - 9} = (?)$.

(a) 2   (b) 4

(c) 8   (d) 16

In Problems 6 through 10, perform the operations indicated.

6. $2 + 3^2$

7. $23.01 - 6.1(2.0)$

8. $\left(\dfrac{2}{3}\right)^3 + \dfrac{1}{3} - \dfrac{2}{9}$

9. $\sqrt{25 + 144}$

10. $\dfrac{3 \cdot 2^2 - (6 - 3)}{8 - 5}$

11. Find $22\frac{1}{2}\%$ of 60.

14. Simplify $\dfrac{2}{3} \div \dfrac{8}{15} + 2$

12. Find the area of a rectangle with dimensions $2\frac{3}{4}$ in. by $1\frac{1}{5}$ in. (area = length × width).

15. Katy bought $4\frac{1}{2}$ yd of cotton, $2\frac{7}{8}$ yd of polyester, and $6\frac{1}{4}$ yd of denim. How many yards of fabric did she buy?

13. Write $\dfrac{3}{8}$ as a percent.

1 Real Numbers

See Problem Set 1.2, Exercise 39

Connections

For over 6000 years, people have been using numbers and the properties of numbers. The need to record information and transmit knowledge from one place to another forced the development of a written system of numeration. Today, most of the world uses the Hindu-Arabic numeral system, named for the Hindus, who probably invented it, and for the Arabs, who introduced it to Western Europe.

At first, small counting numbers like 1, 2, and 3 were sufficient for barter. As society developed, the growth of commerce dictated a need for larger numbers, negative numbers, and fractions. Numbers like $\frac{5}{8}$ were used by merchants to indicate portions of goods, and numbers like -60 were used to indicate debts.

In today's world, stock market gains and losses are reported using signed numbers like $+5\frac{1}{8}$ and -4, and weather forecasts routinely use negative integers to report temperatures. Scientists and engineers use numbers like $\sqrt{2}$ and π in their daily work.

The system of real numbers discussed in this chapter is a natural extension of the arithmetic of natural numbers discussed in Chapter 0 and provides the basic number system of algebra. We begin by learning the arithmetic of real numbers and their properties. The idea of using a letter for a number is introduced, and we form expressions and equations. These ideas are essential concepts of algebra that are examined in more detail in later chapters.

1.1   Sets of Numbers

DISK 2

A set is a collection of objects. The objects are called **elements,** or **members.** We usually use capital letters to name sets and list the elements in braces. For example, $A = \{a, b, c, d, e\}$ is a set containing the first five letters of the alphabet. $B = \{1, 2, 3, \ldots, 10\}$ is a set containing the counting numbers 1 through 10. (The three dots, called an ellipsis, means "continue on in the pattern that has been established.") $N = \{1, 2, 3, \ldots\}$ is a set containing all the counting numbers.

If the number of elements in a set is a counting number or zero, the set is called a **finite** set. If there are no elements at all in a set, the set is the **empty set.** Otherwise, a set is called an **infinite** set. Of the sets listed above, A and B are finite, and N is infinite.

The empty set is sometimes called the **null set.** It is written $\{ \}$ or $\varnothing$. It would be incorrect to write the empty set as $\{\varnothing\}$, as the set $\{\varnothing\}$ contains one element.

The symbol $\in$ is used to mean **is an element of,** and the symbol $\notin$ is used to mean **is not an element of.** If $D = \{a, b, c, d, e\}$, then we see that $a \in D$, but $2 \notin D$.

Sometimes we don't list the elements of the set. Instead, we use a notation called **set-builder notation.** $\{x \mid x \text{ is a vowel}\}$ describes the "set of all x such that x is a vowel." The bar is read "such that." It is the same set as {a, e, i, o, u}.

Set A is a **subset** of a set B if every element in A is also an element in B. We use the symbol $\subseteq$ to mean **is a subset of.** If $A=\{1, 5, 7\}$ and $B=\{1, 3, 5, 7, 9\}$, then $A \subseteq B$.

► **EXAMPLE 1**

List all the subsets of $\{a, b, c\}$.

Solution

The subsets are

$$\{a, b, c\} \qquad \{a, b\} \qquad \{a\} \qquad \varnothing$$
$$\{a, c\} \qquad \{b\}$$
$$\{b, c\} \qquad \{c\}$$

◄

Notice that the set is a subset of itself and that the empty set is a subset of every set.

> **Subsets of Every Set**
>
> If A is any set, $A \subseteq A$ and $\varnothing \subseteq A$.

Many times we combine sets using the operations of **union** ($\cup$) and **intersection** ($\cap$).

> **Union of Two Sets**
>
> If A and B are sets,
>
> $$A \cup B = \{x \mid x \in A \text{ or } x \in B\}$$

That is, the union of two sets is the set of all elements that belong to either of the two sets (or both of them).

> **Intersection of Two Sets**
>
> If A and B are sets,
>
> $$A \cap B = \{x \mid x \in A \text{ and } x \in B\}$$

The intersection of two sets is the set of all elements that belong to both of the two sets.

► **EXAMPLE 2**

If $A = \{1, 2, 7\}$ and $B = \{1, 2, 3, 4, 5\}$, find $A \cup B$ and $A \cap B$.

Solutions

$$A \cup B = \{1, 2, 3, 4, 5, 7\}$$
$$A \cap B = \{1, 2\}$$
◄

John Venn, an English mathematician, popularized the use of diagrams to help visualize sets. Called **Venn diagrams,** these diagrams use circles to indicate the contents of sets. For example, if circle A indicates the contents of set A and circle B shows the contents of set B, we can use shading to indicate the contents of various sets.

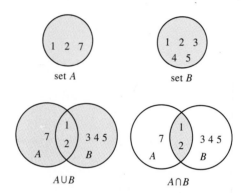

In algebra we are concerned with sets of numbers. We worked with counting numbers, or natural numbers, in Chapter 0, and in the beginning of this section we named the set of counting numbers N.

> **Natural Numbers or Counting Numbers**
> $$N = \{1, 2, 3, \ldots\}$$

If we put the number 0 into this set, we have a set called the set of whole numbers. We name the set W and write it as follows.

> **Whole Numbers**
> $$W = \{0, 1, 2, 3, \ldots\}$$

It is much easier to visualize sets of numbers if we draw a picture of them. This is done by using a **number line.** We draw a line and pick a starting point called the **origin** and label it with the number 0. We move to the right of 0 and represent natural numbers in increasing order. We space the numbers at equal distances with a convenient unit of measure.

The number associated with a point on a number line is called the point's **coordinate.** To **graph** a set of numbers means to locate them on a number line.

There are points to the left of 0 as well as to the right. If we move to the left of 0 with the same unit of measure, we name the points associated with negative numbers. The number one unit to the left of 0 is called -1 (read "negative one"). The number two units to the left of 0 is called -2, and so on.

The whole numbers and these negative numbers form another set of numbers. We call this set of numbers the integers and name it Z.

Integers

$$Z = \{\ldots, -3, -2, -1, 0, 1, 2, 3, \ldots\}$$

The digits 0, 2, 4, 6 and 8 are called **even.** The digits 1, 3, 5, 7, and 9 are called **odd.** An **even integer** is an integer that ends in an even digit, whereas an **odd integer** is one that ends in an odd digit.

Fractions are placed on the number line using the same unit of measure.

This new set of numbers is called the **rational numbers.** We name it Q and write it in set notation as follows.

Rational Numbers

$$Q = \left\{x \mid x \text{ can be written as } \frac{p}{q}, \text{ where } p \text{ and } q \text{ are integers and } q \neq 0\right\}$$

Notice that, since $5 = \frac{5}{1}$, the integer 5 is a rational number. Likewise, all integers are rational numbers.

Since a rational number is the quotient of two integers, we often refer to rational numbers as fractions. We saw in Chapter 0 that every quotient of natural numbers can be written as a decimal that repeats or terminates. So every rational number can be written as a fraction or as a terminating or repeating decimal.

There are other points on the number line that we have not named. When written in decimal form, they do not repeat or terminate. They are called the **irrational numbers.** Some examples of irrational numbers are $\sqrt{2}$, $\sqrt{3}$, and π. In fact, $\sqrt{x}$, where x is a positive integer but not a perfect square, is an irrational number.

Irrational Numbers

$I = \{x \mid x$ is on the number line but is not a rational number$\}$

We have now named all the numbers on the number line. We call all these numbers the **real numbers.** We name the set of real numbers R and with set notation write them as follows.

Real Numbers

$R = \{x \mid x$ is a rational or an irrational number$\}$

The following diagram shows how these sets are related to each other.

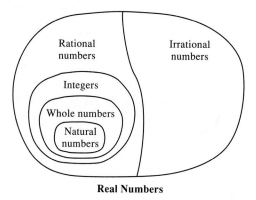

Real Numbers

► **EXAMPLE 3**

Classify each number as rational or irrational.

(a) $\dfrac{3}{7}$ (b) $0.\overline{78}$ (c) $\sqrt{7}$ (d) 1.7

Solutions

(a) $\dfrac{3}{7}$ is rational.

(b) $0.\overline{78}$ is a repeating decimal; thus it is rational.

(c) $\sqrt{7}$ is irrational, since 7 is not a perfect square.

(d) 1.7 is rational because $1.7 = \dfrac{17}{10}$.   ◄

The number line helps us to understand many concepts of numbers. If a number p lies to the left of another number q on a number line, we say that p is **less than** q.

Less Than

If p lies to the left of q on a number line,

$$p < q$$

We read this "p is less than q."

Likewise, if a number r lies to the right of a number s on a number line, we say that r is **greater than** s, which we write as $r > s$.

Greater Than

If r lies to the right of s on a number line,

$$r > s$$

We read this "r is greater than s."

Sometimes we combine equals with less than or greater than, using the symbols $\leq$ for **less than or equal to** and $\geq$ for **greater than or equal to.**

Less Than or Equal To; Greater Than or Equal To

$p \leq q$ means that p is less than or equal to q.
$r \geq s$ means that r is greater than or equal to s.

► **EXAMPLE 4**

Indicate whether each statement is true or false.

(a) $-\dfrac{1}{2} < 0$

(b) $\dfrac{3}{5} < \dfrac{2}{3}$

(c) $-1.4 > -1.2$

(d) $5 \leq 5$

(e) $-\dfrac{1}{2} \leq \dfrac{1}{2}$

Solutions

(a) $-\dfrac{1}{2} < 0$

Because $-\frac{1}{2}$ lies to the left of 0, the statement is true.

(b) $\dfrac{3}{5} < \dfrac{2}{3}$

To see where these numbers fall on a number line, we write them with the same denominator.

$\dfrac{3}{5} = \dfrac{9}{15}$ and $\dfrac{2}{3} = \dfrac{10}{15}$

So, $\dfrac{9}{15}$ lies to the left of $\dfrac{10}{15}$. Thus $\dfrac{3}{5} < \dfrac{2}{3}$ is a true statement.

(c) $-1.4 > -1.2$

Since -1.4 lies to the left of -1.2 on a number line, the statement is false.

(d) $5 \le 5$

Since $5 = 5$, the statement is true.

(e) $-\dfrac{1}{2} \le \dfrac{1}{2}$

Because $-\dfrac{1}{2}$ lies to the left of $\dfrac{1}{2}$ on a number line, the statement is true.

◀

Suppose that the numbers p and q are graphed on a number line:

We can see that p is less than q and also that q is greater than p. When we write $p < q$ and read it as p is less than q, we are reading from left to right. If we read $p < q$ from right to left, it says q is greater than p. We usually read from left to right.

The numbers to the right of 0 are called the **positive** numbers, and the numbers to the left of 0 are called the **negative** numbers. Zero is neither positive nor negative.

```
      Negative            Positive
    real numbers        real numbers
    ←————————————+————————————→
                 0
```

In set notation we write these sets as follows.

Positive Real Numbers

$$\{x \mid x > 0\}$$

Negative Real Numbers

$$\{x \mid x < 0\}$$

► **EXAMPLE 5**

Locate each number on a number line.

(a) x, where x is a positive real number.

(b) y, where y is a negative real number.

(c) x and y, where $x>0$ and $y<0$.

Solutions

(a) x lies to the *right* of 0

(b) y lies to the *left* of 0

(c)

◄

The points on the number line whose coordinates are 3 and -3 are on opposite sides of 0 but are the same distance from 0.

We call 3 and -3 *opposites*.

> **Additive Inverse or Opposite**
>
> If p is a real number, then $-p$ is the real number that is on the opposite side of 0 and the same distance from 0 as p. We call $-p$ the **additive inverse,** or **opposite,** of p.

► **EXAMPLE 6**

Find the additive inverse of each number.

(a) $\dfrac{2}{3}$ (b) -0.2 (c) 0 (d) x

Solutions

(a) The additive inverse of $\dfrac{2}{3}$ is $-\dfrac{2}{3}$.

(b) The additive inverse of -0.2 is 0.2.

(c) The additive inverse of 0 is 0.

(d) The additive inverse of x is $-x$.

◄

If p is a positive number, then p is to the right of 0. The opposite of p, which we call $-p$, is to the left of 0.

If q is a negative number, then q lies to the left of 0. The opposite of q, which we call $-q$, is to the right of 0. Notice that $-q$ is a positive number because q is negative.

The opposite of -3 is 3. We can write this by saying

$$-(-3) = 3$$

What is the opposite of the opposite of a number p? This question sounds confusing. However, if we write it in symbols, it is less confusing. The opposite of p is $-p$. So the opposite of the opposite of p is $-(-p)$. What number lies on the opposite side of 0 from $-p$? It is p.

Double-negative Property

If p is a real number, $-(-p) = p$.

► **EXAMPLE 7**

Simplify each of the following.

(a) $-(-5)$ (b) $-\left(-\dfrac{3}{4}\right)$ (c) $-(-z)$

Solutions

(a) $-(-5) = 5$ (b) $-\left(-\dfrac{3}{4}\right) = \dfrac{3}{4}$ (c) $-(-z) = z$ ◄

We have seen three uses of the symbol $-$. We have used it to indicate the *operation of subtraction,* as part of the *name of a number,* and to indicate the *opposite,* or *additive inverse,* of a number. Sometimes we read it as "minus," sometimes as "subtract," sometimes as "negative," and sometimes as "the opposite of."

$$\boxed{1}\ 6-2 \qquad \boxed{2}\ -5 \qquad \boxed{3}\ -p$$

In $\boxed{1}$ the $-$ symbol indicates subtraction and we read it "6 minus 2" or "6 subtract 2." In $\boxed{2}$ it indicates the name of a number, negative 5. In $\boxed{3}$ it indicates the opposite of p and is read "the opposite of p" or "negative p."

A number and its opposite have the common property that each is the same distance from 0. Both 3 and -3 are 3 units from 0. We name this distance from 0 the **absolute value** of a number and indicate it by vertical bars, $|\ \ |$. $|\,3\,|$ is the distance between 0 and 3. So $|\,3\,| = 3$. Likewise, $|-3|$ is the distance between 0 and -3. So $|-3| = 3$. Thus we see that $|\,3\,| = |-3|$.

Absolute Value

If p is a real number, the absolute value of p, $|\,p\,|$, is its distance from 0.

► **EXAMPLE 8**

Simplify each of the following.

(a) $|-5|$ (b) $2|17|$ (c) $-|-7|$ (d) $|-\pi|$

Solutions

(a) On the number line we see that -5 is 5 units from 0.

|←5 units→|

-5 0

Thus $|-5| = 5$.

(b) Again we look at the number line.

|————17 units————|

0 17

So $|17| = 17$; then

$$2|17| = 2 \cdot 17$$
$$= 34$$

(c) |←—7 units—→|

-7 0

Thus $|-7| = 7$, and so $-|-7| = -7$.

(d) π units
       |←——→|

$-\pi$ 0

And so $|-\pi| = \pi$.   ◄

PROBLEM SET 1.1

Warm-Ups

In Problems 1 through 6, A = {2, 4, 6, 8}, B = {1, 4}, and C = {−2, 0, 2}. For Problems 1 and 2, see Example 1.

1. List all the subsets of B.

2. List all the subsets of C.

For Problems 3 through 6, see Example 2.

3. List the elements in $A \cup B$.

4. List the elements in $A \cup C$.

5. List the elements in $A \cap B$.

6. List the elements in $A \cap C$.

In Problems 7 through 10, classify each number as rational or irrational. See Example 3.

7. 5.41

8. $\dfrac{3}{8}$

9. $\sqrt{11}$

10. $2.\overline{12}$

In Problems 11 through 16, indicate whether each statement is true or false. See Example 4.

11. $-\dfrac{1}{3} < -\dfrac{1}{4}$

12. $1.1 \geq 1.1$

13. $-5 \leq -10$

14. $5.7 > 0$

15. $-1.12 \leq -1.25$

16. $-0.111 > 2.333$

In Problems 17 through 20, graph each number on a number line. Assume that x is a positive real number and that y is a negative real number. See Example 5.

17. x

18. y

19. $-x$

20. $-y$

In Problems 21 through 24, find the additive inverse (opposite) of each number. See Example 6.

21. -3.5

22. $\sqrt{5}$

23. $\dfrac{11}{13}$

24. z

Simplify each of the following. See Examples 7 and 8.

25. $-(-6)$

26. $-(-35)$

27. $-(-\pi)$

28. $-(-q)$

29. $|-12|$

30. $|29|$

31. $2|-8|$

32. $3|6|$

33. $-|4|$

34. $-|-5|$

35. $-|-15|$

36. $-|-3|$

Practice Exercises

In Problems 37 through 39, $A = \left\{\dfrac{1}{2}, 2, \dfrac{9}{4}, 5\right\}$ and $B = \left\{\dfrac{9}{4}, 5\right\}$.

37. List the elements in $A \cup B$.

38. List the elements in $A \cap B$.

39. List all the subsets of B.

In Problems 40 through 59, indicate whether each statement is true or false.

40. The set of natural numbers is a subset of the set of whole numbers.

41. Every negative integer is a rational number.

42. $\frac{1}{2}$ is a real number.

43. 0 is a rational number.

44. The positive integers are a subset of the whole numbers.

45. A whole number is a nonnegative integer.

46. $\sqrt{16}$ is an irrational number.

47. The empty set is written as $\{\emptyset\}$.

48. Every rational number is an integer.

49. $\sqrt{11}$ is real.

50.

$a > 0$

51.

$b < 0$

52.

$-c < 0$

53.

$-d < 0$

54. $-\dfrac{2}{3} < -\dfrac{5}{8}$

55. $7 \leq 7$

56. $-5 \leq -1$

57. $-5.7 > 0$

58. $-3.415 < 2.333$

59. $-3.14 \leq -3.18$

60. Fill in the chart to indicate the set(s) to which each number belongs. N = natural numbers, W = whole numbers, Z = integers, Q = rational numbers, I = irrational numbers, R = real numbers.

Number	N	W	Z	Q	I	R
$\dfrac{4}{3}$						
-1						
$\sqrt{7}$						
π						
$\sqrt{25}$						
10						
$1.\overline{32}$						
-2.3						
$\dfrac{20}{10}$						

In Problems 61 through 72, simplify each.

61. $|-2|$

62. $|29|$

63. $2|-4|$

64. $|9|$

65. $-(-5)$

66. $-(-98)$

67. $-(-3.14)$

68. $-(-k)$

69. $-|8|$

70. $-|-11|$

71. $-|-37|$

72. $-|-6|$

73. Write three consecutive integers if the smallest is 48.

74. Write three consecutive *even* integers if the smallest is 88.

75. Write three consecutive *odd* integers if the smallest is 37.

76. Write three consecutive integers if the largest is 20.

Challenge Problems

77. Express 0.125 as a quotient of two integers.

78. Express 1.2 as a quotient of two integers.

In Problems 79 through 82, list the elements of A.

79. $A = \{x \mid x$ is a natural number less than 10$\}$

80. $A = \{x \mid x$ is an integer between -5 and 2$\}$

81. $A = \{x \mid x$ is a natural number divisible by 5$\}$

82. $A = \{x \mid x$ is an integer divisible by 5$\}$

In Problems 83 through 86, write A using set-builder notation.

83. $A = \{2, 4, 6, \ldots\}$

84. $A = \{2, 4, 6, 8, 10, 12\}$

85. $A = \{-3, -2, -1, 0, 1, 2, 3\}$

86. $A = \{3, 4, 5, 6, 7, 8\}$

In Your Own Words

87. Describe the set of integers.

88. What does it mean to say that the number x is less than 4?

89. What do we mean by the absolute value of a number?

90. What do we mean by the opposite of a number?

1.2 Addition

DISK 2

In this section we will study addition of real numbers. In arithmetic we add natural numbers. In algebra we must learn how to add real numbers. We know that $2 + 3 = 5$, but what about $-2 + 3$?

To help us understand how the rules for addition work, imagine that we are statisticians for a football game. We use a positive number to represent yards gained and a negative number to represent yards lost. If two plays gained 2 yd and 3 yd, then the total yardage gained would be 5 yd, which is the sum of 2 and 3. Now suppose that the first play lost 2 yd and the second play gained 3 yd. The total yardage gained would be $-2 + 3$, which is 1 yd. So,

$$-2 + 3 = 1$$

Likewise, a gain of 2 yd followed by a loss of 3 yd would give a total loss of 1 yd. So, $2 + -3$ must equal -1. That is,

$$2 + -3 = -1$$

This looks awkward, so we use parentheses around the -3 to separate the two symbols $+$ and $-$.

$$2 + (-3) = -1$$

A loss of 2 yd and another loss of 3 yd would give a total loss of 5 yd. So

$$-2 + (-3) = -5$$

These ideas can be pictured using a number line. We think of moving to the right as a gain and moving to the left as a loss. So, to illustrate that $-2 + 3 = 1$, we start at 0 and move to the left two units. This puts us at -2.

From there we move 3 units to the right.

We arrive at 1.

We have two rules for the addition of real numbers. One rule is for addition of numbers that are both positive or both negative and the other is for numbers that are not both positive or not both negative. We must use the idea of absolute value to write the rules.

Addition of Real Numbers

1. To add two numbers with *like signs* (both positive or both negative), add their absolute values and keep the common sign.
2. To add two numbers with *unlike* signs (one positive and one negative), subtract the smaller absolute value from the larger absolute value and keep the sign of the number with the greater absolute value.

Using the rules is easier than reading the rules. Practicing with many problems is the secret to success.

The next example uses the first rule. Notice how the rule is used. Each problem can be verified using a number line.

▶ **EXAMPLE 1**

Find the following sums.

(a) $-5 + (-9)$ (b) $-7 + (-12)$

Solutions

(a) We use the first rule, since both numbers are negative. $|-5| = 5$ and $|-9| = 9$. We add 5 and 9 and keep the negative sign.

$$-5 + (-9) = -14.$$

(b) Again, we use the first rule. $|-7| = 7$ and $|-12| = 12$. We add 7 and 12 and keep the negative sign.

$$-7 + (-12) = -19$$ ◀

The next example uses the second rule.

▶ **EXAMPLE 2**

Find the sums.

(a) $-5 + 9$ (b) $5 + (-9)$

Solutions

(a) We use the second rule, since one number is positive and the other is negative.

$$|-5| = 5 \quad \text{and} \quad |9| = 9$$

So we subtract 5 from 9 and keep the positive sign, since $|9|$ is greater than $|-5|$.

$$-5 + 9 = 4$$

(b) Again we use the second rule.

$$|5| = 5 \quad \text{and} \quad |-9| = 9$$

We subtract 5 from 9 and keep the negative sign, since $|-9|$ is greater than $|5|$.

$$5 + (-9) = -4$$ ◀

▶ **EXAMPLE 3**

Find each sum.

(a) $4 + (-8)$ (b) $-4 + (-7)$ (c) $-7 + 5$

Solutions

(a) $4 + (-8) = -4$

(b) $-4 + (-7) = -11$

(c) $-7 + 5 = -2$ ◀

When we add 0 to any real number, we get the same number.

Addition Property of Zero

If p is a real number,

$$p + 0 = p \quad \text{and} \quad 0 + p = p$$

The order of operations that we learned in Chapter 0 applies to real numbers as well as natural numbers.

Order of Operations

If grouping symbols are present, perform operations inside them, starting with the innermost grouping symbol, in the following order.

1. Perform any exponentiations.
2. Perform all multiplications and divisions in order from left to right.
3. Perform all additions and subtractions in order from left to right.

If grouping symbols are *not* present, perform operations in the above order.

► **EXAMPLE 4**

Perform the operations indicated.

(a) $5 + (-8 + 6)$ (b) $[5 + (-13)] + [-6 + (-8)]$
(c) $\{-1.2 + [2.7 + (-6.1)]\}$

Solutions

(a) We work inside the parentheses first.

$$5 + (\boxed{-8 + 6}) = 5 + (\boxed{-2})$$
$$= 3$$

(b) We work inside the brackets first.

$$[5 + (-13)] + [-6 + (-8)] = [-8] + [-14]$$
$$= -22$$

(c) We begin with the brackets.

$$\{-1.2 + \boxed{[2.7 + (-6.1)]} \} = \{-1.2 + \boxed{(-3.4)} \}$$
$$= -4.6 \qquad ◄$$

Fraction bars, absolute value bars, and square root symbols are also grouping symbols. That is, we perform operations *over* and *under* the fraction bar, *inside* the absolute value symbols, and *inside* the square root symbol *before* we evaluate further.

► **EXAMPLE 5**

Evaluate the following:

(a) $|-4 + 9|$ (b) $|-4| + |9|$ (c) $\sqrt{25 + 144}$

(d) $\sqrt{25} + \sqrt{144}$ (e) $\dfrac{9 + 5}{9 - 2}$

Solutions

(a) Since absolute value bars are a grouping symbol, we must perform the operations inside first.

$$|-4 + 9| = |5|$$
$$= 5$$

(b) $|-4| + |9| = 4 + 9$
$$= 13$$

(c) The $\sqrt{}$ is another grouping symbol.

$$\sqrt{25 + 144} = \sqrt{169}$$
$$= 13$$

(d) $\sqrt{25} + \sqrt{144} = 5 + 12$
$$= 17$$

(e) The fraction bar is also a grouping symbol. We evaluate both the numerator and the denominator *before* we divide.

$$\frac{9 + 5}{9 - 2} = \frac{14}{7}$$
$$= 2$$ ◄

► **EXAMPLE 6**

Answer each of the following questions.

(a) What is the sum of 12 and -5 increased by 9?

(b) The low temperature this morning was $-3°$. If the temperature rose $21°$ by noon, what was the noontime temperature?

Solutions

(a) To increase a number by 9, we add 9 to it.

$$[12 + (-5)] + 9 = 7 + 9$$
$$= 16$$

(b) Since the temperature *rose* $21°$ from $-3°$, we must add 21 to -3.

$$-3 + 21 = 18$$

The noontime temperature was $18°$.   ◄

Calculator Box

Entering Negative Numbers on a Calculator

We must be careful when entering a negative number on a calculator. The $\boxed{-}$ key is for *subtraction*. Look for a key marked $\boxed{+/-}$, $\boxed{(-)}$, or $\boxed{\text{CHS}}$. Let's calculate $-17 \cdot 29$.

Scientific Calculator:

Press $\boxed{17}$ $\boxed{+/-}$ $\boxed{\times}$ $\boxed{29}$ $\boxed{=}$ and read $\boxed{-493}$ on the display.

Graphing Calculator:
The negation key may look like $\boxed{(-)}$.
Press

$$\boxed{(-)}\ \boxed{17}\ \boxed{\times}\ \boxed{29}\ \boxed{=}\ (\text{or } \boxed{\text{ENTER}} \text{ or } \boxed{\text{EXE}})$$

Calculator Exercises

Perform the following computations on a calculator and check the results with pencil and paper.

1. $-221 \cdot 19$ **2.** $-11(41 - 53)$ **3.** $-\sqrt{633 - 584}$

4. $\dfrac{-7 - 4}{-2(-3 - 8)}$ **5.** $\dfrac{3(658 - 576)}{-\sqrt{1808 - 127}}$ **6.** $\sqrt{\dfrac{-4 \cdot 144}{155 - 219}}$

Answers

1. -4199 **2.** 132 **3.** -7 **4.** -0.5 **5.** -6 **6.** 3

Problem Set 1.2

Warm-Ups

In Problems 1 through 19, perform the operations indicated. See Examples 1, 2, and 3.

1. $5 + 3$ **2.** $5 + (-8)$ **3.** $-3 + 7$

4. $6 + 0$ **5.** $-5 + (-8)$ **6.** $0 + (-1)$

7. $-1 + (-1)$ **8.** $-11 + (-3)$ **9.** $7 + (-7)$

10. $\dfrac{1}{4} + \left(-\dfrac{1}{3}\right)$ **11.** $-\dfrac{1}{2} + \left(-\dfrac{1}{2}\right)$ **12.** $2 + \left(-\dfrac{1}{8}\right)$

13. $-\dfrac{1}{8} + \dfrac{1}{3}$ **14.** $-\dfrac{5}{6} + \dfrac{5}{6}$ **15.** $-\dfrac{2}{3} + (-1)$

16. $-\dfrac{3}{5} + \left(-\dfrac{1}{10}\right)$ **17.** $-1.2 + 3.4$ **18.** $-9.8 + (-2.7)$

19. $5.3 + 8.7$

In Problems 20 through 31, perform the indicated operations. See Example 4.

20. $-7 + (-5 + 9)$

21. $(-9 + 12) + (-5 + 3)$

22. $\{-9 + [-8 + (-3)]\} - 3$

23. $[13 + (-8)] + [-2 + (-7)]$

24. $\dfrac{1}{4} + \left(-\dfrac{1}{2} + \dfrac{1}{8}\right)$

25. $[1.2 + (-3.6)] + (-2.5)$

26. $7 + \{-12 + [-5 + (-3)]\}$

27. $-3 + \{[-2 + 4] + [5 + (-8)]\}$

28. $-3.6 + [-2.4 + (-3.1)]$

29. $\left(-\dfrac{2}{3} + \dfrac{1}{6}\right) + \left(-\dfrac{3}{8}\right)$

30. $-3 + \{[-5 + (-5)] + [7 + (-13)]\}$

31. $-\dfrac{6}{5}\left[-\dfrac{3}{4} + \left(-\dfrac{1}{2}\right)\right]$

In Problems 32 and 33, evaluate the given expressions. See Example 5.

32. (a) $|-2 + 7|$ (b) $|-2| + |7|$

33. (a) $\sqrt{225 + 64}$ (b) $\sqrt{225} + \sqrt{64}$

In Problems 34 through 39, answer each question. See Example 6.

34. Find the sum of -6 and 3.

35. What is the sum of -9 and 6 increased by 5?

36. Find the sum of 11 and -6 increased by 5.

37. Last night the low temperature was $-8°$. Tonight's low is predicted to be $12°$ higher than last night's low. What is the predicted low temperature for tonight?

38. The high temperature for today was $-3°$. If the temperature dropped $27°$ to reach the low temperature, find the low temperature for the day.

39. A football team had the following yardage on its first four downs, a gain of 1 yd, a loss of 5 yd, a loss of 6 yd, and a gain of 15 yd. Did they make a first down? (A first down means that they gained more than 10 yd.)

Practice Exercises

In Problems 40 through 99, perform the operations indicated.

40. $7 + 2$

41. $4 + (-3)$

42. $-2 + 6$

43. $-6 + (-7)$

44. $5 + 0$

45. $3 + (-5)$

46. $-12 + 7$

47. $32 + (-9)$

48. $-9 + 9$

49. $-13 + 5$

50. $21 + (-5)$

51. $-8 + 8$

52. $6 + (-2)$

53. $-3 + (-3)$

54. $8 + (-5)$

55. $5.5 + (-5.8)$

56. $-6.3 + 6.3$

57. $-5.0 + 1.0$

58. $25 + 25$

59. $-30 + 20$

60. $14 + (-17)$

61. $-13 + (-5)$

62. $8 + (-8)$

63. $\dfrac{1}{2} + \left(-\dfrac{1}{3}\right)$

64. $-\dfrac{1}{5} + \left(-\dfrac{1}{5}\right)$

65. $3 + \left(-\dfrac{1}{4}\right)$

66. $-\dfrac{2}{5} + \dfrac{1}{3}$

67. $-\dfrac{7}{8} + \dfrac{7}{8}$

68. $-\dfrac{3}{4} + (-1)$

69. $-\dfrac{3}{7} + \left(-\dfrac{1}{14}\right)$

70. $-1.5 + 2.1$

71. $-7.7 + (-2.3)$

72. $6.8 + 3.7$

73. $6.5 + (-6.8)$

74. $-4.3 + 4.3$

75. $-7.0 + 3.0$

76. $-5 + (-7 + 8)$

77. $(-5 + 11) + (-7 + 2)$

78. $-8 + [-9 + (-2)]$

79. $[17 + (-9)] + [-3 + (-6)]$

80. $\dfrac{1}{3} + \left(-\dfrac{1}{2} + \dfrac{5}{6}\right)$

81. $[7.2 + (-4.6)] + (-8.5)$

82. $9 + \{-10 + [-3 + (-3)]\}$

83. $-1 + \{[-3 + 7] + [1 + (-1)]\}$

84. $[-4 + (-8)] + (-6)$

85. $[2 + (-22)] + (-12)$

86. $-5.6 + [-1.4 + (-3.1)]$

87. $\left(-\dfrac{2}{3} + \dfrac{1}{2}\right) + \left(-\dfrac{5}{8}\right)$

88. $5 + [3 + (-7)]$

89. $[-1 + (-1)] + (-6 + 7)$

90. $-1 + \{[-4 + (-4)] + [7 + (-12)]\}$

91. $[-5 + (-3)] + [3 + (-8)]$

92. $[-8 + (-3)] + \{7 + [1 + (-9)]\}$

93. $-\dfrac{7}{15} + \left[-\dfrac{3}{5} + \left(-\dfrac{2}{3}\right)\right]$

94. $[1 + (-2)] + (-3 + 2)$

95. $-0.9 + [-2.4 + (-4.6)]$

96. (a) $|-6 + 17|$ (b) $|-6| + |17|$

97. (a) $\sqrt{15^2 + 8^2}$ (b) $\sqrt{15^2} + \sqrt{8^2}$

98. (a) $|5 - (-3)|$ (b) $|-5| - |-3|$

99. (a) $\sqrt{289 - 64}$ (b) $\sqrt{289} - \sqrt{64}$

In Problems 100 through 110, answer each question.

100. Find the sum of -11 and -11.

101. Find the sum of 16 and -8 decreased by 9.

102. What is the sum of -25 and -10 increased by 50?

103. The high temperature yesterday was $-10°$. If today's high is $5°$ more than yesterday's high, what is today's high?

104. If the temperature is $-15°$ at 5 A.M. and rises at the rate of $3°$ per hour, what will the temperature be at noon?

105. Bill Barber got a raise of $200 a month. If his federal income tax increased by $59 monthly and his state income tax rose by $12 monthly, how much was his net raise?

106. Chuck makes a deposit of $450 and withdrawals of $235 and $350. If he started with a balance of $195, what is his new balance?

107. A football team gains 3 yd, loses 12 yd, gains 9 yd, and gains 5 yd. What is the team's total yardage?

108. Julien bought 10 shares in Sun Trust for $22.50 each one week ago. The following table shows the change in the price of a share of Sun Trust for the next seven days.

T	W	Th	F	M	T	W
$-\dfrac{1}{2}$	$+\dfrac{3}{8}$	$+\dfrac{1}{8}$	$-\dfrac{1}{4}$	$+\dfrac{1}{8}$	$-\dfrac{1}{2}$	$+\dfrac{1}{8}$

(a) Is a share of Sun Trust worth more or less than $22.50 at the end of the seventh day?

(b) What is a share of Sun Trust worth seven days after Julien bought his stock?

(c) How much money has Julien made or lost in seven days?

109. David Werner made a time machine. The controls are designed so that he enters a negative integer to go back from 1996 and a positive integer to go forward. If he enters -3, $+7$, -11, and $+1$ in succession, in what year will he find himself?

110. The change in weight measured in grams for a laboratory rat weighing 265.3 grams recorded in a laboratory experiment over a week is shown in the following table:

M	T	W	Th	F	Sa	S
+3.2	−1.2	−0.09	+1.25	−0.9	−0.3	+0.85

What is the weight of the rat at the end of this week?

Challenge Problems

111. If $a > 0$ and $b > 0$, what can be said about $a + b$?

112. If $a < 0$ and $b < 0$, what can be said about $a + b$?

In Your Own Words

113. How do we add real numbers?

Write in words the meaning of each of the following expressions.

114. $7 + (-3)$

115. $-4 + 0$

116. $-11 + 8$

117. $-2 + (-2)$

1.3   Subtraction

DISK 2

We remember from arithmetic that $5 - 3$ is 2. In the last section we noticed that $5 + (-3)$ is also 2. That is, subtracting 3 is the same as adding the opposite of 3. We use this relationship for our definition of subtraction.

Subtraction of Real Numbers

$$p - q = p + (-q)$$

1. Rewrite the problem as an addition problem by adding the opposite of the number being subtracted.
2. Follow the rules for addition.

▶ **EXAMPLE 1**

Perform the subtractions indicated.

(a) $7 - 2$ (b) $-7 - 2$ (c) $-7 - (-2)$

Solutions

(a) $7 - 2 = 7 + (-2)$ Add the opposite of 2.
          $= 5$

(b) $-7 - 2 = -7 + (-2)$ Add the opposite of 2.
          $= -9$

(c) $-7 - (-2) = -7 + 2$ Add the opposite of -2.
          $= -5$ ◀

▶ **EXAMPLE 2**

Perform the operations indicated.

(a) $[(9 - 4) - (14 - 18)]$ (b) $\dfrac{1}{2} - \left(\dfrac{1}{3} - 2\right)$

(c) $14 - |3 - 8|$ (d) $\sqrt{225 + 8^2} - 8^2$

Solutions

(a) We begin with the innermost grouping symbol.

$$[(9 - 4) - (14 - 18)] = [5 - (-4)]$$
$$= 5 + 4$$
$$= 9$$

(b) $\dfrac{1}{2} - \left(\dfrac{1}{3} - 2\right) = \dfrac{1}{2} - \left(\dfrac{1}{3} - \dfrac{6}{3}\right)$

$= \dfrac{1}{2} - \left(-\dfrac{5}{3}\right)$

$= \dfrac{1}{2} + \dfrac{5}{3}$

$= \dfrac{3}{6} + \dfrac{10}{6}$

$= \dfrac{13}{6}$

(c) Absolute value bars are a grouping symbol.

$$14 - |3 - 8| = 14 - |-5|$$
$$= 14 - 5$$
$$= 9$$

(d) The $\sqrt{}$ symbol is another grouping symbol.

$$\sqrt{225 + 8^2} - 8^2 = \sqrt{225 + 64} - 64$$
$$= \sqrt{289} - 64$$
$$= 17 - 64$$
$$= -47$$

◄

► **EXAMPLE 3**

Answer each question.

(a) Subtract 8 from 5.

(b) Subtract -11 from 2.

(c) What is -8 decreased by the sum of 3 and -5?

(d) If today's high temperature is 7° and the temperature falls 24° for the nighttime low, what is the nighttime low?

Solutions

(a) To subtract 8 from 5, we add the opposite of 8 to 5.

$$5 - 8 = 5 + (-8)$$
$$= -3$$

(b) To subtract -11 from 2, we add the opposite of -11 to 2.

$$2 - (-11) = 2 + 11$$
$$= 13$$

(c) To *decrease* -8 by a quantity, we subtract the quantity from -8. So our problem is to evaluate

$$-8 - [3 + (-5)]$$

The order of operations tells us to work inside the brackets first.

$$-8 - [3 + (-5)] = -8 - (-2)$$
$$= -8 + 2$$
$$= -6$$

(d) If today's high temperature is $7°$ and the temperature falls $24°$, we must subtract 24 from 7 to obtain the nighttime low.

$$7 - 24 = 7 + (-24)$$
$$= -17$$

The nighttime low is $-17°$. ◀

Problem Set 1.3

Warm-Ups

In Problems 1 through 24, perform the operations indicated. See Example 1.

1. $5 - 3$

2. $5 - (-8)$

3. $-3 - 7$

4. $-13 - 5$

5. $21 - (-5)$

6. $-8 - 8$

7. $-5 - (-8)$

8. $6 - 0$

9. $0 - 1$

10. $5 - (-1)$

11. $-1 - (-1)$

12. $8 - (-4)$

13. $-15 - 10$

14. $-20 - 10$

15. $18 - (-3)$

16. $-\dfrac{1}{2} - \left(-\dfrac{1}{2}\right)$

17. $2 - \left(-\dfrac{1}{8}\right)$

18. $-\dfrac{1}{8} - \dfrac{1}{3}$

19. $-\dfrac{5}{6} - \dfrac{5}{6}$

20. $-\dfrac{2}{3} - (-1)$

21. $-\dfrac{3}{5} - \left(-\dfrac{1}{10}\right)$

22. $-1.2 - 3.4$

23. $-9.8 - (-2.7)$

24. $5.3 - 8.7$

In Problems 25 through 38, perform the indicated operations. See Example 2.

25. $-7 - (-5 - 9)$

26. $(-9 - 12) - (-5 - 3)$

27. $-9 - [-8 - (-3)]$

28. $[13 - (-8)] - [-2 - (-7)]$

29. $\dfrac{1}{4} - \left(-\dfrac{1}{2} - \dfrac{1}{8}\right)$

30. $[1.2 - (-3.6)] - (-2.5)$

31. $2 - [(-8 - 7) - (-3 + 1)]$

32. $[-11 + (-3)] - 5$

33. $\dfrac{2}{5} + \left(\dfrac{1}{3} - \dfrac{3}{5}\right)$

34. $(-2.7 + 5.1) - 4.8$

35. $-5 - \sqrt{-3 - (-7)}$

36. $[-6 - 7] - |-6 + 13|$

37. $12 - |7 - 11| - 7$

38. $23 - \sqrt{11 - (3 - 8)}$

In Problems 39 through 44, answer the question in each. See Example 3.

39. Subtract -6 from 6.

40. Subtract 8 from the sum of -11 and -4.

41. What is -12 decreased by 5?

42. What is the sum of 4 and -9, decreased by 1?

43. A grocer set his freezer at $-5°$. If he decreases the temperature by $10°$, what will the temperature be?

44. The high temperature for today was $15°$. If the temperature dropped $27°$ to reach the low temperature, find the low temperature for the day.

Practice Exercises

In Problems 45 through 92, perform the operations indicated.

45. $8 - 5$

46. $-7 - 1$

47. $-8 - (-3)$

48. $-6 - 2$

49. $3 - (-1)$

50. $5 - (-5)$

51. $-5 - (-5)$

52. $4 - (-6)$

53. $-3 - 1$

54. $-9 - (-7)$

55. $-3 - (-3)$

56. $-5 - (-8)$

57. $6 - (-6)$

58. $-6 - (-6)$

59. $6 - (-1)$

60. $-10 - 10$

61. $-12 - (-7)$

62. $13 - 7$

63. $15 - (-12)$

64. $-17 - 17$

65. $-17 - (-17)$

66. $18 - 32$

67. $18 - (-32)$

68. $-18 - 32$

69. $-18 - (-32)$

70. $-32 - 18$

71. $-32 - (-18)$

72. $-\dfrac{2}{5} - \dfrac{2}{3}$

73. $\dfrac{7}{8} - \dfrac{3}{4}$

74. $-\dfrac{1}{6} - \left(-\dfrac{1}{3}\right)$

75. $-\dfrac{4}{5} - \dfrac{4}{5}$

76. $3 - \dfrac{3}{2}$

77. $\dfrac{1}{2} - 1$

78. $1.2 - 2.4$

79. $-2.5 - 3.4$

80. $-3.3 - (-6.6)$

81. $7 - (6 - 12)$

82. $(-7 - 9) - 3$

83. $[(7 - 14) - (-5 - 8)]$

84. $[11 - (-6 - 6)]$

85. $[-9 - (-6)] - [13 - (-7)]$

86. $-\dfrac{2}{3} - \left(-\dfrac{5}{6} - \dfrac{1}{2}\right)$

87. $(-4 + |-4 - 15|) + \sqrt{16 - (-9)}$

88. $\{[8 + (-12)] - (4 - 7)\} - 8$

89. $12 - [(-3 - 20) + (-33 + 10)]$

90. $\{-11 + [-9 - (-20)]\} - \{9 + [-3 - (-6)]\}$

91. $\sqrt{|-7| - (-1 + 7)}$

92. $|-3 - \sqrt{10 - (2 - 8)}|$

In Problems 93 through 98, answer each question.

93. Find the difference when -6 is subtracted from 3.

94. Subtract -5 from -2.

95. Subtract -7 from the sum of -6 and -5.

96. Last night the low temperature was 8°. Tonight's low is predicted to be 12° less than last night's low. What is the predicted low temperature for tonight?

97. The temperature was $-3°$ at 7:00 P.M. If the temperature dropped 15° by midnight, what was the temperature at midnight?

98. Wan Moon had his annual physical in mid-January and discovered his blood pressure to be 178/97. (The first number is called systolic pressure and the second is called diastolic pressure.) Using diet and exercise, he lowered his systolic pressure by 10 units each a month and his diastolic pressure by 3 units each a month for the following 3 months. What was Wan's blood pressure in mid-April?

Problem 98

Challenge Problems

99. Evaluate $-(-\{-[-(-3-(-8))]\})$.

100. Evaluate
$$-\{-[-(-11+7)]-[-(-10-(-6))]\}.$$

In Your Own Words

101. How is subtraction of real numbers done?

Write the meaning of the following expressions.

102. $3-7$

103. $7-(-3)$

104. $-7-(-3)$

105. $16-(4+10)$

1.4 Multiplication and Division

TAPE 2

Multiplication and division with real numbers follow two rules. The rules are the same for both operations, since division is just a form of multiplication. Writing $\frac{6}{3}$ is the same as writing $6 \cdot \frac{1}{3}$. That is, a division problem can be written as a multiplication problem, as long as the divisor is not 0.

Division

$$\frac{p}{q} = p \cdot \frac{1}{q}, \qquad q \neq 0$$

If we look at all the possibilities for multiplying two real numbers, we see that both numbers can be positive, both numbers can be negative, or one number can be positive and the other number negative.

We already know how to multiply two positive numbers. For example,

$$2 \cdot 3 = 6$$

That is, a positive number times a positive number is a positive number.

To understand what $2(-3)$ is, remember that multiplication of -3 by 2 is the same as adding $-3 + (-3)$, which is -6. This gives us the idea that a positive number times a negative number is a negative number.

The product of two negative numbers is a positive number. To gain some understanding why this must be so, consider the following.

$$(-2) \cdot 3 = -6$$
$$(-2) \cdot 2 = -4$$
$$(-2) \cdot 1 = -2$$
$$(-2) \cdot 0 = 0$$
$$(-2) \cdot (-1) = 2$$
$$(-2) \cdot (-2) = 4$$

Notice that the answer increases by 2 each time. So the product of two negative numbers must be positive.

> **Multiplication and Division of Real Numbers**
>
> **1.** To multiply or divide two numbers with *like* signs (both positive or both negative), multiply or divide their absolute values. The sign will be positive.
>
> **2.** To multiply or divide two numbers with *unlike* signs (one positive and one negative), multiply or divide their absolute values. The sign will be negative.

Remember, we **cannot** divide by 0.

► **EXAMPLE 1**

Perform the multiplications indicated.

(a) $(-3)(5)$ (b) $(-4)(-2)$

(c) $\dfrac{1}{3}\left(-\dfrac{3}{2}\right)$ (d) $(-2)^2$

Solutions

(a) $(-3)(5) = -15$ (b) $(-4)(-2) = 8$

(c) $\dfrac{1}{3}\left(-\dfrac{3}{2}\right) = -\dfrac{1}{2}$ (d) $(-2)^2 = (-2)(-2)$

$$= 4$$ ◄

► **EXAMPLE 2**

Perform the divisions indicated.

(a) $-6 \div 3$ (b) $\dfrac{15}{-5}$ (c) $10 \div (-2)$

Solutions

(a) $-6 \div 3 = -2$ (b) $\dfrac{15}{-5} = -3$ (c) $10 \div (-2) = -5$ ◄

The three real numbers 1, -1, and 0 have special properties. If we multiply a real number by 1, we get the real number itself. If we multiply a real number by -1, we get the opposite of the real number. If we multiply a real number by 0, we get 0.

> **Multiplication with 1, -1, and 0**
>
> If p is any real number,
>
> $$1 \cdot p = p \qquad p \cdot 1 = p$$
> $$-1 \cdot p = -p \qquad p \cdot (-1) = -p$$
> $$0 \cdot p = 0 \qquad p \cdot 0 = 0$$

Using 0 in quotients can be troublesome. Remember, we know that $\frac{6}{2} = 3$ because $3 \cdot 2 = 6$. In the same manner, $\frac{0}{6} = 0$ because $0 \cdot 6 = 0$. What about $\frac{6}{0}$? Is $\frac{6}{0} = 0$? This would mean that $0 \cdot 0 = 6$, which is not true. So it is not 0. If we try any other real number, we see that $\frac{6}{0} = ?$ would mean that $? \cdot 0$ must be 6. This is not possible. Thus we cannot divide 0 into a real number.

Quotients with Zero

If p and q are real numbers and p is not zero,

1. $\dfrac{0}{p} = 0$

2. $\dfrac{q}{0}$ is undefined.

► **EXAMPLE 3**

Perform the operations indicated.

(a) $-1 \cdot 7$ (b) $6 \cdot 0$ (c) $1 \cdot x$

(d) $-1(-4)$ (e) $-1(-x)$ (f) $\dfrac{-1}{0}$

Solutions

(a) $-1 \cdot 7 = -7$

(b) $6 \cdot 0 = 0$

(c) $1 \cdot x = x$

(d) $-1(-4) = 4$ Notice 4 is the opposite of -4

(e) $-1(-x) = x$ Notice x is the opposite of $-x$

(f) $\dfrac{-1}{0}$ is undefined.

An important property of fractions concerns signs. Notice the signs in the following computations.

☐1 $\dfrac{-6}{2} = -3$ A negative number divided by a positive number

☐2 $\dfrac{6}{-2} = -3$ A positive number divided by a negative number

☐3 $-\dfrac{6}{2} = -3$ The opposite of a positive number

That is, because of the sign rules for multiplication and division, we have

$$\frac{-6}{2} = \frac{6}{-2} = -\frac{6}{2}$$

This idea holds true in general.

> **A Sign Property of Fractions**
>
> For p and q integers, $q \neq 0$
>
> $$\frac{-p}{q} = \frac{p}{-q} = -\frac{p}{q}$$

The order of operations still applies, as the next example illustrates.

▶ **EXAMPLE 4**

Perform the operations indicated.

(a) $-3(-2)^2$ (b) $\dfrac{-4(5)}{10}$ (c) $8 \div 4(-3)$

Solutions

(a) We must square -2 first.

$$-3\ (-2)^2 = -3\ (4)$$
$$= -12$$

(b) We must multiply in the numerator first.

$$\frac{-4(5)}{10} = \frac{-20}{10}$$
$$= -2$$

(c) We perform multiplications and divisions in order from left to right. So, first, we divide 8 by 4.

$$8 \div 4\ (-3) = 2\ (-3)$$
$$= -6$$

◀

▶ **EXAMPLE 5**

A 22-gallon (gal) solution is 10% molasses by volume. How many gallons of molasses does it contain?

Solution

The solution is 10%, or $\frac{10}{100}$, molasses. Thus it contains $\frac{10}{100} \cdot 22$ gal of molasses.

$$\frac{10}{100} \cdot 22 = \frac{1}{10} \cdot 22$$

$$= \frac{22}{10} = 2.2$$

There are 2.2 gal of molasses in the solution.

◀

► **EXAMPLE 6**

Eighteen-carat gold is 75% pure gold by weight. How much pure gold is contained in an 18-carat ring that weighs 120 grams (g)?

Solution

75% of 120 is

$$\frac{75}{100} \cdot 120 = \frac{3}{4} \cdot 120$$

$$= \frac{3 \cdot 120}{4}$$

$$= 3 \cdot 30 = 90$$

The ring contains 90 g of pure gold.   ◄

Problem Set 1.4

Warm-Ups

In Problems 1 through 12, find each product. See Examples 1 and 3.

1. $(-6)(-1)$

2. $3(0)$

3. $\left(\frac{1}{2}\right)\left(-\frac{1}{2}\right)$

4. $(-5)(-7)$

5. $(1.1)(-10)$

6. $-1(-3)$

7. $(-5)^2$

8. $(-3)^2$

9. $(-5)(-4)$

10. $4(-7)$

11. $0(-4)$

12. $\left(-\frac{3}{4}\right)^2$

In Problems 13 through 28, find each quotient. See Examples 2 and 3.

13. $\frac{-5}{-1}$

14. $9/(-3)$

15. $8 \div 0$

16. $\frac{18}{-9}$

17. $\frac{-35}{7}$

18. $-28/(-4)$

19. $0/(-1)$

20. $12 \div (-4)$

21. $\frac{14}{-7}$

22. $-6 \div (-6)$

23. $\frac{15}{-1}$

24. $\frac{-63}{-7}$

25. $25 \div (-5)$

26. $\frac{3}{5} \div \left(-\frac{9}{10}\right)$

27. $5 \div 0$

28. $-48 \div 8$

In Problems 29 through 39, perform the indicated operations. See Example 4.

29. $\frac{2(-3)}{3}$

30. $2(3)^2$

31. $(-2)^2(3)$

32. $(-4)^2 \div (-8)$

33. $\frac{(-7)^2}{7^2}$

34. $\frac{(-4)(3)}{2^2}$

35. $(-3) \cdot 2^2$

36. $(-1)^2(-6)^2$

37. $16(-2) \div 8$

38. $12 \div (-2)(3)$

39. $(-3)^2 \div 3(2)$

In Problems 40 and 41, answer the question. See Example 5.

40. A gold coin is 95% pure gold by weight. If the coin weighs 2 oz, how much pure gold does it contain?

41. A solution contains 20% acid by volume. How much acid is there in 485 L of the solution?

Practice Exercises

In Problems 42 through 92, perform the operations indicated.

42. $(-8)(-1)$

43. $7(0)$

44. $\left(\dfrac{2}{3}\right)\left(-\dfrac{2}{3}\right)$

45. $(-3)(8)$

46. $(-2.2)(10)$

47. $(-7)^2$

48. $-1(-9)$

49. $(-6)^2$

50. $1(-2)$

51. $(-8)(-4)$

52. $-1 \cdot z$

53. $(-4)(-6)$

54. $(-9)(-8)$

55. $-1(-d)$

56. $(-8)(8)$

57. $(-10)^2$

58. $(-6)(-3)$

59. $5(-7)$

60. $0(-8)$

61. $\left(-\dfrac{5}{6}\right)^2$

62. $\dfrac{-7}{-1}$

63. $8/(-4)$

64. $7 \div 0$

65. $\dfrac{28}{-7}$

66. $22 \div (-2)$

67. $\dfrac{-45}{5}$

68. $-36/(-4)$

69. $0/(-7)$

70. $\dfrac{42}{-7}$

71. $-4/(-4)$

72. $\dfrac{17}{-1}$

73. $\dfrac{-56}{-7}$

74. $36 \div (-6)$

75. $\dfrac{0}{9}$

76. $4/0$

77. $-72 \div 8$

78. $\dfrac{-2}{2}$

79. $\dfrac{3}{-3}$

80. $-64/(-8)$

81. $\dfrac{2}{0}$

82. $\dfrac{2(-5)}{5}$

83. $2(6)^2$

84. $(-3)^2(3)$

85. $(-8)^2 \div (-4)$

86. $\dfrac{(-9)^2}{9^2}$

87. $\dfrac{(-4)(5)}{2^2}$

88. $(-4) \cdot 2^2$

89. $(-1)^2(-5)^2$

90. $(-2)(-9) \div (-3)$

91. $18 \div (-6)(4)$

92. $45 \div (-3)/5$

93. If Ivory soap is $99\frac{44}{100}\%$ pure, how much impurity is there in a ton of it? (A ton is 2000 lb.)

94. During the past week, Juanita Estez earned \$11.28 an hour for $38\frac{1}{2}$ hours. What were her earnings last week?

95. Suppose Juanita earned *time and a half* ($1\frac{1}{2}$ times her regular hourly rate) for overtime (see Problem 94). If her basic work week is 35 hours, how much were her earnings last week?

Problems 96 and 97 use the fact that distance traveled at a constant speed equals rate times time and that sound travels at approximately 1090 ft/sec in air.

96. An echo returns to a person from a cliff in 2.5 seconds. How far away is the cliff?

97. A flash of lightning is seen by an observer, and 7 seconds later there is a clap of thunder. Neglecting the speed of light, how far away from the observer did the lightning strike?

Challenge Problems

98. Are $12 \div 3^2$ and $12 \div 3 \cdot 3$ the same?

99. Write $\dfrac{14}{5 \cdot 11}$ using the $\div$ symbol.

In Your Own Words

100. State the rules that govern the *signs* in multiplication and division.

Write in words the meaning of the following expressions.

101. $3(-5)$

102. $\dfrac{-7}{3}$

103. $6 \cdot \dfrac{10}{11} + 5$

104. $6\left(\dfrac{10}{11} + 5\right)$

1.5 Expressions

TAPE 3

In this section we focus on combining the operations of addition, subtraction, multiplication, and division. Such combinations are called **numerical expressions,** or just **expressions.** We simplify such expressions with the order of operations that we have already learned.

► **EXAMPLE 1**

Simplify the following expressions.

(a) $(-4)(3) - (5)(-6)$ (b) $-6/(-2) + 2(-3)^2 - 2^2$

Solutions

(a) We must perform the multiplications first.
$$(-4)(3) - (5)(-6) = -12 - (-30)$$

Next we do the subtraction
$$-12 - (-30) = -12 + 30$$
$$= 18$$

(b) We must perform the exponentiation first and then the division and multiplications before the addition.
$$-6/(-2) + 2(-3)^2 - 2^2 = -6/(-2) + 2(9) - 4$$
$$= 3 + 18 - 4$$
$$= 17 \qquad \blacktriangleleft$$

We must begin inside grouping symbols, if they are present. Remember that fraction bars, absolute value bars, and square root symbols are grouping symbols.

▶ **EXAMPLE 2**

Simplify each expression.

(a) $\dfrac{8(-3) - 6^2}{8 - 3(-1) + (-6)}$

(b) $-6 - (5 - \sqrt{25 - 16} + 1)$

(c) $|5 - 8| \div (-3) - 4(-2)$

Solutions

(a) We must simplify the numerator and the denominator before we divide.

$$\frac{8(-3) - 6^2}{8 - 3(-1) + (-6)} = \frac{-24 - 36}{8 - (-3) + (-6)}$$

$$= \frac{-60}{8 + 3 + (-6)}$$

$$= \frac{-60}{5}$$

$$= -12$$

(b) We start with the innermost grouping symbol.

$$-6 - (5 - \sqrt{25 - 16} + 1) = -6 - (5 - \sqrt{9} + 1)$$

$$= -6 - (5 - 3 + 1)$$

$$= -6 - 3$$

$$= -9$$

(c) Remember that absolute value bars are grouping symbols.

$$|5 - 8| \div (-3) - 4(-2) = |-3| \div (-3) - 4(-2)$$

$$= 3 \div (-3) - 4(-2)$$

$$= -1 - (-8)$$

$$= -1 + 8$$

$$= 7$$ ◀

It is common in algebra to use letters to name numbers. We write such things as

$$2x + 3$$

to mean

2 times *some number* plus 3

The value of $2x + 3$ depends on the number x. This provides us with a way to write many statements with just one expression.

In algebra, as in arithmetic, we perform four operations (addition, subtraction, multiplication, and division) with numbers. The following chart indicates how we read and write these operations with the numbers x and y.

Operation	Symbol(s)	Read
Addition	$x + y$	x plus y
Subtraction	$x - y$	x minus y
Multiplication	xy	x times y
	$x \cdot y$	
	$x(y)$	
	$(x)y$	
	$(x)(y)$	
Division	$x \div y$	x divided by y
	$\dfrac{x}{y}$	
	x/y	
	$y\overline{)x}$	

Meaningful collections of sums, differences, products, quotients, exponents, and grouping symbols that contain letters are called **algebraic expressions** or simply **expressions.** For example,

$$6x + 2(x - 3)$$

and

$$\frac{3x}{5 + 4x}$$

are algebraic expressions.

In a sum or difference, the numbers being added or subtracted are called **terms.** In a product the numbers being multiplied are called **factors.** These are *important* words in algebra as they are used very often, and their meanings are quite different.

► **EXAMPLE 3**

In each of the following expressions, state if 2 is a *term* or a *factor*.

(a) $5x - 2$ (b) $2x$ (c) $2(x - 5)$

Solutions

(a) The expression $5x - 2$ is a difference. It has two *terms,* $5x$ and 2. Therefore, 2 is a term of this expression.

(b) The expression $2x$ is a product. It has one *term,* $2x$. This term has two *factors,* 2 and x. Thus 2 is a factor of this expression.

(c) The expression $2(x - 5)$ is a product of two *factors,* 2 and $(x - 5)$. Therefore, 2 is a factor.   ◄

The order of operations comes up in algebra when working with expressions. Consider the expression $2x - 1$. It has many different values, depending on the number x. If x is 2,

$$2x - 1 = 2(2) - 1$$
$$= 4 - 1$$
$$= 3$$

So $2x - 1$ has a value of 3 when x is 2.

If x is 3,

$$2x - 1 = 2(3) - 1$$
$$= 6 - 1$$
$$= 5$$

Thus $2x - 1$ has a value of 5 when x is 3.

When we find the value of an expression, we say that we **evaluate** the expression.

► ## EXAMPLE 4

Evaluate the following expressions when x is -2.

(a) $10 + x$ (b) $10 - x$ (c) $\dfrac{10 - 3x}{x}$

(d) $5 + 3x + x^2$ (e) $x\sqrt{7 - x}$ (f) $|3 + 7x| + 21$

Solutions

(a) It is convenient to think of the expression as

$$10 + x = 10 + (\;\;)$$

and then fill the parentheses with -2.

$$10 + x = 10 + (-2)$$
$$= 8$$

(b) Replace x with -2.

$$10 - x = 10 - (\;\;)$$
$$= 10 - (-2)$$
$$= 10 + 2$$
$$= 12$$

(c) When replacing a variable in an expression with a number, we must be sure to replace the variable everywhere it occurs.

$$\frac{10 - 3x}{x} = \frac{10 - 3(\;\;)}{(\;\;)}$$
$$= \frac{10 - 3(-2)}{(-2)}$$
$$= \frac{10 + 6}{-2}$$
$$= \frac{16}{-2}$$
$$= -8$$

(d) $5 + 3x + x^2 = 5 + 3(\ \ ) + (\ \ )^2$

$$= 5 + 3(-2) + (-2)^2$$

The order of operations tells us we must evaluate the exponent first.

$$5 + 3x + x^2 = 5 + 3(-2) + 4$$

Then we do multiplication.

$$5 + 3x + x^2 = 5 + (-6) + 4$$
$$= 3$$

(e) Square root symbols are grouping symbols. We must perform operations inside them first.

$$x\sqrt{7 - x} = (-2)\sqrt{7 - (-2)}$$
$$= (-2)\sqrt{7 + 2}$$
$$= (-2)\sqrt{9}$$
$$= (-2)3$$
$$= -6$$

(f) Absolute value bars are also grouping symbols.

$$|3 + 7x| + 21 = |3 + 7(-2)| + 21$$
$$= |3 - 14| + 21$$
$$= |-11| + 21$$
$$= 11 + 21$$
$$= 32 \qquad \blacktriangleleft$$

Problem Set 1.5

Warm-Ups

In Problems 1 through 8, simplify each expression. See Example 1.

1. $3(-2) + (-3)(-1)$

2. $\dfrac{-18}{-2} + \dfrac{6}{-3}$

3. $2^2 - 6/(6 - 3)$

4. $-7 + 3(-3)^2 + 5$

5. $|-3| + (-6)/(-2) - 3$

6. $2(-3)^2 - \dfrac{-60}{15} + (-8)$

7. $-\dfrac{8}{9} \div \dfrac{2}{3} + \left(-\dfrac{5}{6}\right)(-5)(-2)$

8. $(-3)(-2)(-1) - (-8) \div (-8)$

In Problems 9 through 19, simplify each expression. See Example 2.

9. $\dfrac{2(-3) + 4(-2)}{5 \cdot 2 - 3}$

10. $-5 - 2[8 - 2(4 - 6) - 7]$

11. $2|7 - 5| \div (-2 - 2)$

12. $\dfrac{3 - |0 - 3|}{4^2}$

13. $8 - 6 - (6 - 4)$

14. $\dfrac{3(-3) - 3^2 + (-12)}{(-3)^2 - 2(-3)}$

15. $\dfrac{8 - (-5) - (7 - 9)}{3(2^2 - 3^2)}$

16. $\dfrac{\sqrt{64 + 36}}{4}$

17. $14 - [2(4 - 7) - 3(8) - (5 - 11)]$

18. $10 - \sqrt{|16 - 25|} + 16$

19. $5 + 3^2 \div \sqrt{4 + 5}$

In Problems 20 through 25, state whether 7 is a term or a factor. See Example 3.

20. $7 + 3x - 5x^3$

21. $8x - 7$

22. $(1 - 3s)7$

23. $x^2 + 13x - 7$

24. $\dfrac{ax}{by} \cdot 7$

25. $7\sqrt{x + 3}$

In Problems 26 through 31, evaluate each expression when x is 3. See Example 4.

26. $15x$

27. $15 + x$

28. $\dfrac{15}{x}$

29. $\dfrac{5 + x}{5 - x}$

30. $\sqrt{82 + 6x}$

31. $x - |3x - 15|$

In Problems 32 through 37, evaluate each expression when x is −3.

32. $15x$

33. $15 + x$

34. $\dfrac{15}{x}$

35. $\dfrac{5 + x}{5 - x}$

36. $\sqrt{82 + 6x}$

37. $x - |3x - 15|$

Practice Exercises

In Problems 38 through 59, simplify each expression.

38. $3(-5) - (-3)(-1)$

39. $\dfrac{-28}{4} + \dfrac{-9}{-3}$

40. $3^2 - 8/(6 - 3)$

41. $-8 + 2(-2)^2 + 7$

42. $|-7| + (-14)/(-2) - 14$

43. $3(-3)^2 - \dfrac{-30}{-6} + (-8)$

44. $3(-2) - (-8)(-4) - (-3)(-2)$

45. $3|-5| \div (-5) + (-3)^2$

46. $-\dfrac{6}{7} \div \dfrac{2}{7} + \left(-\dfrac{7}{9}\right)(-5)(-9)$

47. $(-4)(-3)(-1) - [(-6) \div (-6)]$

48. $\dfrac{2[(-3) + 4(2)]}{5(2) - 5}$

49. $-1 - 2[1 - 2(4 - 7) - 8]$

50. $3|5 - 7| \div (-3 - 3)$

51. $\dfrac{3 + |0 - 3|}{4^2 - 4(5) + 1}$

52. $9 - 4 - (4 - 2)$

53. $\dfrac{5(-6) + 2 \cdot 3^2 + (-12)}{(-3)^2 - 2 - 3}$

54. $\dfrac{3}{4}\left(-\dfrac{4}{3}\right) - \dfrac{1}{5}(-5) + \dfrac{5}{7}\left(-\dfrac{7}{15}\right)$

55. $|6 - 9| - 2(-3) \div (-3)$

56. $\dfrac{9 - (-7) - (4 - 9)}{3(4^2 - 3^2)}$

57. $4 - [2(4 - 8) - 3(4) - 3 - 7)]$

58. $\sqrt{169 - 25} + 11^2$

59. $\dfrac{2 + |2 - 8|}{2\sqrt{5^2 - 3^2}}$

In Problems 60 through 65, state whether x is a term *or a* factor.

60. $7 + x$

61. $x - 7y$

62. $x(1 - 3s)$

63. $x - 7$

64. $7ax$

65. $7\sqrt{y + 3} - x$

In Problems 66 through 74, evaluate each expression when x is 2 and again when x is -2.

66. $21x$

67. $21 + x$

68. $\dfrac{144}{x}$

69. $5x^2$

70. $(5x)^2$

71. $-x^2$

72. $\dfrac{1 - x}{1 + x}$

73. $6x|x - 6|$

74. $\sqrt{4x + 17} - 5$

75. The distance in feet traveled by objects in a free fall with no initial velocity is given by the expression $16.1t^2$, where t is the elapsed time in seconds. Sallie Paschal drops her sunglasses while washing windows on the IBM building. Mr. Lower observes from his office window that the sunglasses hit the ground in 3.5 sec. How high up is Sallie?

76. A rocket fired upward with an initial velocity of 128 ft/sec travels at a velocity of $128 - 32t$ feet per second, where t is the time of flight in seconds. Calculate the velocity after (a) 1 sec, (b) 2 sec, (c) 4 sec, and (d) 6 sec. Explain the answers.

77. The low temperatures for a city recorded by a meteorologist for the week between Christmas and New Year's Day are given in the following table:

S	M	T	W	Th	F	Sa
22°	18°	5°	−3°	−7°	−2°	0°

What was the average low temperature for that week?

Challenge Problems

Problems 78 through 81 each contain two expressions. Decide whether they are equal or not; then test by replacing x with -6.

78. $|x - 3|$ and $|x| - |3|$

79. $|x - 3|$ and $|3 - x|$

80. $|2x|$ and $|2||x|$

81. $\left|\dfrac{x}{3}\right|$ and $\dfrac{|x|}{|3|}$

In Your Own Words

82. What do the expressions $20 \div 4 + x$ and $20 \div (4 + x)$ mean?

1.6   Properties of Real Numbers

TAPE 3

In our work with real numbers in this chapter, we have not discussed their properties. Many of them are so natural to us that we completely overlook them. For example, we know that

$$2 + 3 = 3 + 2$$

and that

$$2 \cdot 3 = 3 \cdot 2$$

These examples illustrate what we call the **Commutative Properties.** If we add or multiply two real numbers, the *order* in which we do the operations does not matter.

Commutative Properties

If p and q are real numbers,

Addition: $p + q = q + p$

Multiplication: $pq = qp$

We also know that

$$(2 + 3) + 4 = 2 + (3 + 4)$$

and that

$$(2 \cdot 3) \cdot 4 = 2 \cdot (3 \cdot 4)$$

These examples illustrate the **Associative Properties**. If we add or multiply three numbers, we may group the numbers in different ways to add or multiply.

Associative Properties

If p, q, and r are real numbers,

Addition: $(p + q) + r = p + (q + r)$

Multiplication: $(pq)r = p(qr)$

► **EXAMPLE 1**

Name the property used in each of the following.

(a) $3(4) = 4(3)$ (b) $(x + 3) + 4 = x + (3 + 4)$
(c) $x(3 + y) = x(y + 3)$ (d) $5 \cdot (y + z) = (y + z) \cdot 5$

Solutions

(a) $3(4) = 4(3)$
   This is the Commutative Property of Multiplication.
(b) $(x + 3) + 4 = x + (3 + 4)$
   This is the Associative Property of Addition.
(c) $x(3 + y) = x(y + 3)$
   Notice that the order of 3 and y is all that has changed. This is the Commutative Property of Addition.
(d) $5 \cdot (y + z) = (y + z) \cdot 5$
   All that is changed is the order in which the two numbers 5 and $(y + z)$ are written. This is the Commutative Property of Multiplication.   ◄

If we are given more than two numbers to add with no symbols of grouping, the Commutative and Associative Properties allow us to choose which two numbers to add first. Suppose we want to add $2 + 3 + 4$. We may choose which two numbers to add first.

$$2 + 3 + 4 = (2 + 3) + 4 \qquad \text{Associative Property}$$
$$= 5 + 4 = 9$$

or

$$2 + 3 + 4 = 2 + (3 + 4) \qquad \text{Associative Property}$$
$$= 2 + 7 = 9$$

or

$$2 + 3 + 4 = 2 + 4 + 3 \qquad \text{Commutative Property}$$
$$= (2 + 4) + 3 \qquad \text{Associative Property}$$
$$= 6 + 3 = 9$$

The same idea applies to multiplication.

The Commutative and Associative Properties can be useful for working problems efficiently. Notice how they work in the next example.

► **EXAMPLE 2**

Perform the operations indicated.

(a) $-1345 + 23{,}456 + 1345$ (b) $\dfrac{1}{8} \cdot 15 \cdot 16$

Solutions

(a) First we apply the Commutative Property to the last two terms.

$$-1345 + 23{,}456 + 1345 = -1345 + 1345 + 23{,}456$$

Now we use the Associative Property.

$$-1345 + 1345 + 23{,}456 = (-1345 + 1345) + 23{,}456$$
$$= 0 + 23{,}456$$
$$= 23{,}456$$

(b) $\dfrac{1}{8} \cdot 15 \cdot 16 = \dfrac{1}{8} \cdot 16 \cdot 15 \qquad\qquad$ Commutative Property

$$= \left(\dfrac{1}{8} \cdot 16\right) \cdot 15 \qquad\qquad \text{Associative Property}$$
$$= 2 \cdot 15$$
$$= 30 \qquad\qquad\qquad\qquad\qquad ◄$$

The Distributive Property combines addition and multiplication. For example,

$$2(3 + 4) = 2 \cdot 7$$
$$= 14$$

Notice also that

$$2 \cdot 3 + 2 \cdot 4 = 6 + 8$$
$$= 14$$

So

$$2(3 + 4) = 2 \cdot 3 + 2 \cdot 4$$

Distributive Properties

If p, q, and r are real numbers,

$$p(q + r) = p \cdot q + p \cdot r$$
$$p(q - r) = p \cdot q - p \cdot r$$

Example 3 shows how to use the Distributive Property.

▶ **EXAMPLE 3**

Multiply each of the following using the Distributive Property.

(a) $3(5 + 8)$ (b) $2(x - 3)$ (c) $3\left(\dfrac{2}{3} + y\right)$

Solutions

(a) $3(5 + 8) = 3 \cdot 5 + 3 \cdot 8$
$$= 15 + 24$$
$$= 39$$

(b) $2(x - 3) = 2 \cdot x - 2 \cdot 3$
$$= 2x - 6$$

(c) $3\left(\dfrac{2}{3} + y\right) = 3 \cdot \dfrac{2}{3} + 3 \cdot y$
$$= 2 + 3y$$ ◀

There are two real numbers, 0 and 1, that have special properties. Since the sum of 0 and any real number is the real number itself, we call 0 the identity for addition. Likewise, since 1 times any real number is the real number itself, we call 1 the multiplicative identity.

Identities

If p is a real number, then

1. $p + 0 = p$ and $0 + p = p$ Identity for addition
2. $p \cdot 1 = p$ and $1 \cdot p = p$ Identity for multiplication

---

We say that 0 is the **additive identity** and that 1 is the **multiplicative identity.** If we add opposites, we get the additive identity.

$$3 + (-3) = 0$$

The product of 2 and $\frac{1}{2}$ is the multiplicative identity.

$$2 \cdot \frac{1}{2} = 1$$

Inverses

1. *Additive Inverse* (*opposite*)
If p is a real number, then there is a real number $-p$ such that

$$p + (-p) = 0 \text{ and } (-p) + p = 0$$

$-p$ is called the **additive inverse,** or **opposite,** of p. (We read $-p$ as "negative p" or "the opposite of p.")

2. *Multiplicative Inverse* (*reciprocal*)
If p is a nonzero real number, then there is a real number $\frac{1}{p}$ such that

$$p \cdot \frac{1}{p} = 1 \text{ and } \frac{1}{p} \cdot p = 1$$

$\frac{1}{p}$ is called the **multiplicative inverse,** or **reciprocal,** of p.

► **EXAMPLE 4**

Give the additive inverse (opposite) and the multiplicative inverse (reciprocal) for each.

(a) -10 (b) $-\dfrac{3}{4}$ (c) $\dfrac{2}{x}$

Solutions

(a) The additive inverse is 10 because $-10 + 10 = 0$.

The multiplicative inverse is $-\dfrac{1}{10}$ because

$$(-10)\left(-\frac{1}{10}\right) = 1$$

(b) The additive inverse is $\dfrac{3}{4}$ because $-\dfrac{3}{4} + \dfrac{3}{4} = 0$.

The multiplicative inverse is $-\dfrac{4}{3}$ because

$$\left(-\frac{3}{4}\right)\left(-\frac{4}{3}\right) = 1$$

(c) The additive inverse is $-\dfrac{2}{x}$ because

$$\frac{2}{x} + \left(-\frac{2}{x}\right) = 0$$

The multiplicative inverse is $\dfrac{x}{2}$ because

$$\frac{2}{x} \cdot \frac{x}{2} = 1$$

◄

► **EXAMPLE 5**

Name the property that is illustrated in each.

(a) $4 + 0 = 4$ (b) $5 + (-5) = 0$ (c) $1 \cdot x = x$ (d) $\dfrac{7}{x} \cdot \dfrac{x}{7} = 1$

Solutions

(a) $4 + 0 = 4$ Additive identity

(b) $5 + (-5) = 0$ Additive inverse

(c) $1 \cdot x = x$ Multiplicative identity

(d) $\dfrac{7}{x} \cdot \dfrac{x}{7} = 1$ Multiplicative inverse

◄

Problem Set 1.6

Warm-Ups

In Problems 1 through 15, name the property illustrated by each statement. See Examples 1 and 5.

1. $4 \cdot 7 = 7 \cdot 4$

2. $(3 + 7) + 4 = (7 + 3) + 4$

3. $(x + 7) + 4 = x + (7 + 4)$

4. $8 + n = n + 8$

5. $2(5 + y) = 2 \cdot 5 + 2y$

6. $x + 0 = x$

7. $-\dfrac{1}{3}(-3) = 1$

8. $-\sqrt{6} + \sqrt{6} = 0$

9. $-x + y = y + (-x)$

10. $\dfrac{1}{t} \cdot 1 = \dfrac{1}{t}$

11. $1 \cdot 14 = 14$

12. $9(7 - \sqrt{3}) = 9 \cdot 7 - 9 \cdot \sqrt{3}$

13. $-\dfrac{1}{2} + \dfrac{1}{2} = 0$

14. $(1 + x) + z = 1 + (x + z)$

15. $\dfrac{2}{3} \cdot \dfrac{3}{2} = 1$

In Problems 16 through 19, use the Commutative and Associative Properties to perform the operations indicated efficiently. See Example 2.

16. $367 + 874 + (-367)$

17. $-12 + 48 + 8 + 4$

18. $-6(4)(1/6)$

19. $\left(\dfrac{1}{3}\right)\left(\dfrac{2}{5}\right)(6)(10)$

In Problems 20 through 25, use the Distributive Property to multiply each. See Example 3.

20. $2(1 + x)$

21. $3(x - 4)$

22. $-2(x + 9)$

23. $-4(x - 2)$

24. $x(y + z)$

25. $x(y - z)$

In Problems 26 through 30, give the additive inverse (opposite) and the multiplicative inverse (reciprocal) of each number. Assume x is not zero. See Example 4.

26. x

27. $-\dfrac{2}{7}$

28. π

29. $-5x$

30. 1

Practice Exercises

In Problems 31 through 35, match expressions that have the same value and name the property that justifies the match.

31. $2 \cdot 5 - 2 \cdot 8$ **a.** $3 + 7$

32. $7 + 3$ **b.** 9

33. $(4 + 5) + 6$ **c.** -5

34. $-5 + 0$ **d.** $2(5 - 8)$

35. $1 \cdot 9$ **e.** $(5 + 4) + 6$

In Problems 36 through 49, name the property illustrated by each statement.

36. $2 \cdot 3 \cdot 4 = 3 \cdot 2 \cdot 4$

37. $(3 + 7) + 2 = 3 + (7 + 2)$

38. $(g + 7) + 2 = (7 + g) + 2$

39. $5(6x) = (5 \cdot 6)x$

40. $9(r + 2) = 9r + 9 \cdot 2$

41. $y \cdot 1 = y$

42. $-\dfrac{1}{3} + \dfrac{1}{3} = 0$

43. $x \cdot \dfrac{1}{x} = 1$

44. $-p + p = 0$

45. $\dfrac{1}{j} + 0 = \dfrac{1}{j}$

46. $4q + 4 \cdot 7 = 4(q + 7)$

47. $0 + \sqrt{7u} = \sqrt{7u}$

48. $\dfrac{x}{y} \cdot 1 = \dfrac{x}{y}$

49. $\dfrac{1}{2} \cdot 2 = 1$

In Problems 50 through 57, simplify each expression.

50. $9\left(\dfrac{1}{3}x\right)$

51. $\dfrac{1}{12}(4x)$

52. $251 + 976 + (-511)$

53. $-52 + 148 + 52$

54. $-14(4)\left(\dfrac{1}{7}\right)$

55. $\left(\dfrac{1}{3}\right)\left(\dfrac{2}{7}\right)(9)(14)$

56. $-5\left(-\dfrac{1}{15}y\right)$

57. $\dfrac{2}{5}\left(\dfrac{15}{8}z\right)$

In Problems 58 through 62, give the additive inverse (opposite) and the multiplicative inverse (reciprocal) of each number.

58. 14   **59.** $\dfrac{5}{3}$ **60.** $-\pi$ **61.** 0   **62.** -1

In Problems 63 through 76, use the distributive property to multiply.

63. $2(4 + x)$ **64.** $3(x - 8)$ **65.** $-2(6 + y)$

66. $-4(y - 1)$ **67.** $x(y + 3)$ **68.** $x(3 + z)$

69. $x(y - 5)$ **70.** $x(4 - y)$ **71.** $t(3 - x)$

72. $x(y + 2z)$ **73.** $n(3 + 2x)$ **74.** $k(x - 9)$

75. $c(-9 + x)$ **76.** $-5(t + x)$

Challenge Problems

77. Is subtraction commutative? (Is $p - q$ the same as $q - p$?)

78. Is division commutative? (Is $p \div q$ the same as $q \div p$?)

In Problems 79 through 82, use the properties to carry out the multiplications indicated.

79. $x(x + 3)$ **80.** $5(x + y + z)$ **81.** $2x \cdot 3x^2$

82. $2x(3x^2 + z)$

In Your Own Words

State the following properties.

83. Commutative Properties   **84.** Associative Properties   **85.** Distributive Property

---

1.7   The Language of Algebra

TAPE 3

We have noticed that in algebra letters are used to name numbers. We call such letters constants or variables.

A **constant** is a number whose value does not change. For example, 2 and π are constants. They always have the same fixed value. When we use letters to name constants, they have one fixed value.

A **variable** is a letter that may have many different values. For example, the circumference of a circle is given by the product

$$2\pi r$$

The letter r stands for the radius of the circle and can be *any* positive number. It is a variable. The numeral 2 and the Greek letter π stand for fixed numbers and so they are constants. (The constant π is an irrational number whose approximate value is 3.1416.)

Most problems from the fields of engineering and the sciences are solved by the use of mathematics. Real problems are usually modeled with a mathematical statement; then the laws and properties of mathematics are used to find a solution to the problem. An important step in this direction is translating from the English language to the language of mathematics.

In this section we will examine the problem of translating verbal expressions into algebraic expressions. We must learn to recognize certain English language expressions that are commonly used to indicate mathematical operations.

A list of some common phrases that indicate the various operations is given next.

Phrase	Example	Algebraic Expression
Addition:		
Added to	6 *added to* 9	$9 + 6$
The sum of	*The sum of* 5 and x	$5 + x$
More than	9 *more than* t	$t + 9$
Greater than	15 *greater than* y	$y + 15$
Increased by	31 *increased by* x	$31 + x$
The total of	*The total of* 3 and t	$3 + t$
Subtraction:		
Subtracted from	11 *subtracted from* 100	$100 - 11$
Minus	x *minus* 22	$x - 22$
Less than	y *less than* 10	$10 - y$
Fewer than	17 *fewer than* t	$t - 17$
Decreased by	7 *decreased by* x	$7 - x$
The difference	*The difference* when k is subtracted from 5	$5 - k$
Multiplication:		
Multiplied by	3 *multiplied by* y	$3y$
The product of	*The product of* x and y	xy
Times	4 *times* 7	$4 \cdot 7$
Twice	*Twice* x	$2x$
Of	$\frac{1}{2}$ *of* x	$\frac{1}{2}x$
Division:		
Divided by	16 *divided by* 23	$\frac{16}{23}$
Quotient of	*The quotient of* 9 and 5	$\frac{9}{5}$
The ratio of	*The ratio of* 7 and 11	$\frac{7}{11}$

► **EXAMPLE 1**

Translate each phrase into an algebraic expression.

(a) The sum of $3x$ and 16

(b) 4 less than twice y

(c) The product of 17 and x, divided by the sum of 17 and x

(d) x divided by the quantity 5 less than x, subtracted from 17

Solutions

(a) The word *sum* implies addition:
$$3x + 16$$

(b) Twice y is $2y$.
$$2y - 4$$

(c) $\dfrac{17x}{17 + x}$

(d) $17 - \dfrac{x}{x - 5}$ ◄

It is often necessary to develop an expression in mathematics to represent, or model, a real-life situation.

► EXAMPLE 2

Answer each question.

(a) If 5 ft are cut off the end of a board that is 16 ft long, how much is left?

(b) If x feet are cut off the end of a board that is 16 ft long, how much is left?

Solutions

(a) To find out how much of the board remains, we subtract 5 from 16.
$$16 - 5 = 11$$
There are 11 ft of board left.

(b) We do this exactly like we did part (a). We subtract x from 16.
There are $16 - x$ feet of board left. ◄

► EXAMPLE 3

Answer each question.

(a) If Harold has 20 nickels and dimes in his pocket and 7 of them are dimes, how many nickels does Harold have in his pocket?

(b) If Harold has 20 nickels and dimes in his pocket and x of them are dimes, how many nickels does Harold have in his pocket?

Solutions

(a) If we take away the number of dimes from the total number of coins, we have the number of nickels.
$$20 - 7 = 13$$
He has 13 nickels in his pocket.

(b) In the same manner, we subtract the number of dimes, x, from the total. He has $20 - x$ nickels in his pocket. ◄

► EXAMPLE 4

Answer each question.

(a) If Maria has 7 dimes in her pocket, how much are they worth?

(b) If Maria has x dimes in her pocket, how much are they worth?

(c) If Maria has $20 - x$ nickels in her pocket, how much are they worth?

(d) Maria has x dimes and $20 - x$ nickels in her pocket. What is the value of the money Maria has in her pocket?

Solutions

(a) Dimes are worth 10¢ each, and Maria has 7. Their value is $10 \cdot 7$ or 70 cents.

(b) By the same reasoning, x dimes are worth 10 times x cents. The x dimes are worth $10x$ cents.

(c) As nickels are worth 5¢ each, the nickels in her pocket are worth $5(20 - x)$ cents.

(d) The total value of the collection of coins is the value of the dimes plus the value of the nickels. Maria has $10x + 5(20 - x)$ cents in her pocket.   ◄

► **EXAMPLE 5**

Answer each question.

(a) If Sadie is 34 years (yr) old now and she is three years older than her sister, how old will each of them be in 4 years?

(b) If Sadie is x years old now and she is 3 yr older than her sister, how old will each of them be in 4 years?

(c) If Sadie is x years old now and she is 3 yr older than her sister, what will be the sum of their ages 4 years from now?

Solutions

(a) Since Sadie is 3 yr older than her sister, her sister must be $34 - 3 = 31$ yr old *now*. Thus in 4 yr Sadie will be $34 + 4$ yr old and her sister will be $31 + 4$ yr old.

	Now	In 4 Years
Sadie's age	34	$34 + 4$
Sister's age	$34 - 3$	$(34 - 3) + 4$

In 4 yr Sadie will be 38 yr old and her sister will be 35 yr old.

(b) We compute exactly as we did in part (a). Sadie is x years old now and her sister is $x - 3$. After they each age 4 years, Sadie will be $x + 4$ years old and her sister will be $(x - 3) + 4$ years old.

	Now	In 4 Years
Sadie's age	x	$x + 4$
Sister's age	$x - 3$	$(x - 3) + 4$

(c) From part (b) we see that the sum of their ages will be $(x + 4) + [(x - 3) + 4)]$ years.   ◄

The properties of real numbers allow us to simplify certain sums and differences, such as the sum that occurs in the answer to Example 5(c).

The expression $3x + 5x$ is a sum. In it, $3x$ and $5x$ are *terms*. We call them x terms, and we also say that they are *like terms*. In the same manner, in the difference

$$5x^2 - 7x^2$$

we say the x^2 terms, $5x^2$ and $7x^2$, are *like terms*. In the term $3x$, 3 is called the **coefficient** of x, or the numerical coefficient. In $5x^2$, the numerical coefficient is 5.

The sum $6x + 5y$ contains an x term and a y term. They are *not* like terms. Likewise, $9x^2 - 7x^3$ contains two terms that are *not* like terms.

If terms are the same everywhere except for the numerical coefficient, we call them **like terms.**

► EXAMPLE **6**

Determine if each given pair of expressions is a pair of like terms and give the coefficient for each term.

(a) $7y$ and $18y$ (b) x and $-x$ (c) $2xy^2$ and $3x^2y$

(d) $3x^3y$ and $-5x^3y$ (e) 11 and -6 (f) $15\sqrt{x}$ and $\dfrac{2}{3}\sqrt{x}$

Solutions

(a) $7y$ and $18y$ are both y terms. They *are* like terms. The coefficients are 7 and 18.

(b) x and $-x$ are both x terms. They *are* like terms. The coefficient in the term x is 1, as x can be written as $1 \cdot x$. We call 1 an *understood* coefficient. The coefficient in the term $-x$ is -1. It is also an understood coefficient.

(c) $2xy^2$ and $3x^2y$ are *not* like terms. The coefficients are 2 and 3.

(d) $3x^3y$ and $-5x^3y$ are both x^3y terms. They *are* like terms. The coefficients are 3 and -5.

(e) 11 and -6 *are* like terms because they are both constants. The coefficients are 11 and -6.

(f) $15\sqrt{x}$ and $\dfrac{2}{3}\sqrt{x}$ are both $\sqrt{x}$ terms. They *are* like terms. The coefficients are 15 and $\frac{2}{3}$.   ◄

The Commutative and Distributive Properties allow us to simplify sums and differences that contain like terms.

$$3x + 5x = x \cdot 3 + x \cdot 5 \qquad \text{Commutative}$$

$$= x(3 + 5) \qquad \text{Distributive}$$

$$= x \cdot 8$$

$$= 8x \qquad \text{Commutative}$$

So $3x + 5x = 8x$. *Notice that we just added the coefficients.*

$$5x^2 - 7x^2 = x^2 \cdot 5 \ - \ x^2 \cdot 7 \qquad \text{Commutative}$$

$$= x^2(5 - 7) \qquad\quad\ \text{Distributive}$$

$$= x^2(-2)$$

$$= -2x^2 \qquad\qquad\ \text{Commutative}$$

Thus $5x^2 - 7x^2 = -2x^2$. *We just combined the coefficients.*
We call such operations **combining like terms.**

► **EXAMPLE 7**

In each of the following, combine like terms if possible.

(a) $8x^4 + 9x^4$ (b) $6x^2y - x^2y$

(c) $3kx + 4kx - 2kx$ (d) $14x^3y - 11x^3y + 2x^3z$

Solutions

(a) $8x^4 + 9x^4 = 17x^4$

(b) $6x^2y - x^2y = 5x^2y$

(c) $3kx + 4kx - 2kx = 5kx$

(d) $14x^3y - 11x^3y + 2x^3z = 3x^3y + 2x^3z$

     Note that the last term is *not* like the other two.   ◄

► **EXAMPLE 8**

Simplify each expression.

(a) $6x + 1 + 5x - 3$ (b) $4x^2 - 3x + x^2 - 2x$

Solutions

(a) First we rearrange the terms using the commutative property.

$$6x + 1 + 5x - 3 = 6x + 5x + 1 - 3$$

Then we combine like terms.

$$6x + 5x + 1 - 3 = 11x - 2$$

(b) $\ 4x^2 - 3x + x^2 - 2x = 4x^2 + x^2 - 3x - 2x$

$$= 5x^2 - 5x \qquad\qquad\qquad ◄$$

► **EXAMPLE 9**

Answer each of the following questions.

(a) 26 is an even integer. What are the next two *consecutive even* integers?

(b) If x is an even integer, what are the next two *consecutive even* integers?

(c) Suppose x is the smallest of three consecutive even integers. What is their sum?

Solutions

(a) 28 and 30 are the next two consecutive even integers following 26. Notice that 28 is $26 + 2$ and 30 is $26 + 4$.

The integers are 28 and 30.

(b) Using the same reasoning that we did in part (a), we get the next even integer after x by adding 2 to x and the next after that by adding 4.

The integers are $x + 2$, and $x + 4$.

(c) If x is the smallest of three consecutive even integers, then as we saw in part (b) that they must be x, $x + 2$, and $x + 4$, and their sum is

$$x + (x + 2) + (x + 4)$$

However, we can simplify this expression by combining like terms.

$$x + (x + 2) + (x + 4) = x + x + 2 + x + 4$$
$$= x + x + x + 2 + 4$$
$$= 3x + 6$$

The sum of the integers is $3x + 6$.   ◄

Problem Set 1.7

Warm-Ups

In Problems 1 through 9, translate the given phrases into algebraic expressions. See Example 1.

1. The sum of 41 and $6y^2$.

2. 15 fewer than $2x$.

3. x times $8y$.

4. The ratio of y and twice x.

5. 10 divided by x squared.

6. $3t$ increased by 19.

7. Twice x decreased by twice y.

8. One-fifth of the difference when 5 is subtracted from x.

9. The quotient when the sum of x and 4 is divided by the difference when 4 is subtracted from x.

In Problems 10 through 19, answer each question with an English language phrase. For Problems 10 and 11, see Example 2.

10. A 4-in. piece is cut from the end of a 31-in. length of coaxial cable. How much is left?

11. A 4-in. piece is cut from the end of a length of coaxial cable that is x inches long. How much is left?

For Problems 12 and 13, see Example 3.

12. There are 17 pounds of shrimp in a net that contains 56 pounds of shrimp and mixed fish. How many pounds of mixed fish are in the net?

13. There are 17 pounds of shrimp in a net that contains x pounds of shrimp and mixed fish. How many pounds of mixed fish are in the net?

For Problems 14 through 16, see Example 4.

14. The vending machine in the lounge contains 81 nickels and quarters. If it has 70 nickels, how many quarters does it have?

15. If the vending machine in the lounge contains x nickels and quarters and it has 70 nickels, how many quarters does it have?

16. If the vending machine in the lounge contains x nickels and quarters and it has 70 nickels, what is the *value* of the nickels in the bank? What is the value of the quarters? What is the value of the collection?

For Problems 17 through 19, see Example 5.

17. Don is 4 years younger than Barbara. If Barbara is 55, how old is Don?

18. Don is 4 years younger than Barbara. If Barbara is x years old, how old is Don?

19. Don is 4 years younger than Barbara. If Barbara is x years old now, how old will Don be in 10 years?

Problems 14–16

In Problems 20 through 39, simplify the given expressions. See Examples 7 and 8.

20. $3x + 7x$

21. $8x^2 - x^2$

22. $7x - 4x + x$

23. $4abc - 2abc - 5abc$

24. $3x + 2 + 5x + 13$

25. $4xy - 7xy - 5 + 8$

26. $4x^2 + x - 2 - 2x^2 - x - 1$

27. $9x^2y + 4x^2y - 5xy^2 - xy^2$

28. $3x^3 - x^2 + x^3 - 2x^2$

29. $5x^2z + 8z^2 + 31z^2 - 6x^2z$

30. $2x^2 - 3x + 2x - 3 + 2x^2$

31. $4\sqrt{x} - 2\sqrt{x}$

32. $2(x - 7) + 3x$

33. $14x + 5(3x - 7)$

34. $2(x + 3) + 3(x + 4)$

35. $-3(x - 1) + 2(x - 5)$

36. $3(10 - 7y) + 2(15 - 2y)$

37. $3(3b - 1) + 3(5b - 2)$

38. $4(3 - 2a) + 2(3a + 2)$

39. $3\sqrt{x - 1} + 7\sqrt{x - 1}$

In Problems 40 through 45, answer each question with an English language phrase. See Example 9.

40. 71 is an odd integer. What are the next two consecutive odd integers?

41. What is the sum of three consecutive odd integers if 71 is the smallest of the three?

42. Suppose x is an odd integer. What are the next two consecutive odd integers?

43. What is the sum of three consecutive odd integers if x is the smallest of the three?

44. What is the sum of three consecutive even integers if 92 is the largest of the three?

45. What is the sum of three consecutive even integers if x is the largest of the three?

Practice Exercises

In Problems 46 through 52, translate the given phrases into algebraic expressions.

46. 14 added to -11.

47. The total of $6x$ and 23.

48. 10 less than y.

49. One-third of twice t.

50. The ratio of 17 and the difference when 5 is subtracted from x.

51. The product of x cubed and -4.

52. Twice x increased by twice y.

In Problems 53 through 86, simplify the given expressions.

53. $9z + z$

54. $3y^2 - 2y^2$

55. $7x + 4x - x$

56. $3xyz - 7xyz + 5xyz$

57. $x^2 - 5x^2 - 11x^2$

58. $6xy^2z^3 - 11xy^2z^3 + 4xy^2z^3$

59. $5 + 3v + 19 + 8v$

60. $2x^3 + x^3 + y^3 + 12y^3$

61. $4x^2 + x^3 - 2x^2 + 5x^2 - x^3$

62. $1 + 2x - 3y + x + y + 4$

63. $3x^3 + x^2 - x^3 + 2x^2$

64. $5x^2z + 8z^2 - 31z^2 + 6x^2z$

65. $2x^2 - 3x - 2x - 3 - 2x^2$

66. $6\sqrt{y} + 7\sqrt{y}$

67. $3(x + 1) + 3x$

68. $18x + 3(3x - 5)$

69. $x - 4 + 2(x + 3)$

70. $1 + x + 3(x + 2)$

71. $\sqrt{2x} - 6\sqrt{2x} + 16\sqrt{2x}$

72. $2(2 - x) + 3(4 + x)$

73. $3(2x - 7) + 4(x - 1)$

74. $4(6 + x) + 5(x - 3)$

75. $(3y^2 - 4)3 + 8(y^2 - 1)$

76. $2(3x + 1) - 2x - 8$

77. $3x + 3(4x - 4) + 8$

78. $7z + 2(z - 2) - 3$

79. $2(x - 3) + 3(x - 4)$

80. $-2(x - 1) + 3(x + 5)$

81. $4(3 + 2a) + 2(3a - 2)$

82. $6\sqrt{k + 2} + \sqrt{k + 2}$

83. $-5(11 - 2y) + 3(17 - 2y)$

84. $3(3b - a) + 3(5b - a)$

85. $4(3 - 2x) + 2(-1 - 2x)$

86. $23(2x - 3) + (18 - 21x)$

In Problems 87 through 106, answer each question with an English language phrase.

87. Ken has $7700 to invest. If he buys a $2000 certificate of deposit, how much does he have left to invest?

88. Juan has $7700 to invest. If he buys a certificate of deposit for x dollars, how much does he have left to invest?

89. In a given triangle, $\angle A$ has measure $37°$ and $\angle B$ is twice $\angle A$.
What are the measures of $\angle B$ and $\angle C$? (*Hint:* The sum of the angles in a triangle is $180°$.)

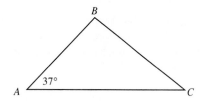

90. In a given triangle, $\angle A$ has measure $x°$ and $\angle B$ is twice $\angle A$. What are the measures of $\angle B$ and $\angle C$?

91. It is 71 miles (mi) from Joo Ran's house to Chin-Ah's. If Joo Ran drives 34 mi toward Chin-Ah's house and then stops for tea, how much farther must she drive?

92. It is 85 mi from Leonard's house to Glenn's. If Leonard drives x miles toward Glenn's house and then stops for coffee, how much farther must he drive?

93. Mr. Maxey removed 102 coins from a broken subway station turnstile before repairing it. If there were 42 quarters and the rest were tokens, how many tokens did Maxey remove?

94. In Problem 93, what is the *value* of the quarters? What is the *value* of the tokens if one token is worth $1.25?

95. Mr. Maxey removed 102 coins from a broken subway station turnstile. If there were x quarters and the rest were tokens, how many tokens did Maxey remove?

96. In Problem 95, what is the *value* of the quarters? What is the *value* of the tokens? What is the *value* of all the coins in the turnstile?

97. If the measure of a certain angle is $56°$, what is the measure of its supplement?

(*Hint:* An angle and its supplement have a sum of $180°$.)

98. If the measure of a certain angle is $x°$, what is the measure of its supplement? What is the measure of its complement? (*Hint:* An angle and its complement have a sum of 90°.)

99. Som is 3 years older than Val. If Val is 22 yr old now, how old is Som?

100. Steve is 3 years older than Angela. If Angela is x years old now, how old is Steve?

101. Steve is 3 years older than Angela. If Angela is x years old now, how old will Steve be in 10 yr?

102. 18 is an even integer. What are the next three consecutive even integers?

103. If x is an even integer, what are the next three consecutive even integers?

104. If x is an odd integer, what are the next three consecutive odd integers?

105. If 18 is the smallest of four consecutive even integers, what is their sum?

106. If x is the smallest of four consecutive even integers, what is their sum?

Challenge Problems

Simplify each of the following.

107. $\sqrt{x^2 + 3} - 3\sqrt{x^2 - 5} - 2\sqrt{x^2 - 5} + 11\sqrt{x^2 + 3}$

108. $\sqrt{x + 1} - 2\sqrt{y + 1} - 2\sqrt{x + 1} + 10\sqrt{y + 1}$

109. $3\sqrt{x + 4} - 2(3\sqrt{x + 4} - \sqrt{2y}) - 4\sqrt{2y}$

110. $13\sqrt{x} - 4(3\sqrt{x} - \sqrt{y}) - 4\sqrt{y}$

In Your Own Words

111. What are constants and variables?

112. Why do we have to use letters in place of numbers in algebra?

113. Make up a question that might have this statement as its answer: There are 22 orange marbles in the bag.

1.8 Equations

TAPE 3

Equations are a central idea of algebra. We define an **equation** to be a statement that two numbers are equal. We use the symbol $=$ to indicate equality.

An equation may be a true statement or it may be a false statement. Consider the statements

$$3 + 14 = 17$$

$$3 + 11 = 17$$

They are both statements that two numbers are equal. The first is *true,* and the second is *false.*

Equations are much more interesting when they contain variables. Consider the equation

$$3 + x = 17$$

Notice that if x has the value 14, the statement is *true,* but if x has the value 11 (or any number other than 14), the statement is *false.* We call 14 a **solution,** or **root,** of the equation. The set of all solutions of an equation is called its **solution set.** Since 14 is the only number that makes the statement true, we write the solution set

$$\{14\}$$

An equation has a **left side,** an equality symbol, and a **right side.**

$$\underbrace{3x - 1}_{\text{Left side}} = \underbrace{4(2 - x)}_{\text{Right side}}$$

Left side   Right side

► **EXAMPLE 1**

Is the given number a solution of the equation?

(a) Equation: $3x + 1 = 5$ (b) Equation: $t^2 - t - 6 = 0$
    Given number: 2   Given number: -2

Solutions

(a) We must decide if replacing x with 2 makes the equation a true or false statement. Since $3(2) + 1$ is 7, the statement is false. So 2 is not a solution of $3x + 1 = 5$.

(b) We replace t with -2 in the left side of the equation

$$t^2 - t - 6 = (-2)^2 - (-2) - 6 \qquad \text{Notice the use of parentheses}$$

$$= 4 + 2 - 6$$

$$= 0$$

Since the right side is also 0, the statement is true. Thus -2 is a solution of $t^2 - t - 6 = 0$.   ◄

To **solve** an equation means to find its solution set. That is, we find the set of numbers that make the statement true. To do this, we have two important tools. The first tool says we can add the *same* number to *both sides* of an equation without changing the solution set. The second tool says we can multiply *both sides* of an equation by the *same* number (as long as it is not zero) without changing the solution set.

Two Tools for Solving Equations

The solution set of an equation will not change if we use either of the following two tools.

Addition Property of Equality

Add (or subtract) the same number on both sides of the equation.

Multiplication Property of Equality

Multiply (or divide) both sides by the same nonzero number.

Our general approach is to use these tools to simplify an equation to the point where it is of the following form: the variable equal to a constant.

► **EXAMPLE 2**

Solve the equation $x - 7 = 11$.

Solution

It may seem that the solution set is obvious as the equation stands. However, as a general practice, we always simplify equations as far as we can before writing the solution set. In this case, we see that if we add 7 to the left side of the equation the 7 and the -7 will have a sum of zero, leaving the x alone. So we add 7 to both sides of the equation.

$$x - 7 = 11$$

$$x - 7 \ +7 \ = 11 \ +7 \qquad \text{Add 7 to both sides}$$

$$x + 0 = 18 \qquad \text{Simplify}$$

$$x = 18$$

This statement is *true* if x has the value 18 and false otherwise. Thus the solution set contains only the number 18, which we write

$$\{18\} \qquad \blacktriangleleft$$

A solution of an equation is a number that makes the equation a *true* statement. We can decide if a number is in fact a solution by **checking.** The type of equations we are solving now do not *require* checking as a part of the solving procedure. However, they can be checked for arithmetic errors. In checking a possible solution, *be sure to use a correct procedure.*

We recommend a checking procedure that evaluates both the left and right sides of the original equation separately. If the value obtained from the left side is the same as the value obtained from the right side, then the number being checked is in the solution set. Otherwise, it is not.

► **EXAMPLE 3**

Solve the equation $x + 4 = 19$ and check the solution.

Solution

We can add *or subtract* any number on both sides of an equation. In this equation, it is convenient to subtract 4 from both sides.

$$x + 4 = 19$$

$$x + 4 \ -4 \ = 19 \ -4 \qquad \text{Subtract 4 from both sides}$$

$$x + 0 = 15 \qquad \text{Simplify}$$

$$x = 15$$

The instructions require checking the solution. We do this *before* writing the solution set. To check, we replace x by 15 in each side of the original equation and evaluate them *separately*. We show one recommended format.

$$\text{LS (left side):} \quad x + 4 = 15 + 4 = \ 19$$
$$\text{RS (right side):} \quad 19$$

As the left side and the right side yield the same number when x is 15, 15 is the solution, and we write the solution set:

$$\{15\} \qquad \blacktriangleleft$$

► **EXAMPLE 4**

Solve the equation $3x = 21$ and check the solution.

Solution

Adding a number to both sides of this equation will not aid in solving it. However, we can *multiply* both sides by $\frac{1}{3}$, or we can *divide* both sides by 3.

$$3x = 21$$

$$\frac{3x}{3} = \frac{21}{3} \qquad \text{Divide both sides by 3}$$

$$x = 7$$

Again the instructions call for checking.

$$\text{LS:} \quad 3x = 3(7) = \boxed{21}$$

$$\text{RS:} \quad \boxed{21}$$

Thus 7 checks. The solution set is

$$\{7\}$$

◄

► **EXAMPLE 5**

Solve $-\dfrac{1}{2}x = 4$ and check the solution.

Solution

$$-\frac{1}{2}x = 4$$

$$\boxed{-2}\left(-\frac{1}{2}x\right) = \boxed{-2}\,(4) \qquad \text{Multiply both sides by } -2$$

$$\left[(-2)\left(-\frac{1}{2}\right)\right]x = -8 \qquad \text{Associative property}$$

$$1 \cdot x = -8$$

$$x = -8$$

To check -8, we replace x with -8.

$$\text{LS:} \quad -\frac{1}{2}x = -\frac{1}{2}(-8) = \boxed{4}$$

$$\text{RS:} \quad \boxed{4}$$

Since the values of the left side and right side are the same, -8 is a solution.

$$\{-8\}$$

◄

► **EXAMPLE 6**

Solve $\dfrac{3}{4}x = -1$

Solution

We can solve this equation by two methods. We can multiply both sides by the reciprocal of $\dfrac{3}{4}$, which is $\dfrac{4}{3}$, or we can multiply both sides by 4 and then divide both sides by 3. We first solve this equation by multiplying both sides by $\dfrac{4}{3}$, the reciprocal of $\dfrac{3}{4}$.

1

$$\frac{3}{4}x = -1$$

$$\frac{4}{3}\left(\frac{3}{4}x\right) = -1\left(\frac{4}{3}\right)$$

$$\left(\frac{4}{3}\cdot\frac{3}{4}\right)x = -1\left(\frac{4}{3}\right) \qquad \text{Associative property}$$

$$x = -\frac{4}{3}$$

$$\left\{-\frac{4}{3}\right\}$$

Compare method 1 with the following method.

2

$$\frac{3}{4}x = -1$$

$$4\left(\frac{3}{4}x\right) = 4(-1) \qquad \text{Multiply both sides by 4}$$

$$\left(4\cdot\frac{3}{4}\right)x = 4(-1) \qquad \text{Associative property}$$

$$3x = -4$$

Notice that the equation now contains no fractions. We often call this step **clearing fractions.**

$$\frac{3x}{3} = \frac{-4}{3} \qquad \text{Divide both sides by 3}$$

$$x = -\frac{4}{3}$$

$$\left\{-\frac{4}{3}\right\}$$

The second method is longer, but it is good to see how to clear an equation of fractions. This will be useful in Chapter 3. ◄

The next example illustrates a situation that arises often in solving equations.

► **EXAMPLE 7**

Solve $-x = 3$.

Solution

If we multiply (or divide) both sides by -1, we can write the solution set.

$$-x = 3$$
$$(\boxed{-1})(-x) = (\boxed{-1})3 \qquad \text{Multiply both sides by } -1$$
$$1x = -3$$
$$x = -3$$
$$\{-3\}$$
◄

Equations are used in the real world to solve problems as diverse as finding interest and designing rocket boosters. They are the heart of engineering and the sciences. In an algebra course we use word problems to imitate such real-world applications.

We solve a word problem by first translating it into an equation. This equation is often called a **mathematical model** of the problem. Then we solve the equation and answer the question.

► **EXAMPLE 8**

The sum of a number x and 8 is 12. What is the number?

(a) Translate the statement into an equation.

(b) Solve the equation and find the number.

(c) Answer the question.

Solution

(a) $x + 8 = 12$

(b) $x + 8 \boxed{-8} = 12 \boxed{-8} \qquad \text{Subtract 8 from both sides}$

$\qquad x = 4$

(c) The number is 4.

We can check our answer by reading the problem and replacing x with 4. The sum of a number 4 and 8 is 12.
◄

Problem Set 1.8

Warm-Ups

In Problems 1 through 8, decide if the given number is a solution of the equation. See Example 1.

1. Equation: $2x - 5 = -8$

   Given number: -1

2. Equation: $6x + 3 = 5$

   Given number: $\dfrac{1}{3}$

3. Equation: $1.5x - 1.25 = 0.5$
Given number: 0.5

4. Equation: $-3 - t = -2$
Given number: -1

5. Equation: $\dfrac{w - 1}{w - 4} = 2$
Given number: 6

6. Equation: $|x| = 3$
Given number: -3

7. Equation: $x^2 - 3x - 10 = 0$
Given number: -2

8. Equation: $\sqrt{x} - 3 = -2$
Given number: 25

In Problems 9 through 18, solve each equation. See Examples 2 and 3.

9. $x - 3 = 2$

10. $x - 7 = 1$

11. $t + 3 = 8$

12. $x + 1 = 0$

13. $15 + x = 22$

14. $-11 + z = 5$

15. $x + 3 = -2$

16. $x + 21 = -4$

17. $x - 15 = -15$

18. $y - 7 = -11$

In Problems 19 through 22, solve each equation. See Example 4.

19. $2z = 12$

20. $16 = 4x$

21. $7x = 14$

22. $3x = 0$

In Problems 23 through 26, solve each equation. See Examples 5 and 6.

23. $\dfrac{1}{2}x = 13$

24. $\dfrac{1}{7}x - 3 = 0$

25. $\dfrac{2}{3}x = 9$

26. $\dfrac{1}{5}t = \dfrac{1}{6}$

In Problems 27 through 32, solve each equation. See Example 7.

27. $-y = 5$

28. $-x = -7$

29. $-x = \dfrac{1}{2}$

30. $-\dfrac{2}{5} = -x$

31. $-\dfrac{3}{2}s = \dfrac{5}{2}$

32. $-\dfrac{2}{7}x = -11$

In Problems 33 through 40, translate each sentence into an equation, solve the equation, and find the number. See Example 8.

33. The sum of a number y and 6 is -12.

34. A number w decreased by 1 is -3.

35. A number x increased by 5 is 0.

36. One-half of a number t is -12.

37. If -7 is subtracted from a number y, the result is 13.

38. The product of a number s and 5 is -35.

39. If a number t is divided by 5, the result is 15.

40. If the reciprocal of $\dfrac{2}{3}$ is multiplied by a number z, the result is -12.

Practice Exercises

In Problems 41 through 48, decide if the given number is a solution of the equation.

41. Equation: $2x - 5 = x - 8$
Given number: -3

42. Equation: $5x - 3 = 5$
Given number: $\dfrac{1}{5}$

43. Equation: $6 - x = 0$
Given number: -6

44. Equation: $0.8x - 1.0 = -0.4$
Given number: 1.2

45. Equation: $\dfrac{w - 1}{w + 8} = \dfrac{1}{2}$
Given number: -2

46. Equation: $\dfrac{1}{2}y - 2 = \dfrac{1}{3}y + 4$
Given number: 12

47. Equation: $|x| = -3$
Given number: -3

48. Equation: $\sqrt{x} - 3 = 1$
Given number: 16

In Problems 49 through 62, solve the equations.

49. $x + 3 = 2$

50. $x + 7 = -1$

51. $12 + x = 27$

52. $17 + x = 4$

53. $8x = 72$

54. $11x = 121$

55. $x - 14 = -8$

56. $x - 9 = -31$

57. $-z = -20$

58. $-x = \dfrac{2}{3}$

59. $\dfrac{2}{5}w = 1$

60. $-\dfrac{5}{4}s = \dfrac{3}{4}$

61. $\dfrac{1}{7}x = \dfrac{1}{4}$

62. $-\dfrac{3}{2}x = -18$

In Problems 63 through 77, translate each sentence into an equation, solve the equation, and find the number.

63. The product of a number y and 6 is -12.

64. A number w decreased by 3 is -3.

65. A number x increased by 7 is 0.

66. One-fourth of a number t is -12.

67. If -5 is subtracted from a number y, the result is 28.

68. The sum of a number s and 5 is -35.

69. If a number t is divided by 7, the result is -5.

70. If the reciprocal of $\frac{4}{3}$ is multiplied by a number z, the result is -12.

71. Find the width of a rectangle with side lengths x and $2x - 1$ if the perimeter is 40 meters.

72. Find the height of a triangle whose base is 12 in. and whose area is 56 sq in.

73. The relationship between Celsius degrees and Fahrenheit degrees is given by the formula,

$$C = \frac{5}{9}(F - 32).$$

Convert the Fahrenheit degrees in the following chart to Celsius degrees.

F	32°	212°	−13°	−76°
C				

74. The width of a rectangular strip of plastic is $2\frac{1}{3}$ inches and its area is $14\frac{7}{12}$ square inches. Find the length of the strip. The area of a rectangle is length times width.

75. The perimeter of a square is 25 feet. Find the length of a side of the square.

76. The area of the following triangle is $77\frac{1}{2}$ square centimeters. Use the formula for area of a triangle, $A = \frac{1}{2}$ base $\cdot$ height, to find the length of the base.

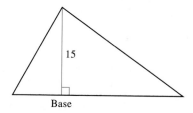

77. The volume of the following rectangular box is 30 cubic inches. Use the formula $V = $ length $\times$ width $\times$ height to find the height of the box.

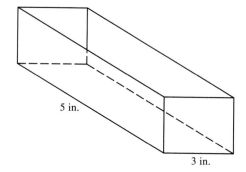

Challenge Problems

In Problems 78 through 83, x is the variable and p is a constant. Solve each equation.

78. $x - 2 = p$

79. $x + p = 2$

80. $5x = p$

81. $\dfrac{2}{3}x = p$

82. $px = 6$ (What restriction must we place on the value of p?)

83. A rectangular pasture is twice as long as it is wide. Find the dimensions of the pasture if it took 1 mile of fencing to completely enclose it.

In Your Own Words

84. What is an equation?

85. What does it mean to solve an equation?

86. What determines whether a number is a solution of an equation?

87. Make up a word problem that has the equation $2x = 8$ as its mathematical model.

Chapter Summary

GLOSSARY

Set   A collection of objects.

Member or **element**   One of the objects in a set.

Empty or **null set**   A set with no members.

Subset   A set whose members are also members of a given set.

Union of two sets   The set of all elements that belong to *either* of the two sets.

Intersection of two sets   The set of all elements that belongs to *both* of the two sets.

Natural numbers   1, 2, 3, . . .

Whole numbers   0, 1, 2, 3, . . .

Integers   . . . , $-3, -2, -1, 0, 1, 2, 3,$ . . .

Rational numbers   All possible quotients of integers. Numbers that can be written in the form $\frac{p}{q}$, where p and q are integers and $q \neq 0$.

Irrational numbers   Numbers on the number line that are not rational, such as $\sqrt{2}$ and π.

Real numbers   The rational and irrational numbers.

Positive real numbers   All real numbers greater than zero.

Negative real numbers   All real numbers less than zero.

Absolute value   The distance of a number from zero on the number line.

Opposite   The number on the opposite side of zero on the number line from a given number, but the same distance from zero.

Reciprocal   If p is a nonzero number, then $\frac{1}{p}$ is its reciprocal.

Equation   A statement that two numbers are equal.

Variable   A letter that stands for a number.

Solution   A number that when substituted for the variable makes an equation a true statement.

The Set of Real Numbers

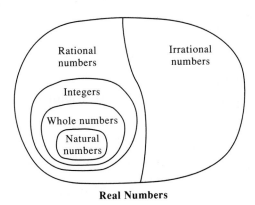

Real Numbers

Addition of Real Numbers

1. To add two numbers with *like signs* (both positive or both negative), add their absolute values and keep the common sign.

2. To add two numbers with *unlike signs* (one positive and one negative), subtract the smaller absolute value from the larger absolute value and keep the sign of the number with the greater absolute value.

Subtraction of Real Numbers

1. To subtract two real numbers, rewrite the problem as an addition problem.

$$p - q = p + (-q)$$

2. Follow the rules for addition.

Multiplication and Division of Real Numbers

1. To multiply or divide two numbers with *like signs* (both positive or both negative), multiply or divide their absolute values. The sign will be positive.

2. To multiply or divide two numbers with *unlike signs* (one positive and one negative), multiply or divide their absolute values. The sign will be negative.

Division Involving Zero

If p and q are real numbers with $p \neq 0$, then

1. $\dfrac{0}{p} = 0$

2. $\dfrac{q}{0}$ is undefined.

Properties of Real Numbers

Addition

1. $p + 0 = 0$ and $0 + p = p$	Identity
2. $p + (-p) = 0$ and $(-p) + p = 0$	Inverse
3. $p + q = q + p$	Commutative
4. $(p + q) + r = p + (q + r)$	Associative

Multiplication

1. $p \cdot 1 = p$ and $1 \cdot p = p$	Identity	
2. $p \cdot \dfrac{1}{p} = 1$ and $\dfrac{1}{p} \cdot p = 1$	Inverse	$(p \neq 0)$
3. $pq = qp$	Commutative	
4. $(pq)r = p(qr)$	Associative	
5. $0(p) = p(0) = 0$		
6. $-1(p) = p(-1) = -p$		

Distributive Properties

1. $p(q + r) = pq + pr$
2. $p(q - r) = pq - pr$

Equations

Two Tools for Solving Equations

1. Add or subtract the same number on both sides of the equation.
2. Multiply or divide both sides by the same nonzero number.

CHECK UPS

1. Classify each as rational or irrational:

   a. $\dfrac{3}{7}$ **b.** $0.\overline{78}$ **c.** $\sqrt{7}$ **d.** 1.7 Section 1.1; Example 3

2. Simplify each of the following: Section 1.1; Example 8
   a. $|-5|$ **b.** $2|17|$ **c.** $-|-7|$ **d.** $|-\pi|$

3. Perform the operations indicated.
   a. $4 + (-8)$ — Section 1.2; Example 3a
   b. $-7 - (-2)$ — Section 1.3; Example 1c
   c. $\frac{1}{3}\left(-\frac{3}{2}\right)$ — Section 1.4; Example 1c
   d. $10/(-2)$ — Section 1.4; Example 2c
   e. $-6 - (5 - \sqrt{25 - 16} + 1)$ — Section 1.5; Example 2b

4. Simplify the following expression.
$$-6/(-2) + 2(-3)^2 - 2^2$$
Section 1.5; Example 1b

5. Evaluate the following expression when x is -2.
$$|3 + 7x| + 21$$
Section 1.5; Example 4f

6. Name the property that is illustrated in each of the following:
   a. $4 + 0 = 4$ b. $5 + (-5) = 0$ — Section 1.6; Example 5

7. Write the opposite and reciprocal of each of the following:
   a. -10 b. $-\frac{3}{4}$ c. $\frac{2}{x}$ — Section 1.6; Example 4

8. If Harold has 20 nickels and dimes in his pocket and x of them are dimes, how many nickels does Harold have in his pocket? — Section 1.7; Example 3b

9. Simplify each expression. — Section 1.7; Example 8
   a. $6x + 1 + 5x - 3$
   b. $4x^2 - 3x + x^2 - 2x$ — Section 1.7; Example 8

10. Solve the equation, $x + 4 = 19$. — Section 1.8; Example 3

11. Solve the equation $-\frac{1}{2}x = 4$ and check the solution. — Section 1.8; Example 5

12. The sum of a number and 8 is 12. Find the number. — Section 1.8; Example 8

► Review Problems

In Problems 1 through 20, perform the operations indicated.

1. $-5 + 3$
2. $-\frac{3}{2} - \frac{1}{3} \div \frac{1}{6}$
3. $-8.1 + (-3.2 + 2.1)$
4. $5 + \{2(6 - 8) + 3[5 + (-7)]\}$
5. $\frac{3^2(4 - 6) - 6}{2^3}$
6. $-(-5)$
7. $|6 - 11| - 3[4 + (-8)]$
8. $(-5 - 6) - 4$
9. $16/4(-2)$
10. $[-9 - (-5)] - (-7 - 2)$
11. $2|-3| \div (-6)$
12. $8 \div 4 + (-4)(-4)$

13. $\dfrac{0}{-4}$

14. $\dfrac{-4}{0}$

15. $2|0 - 5| \div (-1 - 1)$

16. $\dfrac{3(-1) - 3 \cdot 2^2 + (-9)}{2^3 - 3^2}$

17. $-\dfrac{1}{3}\left(-\dfrac{1}{5}\right)(18)(-15)$

18. $487 + 28 - 487$

19. $-3(-8) - 2(4 - 8) + (3 - 6)$

20. $8.1/(0.09) - (-10)^2$

In Problems 21 through 25, answer each question.

21. Subtract -3 from the sum of -8 and -2.

22. Divide the product of -8 and -1 by 2.

23. Subtract the sum of -15 and 10 from 25.

24. The temperature at 5 A.M. was $-10°$. If the temperature rose $2°$ per hour, what was the temperature at noon?

25. On three consecutive plays, the Atlanta Falcons football team gained 8 yd, lost 5 yd, and gained 6 yd. What was their total gain for the three plays?

In Problems 26 through 30, write an algebraic expression for each phrase.

26. The sum of 2 and 6 multiplied by the opposite of 5.

27. The absolute value of the sum of 8 and 11.

28. One less than 9 divided by one more than 9.

29. The cube of twice 7.

30. One-half of 11 times one-third of 11.

In Problems 31 through 35, find the solution set for each equation.

31. $x - 4 = 6$

32. $x + 8 = 11$

33. $9 + x = 24$

34. $3x = 12$

35. $-\dfrac{3}{4}y = 6$

In Problems 36 through 45, indicate whether each statement is true or false.

36. Every whole number is an integer.

37. $\dfrac{2}{3} \le 0.\overline{6}$

38. $\sqrt{7}$ is a real number.

39. $\{2, 3\} \subseteq \{1, 2, 3\}$

40. $3 \in \{1, 2, 3\}$

41. $\dfrac{2}{3}$ is an irrational number.

42. An integer is a real number.

43. $-5.4 > -5.3$

44. $|-3| \le 0$

45. $2.\overline{3}$ is a rational number.

46. Alex has 250 feet of chain link fencing. He plans to fence his rectangular garden plot with dimensions $30\dfrac{1}{2}$ feet by $85\dfrac{3}{4}$ feet. Does Alex have enough fencing?

47. Allen is painting the ceiling of his den, which measures $15\dfrac{1}{2}$ feet by 20 feet. How many square feet must he paint?

48. The area of the following parallelogram is 2584 sq ft. Use the formula for the area of a parallelogram to find the base x. (Area = base $\times$ height)

17 ft

49. The volume of the following right circular cone is 900π cu ft. If r is 10 ft, find the height h. $\left(\text{Volume} = \dfrac{1}{3}\pi r^2 h\right)$

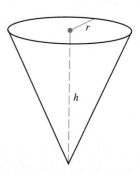

Let's Not Forget . . .

In Problems 50 through 52, find each sum or difference.

50. $5 + (-4)$

51. $5 - (-4)$

52. $-5 - (-4)$

In Problems 53 through 56, find each product or quotient.

53. $\dfrac{0}{3}$

54. $(-1)(5)$

55. $(-1)(0)$

56. $(-1)(-6)$

In Problems 57 through 59, simplify each expression. Watch the role of the parentheses!

57. $(-3 - 5)(3 + 7)$

58. $-3 - 5(3 + 7)$

59. $(-3 - 5)3 + 7$

In Problems 60 through 62, multiply each expression by 2.

60. $\dfrac{1}{2} \cdot 7$

61. $\dfrac{1}{2}(\pi - 2)$

62. $\pi - \dfrac{1}{2}$

In Problems 63 through 68, label each as an expression *or as an* equation. *Solve the equations.*

63. $x + 6$

64. $x + 6 = 0$

65. $2x = 10$

66. $2x - 10$

67. $x - 4 + 7$

68. $x - 4 = 7$

► You Decide

Your new diet allows 320 calories for the entree portion of your evening meal tonight. You have the following choices:

A. Grilled chicken   40 calories per ounce
B. Grilled steak   120 calories per ounce
C. Grilled fish   30 calories per ounce

How many ounces of each do you choose and why?

► Chapter 1 Test

In Problems 1 through 5, choose the correct letter.

1. $\dfrac{4(-1) + 3(-2)}{-2 - 3} = (?)$
   (a) -10 (b) -4
   (c) 4 (d) 2

2. $10 - (4)(8) + 6 = (?)$
   (a) 52   (b) -32
   (c) 84   (d) -16

3. When expressed as a decimal, $\dfrac{11}{6} = (?)$.
   (a) $1.8\overline{3}$ (b) 1.83
   (c) $.18\overline{3}$ (d) .183

4. $2\{3 + [8 - (6 - 8)]\} = (?)$
   (a) 18   (b) 26
   (c) 9   (d) 13

5. $2(-3)^2 = (?)$
   (a) -18 (b) -12
   (c) 12   (d) 18

In Problems 6 through 10, indicate whether each statement is true or false.

6. $-\dfrac{3}{5} \le -\dfrac{1}{2}$

7. $-(-5) > 0$

8. All rational numbers are real numbers.

9. The reciprocal of -3 is $\dfrac{1}{3}$.

10. $\dfrac{-15}{0} = 0.$

In Problems 11 and 12, find the value of each expression when x is 2 and y is −1.

11. $2x^2 - y$

12. $\dfrac{5x + 3y + 1}{2x}$

In Problems 13 and 14, find each quantity.

13. The product of 3 and -4 is added to 5.

14. The sum of 3 cubed and 10.

In Problems 15 through 20, perform the operations indicated.

15. $7 + (-8)$

16. $(-4)(-3)$

17. $|11| - |-5|$

18. $6 + 6 \cdot 6 - \dfrac{6}{6}$

19. $\dfrac{(-4)(-9)}{(-3)^2}$

20. $-5 + [6 + (-8)]$

21. Match each statement with the property that justifies it.

   ___ a. $3(8 + 1) = 24 + 3$
   ___ b. $(-5)\left(-\dfrac{1}{5}\right) = 1$
   ___ c. $-3 + 0 = -3$
   ___ d. $3 + (-3) = 0$
   ___ e. $7 + (2 + 3) = (7 + 2) + 3$
   ___ f. $(4 + 3) + 1 = 1 + (4 + 3)$

   1. Commutative
   2. Associative
   3. Identity
   4. Inverse
   5. Distributive

22. Using the following set of numbers,
   $$\left\{-5, -\dfrac{3}{5}, 0, \sqrt{3}, \dfrac{17}{3}, 12.7, 102\right\}$$
   list the numbers that belong to each set.
   (a) Integers
   (b) Rational numbers
   (c) Irrational numbers

In Problems 23 through 25, find the solution set.

23. $x - 12 = 13$

24. $-5x = 20$

25. $16 + z = 22$

2 Polynomials

See Problem Set 2.2, Exercise 136.

◤ **Connections**

A cardboard box manufacturer is making open boxes from a 12-in.-square piece of cardboard by cutting equal squares from each of the corners and turning up the sides. She wishes to make a box of the largest possible volume. To figure out the size of the squares, she would use the expression $4x^3 - 48x^2 + 144x$. Such expressions are called polynomials. This chapter examines polynomials and their properties. Operations of addition, subtraction, multiplication, and division are performed with polynomials just as with real numbers.

We begin the study of polynomials by looking at the definitions and properties of different kinds of exponents. We then learn how to classify, evaluate, and perform simple algebra with polynomials. Understanding how to add, subtract, multiply, and divide polynomials is essential to the study of more advanced topics in algebra.

2.1 **Whole Number Exponents**

TAPE 4

We use exponents in algebra just as we use them in arithmetic. Thus x^3 means $x \cdot x \cdot x$ just as 2^3 means $2 \cdot 2 \cdot 2$. We read x^3 as "x to the third power," or "x cubed." We now look at a definition of exponents that are natural numbers.

Natural Number Exponent

If x is a real number and n is a natural number, then

$$x^n = \underbrace{x \cdot x \cdot x \cdots x}_{n \text{ factors of } x}$$

So we see that

$$x^1 = x$$
$$x^2 = x \cdot x$$
$$x^{10} = x \cdot x \cdot x \cdot x \cdot x \cdot x \cdot x \cdot x \cdot x \cdot x$$

In writing x^n, we call x the **base** and n the **exponent.** The exponent gives the number of times the base is used as a factor. Sometimes the exponent is called the **power.** We read x^n as x to the nth power.

Be Careful!   *Note:* x^n does *not* mean n times x. Don't confuse $x^2 = x \cdot x$ with $2x = x + x$.

▶ **EXAMPLE 1**

Identify the base and simplify each of the following.

(a) 3^2 (b) $(-3)^2$ (c) -3^2 (d) $(-2)^3$ (e) -2^3

Solutions

(a) 3^2 is 3 squared. The base is 3.

$$3^2 = 3 \cdot 3 = 9$$

(b) $(-3)^2$ is -3 squared. The base is -3.

$$(-3)^2 = (-3)(-3) = 9$$

Be Careful!

(c) -3^2 is the opposite of 3 squared. The base is 3.

$$-3^2 = -(3 \cdot 3) = -9$$

(d) $(-2)^3$ is -2 cubed. The base is -2.

$$(-2)^3 = (-2)(-2)(-2) = -8$$

(e) -2^3 is the opposite of 2 cubed. The base is 2.

$$-2^3 = -(2 \cdot 2 \cdot 2) = -8$$ ◀

As Example 1 indicates, $(-p)^n$ and $-p^n$ are *not* the same. The distinguishing feature is the base.

Identifying the Base

$(-p)^n$	$-p^n$
The base is $-p$.	The base is p.
$(-4)^2 = (-4)(-4) = 16$	$-4^2 = -(4 \cdot 4) = -16$

Looking closer at $(-p)^n$, we notice that if n is *odd* then we have

$$(-p)^n = (-p)(-p) \cdots (-p) \qquad (n \text{ factors of } -p, \text{ and } n \text{ is odd})$$
$$= -p^n$$

However, if n is *even,* we have

$$(-p)^n = (-p)(-p) \cdots (-p) \qquad (n \text{ factors of } -p, \text{ and } n \text{ is even})$$
$$= p^n$$

Simplifying $(-p)^n$

If p is a real number and n is a natural number, then

$$(-p)^n = p^n \qquad \text{if } n \text{ is even}$$
$$(-p)^n = -p^n \qquad \text{if } n \text{ is odd}$$

► **EXAMPLE 2**

Simplify each of the following.

(a) $(-x)^4$ (b) $(-y)^7$ (c) $-z^6$

Solutions

(a) The base is $-x$ and the exponent is *even*.

$$(-x)^4 = x^4$$

(b) The base is $-y$ and the exponent is *odd*.

$$(-y)^7 = -y^7$$

(c) The base is z, and this expression *cannot* be simplified. *Be sure to notice* the difference between this expression and the expression in part (a).   ◄

Not only is exponential notation convenient to use, but it has some very nice properties. Suppose we want to multiply $x^3 \cdot x^2$. We use the definition of exponent and write

$$x^3 \cdot x^2 = (x \cdot x \cdot x)(x \cdot x) = x \cdot x \cdot x \cdot x \cdot x \cdot = x^5$$

This leads us to the first property of exponents.

Product with Same Base

$$x^m \cdot x^n = x^{m+n}$$

To multiply powers of the *same* base, keep the base and *add* the exponents.

► **EXAMPLE 3**

Simplify each expression.

(a) $\left(\dfrac{1}{2}\right)^2 \cdot \left(\dfrac{1}{2}\right)^3$ (b) $x^4 \cdot x^7$ (c) $x^2 y^3$ (d) $(2x)(3x^2)$

Solutions

(a) $\left(\dfrac{1}{2}\right)^2 \cdot \left(\dfrac{1}{2}\right)^3 = \left(\dfrac{1}{2}\right)^{2+3}$ Keep the base and add exponents

$$= \left(\dfrac{1}{2}\right)^5$$

We can leave this answer or we can write

$$\left(\dfrac{1}{2}\right)^5 = \dfrac{1}{32}$$

(b) $x^4 \cdot x^7 = x^{4+7}$ Product with same base

$$= x^{11}$$

(c) x^2y^3 cannot be simplified because the bases are not the same.

(d) $(2x)(3x^2) = (2 \cdot 3)(x \cdot x^2)$ Commutative and Associative Properties

$$= 6x^{1+2} \qquad\qquad \text{Product with same base}$$

$$= 6x^3 \qquad\qquad\qquad\qquad\qquad\qquad \blacktriangleleft$$

Be Careful! It is *important* to note that to use this property, both *bases* must be the same.

Let's look at $(x^2)^3$. The exponent 3 has a base of x^2. We use the definition of exponents and write

$$(x^2)^3 = (x^2)(x^2)(x^2)$$

Next we use the first property and add exponents.

$$(x^2)(x^2)(x^2) = x^{2+2+2} = x^{3(2)}$$

$$= x^6$$

This gives the second property of exponents.

Power to a Power

$$(x^m)^n = x^{mn}$$

To raise a power to a power, keep the base and *multiply* exponents.

▶ **EXAMPLE 4**

Simplify each.

(a) $(2^2)^6$ (b) $(y^3)^5$

Solutions

(a) $(2^2)^6 = 2^{2 \cdot 6}$ Keep the base and multiply exponents

$$= 2^{12}$$

(b) $(y^3)^5 = y^{3 \cdot 5}$ Power to a power

$$= y^{15} \qquad\qquad\qquad\qquad\qquad\qquad \blacktriangleleft$$

Now let's consider a product raised to a power, such as $(xy)^2$. The base is xy, so we use the definition of exponent and write

$$(xy)^2 = (xy)(xy)$$

Using the Commutative and Associative Properties, we can write

$$(xy)(xy) = x \cdot x \cdot y \cdot y$$

$$= x^{1+1}y^{1+1} \qquad \text{Add exponents}$$

$$= x^2y^2$$

So, $(xy)^2 = x^2y^2$. Now we have the third property of exponents.

> **Product to a Power**
>
> $$(xy)^n = x^n y^n$$

To raise a product to a power, raise each factor in the product to the power.

► **EXAMPLE 5**

Simplify each.

(a) $(3x)^4$ (b) $(-xy^2)^2$

Solutions

(a) $(3x)^4 = 3^4 x^4$ Raise each factor to the fourth power

   $\quad\quad = 81x^4$ $(3^4 = 3 \cdot 3 \cdot 3 \cdot 3)$

(b) $(-xy^2)^2 = (-x)^2(y^2)^2$ Product to a power

   $\quad\quad\quad = x^2 y^4$ ◄

The next property is just like the previous one, except that we use a quotient instead of a product. Consider $\left(\dfrac{x}{y}\right)^3$. The base is $\dfrac{x}{y}$. We use the definition of exponent and write

$$\left(\frac{x}{y}\right)^3 = \frac{x}{y} \cdot \frac{x}{y} \cdot \frac{x}{y}$$

$$= \frac{x \cdot x \cdot x}{y \cdot y \cdot y}$$

$$= \frac{x^3}{y^3} \quad\quad\quad \text{Product with same base}$$

This leads to the fourth property of exponents.

> **Quotient to a Power**
>
> $$\left(\frac{x}{y}\right)^n = \frac{x^n}{y^n}, \quad\quad y \neq 0$$

To raise a quotient to a power, raise the numerator and the denominator to the power.

► **EXAMPLE 6**

Simplify each.

(a) $\left(\dfrac{2}{k}\right)^6$ (b) $\left(\dfrac{-1}{x}\right)^3$

Solutions

(a) $\left(\dfrac{2}{k}\right)^6 = \dfrac{2^6}{k^6}$ Raise the numerator and the
                                       denominator to the sixth power

$\qquad\quad = \dfrac{64}{k^6}$

(b) $\left(\dfrac{-1}{x}\right)^3 = \dfrac{(-1)^3}{x^3}$ Quotient to a power

$\qquad\quad = \dfrac{-1}{x^3}$ ◄

The final property of exponents deals with quotients and the *same* base. Let's look at three examples of this situation: $\dfrac{x^5}{x^3}, \dfrac{x^3}{x^5}$, and $\dfrac{x^4}{x^4}$.

$$\frac{x^5}{x^3} = \frac{x \cdot x \cdot x \cdot x \cdot x}{x \cdot x \cdot x} = \frac{x \cdot x}{1} = x^2, \qquad \text{so} \qquad \frac{x^5}{x^3} = x^{5-3} = x^2$$

$$\frac{x^3}{x^5} = \frac{x \cdot x \cdot x}{x \cdot x \cdot x \cdot x \cdot x} = \frac{1}{x \cdot x} = \frac{1}{x^2}, \qquad \text{so} \qquad \frac{x^3}{x^5} = \frac{1}{x^{5-3}} = \frac{1}{x^2}$$

$$\frac{x^4}{x^4} = \frac{x \cdot x \cdot x \cdot x}{x \cdot x \cdot x \cdot x} = 1, \qquad \text{so} \qquad \frac{x^4}{x^4} = 1$$

Putting these ideas together gives us the fifth property of exponents.

Quotient with Same Base

$$\frac{x^m}{x^n} = \begin{cases} x^{m-n} & \text{if } m > n \\[2mm] 1 & \text{if } m = n, \quad x \neq 0 \text{ for each case} \\[2mm] \dfrac{1}{x^{n-m}} & \text{if } n > m \end{cases}$$

► EXAMPLE 7

Simplify each of the following.

(a) $\dfrac{x^{11}}{x^8}$ (b) $\dfrac{(x + y)^3}{(x + y)^{17}}$ (c) $\dfrac{2r^9}{r^9}$

Solutions

(a) Since $11 > 8$, we have

$$\frac{x^{11}}{x^8} = x^{11-8} \qquad \text{Quotient with same base}$$

$$= x^3$$

(b) Since $17 > 3$, we have

$$\frac{(x + y)^3}{(x + y)^{17}} = \frac{1}{(x + y)^{17-3}} \qquad \text{Quotient with same base}$$

$$= \frac{1}{(x + y)^{14}}$$

(c) Since $9 = 9$, we have

$$\frac{2r^9}{r^9} = 2 \qquad \text{Quotient with same base} \qquad \blacktriangleleft$$

The properties of exponents are summarized in the following box.

Properties of Exponents

If m and n are natural numbers and x and y are real numbers,

1. $x^m x^n = x^{m+n}$ Product with same base.

2. $(x^m)^n = x^{mn}$ Power to a power.

3. $(xy)^n = x^n y^n$ Product to a power.

4. $\left(\dfrac{x}{y}\right)^n = \dfrac{x^n}{y^n},$ $y \neq 0$ Quotient to a power.

5. $\dfrac{x^m}{x^n} = \begin{cases} x^{m-n} & \text{if } m > n \\ 1 & \text{if } m = n, \quad x \neq 0 \text{ for each case} \\ \dfrac{1}{x^{n-m}} & \text{if } n > m \end{cases}$ Quotient with same base.

► **EXAMPLE 8**

Use the properties of exponents to simplify.

(a) $x^4(y^3)^5$ (b) $(3x)^2(3x)^2$

(c) $\left(\dfrac{2x}{k^2}\right)^3$ (d) $\dfrac{(3b)^5}{(3b)^3}$

Solutions

(a) We begin with the parentheses, a power to a power.

$$x^4(y^3)^5 = x^4 y^{3 \cdot 5} \qquad \text{Power to a power}$$

$$= x^4 y^{15}$$

This will not simplify further as x and y are *different* bases.

(b) We notice that the base is $3x$ in both expressions. So we add exponents.

$$(3x)^2(3x)^2 = (3x)^{2+2} \qquad \text{Product with same base}$$

$$= (3x)^4$$

Now we have a product to a power. So we raise each factor in the product to the fourth power.

$$(3x)^4 = 3^4 x^4 \qquad \text{Power to a power}$$

$$= 81x^4$$

(c) This is a quotient to power.

$$\left(\frac{2x}{k^2}\right)^3 = \frac{(2x)^3}{(k^2)^3} \qquad \text{Quotient to a power}$$

This will simplify further. The numerator is a product to a power, and the denominator is a power to a power.

$$\frac{(2x)^3}{(k^2)^3} = \frac{2^3 \cdot x^3}{k^{2 \cdot 3}}$$

$$= \frac{8x^3}{k^6}$$

(d) The base is $3b$ in both the numerator and denominator. We subtract the exponents.

$$\frac{(3b)^5}{(3b)^3} = (3b)^{5-3} \qquad \text{Quotient with same base}$$

$$= (3b)^2$$

$$= 3^2 b^2 \qquad \text{Product to a power}$$

$$= 9b^2 \qquad \blacktriangleleft$$

Properties 2, 3, and 4 tell us when to *multiply* exponents. They are sometimes called power rules and are used in problems in the form,

$$(\text{Expression})^{\text{power}}$$

We must be very careful in simplifying problems of this type. The properties of exponents allow products and quotients that are raised to a power to be simplified. However, the properties *do not* allow us to simplify sums and differences that are raised to powers.

▶ **EXAMPLE 9**

Use the properties of exponents to simplify each expression to a power if possible.

(a) $(xy^5)^2$ (b) $(x + y)^2$

Solutions

(a) This is a product to a power. The properties of exponents can be used to simplify.

$$(xy^5)^2 = x^2(y^5)^2 \qquad \text{Product to a power}$$

$$= x^2 y^{10} \qquad \text{Power to a power}$$

Be Careful!

(b) This is a *sum* to a power. It cannot be simplified using the properties of exponents. $(x + y)^2 = (x + y)(x + y)$, which is *not* $x^2 + y^2$. We will study how to find this product in Section 2.5. $\blacktriangleleft$

► **EXAMPLE 10**

Use the properties of exponents to simplify each expression to a power if possible.

(a) $\left(\dfrac{xy^4}{z}\right)^3$ (b) $(x - 2y)^3$

Solutions

(a) This is a quotient to a power. The properties of exponents allow us to simplify.

$$\left(\frac{xy^4}{z}\right)^3 = \frac{x^3(y^4)^3}{z^3} \qquad \text{Product to a power}$$

$$= \frac{x^3y^{12}}{z^3} \qquad \text{Power to a power}$$

*Be
Careful!*
(b) This is a *difference* to a power. It *cannot* be simplified using properties of exponents. $(x - 2y)^3 = (x - 2y)(x - 2y)(x - 2y)$, which is *not* $x^3 - 8y^3$.   ◄

► **EXAMPLE 11**

Simplify each expression.

(a) $(-3x^3y)^2$ (b) $\left(\dfrac{-2r^3s}{t^2}\right)^3$ (c) $(3mn^3)^2(2m)^3$ (d) $\dfrac{(-xy^3)^2}{(-x^3y)^3}$

Solutions

(a) We notice that this is an expression in parentheses to a power and that the expression is a *product*.

$$(-3x^3y)^2 = (-3)^2(x^3)^2y^2 \qquad \text{Product to a power}$$

$$= (-3)^2x^{3 \cdot 2}y^2 \qquad \text{Power to a power}$$

$$= 9x^6y^2$$

(b) This is an expression to a power and the expression is a *quotient*.

$$\left(\frac{-2r^3s}{t^2}\right)^3 = \frac{(-2r^3s)^3}{(t^2)^3} \qquad \text{Quotient to a power}$$

$$= \frac{(-2)^3(r^3)^3s^3}{(t^2)^3} \qquad \text{Product to a power}$$

$$= \frac{-8r^9s^3}{t^6} \qquad \text{Power to a power}$$

(c) We use property 3 to simplify each expression to a power.

$$(3mn^3)^2(2m)^3 = 3^2m^2(n^3)^2 \cdot 2^3m^3 \qquad \text{Product to a power}$$

$$= 9m^2n^6 \cdot 8m^3 \qquad \text{Power to a power}$$

$$= 9 \cdot 8m^2m^3n^6 \qquad \text{Rearrange factors}$$

$$= 72m^{2+3}n^6 \qquad \text{Product with same base}$$

$$= 72m^5n^6$$

(d) Both the numerator and the denominator are expressions raised to powers that can be simplified by using properties 2, 3, and 5.

$$\frac{(-xy^3)^2}{(-x^3y)^3} = \frac{(-x)^2(y^3)^2}{(-x^3)^3y^3} \qquad \text{Product to a power}$$

$$= \frac{x^2y^6}{-x^9y^3} \qquad \text{Power to a power}$$

$$= \frac{y^3}{-x^7} \qquad \text{Quotient with same base}$$

which can be written as $-\dfrac{y^3}{x^7}$ or $\dfrac{-y^3}{x^7}$.   ◄

Next, let's consider what we might mean if we use zero as an exponent. For example, what is 2^0? Since we want the properties of exponents to be valid,

$$2^5 \cdot 2^0 = 2^{5+0}$$

$$2^5 \cdot 2^0 = 2^5$$

Thus, 2^0 must be 1. This leads us to our second definition.

Zero Exponent

If x is any nonzero real number, then

$$x^0 = 1$$

► **EXAMPLE 12**

Simplify.

(a) 3^0 (b) $(23x^3y^5)^0$
(c) $(-5)^0$ (d) $(x + y)^0$
(e) $\left(\dfrac{s}{t}\right)^0$ (f) $5x^0$
(g) -6^0 (h) $4(2a + 3b)^0$

Solutions

(a) $3^0 = 1$
(b) $(23x^3y^5)^0 = 1$
(c) $(-5)^0 = 1$ (The base is -5.)
(d) $(x + y)^0 = 1$ (The base is $x + y$).
(e) $\left(\dfrac{s}{t}\right)^0 = 1$
(f) $5x^0 = 5(1) = 5$ (The base is x.)
(g) $-6^0 = -1$ (The base is 6.)
(h) $4(2a + 3b)^0 = 4(1) = 4$ ◄

Calculator Box

Exponents on a Calculator

Let's find $(1.302 - 0.877)^2$.

Scientific Calculator:

Press $\boxed{1.302}$ $\boxed{-}$ $\boxed{0.877}$ $\boxed{=}$ $\boxed{x^2}$

and read $\boxed{0.180625}$ on the display.

Graphing Calculator:

Press $\boxed{(}$ $\boxed{1.302}$ $\boxed{-}$ $\boxed{0.877}$ $\boxed{)}$ $\boxed{x^2}$ $\boxed{\text{ENTER}}$

To find 4^7, press $\boxed{4}$ $\boxed{y^x}$ $\boxed{7}$ $\boxed{=}$ and read $\boxed{16384}$ on the display. Some calculators label the exponent key $\boxed{x^y}$ or $\boxed{a^b}$ instead of $\boxed{y^x}$ and the graphing TI calculator uses $\boxed{\wedge}$.

Calculator Exercises

Find the value of each expression. Approximate to two decimals if necessary.
1. 3^2 **2.** 2^3 **3.** 3^{10} **4.** $(1.24)^4$ **5.** $(\sqrt{2})^3$ **6.** $(1.33 - 0.84)^2$

Answers

1. 9   **2.** 8   **3.** 59049   **4.** ≈ 2.36 **5.** ≈ 2.83 **6.** ≈ 0.24

Problem Set 2.1

Warm-Ups

In Problems 1 through 6, identify the base and simplify. See Example 1.

1. 4^2

2. -4^2

3. $(-4)^2$

4. $(-4)^3$

5. -4^3

6. $\left(\dfrac{2}{3}\right)^3$

In Problems 7 through 9, simplify. See Example 2.

7. $(-x)^4$

8. $(-x)^9$

9. $-x^2$

In Problems 10 through 15, simplify each if possible. See Example 3.

10. $x^4 x^5$

11. $y^{11} y^{14}$

12. $6^9 \cdot 6^9$

13. $(3z)(-2z^5)$

14. $2^{10} \cdot 2$

15. $x^2 y^3$

In Problems 16 through 18, simplify each. See Example 4.

16. $(x^4)^5$

17. $(y^2)^3$

18. $(5^4)^4$

In Problems 19 through 24, simplify each. See Example 5.

19. $(xy)^7$

20. $(3s)^3$

21. $(11k)^2$

22. $(-2z^3)^2$

23. $(-3t)^3$

24. $(-xy^2)^5$

In Problems 25 through 30, simplify each. See Example 6.

25. $\left(\dfrac{x}{y}\right)^8$

26. $\left(\dfrac{2}{t}\right)^5$

27. $\left(\dfrac{x}{5}\right)^3$

28. $\left(\dfrac{-3}{x}\right)^2$

29. $\left(\dfrac{-5}{p}\right)^3$

30. $\left(\dfrac{-1}{x}\right)^4$

In Problems 31 through 36, simplify each. See Example 7.

31. $\dfrac{x^{16}}{x^8}$

32. $\dfrac{y^4}{y^7}$

33. $\dfrac{3k^{22}}{k}$

34. $\dfrac{7^4}{7^4}$

35. $\dfrac{(x+y)^6}{(x+y)^6}$

36. $\dfrac{(2a+1)^7}{(2a+1)^6}$

In Problems 37 through 42, simplify each. See Example 8.

37. $x(z^7)^2$

38. $(2x^4)^2(2x^4)^3$

39. $\left(\dfrac{5}{2t}\right)^3$

40. $x^3(x^4)^2$

41. $\dfrac{(7a^3)^4}{(7a^3)^6}$

42. $\left(\dfrac{4v^9}{3k}\right)^2$

In Problems 43 through 48, use the properties of exponents to simplify if possible. See Examples 9 and 10.

43. $(2xy)^2$

44. $\left(\dfrac{2x}{y}\right)^2$

45. $(2x+y)^2$

46. $(2x-y)^2$

47. $(2x^2y^3)^2$

48. $(a-b)^3$

In Problems 49 through 54, simplify each. See Example 11.

49. $(-2x^2y^3)^2$

50. $(-3x^3y^2)^3$

51. $(2pq)^3(2p^2)^3$

52. $\dfrac{(x^2y^3)^2}{(xz)^3}$

53. $\left(\dfrac{-6x^7t}{s^6}\right)^2$

54. $\left(\dfrac{abc}{-2d}\right)^3$

In Problems 55 through 60, simplify each. See Example 12.

55. 9^0

56. $4x^0yx^2$

57. $\left(\dfrac{5x^6y^7}{-3z^8}\right)^0$

58. $-x^0$

59. $-12y^0$

60. $(-8)^0$

Practice Exercises

In Problems 61 through 66, find the value of each expression.

61. $2^2+3^3-4^2$

62. 2^0-2-2^2

63. -3^2-2^2

64. $(-3)^2-(-2)^3$

65. $-7^0+(-7)^0$

66. $(-1)^3-1^3$

In Problems 67 through 102, simplify each, if possible.

67. $(-k)^3$

68. $(-k)^8$

69. $(-k)^0$

70. $x^5 x^4$

71. $y^{12} y^{24}$

72. $4^7 \cdot 4^7$

73. $p \cdot p^8$

74. $3^{12} \cdot 3$

75. $s^3 t^9$

76. $(h^3)^3$

77. $(2^3)^2$

78. $(9^9)^3$

79. $(st)^4$

80. $(4 + x)^3$

81. $(13x)^2$

82. $\left(\dfrac{y}{x}\right)^5$

83. $\left(\dfrac{r}{2}\right)^6$

84. $\left(\dfrac{-6}{x}\right)^3$

85. $\dfrac{5x^{10}}{2x^8}$

86. $\dfrac{3t^5}{t^6}$

87. $\dfrac{q^2}{q}$

88. $\dfrac{3^5}{3^5}$

89. $\dfrac{(x-y)^4}{(x-y)^4}$

90. $\dfrac{(2a+1)^6}{(2a+1)^7}$

91. $a(ab^4)^9$

92. $(3x^5)^5$

93. $\left(\dfrac{4r}{3}\right)^3$

94. $y^2(y^3)^5$

95. $(4x^3y^4)^4$

96. $\left(\dfrac{3v^7}{2k^2}\right)^3$

97. $(-2x^2y^3)^3$

98. $(-3x^3y^2)^2$

99. $\dfrac{(7a)^2}{(7a)^3}$

100. $\left(\dfrac{s^3 t^4}{v^5}\right)^6$

101. $\left(\dfrac{-6x^7 t}{s^6}\right)^3$

102. $\dfrac{(2xy)^2}{(x^2 y)^3}$

Problem 103

103. Dr. Lao calculates the concentration in milliliters per cubic centimeter (ml/cc) of a heart drug in the patient's bloodstream using the expression

$$\left(\frac{1}{2t+1}\right)^2$$

where t is the number of hours after the drug is administered. Fill in the following chart and describe what is happening to the concentration of the drug.

t	10 sec	20 sec	30 sec	1 min	5 min
$\left(\dfrac{1}{2t+1}\right)^2$					

104. Fill in the following table and describe the relationship between the value of n^2 and 2^n.

n	1	2	3	4	5	6	7
n^2							
2^n							

Challenge Problems

105. (a) Simplify each of the following:
$(-1)^0$; $(-1)^1$; $(-1)^2$; $(-1)^3$;
$(-1)^4$;
(b) Make a rule for simplifying $(-1)^n$, where n is any whole number.

106. Simplify $\dfrac{9^5}{3^5}$. $\left(\textit{Hint: Use property 4, } \left(\dfrac{x}{y}\right)^n = \dfrac{x^n}{y^n}.\right)$

107. Simplify $\dfrac{32^3}{8^5}$. (*Hint:* Use powers of 2.)

108. What is wrong with saying $\dfrac{x^8}{x^4} = x^2$?

In Your Own Words

109. What does x^n mean when n is a natural number?

110. What are the five properties of exponents?

111. What do the expressions -3^2 and $(-3)^2$ mean?

112. Make up an expression raised to a power that can be simplified using the properties of exponents and one that cannot be simplified using the properties of exponents.

2.2 Negative Integer Exponents

TAPE 4

In the previous section we worked with exponents that were natural numbers or zero. We now examine negative integer exponents. Recall the five properties of exponents.

Properties of Exponents

For m and n natural numbers and x and y real numbers,

1. $x^m \cdot x^n = x^{m+n}$ Product with same base.

2. $(x^m)^n = x^{m \cdot n}$ Power to a power.

3. $(x \cdot y)^n = x^n y^n$ Product to a power.

4. $\left(\dfrac{x}{y}\right)^n = \dfrac{x^n}{y^n}, \quad y \neq 0$ Quotient to a power.

5. $\dfrac{x^m}{x^n} = \begin{cases} x^{m-n} & \text{if } m > n \\ 1 & \text{if } m = n \\ \dfrac{1}{x^{n-m}} & \text{if } n > m \end{cases}$ $x \neq 0$ for each case   Quotient with same base.

Now suppose we have a definition for 2^{-5} such that the five properties of exponents are still valid. If so, the following must be true:

$$2^5 \cdot 2^{-5} = 2^{5+(-5)} \qquad \text{Add exponents}$$

$$= 2^0$$

$$= 1 \qquad \text{Definition of 0 exponent}$$

If $2^5 \cdot 2^{-5} = 1$, then 2^5 and 2^{-5} are reciprocals. Therefore,

$$2^{-5} = \frac{1}{2^5}$$

Negative Exponent

For x any nonzero real number and n a natural number,

$$x^{-n} = \frac{1}{x^n}$$

This definition is consistent with all five of the properties of exponents.

► **EXAMPLE 1**

Rewrite each expression without negative exponents and in simplest form.

(a) 5^{-3} (b) x^{-11} (c) $2^4 x^{-5}$ (d) $(-5)^{-2}$

(e) -3^{-2} (f) $7x^{-3}$

Solutions

(a) $5^{-3} = \dfrac{1}{5^3}$ Definition of negative exponent

$\quad\quad = \dfrac{1}{5 \cdot 5 \cdot 5}$

$\quad\quad = \dfrac{1}{125}$

Notice that the negative exponent did not give a negative answer. 5^{-3} is the *reciprocal* of 5^3.

(b) $x^{-11} = \dfrac{1}{x^{11}}$ Definition of negative exponent

(c) $2^4 x^{-5} = 2^4 \cdot \dfrac{1}{x^5}$ Definition of negative exponent

$\quad\quad = 16 \cdot \dfrac{1}{x^5}$

$\quad\quad = \dfrac{16}{x^5}$

(d) $(-5)^{-2} = \dfrac{1}{(-5)^2}$ Definition of negative exponent

$\quad\quad = \dfrac{1}{25}$

Note how the parentheses show the base for the exponent is -5 and not 5.

Be Careful!   (e) $-3^{-2} = -\dfrac{1}{3^2}$ Definition of negative exponent

$\quad\quad = -\dfrac{1}{9}$

Notice that in this case the base for the exponent is 3 and *not* -3.

Be Careful!   (f) $7x^{-3} = 7 \cdot \dfrac{1}{x^3}$ Definition of negative exponent

$\quad\quad = \dfrac{7}{x^3}$

Notice that the exponent -3 has a base of x, not $7x$. ◄

Including this last definition, the properties of exponents can now be written as follows.

> ### Properties of Exponents
>
> For m and n any *integers* and x and y real numbers,
>
> **1.** $x^m \cdot x^n = x^{m+n}$ Product with same base.
>
> **2.** $(x^m)^n = x^{m \cdot n}$ Power to a power.
>
> **3.** $(x \cdot y)^n = x^n y^n$ Product to a power.
>
> **4.** $\left(\dfrac{x}{y}\right)^n = \dfrac{x^n}{y^n}, \quad y \neq 0$ Quotient to a power.
>
> **5.** $\dfrac{x^m}{x^n} = x^{m-n}, \quad x \neq 0$ Quotient with same base.

Compare this form of property 5 with that of Section 2.1 and notice the simplification that results in property 5 when negative exponents are allowed. Consider this example:

$$\frac{2^3}{2^5} = 2^{3-5} \qquad \text{Quotient with same base}$$

$$= 2^{-2}$$

$$= \frac{1}{2^2} \qquad \text{Definition}$$

$$= \frac{1}{4}$$

► **EXAMPLE 2**

Use the properties of exponents to simplify each. The answers should not contain negative exponents.

(a) $5^5 \cdot 5^{-3}$ (b) $4^3 \cdot 4^{-7}$ (c) $(2x^{-2})(5x^7)$

Solutions

(a) $5^5 \cdot 5^{-3} = 5^{5+(-3)}$ Add exponents. Product with same base.

$$= 5^2$$

$$= 25$$

(b) $4^3 \cdot 4^{-7} = 4^{3+(-7)}$ Product with same base

$$= 4^{-4}$$

$$= \frac{1}{4^4} \qquad \text{Definition}$$

$$= \frac{1}{256}$$

(c) $(2x^{-2})(5x^7) = (2)(5)x^{-2}x^7$ Commutative and Associative
                                                   Properties for multiplication

$$= 10x^5 \qquad\qquad \text{Product with same base} \blacktriangleleft$$

► EXAMPLE 3

Use the properties of exponents to simplify each. The answers should not contain negative exponents. Assume that all variables represent nonzero real numbers.

(a) $(x^{-2})^{-3}$ (b) $(3x^{-4})^{-2}$ (c) $\left(\dfrac{3a^4}{15a^{-1}}\right)^{-1}$

Solutions

(a) $(x^{-2})^{-3}$ $= x^{(-2)(-3)}$ Multiply exponents. Power to a power.

                 $= x^6$

(b) $(3x^{-4})^{-2}$ $= 3^{-2}(x^{-4})^{-2}$ Raise each factor to the -2 power
                                     Product to a power

                 $= 3^{-2}x^{(-4)(-2)}$ Multiply exponents. Power to a power.

                 $= 3^{-2}x^8$

                 $= \dfrac{1}{3^2} \cdot x^8$ Definition

                 $= \dfrac{x^8}{9}$

(c) $\left(\dfrac{3a^4}{15a^{-1}}\right)^{-1} = \left(\dfrac{a^{4-(-1)}}{5}\right)^{-1}$ Quotient with same base

                 $= \left(\dfrac{a^5}{5}\right)^{-1} = \dfrac{(a^5)^{-1}}{5^{-1}}$ Quotient to a power

                 $= \dfrac{a^{-5}}{5^{-1}}$ Power to a power

                 $= \dfrac{\dfrac{1}{a^5}}{\dfrac{1}{5}}$ Definition of negative exponent

                 $= \dfrac{1}{a^5} \div \dfrac{1}{5}$

                 $= \dfrac{1}{a^5} \cdot \dfrac{5}{1} = \dfrac{5}{a^5}$ $\blacktriangleleft$

► EXAMPLE 4

Simplify each and write answers without negative exponents. Assume that all variables are nonzero real numbers.

(a) $\dfrac{a^{-3}}{a^6}$ (b) $\dfrac{b^6}{b^{-2}}$ (c) $\dfrac{x^{-1}}{x^{-3}}$ (d) $\dfrac{5^{-2}x^4}{5^{-3}x^{-5}}$

Solutions

(a) $\dfrac{a^{-3}}{a^6} = a^{-3-6}$ Subtract exponents. Quotient with same base.

$\quad = a^{-9}$

$\quad = \dfrac{1}{a^9}$ Definition of negative exponents

(b) $\dfrac{b^6}{b^{-2}} = b^{6-(-2)}$ Quotient with same base

$\quad = b^{6+2}$

$\quad = b^8$

(c) $\dfrac{x^{-1}}{x^{-3}} = x^{-1-(-3)}$ Subtract exponents

$\quad = x^{-1+3}$

$\quad = x^2$

(d) $\dfrac{5^{-2}x^4}{5^{-3}x^{-5}} = 5^{-2-(-3)}x^{4-(-5)}$ Subtract exponents

$\quad = 5^{-2+3}x^{4+5}$

$\quad = 5x^9$ ◀

Scientific Notation

If we multiply 123,456 by 123,456 on a calculator, it may display an answer like

$$1.524138394^{10} \quad \text{or} \quad 1.524138394\text{E} + 10$$

These are short for

$$1.524138394 \times 10^{10}$$

a large number written in scientific notation. If we divide 3 by 12,345 on a calculator, it may display an answer like

$$2.430133657^{-04} \quad \text{or} \quad 2.430133657\text{E}-04$$

These are short for

$$2.430133657 \times 10^{-4}$$

a small number written in scientific notation. Each of the numbers is written as a number between 1 and 10 times a power of 10. As is usually the case in calculator computations, these are both approximations. Scientific notation is a convenient way to express very large and very small numbers. Two examples that occur in physics are Planck's constant,

$$h \approx 0.00000000000000000000000000006625 \text{ erg-sec}$$

and the speed of light in a vacuum,

$$c \approx 29,900,000,000 \text{ cm/sec}$$

Written in scientific notation, these become

$$h \approx 6.625 \times 10^{-27} \text{ erg-sec}$$

$$c \approx 2.99 \times 10^{10} \text{ cm/sec}$$

These are both approximations, as the symbol $\approx$ indicates.

Writing a Number in Scientific Notation

1. Starting with the number in decimal format, move the decimal point until the number is between 1 and 10, including 1.

2. Multiply the number formed in step 1 by 10 raised to the number of decimal places moved. If the original number was between 0 and 1, the power of 10 is negative. If the original number was greater than 10, the power is positive.

3. This procedure is for a *positive* number. If the original number is negative, perform steps 1 and 2 on the absolute value of the original number and then append a negative sign to the result.

► ## EXAMPLE 5

Write each number in scientific notation.

(a) 93,000,000   (b) 0.000001554   (c) −254,000

Solutions

(a) We think of 93,000,000 as 93000000.0 and move the decimal point until we have a number between 1 and 10.

STEP 1   9 3 0 0 0 0 0 0.0

We moved the decimal point 7 places. Now we write

STEP 2 9.3×10^{7}

The exponent is *positive* because 93,000,000 is greater than 10.

STEP 1

(b)   0.0 0 0 0 0 1 5 5 4

We moved the decimal point 6 places. We write

STEP 2 1.554×10^{-6}

The exponent is *negative* because 0.000001554 is less than 1.

(c) −254,000 is negative, so we work with its absolute value.

STEP 1   2 5 4 0 0 0.0

We moved the decimal point 5 places. We have now

STEP 2 2.54×10^{5}

The exponent is *positive* because 254,000 is greater than 10. As the original number was negative, we write

STEP 3 -2.54×10^{5} ◄

► EXAMPLE 6

Write the following numbers without exponents.

(a) 8.771×10^{14}

(b) 3.2×10^{-13}

(c) -9.99231×10^{-3}

Solutions

To change scientific notation to standard decimal format, we reverse the steps given earlier.

(a) 8.7 7 1 0 0 0 0 0 0 0 0 0 0 0 0

877,100,000,000,000

Note that we moved the decimal point 14 places to the right (to make a number larger than 10) because 14 is positive.

(b) 0 0 0 0 0 0 0 0 0 0 0 0 0 3.2

0.00000000000032

This time we moved the decimal point 13 places to the left (to make a number smaller than 1) because -13 is negative.

(c) Because the given number is negative, we work with its absolute value.

$$0\ 0\ 0\ 9.9\ 9\ 2\ 3\ 1$$

0.00999231

We moved the decimal point 3 places to the left because the exponent of 10, -3, is negative. Now we append a negative sign because the original number was negative.

$$-0.00999231 \qquad \blacktriangleleft$$

► EXAMPLE 7

Write each number without exponents.

(a) $(2.1 \times 10^{-5})(-2.0 \times 10^{-2})$ (b) $\dfrac{1.6 \times 10^5}{3.2 \times 10^{-3}}$

Solutions

(a) $(2.1 \times 10^{-5})(-2.0 \times 10^{-2}) = (2.1)(-2.0)(10^{-5})(10^{-2})$

$$= -4.2 \times 10^{-7} \qquad \text{Add exponents}$$

$$= -0.00000042$$

(b) $\dfrac{1.6 \times 10^5}{3.2 \times 10^{-3}} = \dfrac{1.6}{3.2} \times 10^{5-(-3)} \qquad \text{Subtract exponents}$

$$= 0.5 \times 10^{5+3}$$

$$= 0.5 \times 10^8$$

$$= 50,000,000$$

Calculator Box

Scientific Notation on a Calculator

To enter 7.2×10^{12} on a calculator, press the keys ⎡7.2⎤ ⎡EE⎤ ⎡12⎤ and the display should show ⎡7.2¹²⎤ or ⎡7.2   12⎤ or ⎡7.2E12⎤, or something similar. Some calculators use ⎡EXP⎤ or simply ⎡E⎤ instead of ⎡EE⎤.

Scientific Calculator:

To enter 1.053×10^{-8}, press ⎡1.053⎤ ⎡EE⎤ ⎡8⎤ ⎡+/−⎤

Graphing Calculator:

press ⎡1.053⎤ ⎡EE⎤ ⎡(−)⎤ ⎡8⎤

The display should show ⎡1.053 ⁻⁸⎤, ⎡1.053   −08⎤, ⎡1.053 E − 8⎤, or something similar.

These representations may be used like any other number entered on the calculator.

Calculator Exercises

In Problems 1 through 4, enter each number on a calculator.

1. 3.58×10^{24}

2. 1.01×10^{-5}

3. 4.3×10^{23}

4. 9.1×10^{-19}

In Problems 5 and 6, evaluate each expression to two decimal places by entering the numbers in scientific notation.

5. $(6.32 \times 10^{11})(1.85 \times 10^{-5})$

6. $\dfrac{2.117 \times 10^{-17}}{5.901 \times 10^{-32}}$

Answers

5. 1.17×10^{7}

6. 3.59×10^{14}

Problem Set 2.2

Assume that all variables are nonzero real numbers.

Warm-Ups

In Problems 1 through 18, rewrite each expression without negative exponents and simplify. See Example 1.

1. 2^{-4}

2. 3^{-2}

3. 2^{-1}

4. x^{-2}

5. y^{-5}

6. A^{-1}

7. $(-2)^{-4}$

8. $(-2)^{-5}$

9. -2^{-2}

10. $(-5)^{-2}$

11. $(-5)^{-3}$

12. 5^{-2}

13. -5^{-2}

14. -5^{-3}

15. $2x^{-3}$

16. $3^{-2}x$

17. $-2x^{-3}$

18. $-7v^{-2}$

In Problems 19 through 26, use the properties of exponents to simplify the expressions. The answers should not contain negative exponents. See Example 2.

19. $2^3 \cdot 2^{-2}$

20. $3^{-3} \cdot 3^5$

21. $4^3 \cdot 4^{-5}$

22. $2^5 \cdot 2^{-9}$

23. $(2x^6)(4x^{-2})$

24. $z^5 \cdot z$

25. $(-2x^{-4})(3x^2)$

26. $y^6 y^{-6}$

In Problems 27 through 38, write each expression without negative exponents. See Example 3.

27. $(2^{-3})^2$

28. $(3^2)^{-2}$

29. $\left(\dfrac{3}{5}\right)^{-1}$

30. $\left(-\dfrac{1}{2}\right)^{-1}$

31. $(x^{-5})^2$

32. $(z^{-3})^{-8}$

33. $(xy)^{-3}$

34. $(2x)^{-4}$

35. $(-3x)^{-2}$

36. $(-2s)^{-4}$

37. $\left(\dfrac{2a^{-1}}{a^{-3}}\right)^{-1}$

38. $\left(\dfrac{x^{-2}}{a^{-4}}\right)^{-1}$

In Problems 39 through 47, simplify the expressions. The answers should not contain negative exponents. See Example 4.

39. $\dfrac{10^{-3}}{10^{-1}}$

40. $\dfrac{6^{-9}}{6^{-7}}$

41. $\dfrac{5^2}{5^{-1}}$

42. $\dfrac{x^4}{x^{-3}}$

43. $\dfrac{r^{-3}}{r^4}$

44. $\dfrac{y^7}{y^{-1}}$

45. $\dfrac{k^{-5}}{k^{-3}}$

46. $\dfrac{6^{-2}x^3}{6^{-3}x^{-2}}$

47. $\dfrac{2^{-4}b^{-5}}{2^{-9}b^{-3}}$

In Problems 48 through 53, write the given numbers in scientific notation. See Example 5.

48. 5280

49. 0.0000012345

50. 0.094

51. 77,722,000,000,000

52. −32,000

53. −0.000000021367

In Problems 54 through 59, write the given numbers without exponents. See Example 6.

54. 1.609×10^5

55. 5.43×10^{-5}

56. 1.1×10^{-8}

57. 6.81×10^2

58. -8.0×10^{11}

59. -4.32101×10^{-2}

In Problems 60 through 63, write the number without exponents. See Example 7.

60. $\dfrac{2.4 \times 10^5}{0.2 \times 10^{-4}}$

61. $\dfrac{4.4 \times 10^{-6}}{2.2 \times 10^{-8}}$

62. $(2.5 \times 10^5)(2.0 \times 10^{-3})$

63. $(1.1 \times 10^{-2})(5.0 \times 10^{-5})$

Practice Exercises

In Problems 64 through 123, use the properties of exponents to simplify the expressions. The answers should not contain negative exponents.

64. 2^{-5}

65. 3^{-3}

66. 3^{-1}

67. x^{-11}

68. k^{-5}

69. T^{-1}

70. $(-3)^{-2}$

71. $(-3)^{-3}$

72. -3^{-3}

73. -2^{-2}

74. $(-6)^{-2}$

75. $(-4)^{-3}$

76. 4^{-3}

77. -4^{-3}

78. -4^{-2}

79. $4x^{-2}$

80. $5^{-3}x$

81. $6x^{-7}$

82. $2r^{-1}$

83. $-8x^{-2}$

84. $-5s^{-3}$

85. bt^{-3}

86. $-cy^{-4}$

87. k^2x^{-2}

88. $a^{-1}x^{-2}$

89. $3a^{-3}t^{-2}$

90. $-q^{-2}x^{-2}$

91. $3^3 \cdot 3^{-2}$

92. $2^{-3} \cdot 2^5$

93. $4^{-3} \cdot 4^4$

94. $5^3 \cdot 5^{-5}$

95. $3^4 \cdot 3^{-8}$

96. $6^{-6} \cdot 6^4$

97. $10^4 \cdot 10^{-2}$

98. $(4x^{11})(6x^{-4})$

99. $(-3z^4)(-4z^{-3})$

100. $x^{-2}x^4$

101. v^5v^{-5}

102. $2u^3 \cdot u^{-3}$

103. $(3^{-2})^3$

104. $(3^2)^{-3}$

105. $\left(\dfrac{3x^{-2}}{x^{-1}}\right)^{-1}$

106. $\left(\dfrac{x^{-1}}{y^{-2}}\right)^{-2}$ **107.** $\left(\dfrac{2}{3}\right)^{-1}$ **108.** $\left(-\dfrac{1}{8}\right)^{-1}$

109. $(x^2)^{-3}$ **110.** $(x^{-4})^3$ **111.** $(x^{-4})^{-9}$

112. $(ab)^{-5}$ **113.** $(2x)^{-5}$ **114.** $(3t)^{-3}$

115. $(-2r)^{-6}$ **116.** $(-3r)^{-2}$ **117.** $-(bw)^{-4}$

118. $\dfrac{y^2}{y^{-2}}$ **119.** $\dfrac{x^{-4}}{x^2}$ **120.** $\dfrac{p^5}{p^{-1}}$

121. $\dfrac{w^{-6}}{w^{-4}}$ **122.** $\dfrac{4^{-8}}{4^{-6}}$ **123.** $\dfrac{3^{-4}}{3^{-5}}$

124. $\dfrac{1}{2^{-2}}$ **125.** $\left(\dfrac{1}{2}\right)^{-2}$ **126.** $2^{-2}-3^{-1}$

127. $\dfrac{2^{-1}+3^{-2}}{6^{-1}}$ **128.** $\left(\dfrac{2}{x}\right)^{-4}$ **129.** $\left(\dfrac{y}{3}\right)^{-3}$

130. $\left(\dfrac{a}{v}\right)^{-4}$

In Problems 131 through 138, write the numbers in scientific notation.

131. McDonald's has served 55 billion hamburgers.
132. The speed of light is 186,000 mi/s.
133. The mass of an electron is 0.0000000000000000000000000009109 kg.
134. The density of oxygen is 0.001429 units.

135. Saturn is 885,000,000 mi from the sun.
136. Mount McKinley is 20,300 ft high.
137. A virus ranges in size from 0.00001 to 0.003 mm.
138. A nanosecond is 1 billionth of a second.

In Problems 139 through 144, write the given numbers without exponents.

139. 3.101×10^4
141. $(5.67 \times 10^3)(2.0 \times 10^{-2})$
143. $(-2.0 \times 10^{-3})(1.5 \times 10^{-2})$

140. 1.23×10^{-3}
142. 9.1×10^{-2}
144. 3.45×10^{10}

Challenge Problems

In Problems 145 through 147, evaluate each expression when x is −2.

145. x^2-x^{-2} **146.** $(2x-1)^{-2}$

147. $\dfrac{x^{-1}+x^{-2}}{x^{-2}}$

148. Evaluate $2^{-n}+3^{-n}$ for $n=0,1,2,3$.
149. Evaluate 2^{1-n} for $n=0,1,2,3$.

In Your Own Words

150. What is the relationship between 3^{-7} and 3^7?
151. Explain what scientific notation is and why it is useful.

2.3   Defining, Evaluating, and Simplifying Polynomials

TAPE 4

The expression bx^n, where b is a nonzero real number and n is a nonnegative integer, is called a **monomial** in one variable. We call b the **coefficient,** or **numerical coefficient,** x the **variable,** and n the **degree.** The following are monomials in one variable.

$$\boxed{1}\ 3x^2 \qquad \boxed{2}\ \frac{1}{2}y^3 \qquad \boxed{3}\ -5z^4$$

	Coefficient	Variable	Degree
$\boxed{1}$	3	x	2
$\boxed{2}$	$\frac{1}{2}$	y	3
$\boxed{3}$	-5	z	4

The following expressions are also monomials.

$$\boxed{4}\ x^5 \qquad \boxed{5}\ -t^7 \qquad \boxed{6}\ 8$$

Since $x^5 = 1 \cdot x^5$, we say that x^5 has an *understood* coefficient of 1. Similarly, $-t^7$ has an understood coefficient of -1. Since $8 = 8 \cdot 1 = 8X^0$ for *any* letter X, we say 8 is a monomial of degree 0. We call such monomials **constants.**

	Coefficient	Variable	Degree
$\boxed{4}$	1	x	5
$\boxed{5}$	-1	t	7
$\boxed{6}$	8	Any letter	0

A **polynomial** is a sum of monomials. Its degree is the highest degree of the monomials.

Consider

$$5x^7 + 3x^3 + 2x^2 + 6$$

This is a polynomial, since it is the sum of four monomials. Its degree is seven. This polynomial is written so that the monomial of highest degree comes first and the degrees of the other monomials decrease in order. This is called **standard form.** The coefficient of the first term is called the **leading coefficient.** The monomials are most often called **terms.** The last term, 6, is called the **constant term.** Another example of a polynomial is

$$3y^5 + (-2y^2) + (-9)$$

We write such polynomials as

$$3y^5 - 2y^2 - 9$$

This is a fifth-degree polynomial with three terms, written in standard form.

▶ EXAMPLE 1

Write each polynomial in standard form and give the degree of each.

(a) $5 + 3x^2 + 2x$ (b) $2x + 3$

(c) $x^5 - x^6 + 2$ (d) $-5 - 2y^2 - 3y + 3y^5$

Solutions

(a) Standard form is $3x^2 + 2x + 5$. The degree is 2.
(b) Standard form is $2x + 3$. The degree is 1.
(c) Standard form is $-x^6 + x^5 + 2$. The degree is 6.
(d) Standard form is $3y^5 - 2y^2 - 3y - 5$. The degree is 5. ◀

- Polynomials that have one term are called **monomials.**
- Polynomials that have two terms are called **binomials.**
- Polynomials that have three terms are called **trinomials.**

Listed next are examples of each type.

Monomials	Binomials	Trinomials
x	$x^2 + x$	$x^5 + 4x^2 + 7$
$5y^4$	$z^3 - 7$	$w^2 - 8w + 6$

    Polynomials have many different values depending on the value of each variable. To **evaluate** a polynomial means to find its value, given some value of the variable. Polynomials are expressions, and we evaluate them in the same manner we evaluated expressions in Section 1.5.

► **EXAMPLE 2**

Evaluate each polynomial for the given values of each variable.

(a) $3x^2 - x + 4$ if x is -2.
(b) $2y^3 + 3y^2 + y + 5$ if y is -2.
(c) $3z^4 - 4z^3 - 2z^2 + z$ if z is -1.

Solutions

(a) $3x^2 - x + 4 = 3(-2)^2 - (-2) + 4$
$$= 3(4) + 2 + 4$$
$$= 12 + 2 + 4$$
$$= 18$$

(b) $2y^3 + 3y^2 + y + 5 = 2(-2)^3 + 3(-2)^2 + (-2) + 5$
$$= 2(-8) + 3(4) - 2 + 5$$
$$= -16 + 12 - 2 + 5$$
$$= -1$$

(c) $3z^4 - 4z^3 - 2z^2 + z = 3(-1)^4 - 4(-1)^3 - 2(-1)^2 + (-1)$
$$= 3 \cdot 1 - 4(-1) - 2 \cdot 1 - 1$$
$$= 3 + 4 - 2 - 1$$
$$= 4$$ ◀

Sometimes we work with polynomials that contain like terms. As we saw in Section 1.5, the commutative, associative, and distributive properties allow us to combine like terms. We often simplify polynomials by combining like terms.

► **EXAMPLE 3**

Simplify each polynomial by combining like terms.

(a) $8x^4 + 5x^4$ (b) $2y + y$

(c) $5z - 3z + 2z$ (d) $2w^3 - w^3 + 3w^3$

Solutions

(a) $8x^4 + 5x^4 = (8 + 5)x^4 = 13x^4$

(b) $2y + y = (2 + 1)y = 3y$

(c) $5z - 3z + 2z = (5 - 3 + 2)z = 4z$

(d) $2w^3 - w^3 + 3w^3 = (2 - 1 + 3)w^3 = 4w^3$ ◄

► **EXAMPLE 4**

Combine like terms in each.

(a) $3a^2 - 2a^2 + 12a + 4a$ (b) $5x^2 - 5 + 3x^2 + 4$ (c) $3y + 7y^2$

Solutions

(a) $3a^2$ and $-2a^2$ are like terms and $12a$ and $4a$ are also like terms.

$$3a^2 - 2a^2 + 12a + 4a = a^2 + 16a$$

(b) $5x^2$ and $3x^2$ are like terms, as are -5 and 4. We use the commutative property and rearrange the terms.

$$5x^2 - 5 + 3x^2 + 4 = 5x^2 + 3x^2 - 5 + 4$$
$$= 8x^2 - 1$$

(c) There are no like terms in this polynomial.   ◄

The distributive property can also be used to simplify a polynomial.

► **EXAMPLE 5**

Simplify each polynomial.

(a) $2(x + 3) + 6x$ (b) $-3(4 - 3y) - 8 + y$

Solutions

(a) The order of operations applies. We must multiply with the distributive property first.

$$2 \, (x + 3) + 6x = \boxed{2} \, (x) + \boxed{2} \, (3) + 6x \qquad \text{Distributive Property}$$
$$= 2x + 6 + 6x$$
$$= 8x + 6 \qquad\qquad\qquad \text{Combine like terms}$$

(b) $\boxed{-3}\ (4 - 3y) - 8 + y = \boxed{-3}\ (4) - \boxed{(-3)}\ (3y) - 8 + y$

$$= -12 + 9y - 8 + y$$

$$= -20 + 10y$$

We can write this in standard form as $10y - 20$. ◀

Calculator Box

Evaluating a Polynomial with a Calculator

Let's evaluate the polynomial $3x^4 - 5x^2 + x - 5$ when x is 1.2. We enter the polynomial with 1.2 substituted for x.

$$\boxed{3}\ \boxed{\times}\ \boxed{1.2}\ \boxed{y^x}\ \boxed{4}\ \boxed{-}\ \boxed{5}\ \boxed{\times}\ \boxed{1.2}\ \boxed{x^2}\ \boxed{+}\ \boxed{1.2}\ \boxed{-}\ \boxed{5}\ \boxed{=}$$

We read $\boxed{-4.7792}$ on the display. Therefore, the value of the polynomial is -4.7792 when x is 1.2. Note that we *did not* write down any intermediate results as we went through the evaluation. The calculator kept the running total. This is an important part of good calculator technique.

Next, let's estimate the value of the polynomial $x^4 - 5x^2 + x + 5$ when x is 1.9526 and then approximate the value to three decimal places with a calculator. Since 1.9526 is close to 2, we estimate the value using 2 for x.

$$2^4 - 5 \cdot 2^2 + 2 + 5 = 16 - 20 + 7 = 3$$

So our estimate is 3. Now, on our calculator we press

$$\boxed{1.9526}\ \boxed{y^x}\ \boxed{4}\ \boxed{-}\ \boxed{5}\ \boxed{\times}\ \boxed{1.9526}\ \boxed{x^2}\ \boxed{+}\ \boxed{1.9526}\ \boxed{+}\ \boxed{5}\ \boxed{=}$$

and see $\boxed{2.425641517}$ on the display. Therefore, the value of the polynomial is approximately 2.426 to three decimal places when x is 1.9526. This approximation is reasonably close to our estimate and thus is a reasonable answer.

Calculator Exercises

Approximate the value of each polynomial to the nearest thousandth when x is 1.0319. Make an estimate to see if the answer is reasonable.

1. $x^3 + 2x^2 + 3x + 4$ **2.** $2x^5 - 2x^4 - 3x + 3$

3. $11x^6 - 12x^5$ **4.** $3x^4 - 2x^3 + 5x^2 - 6x + 1$

Answers

1. 10.324 (Estimate: 10) **2.** −0.023 (Estimate: 0)

3. −0.759 (Estimate: −1) **4.** 1.337 (Estimate: 1)

Problem Set 2.3

Warm-Ups

In Problems 1 through 8, write each polynomial in standard form. Give its degree, and identify those that are monomials, binomials, or trinomials. See Example 1.

1. $x + 6$

2. $4 - 3x + x^2$

3. $\frac{1}{2}x - 3x^3 + 4x^2 - 7$

4. $4y^3 + \frac{2}{3}y - 17y^7 + 89$

5. $x^5 - 25$

6. $3x + x^2 + x^3$

7. $-5y^5 + 6y + y^7 - 1$

8. $9 - z^3$

In Problems 9 through 14, evaluate each polynomial when x is 2. See Example 2.

9. $3x + 7$

10. $1 - x^2$

11. $8 - x$

12. $x^2 - x$

13. $4x^2 - 2x + 1$

14. $2x^3 - x - 2$

In Problems 15 through 20, evaluate each polynomial when x is -2.

15. $3x + 7$

16. $8 - x$

17. $x^2 - x + 1$

18. $3x^3 - 5x^2 + x$

19. $x^4 + 3x^2 + 1$

20. $x^5 + 3x^3 + 5x$

In Problems 21 through 30, simplify by combining like terms. See Example 3.

21. $3x + 7x$

22. $8x^2 - x^2$

23. $7x - 4x + x$

24. $-2t - 8t - t$

25. $-w^3 - 2w^3 + 6w^3$

26. $3y - 2y - y$

27. $4a + a - 2a$

28. $12z - 8z - z$

29. $-11x^2 - 15x^2 + 5x^2$

30. $20y^2 - 25y^2 + y^2$

In Problems 31 through 40, combine like terms. See Example 4.

31. $3x + 2 + 5x + 1$

32. $-2y - 1 - 3y - 4$

33. $4r^3 - 2 + 3r^3 - 3$

34. $4x - 7x - 5 + 3$

35. $5x^2 + x + 2 - 3x^2 - 5x + 4$

36. $x^2 - 3x + 4x + 2x^2 - 5 + 4$

37. $-3y + 2y^2 - 5 - y^2 + 6 - 2y$

38. $5y^2 + 3x^2 - 4y^2 - x^2$

39. $3x^2 + 4y^2 - x^2 - y^2$

40. $3x^3 - x^2 + x^3 - 2x^2$

In Problems 41 through 50, simplify each polynomial. See Example 5.

41. $3(2x - 1) + 2x$

42. $-2(x + 3) - 4x + 3$

43. $5(3 - 2x^3) - 16$

44. $2(3y + 4) - 6y$

45. $-7(1 - t) + 7 - 7t$

47. $\frac{1}{2}(2x^2 + 4) + x^2$

47. $\frac{2}{3}(6 - 15x) - 2$

48. $-\frac{3}{5}\left(\frac{5}{2}x + \frac{1}{3}\right) + \frac{5}{2}x$

49. $1.5(2z - 3) - 1.8$

50. $-3.0(2.1 - 1.1w) + 2.1w$

Practice Exercises

In Problems 51 through 56, write each polynomial in standard form. Give its degree, and identify those that are monomials, binomials, or trinomials.

51. $5 - x$

52. $3 - 2t + t^2$

53. $x^9 - 2x^8$

54. $9z + z^4 + z^3$

55. $5y^5 - 7y - y^7 - 1$

56. $49x$

In Problems 57 through 62, evaluate each polynomial when x is 3.

57. $3x + 7$

58. $2x^3 - 3x - 22$

59. $15 - 2x$

60. $x^3 - x^2$

61. $-5x^2 + 3x - 12$

62. $2x^2 + x - 2$

In Problems 63 through 68, evaluate each polynomial when x is −3 and y is 1.

63. $3x + 7y$

64. $x^2 - x + y$

65. $15 - 2xy$

66. $x^3 + 4x^2 - xy + 5y^2$

67. $x^2 + 3xy + y^2$

68. $x^2 - 3xy^3 + 5x - 3y$

In Problems 69 through 90, simplify each polynomial.

69. $3x + x$

70. $5x^2 - x^2$

71. $6x - 8x + x$

72. $-3t - 5t - t$

73. $-w^3 + 3w^3 - 6w^3$

74. $2(5x - 7) + 3x - 9$

75. $c + 3c - 2c$

76. $-x^2 - 5x^2 + 4x^2$

77. $2z - 7z - z$

78. $10y^2 - 5y^2 - y^2$

79. $4x + 1 + x - 1$

80. $-y - 3 - 5y + 4$

81. $3r^3 - 5 + 2r^3 - 1$

82. $6(y - 7) - 8 + 9y$

83. $4(x + 7) - 2x + 3$

84. $x^2 - x + 3x + 2x^2 - 7 + 1$

85. $-2y + 6y^2 - 8 - y^2 + 3 - 7y$

86. $-7(y^2 + 3) - y^2$

87. $5x^2 + 3y^2 + x^2 + y^2$

88. $\dfrac{1}{2}(x^2 - 6) - \dfrac{3}{2}x^2$

89. $\dfrac{1}{5}\left(5x + \dfrac{5}{2}\right) + \dfrac{3}{2}$

90. $2.5(y - 2.5) + 5.5$

91. A flare is shot upward at a steep angle in a fireworks show. The height of the projectile above the earth (in feet) is given by $160t - 16t^2$, where t is the time after firing (in seconds). How high is the flare 5 seconds after firing? 10 seconds after firing?

92. If the equipment of Problem 91 is mounted on the top of a 100-ft building, the height of the projectile is given by $100 + 160t - 16t^2$. Complete the following table by entering the height corresponding to the given values of t.

Time (sec)	0	1	2	3	4	5	6	7	10
Height (ft)									

Problem 91

Challenge Problems

Evaluate each polynomial when x is −1. Assume that n is a natural number.

93. $x^n + 1$

94. $x^{2n} + 1$

95. The McDonald Company makes boxes of various sizes by cutting squares from each corner of a 12-in. by 18-in. piece of cardboard and then folding up the sides. Government regulations require that the volume of each box be printed on the box. A mathematics consultant has determined that the volume of a box with square cutouts of length x (as shown) is given by the polynomial

$$4x^3 - 60x^2 + 216x$$

Find the volume that must be printed on the boxes if x is 1 in., 2 in., 3 in., 5 in., or 6 in. Guess what size box would have the largest volume.

In Your Own Words

6. What is a polynomial?

97. What do we mean by standard form for a polynomial?

2.4 Addition and Subtraction

TAPE 4

We have seen that the Commutative, Associative, and Distributive Properties enable us to combine like terms. Addition of polynomials is just a matter of combining like terms.

► **EXAMPLE 1**

Perform the addition in each.

(a) $(8x^2 - 3x + 11) + (x^2 + 5x - 4)$

(b) $(4x^3 + 4x^2 - x + 2) + (-7x^2 + x - 5)$

Solutions

(a) We gather the like terms.

$$(8x^2 - 3x + 11) + (x^2 + 5x - 4)$$

$$= (8x^2 + x^2) + (-3x + 5x) + (11 - 4)$$

$$= \quad 9x^2 \quad + \quad 2x \quad + \quad 7$$

(b) $(4x^3 + 4x^2 - x + 2) + (-7x^2 + x - 5)$

$$= 4x^3 + (4x^2 - 7x^2) + (-x + x) + (2 - 5)$$

$$= 4x^3 - 3x^2 - 3 \qquad \blacktriangleleft$$

Subtraction of polynomials is defined in terms of addition, as it was for real numbers. Again we use the idea of the opposite.

> If P is a polynomial, $-P$ is called the **opposite** of P, and
> $$P + (-P) = 0$$

We must learn how to find the opposite of a polynomial before we can subtract. The opposite of $2x$ is $-2x$ because

$$2x + (-2x) = 0$$

The opposite of $x^2 - x + 1$ is $-x^2 + x - 1$ because

$$(x^2 - x + 1) + (-x^2 + x - 1) = 0$$

How do we find the opposite of a polynomial? The preceding discussion suggests that, to find the opposite of a polynomial, we change the sign of each term in the polynomial.

We learned in Chapter 1 that the opposite of a real number is -1 times the real number. That is, $-x = -1x$. The same idea is true for polynomials.

> If P is a polynomial,
> $$-P = -1(P)$$

We can find the opposite of a polynomial by multiplying the polynomial by -1. The Distributive Property allows us to do this.

► **EXAMPLE 2**

Find the opposite of each polynomial.

(a) $4x^2 - 3x - 1$ (b) $x - 1$ (c) $y + 2$

Solutions

(a) $\quad - \quad (4x^2 - 3x - 1) = \quad -1 \quad (4x^2 - 3x - 1)$
$$= -4x^2 + 3x + 1$$

(b) $\quad - \quad (x - 1) = \quad -1 \quad (x - 1)$
$$= -x + 1$$

(c) $\quad - \quad (y + 2) = \quad -1 \quad (y + 2)$
$$= -y - 2 \qquad \blacktriangleleft$$

We define subtraction as we did for real numbers.

> If P and Q are polynomials, then
> $$P - Q = P + (-Q)$$

To subtract Q from P, we add the opposite of Q to P.

► **EXAMPLE 3**

Perform the subtractions indicated.

(a) $(4x^2 - 17) - (x^2 - 11)$
(b) $(3x^3 - 7x + 2) - (x^3 + x - 8)$
(c) $(9x^4 + x^2 - 6) - (-4x^2 + x + 1)$

Solutions

(a) To subtract, we add the opposite. The opposite of $x^2 - 11$ is $-x^2 + 11$.

$$(4x^2 - 17) + (x^2 - 11) = (4x^2 - 17) + (-x^2 + 11)$$
$$= 4x^2 - 17 - x^2 + 11$$
$$= 3x^2 - 6$$

(b) The opposite of $x^3 + x - 8$ is $-x^3 - x + 8$.

$$(3x^3 - 7x + 2) - (x^3 + x - 8)$$
$$= (3x^3 - 7x + 2) + (-x^3 - x + 8)$$
$$= 3x^3 - 7x + 2 - x^3 - x + 8$$
$$= 2x^3 - 8x + 10$$

(c) The opposite of $-4x^2 + x + 1$ is $4x^2 - x - 1$.

$$(9x^4 + x^2 - 6) - (-4x^2 + x + 1)$$
$$= (9x^4 + x^2 - 6) + (4x^2 - x - 1)$$
$$= 9x^4 + x^2 - 6 + 4x^2 - x - 1$$
$$= 9x^4 + 5x^2 - x - 7 \qquad ◄$$

When we combine the operations of addition, subtraction, and multiplication, the order of operations applies.

► **EXAMPLE 4**

Perform the operations indicated.

(a) $[(x^2 - x) - (2x^2 + 3x + 2)] + (6x^2 - x + 1)$
(b) $(x^2 - 1) - 2(x^2 + 4)$

Solutions

(a) We start with the inside parentheses.

$$[(x^2 - x) - (2x^2 + 3x + 2)] + (6x^2 - x + 1)$$
$$= [(x^2 - x) + (-2x^2 - 3x - 2)] + (6x^2 - x + 1)$$
$$= [-x^2 - 4x - 2] + (6x^2 - x + 1)$$
$$= 5x^2 - 5x - 1$$

(b) We perform the multiplication first.

$$(x^2 - 1) - 2(x^2 + 4) = (x^2 - 1) - (2x^2 + 8)$$
$$= (x^2 - 1) + (-2x^2 - 8)$$
$$= -x^2 - 9 \quad \blacktriangleleft$$

Often when we add and subtract polynomials, we think of them in terms of parentheses preceded by a positive or a negative sign.

Look at an addition:

$$(5x^2 + x - 8) + (3x^2 + 4x + 3) = 5x^2 + x - 8 + 3x^2 + 4x + 3$$
$$= 8x^2 + 5x - 5$$

We removed the parentheses. Parentheses preceded by a positive sign can be removed without changing the problem.

Look at a subtraction.

$$(4x^2 - 3x - 7) - (x^2 + 5x - 2)$$
$$= 4x^2 - 3x - 7 - x^2 - 5x + 2$$
$$= 3x^2 - 8x - 5$$

Thus parentheses preceded by a negative sign can be removed if the sign of *each* term inside the parentheses is changed.

▶ **EXAMPLE 5**

Subtract $3y - 4$ from the sum of $6y + 7$ and $-4y + 2$.

Solution

We must add $6y + 7$ and $-4y + 2$ and then subtract $3y - 4$. We use grouping symbols to write this.

$$[(6y + 7) + (-4y + 2)] - (3y - 4) = (2y + 9) - (3y - 4)$$
$$= (2y + 9) + (-3y + 4)$$
$$= -y + 13 \quad \blacktriangleleft$$

In using algebra, polynomials arise in many different situations. The next examples illustrate some of them.

▶ **EXAMPLE 6**

Express the perimeter of the rectangle shown as a polynomial in standard form.

$x - 3$ ft

x ft

Solution

The perimeter is the distance around the rectangle. So, it is $x + (x - 3) + x + (x - 3)$.

$$x + (x - 3) + x + (x - 3) = 4x - 6$$

The perimeter is $4x - 6$ ft. $\quad \blacktriangleleft$

► **EXAMPLE 7**

If n is the smallest of three consecutive odd integers, express the sum of the three integers as a polynomial in standard form.

Solution

The three integers are n, $n + 2$, and $n + 4$.

$$n + (n + 2) + (n + 4) = 3n + 6$$

The sum is $3n + 6$.   ◄

Problem Set 2.4

Warm-Ups

In Problems 1 through 16, perform the operations indicated. See Example 1.

1. $(2x + 3) + (5x + 1)$ **2.** $(7x + 8) + (2x + 5)$
3. $(x^2 - 5) + (3x^2 - 2)$ **4.** $(5x^3 - 2) + (x^3 - 1)$
5. $(4z^2 - 6) + (3 + 4z^2)$ **6.** $(3 + 2x) + (5x + 7)$
7. $(5x - 8) + (7x + 4)$ **8.** $(11x^2 + 15) + (8x^2 - 11)$

9. $(2t^2 - 5) + (-3t^2 - 2)$ **10.** $(-3x + 5) + (-2x - 17)$
11. $(6y + y^2 - 5) + (2y^2 - 7)$ **12.** $(z^2 - 6z) + (2z^2 + 8 - z)$
13. $(x^2 + 4x + 1) + (4x^2 - 7x - 4)$
14. $(5x^3 - 6x^2 + 5x - 7) + (10x^3 + 7x^2 - x - 1)$
15. $(4x^5 - 6x^3 + 8x) + (4x^4 + 7x^3 - x^2 + 2)$
16. $(5x^6 - 3x^4 + 3) + (-4x^5 + 7x^4 + x^3 - 4)$

In Problems 17 through 26, find the opposite of each polynomial. See Example 2.

17. $3x + 1$ **18.** $t^2 - 3$
19. $6x - 9$ **20.** $1 - z$
21. $x^2 - 8x - 7$ **22.** $-x^3 + 2x^2 - x - 3$
23. $5 + 7x^2$ **24.** $x + 5$
25. $4 - x$ **26.** $x - 4$

In Problems 27 through 38, perform the operations indicated. See Example 3.

27. $(3x + 7) - (2x + 4)$ **28.** $(4x + 3) - (x - 3)$
29. $(2x^2 - 8) - (4x^2 + 5)$ **30.** $(-2z^3 - 3) - (-3z^3 + 1)$
31. $6x^4 - (2x^4 + 3)$ **32.** $33 - (x - 19)$
33. $x^3 - (2x^2 - x)$ **34.** $(2y - 5) - (3 + y)$
35. $(2x^2 + 3x - 4) - (x + x^2 - 3)$ **36.** $(y^3 - 6y - 1) - (6y + 3y^3)$
37. $(6x^3 - 5x^2 + 6x - 5) - (2x^3 - 11x^2 - x + 5)$
38. $(-y^4 - y^3 + 6y + 9) - (-5y^4 - y^2 + 9)$

In Problems 39 through 46, perform the operations indicated. See Example 4.

39. $[(x + 7) + (3x - 4)] - (2x - 3)$

40. $(2x^2 + 11) + 3(4x^2 - 1)$

41. $(4x^3 - 9) - 2(6x^3 + 8) + (8x^3 - 2x^2 + 13)$

42. $[(5x + 2) - (2x - 7)] - 2(3x - 1)$

43. $-2(4x - 1) + [(3x + 8) - (x - 1)]$

44. $(z^2 + 3) + (2z^2 - 8) - (4z^2 + 2)$

45. $2(x - 5) + 4(2 - 3x)$

46. $(5w^2 - 4w + 18) + (4w^2 - w + 4) - (2w^2 - 7w - 14)$

In Problems 47 through 52, see Example 5.

47. Add $4x^2 + 3x - 1$ to $x^2 - 1$.

48. Subtract $3s + 2$ from $s^2 + 7s - 6$.

49. Find the sum of $3x^2 - 7x + 5$ and $3x + 4$.

50. Subtract $x^2 + x + 1$ from the sum of $x^2 + 1$ and $x^2 + 2x + 1$.

51. Subtract $x - 3$ from $3 - x$.

52. Find the sum of $4x^2 + 4x + 4$, $-x^2 - x - 1$, and $3x^2 + 3x + 3$.

In Problems 53 through 58, express each quantity as a polynomial in standard form. See Examples 6 and 7.

53. The perimeter of a square with a side of length $2s + 3$ feet.

54. A board is cut into two pieces, as shown. Find the length of the board.

$x^2 + 5$ ft $x^2 - x - 3$ ft

55. The circumference of a circle with radius of $r - 2$ inches.

56. If n is the largest of three consecutive integers, find the sum of the three integers.

57. John has $2x + 5$ dollars and Samuel has $5x - 7$ dollars. How many dollars do they have together?

58. Mary and Dolores run from the same point in opposite directions. If Mary runs a distance of $60x$ mi and Delores runs a distance of $40x$ mi, find the distance between them.

Practice Exercises

In Problems 59 through 81, perform the operations indicated.

59. $(x - 5) + (23x + 9)$

60. $(7x + 6) - (x - 4)$

61. $(4y + 3y^2 - 8) + (3y^2 - 9)$

62. $(z^2 - 3z) - (4z^2 + z)$

63. $(x^2 - 4x - 1) - (3x^2 + 7x - 5)$

64. $(5x^2 - 2x + 4) - (7x^2 - 11 - 5x)$

65. $(3x^3 - 4x^2 + 3x - 9) + (12x^3 + 6x^2 + x + 1)$

66. $(2x^4 + x^2 + 5x^3 - x + 4)$
$- (3x^4 - 8x^3 + 5x^2 - 2x - 1)$

67. $(8x^5 - 4x^3 + 7x) - (3x^4 - 17x^3 + x^2 - 12)$

68. $(4x^6 - 2x^4 + 13) - 2(9x^5 + 4x^4 - x^3 + 3)$

69. $(4x^3 - 2x^2 + 8x - 12) + (2x^3 - 7x^2 + 7x - 1)$

70. $(-3x^4 + 6x^3 + x^2 - 2) - (x^4 - 3x^3 + 2x^2 - 9)$

71. $(13x^5 - 4x^3 - 7x) + (11x^5 + 14x^4 - 3x)$

72. $(17x^4 - 18x^3 - 3x^2 + 19) - 3(2x^3 + 2x - 13)$

73. $(x^2 - 7x - 5) - (x^2 + 6x - 4) + (x^2 - 2x + 11)$

74. $-2(8 + 9y) + 3(10y + 6) - (y - 4)$

75. $(z^2 - 3) - (3 + z^2) + (z^2 - 2z - 1)$

76. $(2x - 3) - 3(x + 4)$

77. $[(3x^2 - 4x + 7) + (2x^2 + 10)] - (7x^2 - x)$

78. $[(2x^3 - x^2 - 3x - 7) - (5x^3 + 3x^2 - 6x)]$
$+ (4x^3 - 1)$

79. $(4v^4 - v^2 + 4) - [(2v^4 - v + 8)$
$+ (2v^3 + 3v^2 - 5v + 8)]$

80. $(t^3 + 27) - [(t^3 - 27) - (27 - t^3)]$

81. $6(x^2 - 4) - (x^2 - 20)$

82. Subtract $5s + 8$ from $2s^2 - 5s + 4$.

83. Find the sum of $6x^2 - 5x + 1$ and $x + 7$.

84. Subtract $x^2 - x - 1$ from the sum of $x^2 - 1$ and $x^2 - 2x - 1$.

85. Subtract $-7t^3 + t^2 + t - 8$ from $t^3 - 2t^2 + t + 4$.

86. Subtract $x - y$ from $y - x$.

87. Find the sum of $2x^2 + 2x + 2$, $-x^2 - 4x - 4$, and $6x^2 + 7x - 3$.

In Problems 88 through 94, express each quantity as a polynomial in standard form.

88. Find the perimeter of a rectangular helicopter pad if the dimensions are $5d - 3$ meters by $3d + 7$ meters. What is the perimeter when d is 10?

89. A length of fire hose is cut into three pieces whose lengths are $5r$, $r - 1$, and $r^2 - 2r + 3$ feet. Write an expression to represent the original length of the fire hose. What was the original length if r is 9?

90. Find the perimeter of a square with side of length s^2 yards.

91. Two planes fly in opposite directions from Atlanta. If one plane travels a distance of $500(x + 4)$ km and the other travels a distance of $400(x - 4)$ km, find the distance between them.

92. Jason has $5d - 3$ compact disks and Steve has $8d + 5$ compact disks. How many disks do they have together?

93. Christy is $2y$ years old and Susannah is $y + 7$ years old. Find the sum of their ages.

94. Find the perimeter of a triangle with sides of $2s + 3$, $s - 6$, and $3s - 1$ in.

Challenge Problems

In Problems 95 and 96, combine like terms.

95. $2x^n + 5x^n$

96. $5x^{2n} - 3x^{2n}$

In Your Own Words

97. How do we find the opposite of a polynomial?

98. How do we add polynomials?

99. How do we subtract polynomials?

2.5   Multiplication

TAPE 5

We can multiply polynomials by using the Distributive, Commutative, and Associative Properties and the laws of exponents.

To multiply two monomials, we use the Commutative and Associative Properties to rearrange the factors.

▶ EXAMPLE **1**

Find the product of $3x^4$ and $5x^2$.

Solution

$$(3x^4) \cdot (5x^2) = 3 \cdot 5x^4x^2 \qquad \text{Associative and Commutative Properties}$$
$$= 15x^6 \qquad\qquad\qquad ◀$$

As illustrated by Example 1, we multiply monomials by multiplying the coefficients and multiplying the variables.

► **EXAMPLE 2**

Find each product.

(a) $(6y^4)(-4y)$ (b) $(-2x^3)(-9x^{11})$ (c) $(-x)(-y)$

(d) $(-a)(a^2)$ (e) $\frac{1}{2}(4x)(x^2)$ (f) $\frac{2}{3}(5t^3)(6t)$

Solutions

(a) $(6y^4)(-4y) = (6)(-4)y^4y$ (b) $(-2x^3)(-9x^{11}) = (-2)(-9)x^3x^{11}$

$= -24y^5$ $= 18x^{14}$

(c) $(-x)(-y) = (-1)(-1)xy$ (d) $(-a)(a^2) = (-1)(1)(a)(a^2)$

$= xy$ $= -1a^3$

$= -a^3$

(e) $\frac{1}{2}(4x)(x^2) = \frac{1}{2}(4)xx^2$ (f) $\frac{2}{3}(5t^3)(6t) = \frac{2}{3}(6)(5)\,t^3t$

$= 2x^3$ $= 4(5)t^4$

$= 20t^4$ ◄

The Distributive Property allows us to multiply a monomial by a polynomial.

► **EXAMPLE 3**

Find each product.

(a) $x^3(x - 2)$
(b) $2x^3(3x^2 - 5x + 3)$

Solutions

(a) $x^3 (x - 2) = x^3 \cdot x - x^3 \cdot 2$

$= x^4 - 2x^3$

(b) $2x^3 (3x^2 - 5x + 3) = 2x^3 (3x^2) - 2x^3 (5x) + 2x^3 (3)$

$= 6x^5 - 10x^4 + 6x^3$ ◄

The product of two binomials is one of the most important products that we learn. Let's look at $(a + b)(c + d)$, which is a product of two binomials.

Using the Distributive and Commutative Properties, we write

$$(a + b)(c + d) = a(c + d) + b(c + d)$$

$$= ac + ad + bc + bd$$

If we examine this closely, we can see that there are four terms in the product and that we get them by multiplying each term in the first binomial by each term in the second binomial.

Product of Two Binomials

$$(a + b)(c + d) = ac + ad + bc + bd$$

A common way to remember this is:

- ac represents the product of the *f*irst terms,
- ad represents the product of the *o*utside terms,
- bc represents the product of the *i*nside terms, and
- bd represents the product of the *l*ast terms.

This is often called the *FOIL* method.

► EXAMPLE 4

Find each product.

(a) $(x + 2)(x + 1)$ (b) $(y - 2)(y + 5)$

(c) $(2x - 1)(x + 3)$ (d) $(z - 7)(2z - 1)$

Solutions

(a) $(x + 2)(x + 1) = x(x) + x(1) + 2(x) + 2(1)$
$$= x^2 + x + 2x + 2$$
$$= x^2 + 3x + 2$$

(b) $(y - 2)(y + 5) = y(y) + y(5) - 2(y) - 2(5)$
$$= y^2 + 5y - 2y - 10$$
$$= y^2 + 3y - 10$$

(c) $(2x - 1)(x + 3) = 2x(x) + 2x(3) - 1(x) - 1(3)$
$$= 2x^2 + 6x - x - 3$$
$$= 2x^2 + 5x - 3$$

(d) $(z - 7)(2z - 1) = z(2z) + z(-1) - 7(2z) - 7(-1)$
$$= 2z^2 - z - 14z + 7$$
$$= 2z^2 - 15z + 7$$ ◄

In multiplying binomials as in Example 4, it is possible to compute the middle term mentally.

The Distributive Property allows us to perform other multiplications, such as

$$(a + b)(c + d + e) = a(c + d + e) + b(c + d + e)$$
$$= ac + ad + ae + bc + bd + be$$

Be Careful! Again, each term in the first polynomial is multiplied by each term in the second polynomial. It is important to note that the FOIL method works only for the product of two binomials.

► **EXAMPLE 5**

Find each product.

(a) $(x - 2)(x^2 - 3x + 4)$

(b) $(x + 2)^3$

(c) $2(x + 3)(x - 4)$

Solutions

(a) $(x - 2)(x^2 - 3x + 4)$

$$= x(x^2) - x(3x) + x(4) - 2(x^2) - 2(-3x) - 2(4)$$
$$= x^3 - 3x^2 + 4x - 2x^2 + 6x - 8$$
$$= x^3 - 5x^2 + 10x - 8$$

(b) $(x + 2)^3 = (x + 2)(x + 2)\ (x + 2)$ $x^2 \quad 2x \ 2x \ 4$

$$= (x^2 + 4x + 4)\ (x + 2)$$
$$= x^2(x) + x^2(2) + 4x(x) + 4x(2) + 4(x) + 4(2)$$
$$= x^3 + 2x^2 + 4x^2 + 8x + 4x + 8$$
$$= x^3 + 6x^2 + 12x + 8$$

(c) The Associative Property tells us that we can multiply any two of the numbers first. We will multiply the binomials first.

$$2\ (x + 3)(x - 4)\ = 2\ (x^2 - 4x + 3x - 12)$$
$$= 2(x^2 - x - 12)$$
$$= 2x^2 - 2x - 24$$ ◄

► **EXAMPLE 6**

Express the area of the rectangle as a polynomial in standard form.

$x + 3$ m

$x + 4$ m

Solution

The area of a rectangle is the product of its length and width. So the area is $(x + 4)(x + 3)$. To write this in standard form, we must multiply the binomials.

$$(x + 4)(x + 3) = x^2 + 3x + 4x + 12$$
$$= x^2 + 7x + 12$$

The area of the rectangle is $x^2 + 7x + 12$ square meters.   ◄

► **EXAMPLE 7**

Express the product of an integer n and the next consecutive integer as a polynomial in standard form.

Solution

The next integer after n is $n + 1$. So the product is $n(n + 1)$.

$$n(n + 1) = n^2 + n$$ ◄

The product is $n^2 + n$.

Problem Set 2.5

Warm-Ups

In Problems 1 through 9, find each product. See Examples 1 and 2.

1. $(2x^2)(5x^3)$

2. $(3x)(7x^8)$

3. $(-4x^3)(6x^4)$

4. $(8t^5)(-3t^2)$

5. $(-z)(-z^5)$

6. $(-9x^9)(-4x^4)$

7. $\frac{1}{2}(-8s)(3s^3)$

8. $\frac{3}{4}(7y^4)(4y^3)$

9. $-\frac{2}{3}(12z)(-z^2)$

In Problems 10 through 15, find each product. See Example 3.

10. $3(x^2 + x - 7)$

11. $(2x + 3)4$

12. $2x(3 - 4x + x^3)$

13. $3x^3(3x^5 - x^2 - 2x)$

14. $-2t(4t^3 + t - 6)$

15. $-5x^4(3 - 7x^3)$

In Problems 16 through 25, find each product. See Example 4.

16. $(x + 3)(x + 1)$

17. $(x - 5)(x + 4)$

18. $(t + 7)(t - 3)$

19. $(y - 2)(y - 6)$

20. $(2x + 3)(x + 9)$

21. $(x + 1)(5x + 8)$

22. $(3t - 5)(t + 1)$

23. $(z + 8)(5z - 9)$

24. $(6x - 3)(x - 5)$

25. $(3x + 1)(2x + 9)$

In Problems 26 through 37, find each product. See Example 5.

26. $(x + 2)(x^2 + x + 1)$

27. $(x - 3)(x^2 + 2x + 4)$

28. $(x + 4)(2x^2 - 3x + 5)$

29. $(2t - 1)(3t^2 - 2t - 6)$

30. $(3x - 2)(2x^2 + 4x - 7)$

31. $(x^2 - 3)(x^2 + 7x - 1)$

32. $(x + 1)^3$

33. $(x - 3)^3$

34. $t(3t - 2)(3t + 2)$

35. $(2y + 3)(y + 3)(y - 3)$

36. $4x(x + 3)(x - 2)$

37. $y^2(2y - 3)(y + 1)$

In Problems 38 through 45, write a polynomial in standard form to represent each quantity. See Examples 6 and 7.

38. Find the area of the square whose side is of length $x - 1$ feet.

39. Find the area of the triangle shown.

2x ft

x + 3 ft

40. Find the area of the rectangle with dimensions of $3x + 2$ meters and $x + 4$ meters.

41. Find the area of the trapezoid.

h + 1 yd

2h yd

2h – 3 yd

42. Find the area of a circle with radius of $r + 2$ centimeters.

43. If x is an even integer, find the product of x and the next even integer.

44. If n is an odd integer, find the product of n and the next two odd integers.

45. If y is an integer, find the product of y and the integer before y.

Practice Exercises

In Problems 46 through 96, find each product.

46. $(3x^2)(2x^4)$

47. $(2r)(9r^5)$

48. $(-5x^2)(7x^7)$

49. $(4w^3)(-6w^2)$

50. $(-8z^{11})(-z^{15})$

51. $(-7x^7)(-3x^3)$

52. $\frac{2}{5}(5x)(2x^2)$

53. $\frac{1}{6}(-x)(12x)$

54. $\frac{2}{7}(3w^2)(7w)$

55. $4(x - 5)$

56. $(4x - 7)5$

57. $2y(3 - 9y + y^2)$

58. $2x^2(x^3 - 3x)$

59. $-3t(7t^2 - 8t + 2)$

60. $-3x^5(3 - 2x + 5x^7)$

61. $(x + 2)(x + 5)$

62. $(x - 3)(x + 2)$

63. $(t + 3)(t - 3)$

64. $(y - 3)(y - 4)$

65. $(3t + 2)(t + 9)$

66. $(x + 2)(4x + 7)$

67. $(2x - 5)(x + 2)$

68. $(x + 6)(6x - 11)$

69. $(4x - 3)(3x + 2)$

70. $(10x - 3)(4x - 1)$

71. $(2r + 7)(3r - 5)$

72. $(3s - 5)(5s - 3)$

73. $(6 + 2x)(3 - x)$

74. $(5 - 9x)(2x + 1)$

75. $(x - 4)(x + 4)$

76. $(3x + 2)(3x - 2)$

77. $(y + 5)(y + 5)$

78. $(z - 4)^2$

79. $(5t - 3)(t - 7)$

80. $(2y + 1)(3y + 8)$

81. $(3x - 5)(5x + 2)$

82. $(12x - 1)(4x - 3)$

83. $(3s + 7)(2s - 5)$

84. $(2y - 7)(2y - 7)$

85. $(3 + 8x)(2 - x)$

86. $(2 - 7x)(3x + 1)$

87. $(x - 1)(x^2 + x + 1)$

88. $(x + 3)(x^2 + 2x + 1)$

89. $(x + 4)(x^2 - 4x + 16)$

90. $(2t - 1)(4t^2 + 2t + 1)$

91. $(2x - 3)(3x^2 + x - 4)$

92. $x(x - 1)(x^2 + 2x - 3)$

93. $(s - 1)^3$

94. $(v + 3)^3$

95. $3t(t - 5)(t + 5)$

96. $(y + 4)(y - 4)(y + 2)$

In Problems 99 through 102, express each quantity as a polynomial in standard form.

97. Find the area of the rectangle shown.

x cm

$2x - 1$ cm

98. Find the area of a square with a side of length $y - 3$ meters.

99. If z is an even integer, find the product of the next two consecutive even integers.

100. Find the product of twice n and 3 more than n.

101. An emergency marker panel is to be made with the dimensions shown. Write an expression to represent the area of the panel. What is the area of the panel if x is 4?

x yd

$2x - 1$ yd

102. Write an expression for the area of a square search pattern with a side of length $y - 3$ miles. What is the area of the search pattern if y is 12 miles?

Challenge Problems

Find the following products.

103. $\left(\dfrac{1}{2}x + 1\right)\left(\dfrac{1}{2}x - 1\right)$

104. $\left(\dfrac{1}{2}y + 1\right)^2$

105. $\left(\dfrac{2}{3}z - 1\right)\left(\dfrac{1}{3}z + 2\right)$

106. $(p - q)(p^2 + pq + q^2)$

107. $(p + q)(p^2 - pq + q^2)$

In Your Own Words

108. How do we multiply two binomials?

109. How do we multiply two polynomials?

2.6 Special Products and Order of Operations

TAPE 5

In this section we will develop some products that occur very often in algebra. They are called special products and should be memorized. The first one is called *squaring a binomial*.

▶ **EXAMPLE 1**

Find each product.

(a) $(p + q)^2$ (b) $(p - q)^2$

Solutions

(a) $(p + q)^2 = (p + q)(p + q)$

$\qquad\qquad = p^2 + pq + qp + q^2$

$\qquad\qquad = p^2 + 2pq + q^2$

(b) $(p - q)^2 = (p - q)(p - q)$

$\qquad\qquad = p^2 - pq - qp + q^2$

$\qquad\qquad = p^2 - 2pq + q^2$ ◀

Square of a Binomial

$$(p + q)^2 = p^2 + 2pq + q^2$$
$$(p - q)^2 = p^2 - 2pq + q^2$$

These should be memorized as a pattern and remembered in words rather than symbols. The square of a binomial is the first term squared plus (or minus) twice the product of the two terms plus the last term squared. Notice that these products are alike everywhere except for the sign in the middle term.

Notice that $(p + q)^2 \neq p^2 + q^2$ and $(p - q)^2 \neq p^2 - q^2$. To illustrate this, let's evaluate each expression when p is 4 and q is 3.

$$(p + q)^2 = (4 + 3)^2 = 7^2 = \boxed{49}$$
$$p^2 + q^2 = 4^2 + 3^2 = 16 + 9 = \boxed{25}$$

So we see that $(p + q)^2 \neq p^2 + q^2$.

$$(p - q)^2 = (4 - 3)^2 = 1^2 = \boxed{1}$$
$$p^2 - q^2 = 4^2 - 3^2 = 16 - 9 = \boxed{7}$$

We also see that $(p - q)^2 \neq p^2 - q^2$.

Be Careful!

$$(p + q)^2 \neq p^2 + q^2$$
$$(p - q)^2 \neq p^2 - q^2$$

► **EXAMPLE 2**

Square each binomial.

(a) $(x + 4)^2$ (b) $(2x - 3y)^2$

Solutions

(a) We note this is the square of a binomial.

$$(x + 4)^2 = x^2 + 2(x)(4) + 4^2$$

(the first term squared plus twice the product of the two terms plus the last term squared)

$$(x + 4)^2 = x^2 + 8x + 16$$

(b) $(2x - 3y)^2 = (2x)^2 - 2(2x)(3y) + (3y)^2$

(the first term squared minus twice the product of the two terms plus the last term squared)

$$(2x - 3y)^2 = 4x^2 - 12xy + 9y^2 \qquad ◄$$

Another special product comes from multiplying the sum of two numbers by the difference of the same two numbers. It is called the difference of two squares.

► **EXAMPLE 3**

Find the product of $(p - q)$ and $(p + q)$.

Solution

$$(p - q)(p + q) = p^2 + pq - qp - q^2$$
$$= p^2 - q^2$$

Note that $(p - q)(p + q) = (p + q)(p - q)$.   ◄

Difference of Two Squares
$$(p - q)(p + q) = p^2 - q^2$$

The product of the sum of two numbers and the difference of the same two numbers is the first number squared minus the second number squared.

► **EXAMPLE 4**

Find each product.

(a) $(x + 2)(x - 2)$ (b) $(4r + 7)(4r - 7)$

Solutions

This is the difference of two squares.

(a) $(x + 2)(x - 2) = x^2 - 2^2$ $x^2 + -2x + 2x - 4$
$$= x^2 - 4$$

(the first number squared minus the second number squared)

(b) $(4r + 7)(4r - 7) = (4r)^2 - 7^2$
$$= 16r^2 - 49$$ ◄

Now that we have studied addition, subtraction, and multiplication of polynomials, we can combine these operations. The order of operations tells us how to simplify such problems.

► **EXAMPLE 5**

Simplify each polynomial.

(a) $(x + 1)(x - 1) - (x^2 + 2x - 1)$ (b) $(x - 2)^2 - (x + 2)(x - 3)$

Solutions

(a) We must multiply $(x + 1)(x - 1)$ first. It is the difference of two squares.

$$(x + 1)(x - 1) - (x^2 + 2x - 1)$$
$$= x^2 - 1 - (x^2 + 2x - 1)$$

Be Careful!

We perform the subtraction next, being *very careful* with the signs!
$$= x^2 + 1 - x^2 - 2x + 1 = -2x$$

(b) We perform the multiplications first.
$$(x - 2)^2 - (x + 2)(x - 3)$$
$$= x^2 - 4x + 4 - (x^2 - x - 6)$$

Be Careful!

Notice that we left the parentheses because the product of $(x + 2)$ and $(x - 3)$ is to be subtracted. Now we remove the parentheses, being careful with signs.
$$= x^2 - 4x + 4 - x^2 + x + 6$$
$$= -3x + 10 \qquad \blacktriangleleft$$

► **EXAMPLE 6**

Express the area of a square with side of length $s - 4$ meters as a polynomial in standard form.

Solution

The area of a square is the square of the length of a side. So the area is $(s - 4)^2$.
$$(s - 4)^2 = s^2 - 2(s)(4) + 4^2$$
$$= s^2 - 8s + 16$$

The area of the square is $s^2 - 8s + 16$ square meters. $\qquad \blacktriangleleft$

Problem Set 2.6

Warm-Ups

In Problems 1 through 10, square each binomial. See Example 2.

1. $(x + 1)^2$

2. $(x - 7)^2$

3. $(x - 1)^2$

4. $(x + 2)^2$

5. $(4s + 3)^2$

6. $(5t - 4)^2$

7. $(2x - 1)^2$

8. $(3x - 2)^2$

9. $(9 + y)^2$

10. $(8 - t)^2$

In Problems 11 through 20, find each product. See Example 4.

11. $(x + 1)(x - 1)$

12. $(x - 6)(x + 6)$

13. $(2x - 3)(2x + 3)$

14. $(3x + 5)(3x - 5)$

15. $(6x + 9)(6x - 9)$

16. $(7 - 11w)(7 + 11w)$

17. $(3x + 7)(3x - 7)$

18. $(10 + 9t)(10 - 9t)$

19. $(9 - 5t)(9 + 5t)$

20. $(1 - t)(1 + t)$

In Problems 21 through 28, simplify each polynomial. See Example 5.

21. $5 - 2(x + 3)$

22. $3 - [4 - (x - 3) + 2(x + 1)]$

23. $6x[5 - 2x(x - 3)] + 4x^2$

24. $2t - 3t[t(t + 1) - t^2]$

25. $x(x - 3) + (x + 3)^2$

26. $(2x - 1)(x + 2) - x(2x - 1)$

27. $(y - 4)(y + 3) + (y - 2)(y + 2)$

28. $2x(x - 8) + (3x + 7)(3x - 7)$

In Problems 29 through 33, write a polynomial in standard form to represent each quantity. See Example 6.

29. The area of a square with side of length $3y + 2$ feet.

30. The area of a circle with radius of $r - 1$ meters.

31. The area of the rectangle shown.

$x - 3$ ft

$x + 3$ ft

32. The square of the next integer after the integer n.

33. The square of the integer before the integer q.

Practice Exercises

In Problems 34 through 53, use special products to find each product.

34. $(x - 1)^2$

35. $(x + 3)^2$

36. $(t + 3)(t - 3)$

37. $(y - 5)(y + 5)$

38. $(3x - 4)^2$

39. $(4t + 5)^2$

40. $(2 - 3x)(2 + 3x)$

41. $(2t - 7)(2t + 7)$

42. $(2x + 6)^2$

43. $(2x - 3)^2$

44. $(3x + 5)^2$

45. $(2z + 1)^2$

46. $(8x + 9)(8x - 9)$

47. $(11q - 5)(11q + 5)$

48. $(7p - 1)^2$

49. $(3w - 4)^2$

50. $(7x - 8)(7x + 8)$

51. $(5 - 12k)(5 + 12k)$

52. $(5x - 1)^2$

53. $(9g + 2)^2$

In Problems 54 through 63, simplify each polynomial.

54. $(x - 7)(x + 7) + (x - 3)^2$

55. $2x - 3x[(x - 5) + 2x(x - 7)]$

56. $(s - 6)^2 - 3s(2s - 3)$

57. $3z(2 - z) - (z + 1)^2$

58. $(y - 3)^2 - (y - 1)(y + 1)$

59. $(w - 5)(w + 3) - (w + 2)(w + 4)$

60. $(z + 3)^2 - 2z(z + 1)$

61. $(t - 4)(t + 4) - (t + 2)(t + 1)$

62. $7y - y(y - 2) + (y + 2)(y - 2)$

63. $7t - t[2(t + 3) + (t + 2)(t - 3)]$

In Problems 64 through 69, express each quantity as a polynomial in standard form.

64. The area of a square with a side of length of $2y - 4$ centimeters.

65. The area of a circle with radius of $r + 7$ inches.

66. The product of the integer before and the integer after the integer p.

67. The area of the following rectangle.

$x - 2$ m

$x + 2$ m

Solution

Both polynomials are in standard form.

$$x + 3 \overline{) x^2 + x - 6}$$

Since $\dfrac{x^2}{x} = x$, we write

$$\begin{array}{r} x \\ x + 3 \overline{) x^2 + x - 6} \end{array}$$

Then we multiply x by $x + 3$ and enter the product.

$$\begin{array}{r} x \\ x + 3 \overline{) x^2 + x - 6} \\ x^2 + 3x \end{array}$$

Now we are *very careful* in subtracting.

$$\begin{array}{r} x \\ x + 3 \overline{) x^2 + x - 6} \\ \ominus \ominus \\ x^2 + 3x \\ \hline -2x \end{array}$$

We bring down the -6.

$$\begin{array}{r} x \\ x + 3 \overline{) x^2 + x - 6} \\ \ominus \ominus \\ x^2 + 3x \\ \hline -2x - 6 \end{array}$$

We repeat the previous steps. Since $\dfrac{-2x}{x} = -2$, we enter the -2 and multiply.

$$\begin{array}{r} x - 2 \\ x + 3 \overline{) x^2 + x - 6} \\ \ominus \ominus \\ x^2 + 3x \\ \hline -2x - 6 \\ -2x - 6 \end{array}$$

Subtract.

$$\begin{array}{r} x - 2 \\ x + 3 \overline{) x^2 + x - 6} \\ \ominus \ominus \\ x^2 + 3x \\ \hline -2x - 6 \\ \oplus \oplus \\ -2x - 6 \\ \hline 0 \end{array}$$

Since the remainder is 0, we are finished with the procedure. We still need to write the answer. Since the remainder is zero, we can write

$$\frac{x^2 + x - 6}{x + 3} = x - 2$$

Since $(x + 3)(x - 2) = x^2 + x - 6$, the answer checks.   ◀

► **EXAMPLE 3**

Divide $2 - 2x + x^3$ by $x + 1$.

Be Careful!

Solution

First, we *must* write both polynomials in standard form.

$$(x^3 - 2x + 2) \div (x + 1)$$

Notice that there is no x^2 term in the dividend. Because we need *every term* in the standard form, we insert $0x^2$ in its proper place and write the display.

$$x + 1 \overline{)x^3 + 0x^2 - 2x + 2}$$

We calculate $\dfrac{x^3}{x} = x^2$, enter it, multiply, and prepare to subtract.

$$
\begin{array}{r}
x^2 \phantom{{} - 2x + 2} \\
x + 1 \overline{)x^3 + 0x^2 - 2x + 2} \\
x^3 + x^2
\end{array}
$$

Now we subtract.

$$
\begin{array}{r}
x^2 \phantom{{} - 2x + 2} \\
x + 1 \overline{)x^3 + 0x^2 - 2x + 2} \\
\underline{x^3 + x^2} \\
-x^2
\end{array}
$$

We bring down the $-2x$. Since $\dfrac{-x^2}{x} = -x$, we enter it and multiply.

$$
\begin{array}{r}
x^2 - x \phantom{{} + 2} \\
x + 1 \overline{)x^3 + 0x^2 - 2x + 2} \\
\underline{x^3 + x^2} \\
-x^2 - 2x \\
-x^2 - x
\end{array}
$$

Subtract, and bring down the 2.

$$
\begin{array}{r}
x^2 - x \phantom{{} + 2} \\
x + 1 \overline{)x^3 + 0x^2 - 2x + 2} \\
\underline{x^3 + x^2} \\
-x^2 - 2x \\
\underline{-x^2 - x} \\
-x + 2
\end{array}
$$

Since $\dfrac{-x}{x} = -1$, we enter -1 and multiply.

$$
\begin{array}{r}
x^2 - x \;\; - 1 \\
x + 1 \overline{)\, x^3 + 0x^2 - 2x + 2} \\
\underline{x^3 + x^2 } \\
-x^2 - 2x \\
\underline{-x^2 - x } \\
-x + 2 \\
-x - 1
\end{array}
$$

We subtract, and get our final display.

$$
\begin{array}{r}
x^2 - x \;\; - 1 \\
x + 1 \overline{)\, x^3 + 0x^2 - 2x + 2} \\
\underline{x^3 + x^2 } \\
-x^2 - 2x \\
\underline{-x^2 - x } \\
-x + 2 \\
\underline{-x - 1} \\
3
\end{array}
$$

The degree of 3 is less than the degree of $x + 1$. We conclude that

$$(2 - 2x + x^3) \div (x + 1) = x^2 - x - 1 + \dfrac{3}{x + 1} \qquad \blacktriangleleft$$

► **EXAMPLE 4**

Divide $(6x^3 - 13x^2 + 4x + 7)$ by $(2x - 3)$.

Solution

$$
\begin{array}{r}
3x^2 - 2x \;\; - 1 \\
2x - 3 \overline{)\, 6x^3 - 13x^2 + 4x + 7} \\
\underline{6x^3 - 9x^2 } \\
-4x^2 + 4x \\
\underline{-4x^2 + 6x } \\
-2x + 7 \\
\underline{-2x + 3} \\
4
\end{array}
$$

So

$$\dfrac{6x^3 - 13x^2 + 4x + 7}{2x - 3} = 3x^2 - 2x - 1 + \dfrac{4}{2x - 3} \qquad \blacktriangleleft$$

Problem Set 2.7

Warm-Ups

Perform the divisions indicated. In Problems 1 through 15, see Example 1.

1. $\dfrac{4a^2 - 4a + 4}{4}$

2. $\dfrac{8 - 4t^3}{-2}$

3. $\dfrac{x^2 - x + 1}{x^2}$

4. $\dfrac{3z^6 - 5z^4 - z^2}{15z^2}$

5. $(6x - 6)/(-6)$

6. $\dfrac{5 - 2y + 5y^2}{5y^2}$

7. $\dfrac{3x^6 - x^4}{x^3}$

8. $\dfrac{-4z^9 + 8z - 16}{-4z^3}$

9. $\dfrac{9x^3 - 3x^2 + 3}{-3x^2}$

10. $\dfrac{x^3 + 2x^4}{2x^4}$

11. $\dfrac{2x^3 + 3x^4}{2x^2}$

12. $\dfrac{t^5 - 7t^3}{t^3}$

13. $(3x^4 + 2x^5) \div 3x^4$

14. $\dfrac{x^2y - xy^2}{xy}$

15. $\dfrac{2x^5 - 3x^7}{6x^5}$

In Problems 16 through 39, see Examples 2, 3, and 4.

16. $\dfrac{x^2 + 3x + 2}{x + 1}$

17. $\dfrac{x^2 + 7x + 12}{x + 3}$

18. $\dfrac{x^2 - 3x - 10}{x + 2}$

19. $\dfrac{x^2 + 4x + 4}{x + 1}$

20. $\dfrac{y^2 + 6y + 11}{y + 2}$

21. $\dfrac{t^2 + t - 4}{t + 3}$

22. $\dfrac{2x^2 + x - 1}{x + 1}$

23. $\dfrac{3x^2 + 8x + 4}{x + 2}$

24. $\dfrac{2x^2 - x - 21}{x + 3}$

25. $\dfrac{s^2 + s - 1}{s - 1}$

26. $\dfrac{2x^2 + 7x - 1}{x + 4}$

27. $\dfrac{3z^2 - 11z + 2}{z - 3}$

28. $\dfrac{x^3 + 2x^2 + 2x + 1}{x + 1}$

29. $\dfrac{x^3 + 4x^2 - 3x - 10}{x + 2}$

30. $\dfrac{t^3 + 2t^2 - 4t - 3}{t + 3}$

31. $\dfrac{3x^2 + x^3 + x - 1}{x + 2}$

32. $\dfrac{x^3 - 3x^2 - x - 1}{x - 1}$

33. $\dfrac{x^3 + 4x^2 + 2x}{x + 2}$

34. $\dfrac{2x^3 + 7x^2 + 7x + 2}{2x + 1}$

35. $\dfrac{4y^3 + 4y^2 - y}{2y + 3}$

36. $\dfrac{x^3 - 26x + 7}{x - 5}$

37. $\dfrac{6v^3 - 7v^2 - 4}{2v - 1}$

38. $(5 - 7x^2 + 3x^3)/(3x + 2)$

39. $(2z^3 - 4z) \div (4 + 2z)$

Practice Exercises

In Problems 40 through 72, perform the divisions indicated.

40. $\dfrac{-3x^5 - 6x}{-3x}$

41. $\dfrac{3y^7 - 2y^3}{2y^3}$

42. $\dfrac{8x^4 - 16x^3 + 4x^2}{-2x^2}$

43. $\dfrac{x^4 + 4x^5}{x}$

44. $\dfrac{3x^5 + 2x^3}{x^2}$

45. $\dfrac{t^6 - 6t^4}{t^4}$

46. $\dfrac{4x^3 + 12x^4 - 4x^5}{4x^3}$

47. $\dfrac{3x^2 - 4x^5}{12x^2}$

48. $\dfrac{sx^5 - tx^7}{stx^5}$

49. $\dfrac{x^2 + 4x + 3}{x + 1}$

50. $\dfrac{x^2 + 6x + 8}{x + 2}$

51. $\dfrac{x^2 + x - 6}{x + 3}$

52. $\dfrac{x^2 + 3x + 3}{x + 1}$

53. $\dfrac{y^2 + 7y + 11}{y + 3}$

54. $\dfrac{t^2 - 3t - 7}{t + 2}$

55. $\dfrac{x^2 + x - 2}{x - 1}$

56. $\dfrac{2x^2 + 7x - 4}{x + 4}$

57. $\dfrac{3x^2 - 11x - 6}{x - 3}$

58. $\dfrac{2s^2 + s + 1}{s + 1}$

59. $\dfrac{3x^2 + 8x + 1}{x + 2}$

60. $\dfrac{2z^2 - z - 16}{z + 3}$

61. $\dfrac{x^3 + 3x^2 + 3x + 2}{x + 2}$

62. $\dfrac{x^3 + 3x^2 - 2x - 4}{x + 1}$

63. $\dfrac{t^3 + 2t^2 - t + 6}{t + 3}$

64. $\dfrac{x^3 + x + 3}{x + 1}$

65. $\dfrac{x^3 - x^2 - 9}{x - 2}$

66. $\dfrac{2x^3 + 5x^2 + 3x + 5}{x + 2}$

67. $\dfrac{3x^2 + 2x^3 - 1 - x}{2x + 1}$

68. $\dfrac{2y^3 - 3y^2 - 5y}{2y - 1}$

69. $\dfrac{x - 11 + 4x^3}{2x - 3}$

70. $\dfrac{6v^3 + 5v^2 - 1}{2v + 3}$

71. $(16x^3 - x) \div (4x - 3)$

72. $(3 + 5t^2 + 6t^3) \div (3t - 2)$

73. Suppose a garden contains $2x^2 - 5x - 3$ hills of Silver Queen sweet corn. If the hills are planted in $x - 3$ equal rows, how many hills of corn are there in each row? How many hills are there in each row if x is 6?

74. The area of a rectangular playground is $6s^2 + 25s + 25$ sq ft. What is the length if the width is $2s + 5$ ft? What is the length if s is 27 feet?

Challenge Problems

Perform the divisions indicated.

75. $\dfrac{4x^4 - 6x^3 - 4x^2 + 13x - 9}{2x - 1}$

76. $\dfrac{x^3 + 3x^2 + x - 1}{x^2 + x - 1}$

77. $(x^4 + 5x^3 + x^2 - 11x + 4) \div (x^2 + 2x - 1)$

78. $(x^3 - 8)/(x - 2)$

79. $\dfrac{x^4 + 2x^3 + 7x^2 + 2x + 5}{x^2 + 1}$

80. Find the value of k so that, if $x^3 + 2x^2 + x + k$ is divided by $x + 3$, the remainder will be zero.

In Your Own Words

81. When do we use long division?

82. What is the procedure for long division?

Chapter Summary

GLOSSARY

Exponent   In the expression x^n, the exponent is n. It tells how many times to use x as a factor.

Base   In the expression x^n, the base is x. It is the number being raised to the power n.

Monomial   An expression of the form bx^n, where n is a whole number and x is a nonzero real number.

Binomial   A sum of two monomials.

Trinomial   A sum of three monomials.

Polynomial   A sum of monomials.

Coefficient   In the expression bx^n, the coefficient of x^n is b.

Degree of a monomial in one variable   The exponent of the variable.

Degree of a polynomial   Highest degree of the terms in the polynomial.

Standard form   A polynomial written so that the degree of the monomials decreases in order from left to right.

Leading coefficient   The coefficient of the first term of a polynomial.

Constant term   A term that has degree 0.

Exponent Definitions

1. If n is a natural number, then $x^n = x \cdot x \cdot x \cdots x$ (n x's).
2. $x^0 = 1; x \neq 0$.
3. If n is a natural number and $x \neq 0$, then $x^{-n} = \dfrac{1}{x^n}$

Properties of Exponents

1. $x^m \cdot x^n = x^{m+n}$ Product with same base.
2. $(x^m)^n = x^{mn}$ Power to a power.
3. $(xy)^n = x^n y^n$ Product to a power.
4. $\left(\dfrac{x}{y}\right)^n = \dfrac{x^n}{y^n}; y \neq 0$ Quotient to a power.
5. $\dfrac{x^m}{x^n} = x^{m-n}; x \neq 0$ Quotient with same base.

Operations with Polynomials

1. To evaluate a polynomial means to replace each variable in the polynomial with the value of the variable.
2. To add polynomials, add like terms.
3. To subtract polynomials, change the subtraction to addition of the opposite.
4. To multiply polynomials, use the distributive property and the properties of exponents.
5. To divide polynomials, use long division if the divisor is not a monomial. If the divisor is a monomial, divide each term in the dividend by the divisor.

To Write a Number in Scientific Notation

1. Starting with the number in decimal format, move the decimal point until the number is between 1 and 10, including 1. Count the number of places moved.

2. Multiply the number formed in step 1 by 10 raised to the number of decimal places moved. If the original number was between 0 and 1, the power of 10 is negative. If the original number was greater than 10, the power is positive.

3. This procedure is for a *positive* number. If the original number is negative, perform steps 1 and 2 on the absolute value of the original number; then attach a negative sign to the result.

Square of a Binomial

1. $(p + q)^2 = p^2 + 2pq + q^2$
2. $(p - q)^2 = p^2 - 2pq + q^2$

Difference of Two Squares

$(p - q)(p + q) = p^2 - q^2$

CHECK UPS

1. Simplify $(-xy^2)^2$	Section 2.1; Example 5b
2. Simplify $\left(\dfrac{2x}{k^2}\right)^3$	Section 2.1; Example 8c
3. Simplify $7x^{-3}$	Section 2.2; Example 1f
4. Simplify $(3x^{-4})^{-2}$	Section 2.2; Example 3b
5. Write 0.000001554 in scientific notation.	Section 2.2; Example 5b
6. Write $5 + 3x^2 + 2x$ in standard form and give its degree.	Section 2.3; Example 1a
7. Evaluate $3x^2 - x + 4$ if x is -2.	Section 2.3; Example 2a

In Problems 8 through 14, perform the indicated operations.

8. $(8x^2 - 3x + 11) + (x^2 + 5x - 4)$	Section 2.4; Example 1a
9. $(3x^3 - 7x + 2) - (x^3 + x - 8)$	Section 2.4; Example 3b
10. $[(x^2 - x) - (2x^2 + 3x + 2)]$ $+ (6x^2 - x + 1)$	Section 2.4; Example 4a
11. $(-a)(a^2)$	Section 2.5; Example 2d
12. $(x + 2)(x + 1)$	Section 2.5; Example 4a
12. $\dfrac{3t^2 - 5t}{2t^2}$	Section 2.7; Example 1b
14. $\dfrac{x^2 + x - 6}{x + 3}$	Section 2.7; Example 2

In Problems 15 and 16, find each product using a special product.

15. $(x + 4)^2$	Section 2.6; Example 2a
16. $(x + 2)(x - 2)$	Section 2.6; Example 4a
17. Express the area of a square with a side of length $s - 4$ meters as a polynomial in standard form.	Section 2.6; Example 6

► Review Problems

In Problems 1 through 9, simplify if possible.

1. $x^5 x^2$

2. $(x^3 t^2)^3$

3. $(2z^4)^{-2}$

4. $\left(\dfrac{-5x^3}{A}\right)^2$

5. -7^{-2}

6. $\dfrac{2y^5}{y^9}$

7. $(t^{-5})^{-2}$

8. $5x^0$

9. $(\pi x^{-5} y^3)^0$

10. Write the following in scientific notation.
   a. 76,500   b. 0.00109

11. Write the following without exponents.
   a. 1.63×10^{-7}
   b. 8.503×10^6

In Problems 12 and 13, evaluate each polynomial when t is −2.

12. $t^2 + 3t + 4$

13. $t^3 - 3t^2 - t - 7$

In Problems 14 through 28, perform the operations indicated.

14. $(2x^3 - 3x + 4) + (x^3 - 2x^2 - 4x - 5)$

15. $(4x^5 - 3x^3 - 6x + 11) + (-x^5 + 2x^4 + 6x + 6)$

16. $(5t^2)(-3t^3)$

17. $-5x^2(1 - 5x^5)$

18. $(2x - 1)(3x + 7)$

19. $(x - 1)(x^2 - x + 1)$

20. $\dfrac{3x^4 + 5x^3}{15x^3}$

21. $\dfrac{3x^2 + x - 10}{x + 2}$

22. $\dfrac{4x^3 + 3x + 2}{2x - 1}$

23. $(x - 2)^3$

24. $x(x - 3)(2x - 5)$

25. $x(x + 5) - (x^2 - 2x + 3)$

26. $(y^2 - y + 7) + y(2y - 1)$

27. $(-xy^2)(-3xy)(-x^2 y)$

28. $(t + 2)(t - 2) - (t + 1)(t + 3)$

In Problems 29 through 31, use special products to multiply.

29. $(x + 7)^2$

30. $(2x + 5)(2x - 5)$

31. $(2t - 3)^2$

In Problems 32 through 35, express each quantity as a polynomial in standard form.

32. The perimeter of a square with side of length $x - 2$ yards.

33. The length of the following rod.

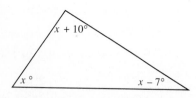

x ft $2x + 5$ ft

34. The sum of the angles of the triangle shown.

35. The area of a rectangle with dimensions of $y - 3$ and $y + 3$ inches.

36. A stone is tossed upward with an initial velocity of 48 feet per second from the top of a tower 800 feet high. The polynomial

$$-16(t^2 - 3t - 50)$$

represents the height of the stone t seconds later. Find the height of the stone when t is 1, 2, 3, and 4 seconds.

37. The Interstate Shipping Company calculates the number of dollars to charge for shipping a package with the polynomial

$$0.05w + 1.25$$

where w is the weight of the package in ounces. Calculate the charge for packages weighing 5, 10, 12, 25, 50, and 100 ounces each.

38. Find the perimeter of the Taco Mac parking lot if it is a rectangle with dimensions of $7k - 5$ meters by $11k + 2$ meters. What is the perimeter when k is 7?

39. Suppose an orchard contains $2x^2 + 11x + 5$ apple trees. If the trees are planted in $x + 5$ equal rows, how many apple trees are in each row?

Let's Not Forget . . .

In Problems 40 through 43, find each sum or difference.

40. $-5 - 11$

41. $(3x - 5) + (6x - 7)$

42. $48 - (x - 42)$

43. $(3x^2 - x + 2) - (2x - 3)$

In Problems 44 through 53, find each product or quotient.

44. $(-x)(-2x^2)$

45. -4^2

46. -4^{-2}

47. $(-4)^{-2}$

48. $(-4)^2$

49. -4^0

50. $\dfrac{0}{-4}$

51. $\dfrac{-4}{0}$

52. $\dfrac{-2x}{x^2}$

53. $\dfrac{4x - 4}{4}$

In Problems 54 through 56, simplify each expression. Watch the role of the parentheses.

54. $(x - 3)(x + 3)$

55. $x - 3(x + 3)$

56. $(x - 3)x + 3$

In Problems 57 through 59, determine if y^2 is a factor or a term.

57. $5y^2$

58. $17 - y^2$

59. $y^2(x - 2)$

In Problems 60 through 62, multiply each expression by 3.

60. $\dfrac{1}{3}x$

61. $\dfrac{1}{3}(x - 1)$

62. $\dfrac{1}{3}x - 1$

-In Problems 63 through 67, label each as an equation *or an* expression. *Solve the equations and perform the operations indicated with the expressions.*

63. $2x = -18$

64. $2(x + 3)$

65. $3x + 5x$

66. $x + 5 = 10$

67. $-11 + 6$

► You Decide

Your backyard swimming pool (20 ft by 40 ft) needs painting. Ronnie from Metro Pool and Chemical Company sells two kinds of paint, rubber based and epoxy. A gallon of rubber-base paint costs $40, covers 250 sq ft, and lasts 2 years while a gallon of epoxy costs $60, covers 108 sq ft, and lasts 3 to 4 years. Ronnie tells you that the total surface area can be approximated by multiplying the area of the pool by 1.65. Color choice is about the same for both kinds of paint, but epoxy is a nicer looking paint. Which kind of paint would you use and why?

► Chapter 2 Test

In Problems 1 through 5, choose the correct answer.

1. $2x^3(9x^2 - x^5) = (?)$
   (a) $18x^5 - 2x^8$ (b) $18x^6 - 2x^{15}$
   (c) $16x^{-6}$ (d) $16x^{10}$

2. The value of the polynomial $3x^3 - 2x^2 + x$ when x is -1 is (?).
   (a) 2   (b) 0
   (c) -2 (d) -6

3. $\left(\dfrac{2a^5}{b^4}\right)^2 = (?)$

   (a) $\dfrac{2a^{10}}{b^8}$ (b) $\dfrac{2a^7}{b^6}$

   (c) $\dfrac{4a^{10}}{b^8}$ (d) $\dfrac{4a^7}{b^6}$

4. $\dfrac{(-3)^4}{(-3)^6} = (?)$

   (a) $\dfrac{1}{6}$ (b) $\dfrac{1}{9}$

   (c) $-\dfrac{1}{6}$ (d) $-\dfrac{1}{9}$

5. $3^{-2} = (?)$

   (a) -9 (b) $\dfrac{1}{9}$

   (c) -6 (d) $\dfrac{1}{6}$

6. Write the following in scientific notation.
   a. 0.00233
   b. 576.3

7. Write the following without exponents.
   a. 9.4×10^3
   b. 4.321×10^{-5}

In Problems 8 through 18, perform the operations indicated.

8. $\dfrac{6x^2 - 4x^3}{x^2}$

9. $(2x - 3)(x + 1)$

10. $(5x^3 - 4x^2 + 3x) + (-3x^3 + x^2 - 7x)$

11. $(3x - 2)^2$

12. $(x + 2)^3$

13. $(6x^3 - 2x^2 + x - 2) \div (x + 1)$

14. $(5x^3 - x^2 + 4) - (x^3 - 2x - 7)$

15. $(2^{-2})^{-1}$

16. $(2 - y)(2 + y)$

17. $x^3(2x^4 - 3x^3 + 3)$

18. $(x + 2)(x - 5) + 2(x - 3)$

In Problems 19 and 20, express each quantity as a polynomial in standard form.

19. The area of a rectangle with dimensions $2y - 1$ and $y + 2$ inches.

20. The length of the pipe shown.

   | 2s yd | s + 22 yd |

3 Linear Equations and Inequalities in One Variable

See Problem Set 3.4, Exercise 21.

► **Connections**

Solving equations has always been a central idea in algebra. The early Egyptians used the rule of false position to solve equations. The idea was to guess a solution and then alter the guess to make it correct. For example, to solve

$$x + \frac{x}{5} = 24$$

guess a solution, say 5, then try it out.

$$5 + \frac{5}{5} = 6$$

Since 24 is 4(6), the correct solution must be 4(5) or 20.

Around the year 830 A.D., a Muslim mathematician named Al-Khowarizmi published a book that gave one of the earliest descriptions on how to solve equations by using the addition and multiplication tools that we studied in Chapter 1. The title of his book contained the Arabic word al-jabr from which our word algebra is derived.

In this chapter we continue the study of equations introduced in Section 1.8. The importance of solving equations lies in solving applied problems. We state a problem in business, economics, or engineering and form an equation as a mathematical model of the real-world problem. If we can solve the equation, we can solve the applied problem. Also, we will study linear inequalities and learn how to solve them.

3.1   Linear Equations

TAPE 6

Equations in which the variable occurs to the first power only are called **linear equations,** or **first-degree equations.** Linear equations in one variable can be written in the form

$$Ax + B = 0, \qquad \text{where } A \text{ is not zero}$$

Three examples of linear equations are

$$2x + 5 = 0$$
$$3(x - 7) = 11$$
$$6t = 4(t - 1) + 2$$

Three examples of **nonlinear** equations are

$$x^2 + 2x + 5 = 0 \qquad \text{Quadratic equation}$$
$$\sqrt{x} + x = 11 \qquad \text{Equation with radicals}$$
$$\frac{1}{x} + \frac{2}{x} = x \qquad \text{Fractional equation}$$

Remember that to **solve** an equation means to find its solution set. We use the addition and multiplication properties that we studied in Section 1.8 to solve linear equations.

Two Tools for Solving Equations

The solution set of an equation will not change if we use either of the following two tools.

Addition Property of Equality

Add (or subtract) the same number on both sides of the equation.

Multiplication Property of Equality

Multiply (or divide) both sides by the same nonzero number.

Our general approach is to use these tools to simplify an equation to the point where it is of the form

$$\text{Variable} = \text{constant}$$

► EXAMPLE 1

Solve each equation. Check the solutions.

(a) $x + 8 = 2$ (b) $-\frac{1}{3}t = 6$

Solutions

(a) We subtract 8 from both sides.

$$x + 8 = 2$$

$$x + 8 - 8 = 2 - 8 \qquad \text{Addition Property}$$

$$x = -6 \qquad \text{Combine like terms}$$

To check -6, we evaluate the left side (LS) and right side (RS) of the original equation.

$$\text{LS:} \quad -6 + 8 = 2$$

$$\text{RS:} \quad 2$$

Since both the left and right sides have the value 2, we say -6 checks and is the solution.

$$\{-6\}$$

(b) We use the Multiplication Property and multiply both sides by -3.

$$-\frac{1}{3}t = 6$$

$$(-3)\left(-\frac{1}{3}t\right) = (-3)\,6$$

$$\left[(-3)\left(-\frac{1}{3}\right)\right]t = (-3)6 \qquad \text{Associative Property}$$

$$t = -18$$

We check -18.

$$\text{LS:} \quad -\frac{1}{3}(-18) = \boxed{6}$$

$$\text{RS:} \quad \boxed{6}$$

The solution is -18.

$$\{-18\} \qquad\qquad\qquad \text{Write the solution set} \quad \blacktriangleleft$$

Often we must use both the Addition and the Multiplication Properties to write an equation in the form *variable = a constant*. It is usually best to use the Addition Property first.

► **EXAMPLE 2**

Solve each equation.

(a) $2x - 3 = 9$

(b) $4 - 5t = 49$

(c) $11 = 6x + 11$

(d) $5 - x = 3$

Solutions

(a) To write the equation in the form $x =$ constant, we must use both the Addition and Multiplication Properties. We use the addition property first.

$$2x - 3 = 9$$

$$2x - 3 \boxed{+ 3} = 9 \boxed{+ 3} \qquad \text{Add 3 to both sides}$$

$$2x = 12 \qquad\qquad \text{Combine like terms}$$

$$\frac{2x}{2} = \frac{12}{2} \qquad\qquad \text{Divide both sides by 2}$$

$$x = 6$$

Check the solution in the original equation.

$$\{6\} \qquad\qquad\qquad \text{Write the solution set}$$

(b)
$$4 - 5t = 49$$

$$4 - 5t \boxed{- 4} = 49 \boxed{- 4} \qquad \text{Subtract 4 from both sides}$$

$$-5t = 45 \qquad\qquad \text{Combine like terms}$$

$$\frac{-5t}{-5} = \frac{45}{-5} \qquad\qquad \text{Divide both sides by } -5$$

$$t = -9$$

Check the solution in the original equation.

$$\{-9\} \qquad\qquad\qquad \text{Write the solution set}$$

(c) We isolate x on the right side.

$$11 = 6x + 11$$

$$11 - 11 = 6x + 11 - 11 \qquad \text{Subtract 11 from both sides}$$

$$0 = 6x \qquad \text{Combine like terms}$$

$$\frac{0}{6} = \frac{6x}{6} \qquad \text{Divide both sides by 6}$$

$$0 = x \qquad \text{Remember } \frac{0}{6} \text{ is } 0$$

Check the solution in the original equation.

$$\{0\} \qquad \text{Write the solution set}$$

(d)

$$5 - x = 3$$

$$5 - x - 5 = 3 - 5 \qquad \text{Subtract 5 from both sides}$$

$$-x = -2 \qquad \text{Combine like terms}$$

$$(-1)\,(-x) = (-1)\,(-2) \qquad \text{Multiply both sides by } -1$$

$$x = 2$$

Check the solution in the original equation.

$$\{2\} \qquad \text{Write the solution set} \qquad \blacktriangleleft$$

Often we must combine like terms on the left side and on the right side before we use our addition and multiplication properties to solve an equation. Notice how this works in the next example.

► **EXAMPLE 3**

Solve each equation and check the solutions.

(a) $3x - 5x = 12 - 8$ (b) $-5 - 5 = 6x - x$

Solutions

(a)

$$3x - 5x = 12 - 8$$

$$-2x = 4 \qquad \text{Combine like terms}$$

$$\frac{-2x}{-2} = \frac{4}{-2} \qquad \text{Divide both sides by } -2$$

$$x = -2$$

Check -2.

$$\text{LS:} \quad 3x - 5x = 3(-2) - 5(-2) = -6 + 10 = 4$$

$$\text{RS:} \quad 12 - 8 = 4$$

Since the left side and the right side have a value of 4, -2 checks and is in the solution set.

$$\{-2\}$$

(b) $-5 - 5 = 6x - x$

$-10 = 5x$ Combine like terms

$\dfrac{-10}{5} = \dfrac{5x}{5}$ Divide both sides by 5

$-2 = x$

Check -2.

LS: $-5 - 5 =$ -10

RS: $6(-2) - (-2) = -12 + 2 =$ -10

The values of both the left side and the right side are the same. Thus -2 checks.

$$\{-2\}$$ ◄

Sometimes the variable may appear on both sides of the equation. We must collect all terms containing the variable on one side of the equation using the addition tool.

► **EXAMPLE 4**

Solve each equation.

(a) $6x = 5x + 7$ (b) $x + 3 = 3x - 5$

Solutions

(a) We must collect all x terms on the same side. We can do this if we subtract $5x$ from both sides.

$6x = 5x + 7$

$6x$ $- 5x$ $= 5x + 7$ $- 5x$ Subtract $5x$

$x = 7$ Combine like terms

$\{7\}$ Write the solution set

(b) We must collect the x terms on one side. We can subtract $3x$ from both sides and collect them on the left side or we can subtract x from both sides and collect them on the right side. Although either would be correct, we often collect them on the side that will yield a *positive* coefficient. So we subtract x from both sides.

$x + 3 = 3x - 5$

$x + 3$ $- x$ $= 3x - 5$ $- x$ Subtract x from both sides

$3 = 2x - 5$ Combine like terms

3 $+ 5$ $= 2x - 5$ $+ 5$ Add 5 to both sides

$8 = 2x$ Combine like terms

$\dfrac{8}{2} = \dfrac{2x}{2}$ Divide both sides by 2

$4 = x$

$\{4\}$ Write the solution set ◄

Equations are used in solving many different kinds of written problems.

▶ **EXAMPLE 5**

Seven more than twice a number x is -9.

(a) Translate the sentence into an equation.

(b) Solve the equation and find the number.

Solutions

(a) Seven more than twice x is -9.

$$2x + 7 = -9$$

(b) $7 + 2x - 7 = -9 - 7$ Subtract 7 from both sides

$$2x = -16$$ Simplify

$$\frac{2x}{2} = \frac{-16}{2}$$ Divide both sides by 2

$$x = -8$$

The number is -8.

Check by replacing x with -8 in the word problem. Twice -8 is -16 and seven more than -16 is -9. ◀

Problems such as these are sometimes called applications, or word problems.

▶ **EXAMPLE 6**

A 36-ft string is cut into two pieces so that the shorter piece is x feet long and the other piece is twice the shorter piece.

(a) Translate the sentence into an equation.

(b) Solve the equation and find the length of the shorter piece of string.

Solutions

(a) A picture helps visualize the situation.

Since the string is 36 ft long, we add the lengths of the two pieces and get 36.

$$x + 2x = 36$$

(b) $3x = 36$ Combine like terms

$$\frac{3x}{3} = \frac{36}{3}$$ Divide both sides by 3

$$x = 12$$

The length of the shorter piece is 12 ft.

We can check by adding the lengths of the two pieces together to see if we get 36. We see that the longer piece is 24 ft long and that $12 + 24 = 36$.   ◄

► **EXAMPLE 7**

The area of the larger triangle is 44 m² more than the area of the smaller triangle. Find the length of the base of the larger triangle.

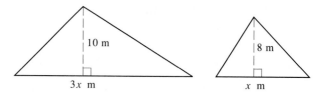

Solution

Remember that the area of a triangle is one-half the product of the base and the height of the triangle.

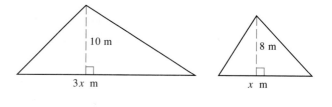

$$\text{Area} = \frac{1}{2} \cdot \text{base} \cdot \text{height} \qquad \text{Area} = \frac{1}{2} \cdot \text{base} \cdot \text{height}$$

$$\text{Area} = \frac{1}{2}(3x)(10) \qquad \text{Area} = \frac{1}{2}(x)(8)$$

$$\text{Area} = 15x \qquad \text{Area} = 4x$$

The area of the larger triangle is 44 more than the area of the smaller triangle. So $15x$ is larger than $4x$. To make them equal, we can add 44 to the smaller number or subtract 44 from the larger number.

Add 44 to smaller number   or   Subtract 44 from the larger number

$$15x = 4x + 44 \qquad\qquad\qquad 15x - 44 = 4x$$

We can solve either equation.

$$15x = 4x + 44$$

$$11x = 44 \qquad \text{Subtract } 4x \text{ from both sides}$$

$$x = 4 \qquad \text{Divide both sides by 11}$$

The base of the larger triangle is $3x$. So the base of the larger triangle is $3(4) = 12$ m.   ◄

Calculator Box

Solving Equations with a Calculator

The proper strategy when approximating the solution(s) of an equation with a calculator is to first write the equation in the form

$$x = \text{an expression (not containing } x)$$

without doing any but the simplest arithmetic. Then complete the approximation with one sequence of keystrokes. Let's approximate to the nearest thousandth the solution to the equation

$$4.5577x - 8.0061 = 6.7442 + 1.3457x$$

We collect the x terms on the left side and the constants on the right.

$$4.5577x - 1.3457x = 6.7442 + 8.0061$$

Next, we *factor out* x on the left side.

$$(4.5577 - 1.3457)x = 6.7442 + 8.0061$$

Then we divide both sides by the coefficient of x.

$$x = \frac{6.7442 + 8.0061}{4.5577 - 1.3457}$$

Finally, we let the calculator do all the remaining work!

| (| 6.7442 | + | 8.0061 |) | ÷ | (| 4.5577 | − | 1.3457 |) | = |

We see ▮4.592247821▮ on the display. Rounding to the nearest thousandth gives the desired approximation $x \approx 4.592$. This logic works for both scientific and graphing calculators.

Calculator Exercises

Use a calculator to approximate the solution of each equation correct to two decimal places. Be sure to indicate that the answers are approximations.

1. $61.45x + 102.09 = 14.22x - 20.93$ **2.** $355 + 211x = 978 + 17x$

3. $0.348t + 0.097 = 3(0.609 - 0.331t)$ **4.** $9\left(x + \dfrac{7}{11}\right) = \pi$

Answers

1. $x \approx -2.60$ **2.** $x \approx 3.21$ **3.** $t \approx 1.29$ **4.** $x \approx -0.29$

Problem Set 3.1

Warm-Ups

In Problems 1 through 8, solve each equation. See Example 1.

1. $y + 8 = -1$

2. $\dfrac{2}{3}x = -12$

3. $z + 8 = 8$

4. $5x = -25$

5. $-w = \dfrac{1}{2}$

6. $-8t = -56$

7. $5 = x - 7$

8. $-3 = 3 + z$

In Problems 9 through 18, solve each equation. See Example 2.

9. $2z - 8 = 22$

10. $3t + 4 = 16$

11. $5x + 2 = 22$

12. $7k - 35 = 0$

13. $6 + 2s = 7$

14. $-5 - 3x = 10$

15. $6 - 3y = 0$

16. $4w - 9 = -9$

17. $8 + 5t = 8$

18. $6 - x = 5$

In Problems 19 through 28, solve each equation. See Example 3.

19. $x + 4x = -35 + 20$

20. $-5x + 2x = 20 - 24$

21. $-x - 2x = -8 - 7$

22. $-2x - 7x + x = -12 - 12$

23. $t - 7t = 12 - 6$

24. $15 - 1 = 3x - 10x$

25. $4x - 5x = 2 - 3$

26. $12 = 9p - 13p$

27. $-y - 2y = -1 - 2$

28. $14 - 18 = 4x - 2x$

In Problems 29 through 42, solve each equation. See Example 4.

29. $5x - 3 = 6x$

30. $7x = 5 + 6x$

31. $10x - 1 = 9x$

32. $13x = 12x - 5$

33. $6x + 5 = x$

34. $3x = x - 4$

35. $4x - 1 = x + 8$

36. $2x + 5 = 3x - 7$

37. $3x + 17 = 5x + 3$

38. $x + 6 = 5 - x$

39. $2 - 3x = 2x - 13$

40. $8 - x = 4 - 3x$

41. $5x - 3 = 4x + 2$

42. $3x + 7 = 4x + 6$

In Problems 43 through 49, translate each sentence into an equation, solve the equation, and find the number. See Example 5.

43. Three times a number y decreased by 5 is 10.

44. The sum of a number z and twice z is -27.

45. Five times a number t increased by three times t is 72.

46. If 8 is added to three times a number x, the result is two more than five times x.

47. If four times a number w is subtracted from 17, the result is the sum of w and 22.

48. Twice a number z is 8 less than six times z.

49. A number y is 3 more than four times y.

In Problems 50 through 53, answer each question. See Example 6.

50. A box marked "miscellaneous hardware" contains only b bolts and $b - 2$ washers. If there is a total of 68 bolts and washers in the box, how many bolts are there in the box?

51. A board 48 in. long is cut into three parts, each x inches long. Find the length of each part.

52. In the following figure, the distance from West Branch to Sunbury is $2d - 3$ miles and the distance from Springdale to Sunbury is $d + 7$ miles. If the distance from West Branch to Springdale is 16 miles, how far is it from West Branch to Sunbury?

53. Mike has 27 coins, which are either nickels or dimes. If he has n nickles and five more dimes than nickles, how many of each coin does he have?

In Problems 54 through 56, answer each question. See Example 7.

54. The perimeter of the larger square is 12 ft more than the perimeter of the smaller square. Find the length of a side of the larger square.

56. The circumference of the larger circle is 2π m more than the circumference of the smaller circle. Find the radius of the larger circle.

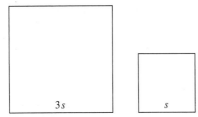

55. The area of the smaller rectangle is 20 cm² less than the area of the larger rectangle. Find the length of the smaller rectangle.

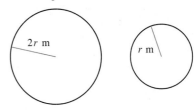

Practice Exercises

In Problems 57 through 86, solve each equation.

57. $3x - 2 = -5$

58. $6 - 5t = 21$

59. $2x - 7 = x$

60. $4x = x - 12$

61. $-x = 3$

62. $1 - x = 0$

63. $3t - 1 = t + 9$

64. $3x - 7 = 4x + 5$

65. $5x - 3x = -17 + 9$

66. $-1 - 5 = -5y - y$

67. $\dfrac{2}{3}x = \dfrac{5}{3}$

68. $7 + x = 7$

69. $2s - 17 = 3s - 3$

70. $x - 9 = 5 - x$

71. $6x - 8 = 5x + 3$

72. $2x + 9 = 3x + 2$

73. $5y - 8 = 3y - 14$

74. $6x - 3x = -5 + 6$

75. $11 - y = 9 + y$

76. $10p - 12 = 7p$

77. $5y - 2y + 3 = 6y - 6$

78. $-x - 1 = x + 1$

79. $13 + 8 - 5 = -7x$

80. $16 = -8z$

81. $15 - 17 = 5x - 8x$

82. $5w + 7w - 2 = 5 + 7 - 2w$

83. $4 - 3y - 8 = 17$

84. $6x - 3 - x - 2 = 40$

85. $x - 8 = -9$

86. $3x - 7 = 7$

In Problems 87 through 91, translate each sentence into an equation, solve the equation, and find the number.

87. Twice a number z increased by 3 is -3.

88. Six times a number t decreased by twice t is 6.

89. If 7 is subtracted from three times a number y, the result is one more than twice y.

90. If seven times a number s is subtracted from 18, the result is twice s.

91. Twice a number x is 5 less than seven times x.

In Problems 92 through 96, answer each question.

92. The perimeter of the triangle on the left is 6 ft more than the perimeter of the triangle on the right. Find the lengths of all sides.

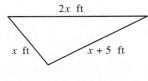

93. The area of the smaller parallelogram is 46 cm² less than the area of the larger parallelogram. Find the length of the base of the smaller parallelogram.

94. A rod 37 in. long is broken into two pieces so that the shorter piece is $2x - 7$ inches long and the other piece is $3x + 4$ inches long. How long is the short piece?

95. Tom has only black and blue socks. He has 12 pairs of socks. If he has b blue socks and twice that many black socks, how many black socks does he have?

96. At a Mystery Night Party, the guests learn that Susie is y years old and Carol is 10 years older than Susie. If the sum of their ages is 50, how old is Susie?

Challenge Problems

Solve the following equations. The variable is x and q is a constant.

97. $2x - q = 3q$ **98.** $3x + q = x - 3q$

In Your Own Words

99. What is a first-degree equation? What is a linear equation?

100. What is the procedure for solving a linear equation?

3.2   Linear Equations with Grouping Symbols or Fractional Coefficients

TAPE 6

In this section we continue our study of equations by examining linear equations that contain grouping symbols and fractional or decimal coefficients.

The properties of real numbers and the tools used in the last section provide all we need to solve these equations. However, it is often necessary to simplify as we solve equations. *Whenever possible, we should combine like terms on either side of an equation.*

> ### A Procedure for Solving Linear Equations
>
> **1.** Clear the equation of fractions (or decimals), if necessary, by multiplying by the least common denominator (LCD) of all the fractions in the equation.
> **2.** Simplify each side by removing grouping symbols and combining like terms, where possible.
> **3.** Collect all terms containing the variable on one side of the equation and all other terms on the other side by using the addition property.
> **4.** Combine like terms.
> **5.** Divide both sides by the coefficient of the variable.
> **6.** Check if required.
> **7.** Write the solution set.

Many times steps 1 and 2 can be interchanged.

► **EXAMPLE 1**

Solve each equation.

(a) $2(x + 3) = 7$ (b) $2(x - 1) = 4x - 5$

Solutions

(a) $2(x + 3) = 7$

STEP 1   Clear fractions.

This equation has no fractions.

STEP 2   Remove grouping symbols.

We must use the Distributive Property to remove the grouping symbols.

$$2(x + 3) = 7$$
$$2 \cdot x + 2 \cdot 3 = 7 \qquad \text{Distributive Property}$$
$$2x + 6 = 7$$

STEP 3   Collect terms.

We subtract 6 from both sides.

$$2x + 6 - 6 = 7 - 6 \qquad \text{Subtract 6}$$

STEP 4   Combine like terms.

$$2x = 1$$

STEP 5   Divide both sides by the coefficient of the variable.

We divide both sides by 2, the coefficient of x.

$$x = \frac{1}{2}$$

STEP 6   Check if required.

$\frac{1}{2}$ may be checked in the original equation.

STEP 7   Write the solution set.

$$\left\{ \frac{1}{2} \right\}$$

(b) $2(x - 1) = 4x - 5$

STEP 2   Remove grouping symbols.

$$2x - 2 = 4x - 5 \qquad \text{Distributive Property}$$

STEP 3   Collect x terms on the right side.

$$2x - 2 - 2x = 4x - 5 - 2x \qquad \text{Subtract } 2x$$

STEP 4   Combine like terms.

$$-2 = 2x - 5$$

STEP 3   Repeat steps 3 and 4 with the constant terms. Collect constant terms on the left side.

$$-2 + 5 = 2x - 5 + 5 \qquad \text{Add 5}$$

STEP 4   Combine like terms.

$$3 = 2x$$

STEP 5   Divide by the coefficient of the variable.

$$\frac{3}{2} = \frac{2x}{2} \qquad \text{Divide by 2}$$

$$\frac{3}{2} = x$$

STEP 6   Check if required.

$\frac{3}{2}$ may be checked in the original equation.

STEP 7   Write the solution set.

$$\left\{ \frac{3}{2} \right\}$$

► EXAMPLE 2

Solve each equation, and *check* the solutions.

(a) $5(x + 2) = 2x + 19$ (b) $3t + 4 = 3(1 + 2t)$

Solutions

(a) $5(x + 2) = 2x + 19$

Remove grouping symbols.

$$5x + 10 = 2x + 19 \qquad \text{Distributive Property}$$

Collect x terms.

$$5x + 10 - 2x = 2x + 19 - 2x \qquad \text{Subtract } 2x$$

Combine like terms.

$$3x + 10 = 19$$

Collect constant terms.

$$3x + 10 - 10 = 19 - 10 \qquad \text{Subtract 10}$$

Combine like terms.

$$3x = 9$$

Divide by the coefficient of x.

$$x = 3 \qquad \text{Divide by 3}$$

Check.

The instructions call for checking our solutions. We must do so before writing the solution set. Check 3 in the original equation.

$$\text{LS:} \quad 5(x + 2) = 5(3 + 2) = 5(5) = \boxed{25}$$

$$\text{RS:} \quad 2x + 19 = 2(3) + 19 = 6 + 19 = \boxed{25}$$

We get the same value in both the left side and the right side. Thus 3 is the solution.

Write the solution set.

$$\{3\}$$

(b) $\quad 3t + 4 = 3(1 + 2t)$

$3t + 4 = 3 + 6t$	Distributive Property
$3t + 4 - 3t = 3 + 6t - 3t$	Subtract $3t$
$4 = 3 + 3t$	Combine like terms.
$4 - 3 = 3 + 3t - 3$	Subtract 3
$1 = 3t$	Combine like terms
$\dfrac{1}{3} = t$	Divide by 3

Check $\dfrac{1}{3}$.

$$\text{LS:} \quad 3t + 4 = 3\left(\frac{1}{3}\right) + 4 = 1 + 4 = \boxed{5}$$

$$\text{RS:} \quad 3(1 + 2t) = 3\left[1 + 2\left(\frac{1}{3}\right)\right] = 3\left[1 + \left(\frac{2}{3}\right)\right] = 3\left(\frac{5}{3}\right) = \boxed{5}$$

The left side has the same value as the right side, so $\dfrac{1}{3}$ checks.

$$\left\{\frac{1}{3}\right\} \qquad \text{Write the solution set} \qquad \blacktriangleleft$$

Linear equations involving fractions can be solved with the addition and multiplication tools. However, arithmetic with fractions is more difficult than arithmetic with integers, and we are more likely to make mistakes. It is wise to clear fractions as a first step in such equations. If we multiply both sides of the equation by the least common denominator (LCD) of all the fractions in the equation, all the fractions will be removed. We call this **clearing fractions.**

When solving equations involving fractions, we frequently encounter expressions like

$$\frac{2}{3}x \quad \text{or} \quad \frac{2x}{3}$$

Because of the rules for multiplying fractions, these have the same value and are just two different ways of writing the same fraction. Note that the left side and the right side of each of the following are equal to each other.

$$\frac{1}{5}x = \frac{x}{5}$$

$$\frac{7}{3}x = \frac{7x}{3}$$

$$\frac{3}{2}(2x - 3) = \frac{3(2x - 3)}{2}$$

► EXAMPLE 3

Solve each equation.

(a) $\dfrac{x}{2} - 1 = \dfrac{2}{5}x + 3$ (b) $\dfrac{1}{2}(x - 2) = 4$

Solutions

(a) $\dfrac{x}{2} - 1 = \dfrac{2}{5}x + 3$

Notice that this equation contains the fractions $\dfrac{x}{2}$ and $\dfrac{2}{5}$. The LCD of the fractions is 10. So we multiply both sides of the equation by 10.

$$10 \cdot \left(\frac{x}{2} - 1\right) = 10 \cdot \left(\frac{2}{5}x + 3\right) \qquad \text{Multiply by 10}$$

$$10 \cdot \left(\frac{x}{2}\right) - 10 \cdot 1 = 10 \cdot \left(\frac{2}{5}x\right) + 10 \cdot 3 \qquad \begin{array}{l}\text{Distributive}\\\text{Property}\end{array}$$

$$10\left(\frac{x}{2}\right) - 10 \cdot 1 = \left(10 \cdot \frac{2}{5}\right)x + 10 \cdot 3 \qquad \begin{array}{l}\text{Associative}\\\text{Property}\end{array}$$

$$5x - 10 = 4x + 30$$

$$5x - 10 - 4x = 4x + 30 - 4x \qquad \text{Subtract } 4x$$

$$x - 10 = 30 \qquad \text{Combine like terms}$$

$$x - 10 + 10 = 30 + 10 \qquad \text{Add 10}$$

$$x = 40 \qquad \text{Combine like terms}$$

$$\{40\} \qquad \text{Write the solution set}$$

(b) $\qquad \dfrac{1}{2}(x - 2) = 4$

$$2\left[\frac{1}{2}(x - 2)\right] = 2\,(4) \qquad \text{Multiply by 2}$$

$$\left(2 \cdot \frac{1}{2}\right)(x - 2) = 8 \qquad \text{Associative Property}$$

$$x - 2 = 8$$

$$x - 2 + 2 = 8 + 2 \qquad \text{Add 2}$$

$$x = 10 \qquad \text{Combine like terms}$$

$$\{10\} \qquad \text{Write the solution set} \qquad ◄$$

If an equation contains decimals, we can work with the decimals or we can clear the decimals. Since decimals can be written as fractions, we will multiply both sides of the equation by 10, 100, 1000 or whatever power of 10 is necessary.

$$0.1 = \frac{1}{10}, \qquad 0.01 = \frac{1}{100}, \qquad 0.001 = \frac{1}{1000}$$

► **EXAMPLE 4**

Solve each equation.

(a) $5.2x = 15.6$ (b) $2.3x - 2.12 = 1.2x + 0.08$

Solutions

(a) $5.2x = 15.6$

We clear decimals by multiplying both sides by 10.

$10\,(5.2x) = 10\,(15.6)$	Multiply by 10
$52x = 156$	Simplify
$\dfrac{52x}{52} = \dfrac{156}{52}$	Divide by 52
$x = 3$	Simplify
$\{3\}$	Write the solution set

(b) $2.3x - 2.12 = 1.2x + 0.08$

We must multiply by 100 to clear the decimals because there are both tenths and hundredths in the original equation.

$100\,(2.3x - 2.12) = 100\,(1.2x + 0.08)$	
$230x - 212 = 120x + 8$	Distributive Property
$230x - 212 - 120x = 120x + 8 - 120x$	Subtract $120x$
$110x - 212 = 8$	
$110x - 212 + 212 = 8 + 212$	Add 212
$110x = 220$	
$x = 2$	Divide by 110
$\{2\}$	

◄

We do not always get one number in our solution set when we solve an equation. Two more equations illustrate this.

► **EXAMPLE 5**

Solve each equation.

(a) $2x + 1 = 2(5 + x)$ (b) $3(2x + 6) = 2(9 + 3x)$

Solutions

(a)
$$2x + 1 = 2(5 + x)$$
$$2x + 1 = 10 + 2x$$
$$2x + 1 - 2x = 10 + 2x - 2x$$
$$1 = 10$$

Oops! What has happened here? Since our work is all correct, the solution set for the original equation is the same as the solution set for the equation

$$1 = 10$$

But there is *no* value of x that will make 1 equal 10, so there are no numbers in the solution set. The solution set is the set with no elements, the empty set. We write the empty set as

$$\{\} \quad \text{or} \quad \varnothing \quad \textit{Never } \{\varnothing\}$$

Be Careful!

(b)
$$3(2x + 6) = 2(9 + 3x)$$
$$6x + 18 = 18 + 6x$$
$$6x + 18 - 6x = 18 + 6x - 6x$$
$$18 = 18$$

Again, the variables disappeared from the equation. However, this time we are left with a *true* statement. That is, 18 equals 18 for *all* values of x. The solution set must be all real numbers. We use set-builder notation (Section 1.1) to write the solution set.

$$\{x \mid x \text{ is a real number}\}$$ ◄

► **EXAMPLE 6**

John has x quarters and $x + 10$ nickels in his pocket. He knows that he has \$5. How many quarters does John have?

Solution

We must find the *value* of x quarters. Since 1 quarter has a value of 25¢, x quarters have a value of $25x$ cents.

The *value* of $x + 10$ nickels is $5(x + 10)$ cents.

The *value* of the money is \$5, or 500¢.

Value of quarters in cents		Value of nickels in cents		Value of money in cents
↓		↓		↓
$25x$	$+$	$5(x + 10)$	$=$	500

We can find how many quarters John has by solving this equation.

$$25x + 5(x + 10) = 500$$
$$25x + 5x + 50 = 500 \qquad \text{Distributive Property}$$
$$30x + 50 = 500$$
$$30x + 50 - 50 = 500 - 50$$
$$30x = 450$$
$$\frac{30x}{30} = \frac{450}{30} \qquad \text{Divide by 30}$$
$$x = 15$$

John has 15 quarters.

We check by noting that $\$.25(15)$ is $\$3.75$. Also, John has 25 nickels worth $\$.05(25)$, or $\$1.25$. Adding the value of this quarters to the value of his nickels, we have

$$\$3.75 + \$1.25 = \$5.00 \qquad \blacktriangleleft$$

► EXAMPLE 7

If x is the smallest of three consecutive integers, twice the largest integer is equal to the sum of twice the smallest integer and the middle integer. Find the integers.

Solution

If x is the smallest integer, then $x + 1$ is the middle integer and $x + 2$ is the largest integer.

We form an equation to find the integers.

Twice the largest	is	the sum of twice smallest and middle.	
↓	↓	↓	↓
$2(x + 2)$	$=$	$2x$	$+ (x + 1)$

Now we solve the equation.

$$2x + 4 = 3x + 1$$
$$2x + 4 - 2x = 3x + 1 - 2x$$
$$4 = x + 1$$
$$4 - 1 = x + 1 - 1$$
$$3 = x$$

Since the integers are x, $x + 1$, and $x + 2$, the numbers are 3, 4, and 5.

Check in the original problem. Twice the largest, $2(5)$, is the same as the sum of twice the smallest, $2(3)$, and the middle, 4. That is,

$$2(5) = 2(3) + 4 \qquad \blacktriangleleft$$

▶ **EXAMPLE 8**

The formula for the volume of a pyramid is one-third the product of the area of its base and its height. If the volume of the following pyramid is 125 m³ and the area of its base is 25 m², find the measure of its height.

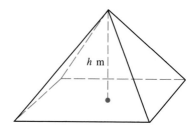

Solution

We use the formula to write an equation.

$$\text{Volume} = \frac{1}{3} \cdot \text{area of base} \cdot \text{height}$$

$$125 = \frac{1}{3}(25)h$$

$$125 = \frac{25}{3}h$$

$$3(125) = 3\left(\frac{25}{3}\right)h \qquad \text{Multiply by 3 to clear the fraction}$$

$$375 = 25h$$

$$\frac{375}{25} = \frac{25h}{25} \qquad \text{Divide by 25}$$

$$15 = h$$

The height is 15 meters.   ◀

▶ **EXAMPLE 9**

If 23% of n is 99.13, find n.

Solution

Since $23\% = 0.23$, we form the equation $0.23n = 99.13$.

$$0.23n = 99.13$$

$$23n = 9913 \qquad \text{Multiply by 100}$$

$$\frac{23n}{23} = \frac{9913}{23} \qquad \text{Divide by 23}$$

$$n = 431$$

n is 431.   ◀

Problem Set 3.2

Warm-Ups

Solve each equation. For Problems 1 through 20, see Examples 1 and 2.

1. $4(x - 2) = 3$

2. $5(x + 1) = 1$

3. $6(3 + x) = 18$

4. $7(1 - x) = 3$

5. $-4(2w - 3) = 6$

6. $-5(3 + 7x) = 20$

7. $6(3 - 4x) - 22 = 0$

8. $15 - 3(2v - 9) = 0$

9. $66 - (12 - 3x) = 0$

10. $13(3 - 2x) + 1 = 1$

11. $2(y - 3) - y = 5$

12. $6x - (5x - 7) = -2$

13. $4(z + 4) - 3(z - 1) = 4$

14. $-2(w + 3) + 3(3w + 2) = -14$

15. $5x - 1 = 2(x + 3)$

16. $2(x - 1) = 6(x + 1)$

17. $2(3 + t) = 3(t - 4)$

18. $x + 3(2x - 1) = 4(2x + 1)$

19. $5(3t - 2) - (12t - 10) = 2(t + 2)$

20. $-2(1 - 3x) = 5(1 - x) - 3x$

In Problems 21 through 32, solve each equation. See Example 3.

21. $\dfrac{x}{2} + \dfrac{1}{2} = \dfrac{3}{2}$

22. $\dfrac{x}{3} - \dfrac{2}{3} = \dfrac{1}{3}$

23. $\dfrac{1}{4} - \dfrac{3x}{4} = 1$

24. $\dfrac{5}{3} - \dfrac{2}{3}x = 1$

25. $\dfrac{3}{2}x - \dfrac{1}{3} = \dfrac{1}{2}x$

26. $\dfrac{5}{6} - \dfrac{x}{2} = \dfrac{x}{3}$

27. $\dfrac{2}{3}x - \dfrac{3}{2}x = -\dfrac{10}{3}$

28. $2x + 3 = \dfrac{1}{2}x - 9$

29. $\dfrac{2}{3}x = x + \dfrac{5}{3}$

30. $\dfrac{1}{3}x = \dfrac{1}{2}x + 1$

31. $\dfrac{1}{3}(x + 2) = 1$

32. $\dfrac{w - 3}{3} = \dfrac{w}{6}$

In Problems 33 through 36, solve each equation. See Example 4.

33. $2.1x = 6.3$

34. $0.49x = 2.45$

35. $1.7x - 8.2 = 2.5x + 3.8$

36. $12.7z - 5.4z = z - 12.6$

In Problems 37 and 38, solve each equation. See Example 5.

37. $4x - 7 = 4(x - 2)$

38. $3(1 - x) = 3 - 3x$

In Problems 39 through 41, answer each question. See Example 6.

39. Tom worked x hours at K-Mart last week and earned $4.50 an hour. How many hours did he work last week if he made $121.50?

40. Jerry bought s 29¢ stamps at a cost of $4.93. How many of these stamps did he buy?

41. Sharon has n nickels and $n + 8$ dimes. If she has $2.60, how many dimes does she have?

In Problems 42 through 44, answer each question. See Example 7.

42. The sum of three consecutive odd integers is 51. If n is the smallest of the integers, find the three integers.

43. The smallest of four consecutive integers is s. If the sum of the smallest one and three times the third one equals the fourth one, find the integers.

44. The smallest of three consecutive even integers is n. If twice the first integer plus three times the second is 82 more than the third integer, find the integers.

In Problems 45 through 48, answer each question. See Example 8.

45. Irma deposited d dollars in her savings account at 6% for one year. If she earned \$34.20 in interest, how much did she deposit? (*Hint:* Interest = principal · rate · time.)

46. Lucille drove 385 mi into the country at a speed of r miles per hour. If the trip took 7 h, find her speed. (*Hint:* Distance = rate · time.)

47. Samson rode his bike for h hours at a speed of 12.5 miles per hour. If he traveled a distance of 100 mi, find how long he rode his bike. (*Hint:* Distance = rate · time.)

48. The volume of the following box is 300 m³. Find its height. (*Hint:* Volume = length · width · height.)

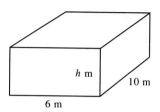

In Problems 49 through 52, answer each question. See Example 9.

49. Fifteen percent of a number q is 75. What is the number?

50. Twenty-five percent of a number x is 36. What is the number?

51. A bottle contains x gallons of a 15% acid solution. If there are 3 gal of acid in the bottle, how many gallons of the solution are in the bottle?

52. A sugar solution contains 4.9 lb of sugar. If the solution is 35% sugar and there are x liters of the sugar solution, how many liters of sugar solution are there?

Practice Exercises

In Problems 53 through 99, solve the following equations.

53. $2(x + 1) = 6$

54. $3(x - 2) = 12$

55. $\dfrac{2x}{3} + \dfrac{1}{3} = \dfrac{5}{3}$

56. $\dfrac{5}{3}x - \dfrac{5}{3} = \dfrac{4}{3}$

57. $3(9 - x) = 18$

58. $(5 - y)7 = -14$

59. $-2(3x - 2) = 10$

60. $-3(5 + 2x) = 7$

61. $\dfrac{5 - 4x}{7} = -1$

62. $\dfrac{2 - x}{8} = \dfrac{1}{2}$

63. $95 - 5(13 - 3x) = 0$

64. $5(4 - 5x) + 5 = 0$

65. $\dfrac{1}{2}(s + 3) = 1$

66. $\dfrac{1}{3}(x - 2) = 1$

67. $3(2 + 3x) = 5$

68. $2(3 - 2x) = 1$

69. $-2(3x + 2) = 10 - 6x$

70. $-3(5 - 2x) + 24 = 9 + 6x$

71. $(3x + 2)2 = 7x - 6$

72. $3(3 - k) = 2(k + 4) - 5k$

73. $7(3x + 4) - 2(14 - 3x) = 54$

74. $6 - 2(3x - 2) = 4 - 2(3 + x)$

75. $6 - 2(3x + 2) = 8 - 2(3 + 3x)$

76. $3(5 - x) + 5(2x - 3) = 3(2x - 1)$

77. $2[3(4 - 3x) - 2x] - 4x = -x - 1$

78. $\dfrac{x}{3} - \dfrac{1}{2} = \dfrac{3}{2}$

79. $\dfrac{1}{4} - \dfrac{x}{2} = \dfrac{1}{8}$

80. $\dfrac{2}{3}x - \dfrac{3}{5} = \dfrac{1}{15}$

81. $\dfrac{3x}{2} + \dfrac{1}{6} = \dfrac{5}{3}$

82. $\dfrac{x}{2} - \dfrac{1}{2} = \dfrac{3}{2}$

83. $3.3t = -1.32$

84. $6(3 - z) = 18$

85. $\dfrac{1 - 3x}{4} = 1$

86. $\dfrac{3 - 7x}{2} = 3$

87. $\dfrac{2x}{3} + \dfrac{3}{5} = -\dfrac{1}{15}$

88. $\dfrac{1}{7} + \dfrac{5w}{3} = \dfrac{8}{21}$

89. $-2.7 = 0.3x$

90. $\dfrac{3}{2}t = t + \dfrac{5}{2}$

91. $\dfrac{x}{2} = \dfrac{x}{3} - 1$

92. $\dfrac{x}{3} + \dfrac{3}{4} = \dfrac{2x}{3}$

93. $\dfrac{1}{3}x + \dfrac{1}{4} = \dfrac{1}{2}x$

94. $\dfrac{x + 3}{3} = \dfrac{x}{6}$

95. $\dfrac{1}{3}(y - 2) = \dfrac{2}{3} - y + \dfrac{4}{3}(y - 1)$

96. $\dfrac{2}{3}(2 - x) = 4 - \dfrac{2}{3}(x + 1)$

97. $\dfrac{1}{2}(x + 5) = \dfrac{3}{2} - x$

98. $3[2(4 + 3y) - 3y] - 5y = 4y + 1$

99. $2(x - 1) - 3x = x - 2(1 + x)$

In Problems 100 through 110, answer each question.

100. James has q quarters and twice that many dimes. If he has $4.05, how many dimes does he have?

101. Thirty percent of s is 144. What is s?

102. The volume of the following cylinder is 1440π in.³ If the radius of the base is 12 in., find the height. (*Hint:* Volume $= \pi \cdot$ (radius)² $\cdot$ height).

103. Ray borrowed x dollars at 14% for one year. If he paid $105 in interest, how much did he borrow? (*Hint:* Interest $=$ principal $\cdot$ rate $\cdot$ time.)

104. Lou bought y apples at 49¢ per apple and she bought twice that many pears at 69¢ per pear. If she spent $9.35, how many apples did she buy?

105. Leonard has a box full of $5 and $10 bills worth $140. If he has t $5 bills and 4 less than t $10 bills, how many $10 bills does he have?

106. Jane walked x miles in the park. If she was gone $3\dfrac{1}{2}$ hr and she walked at a rate of 2 mph, how far did she walk? (*Hint:* Distance $=$ rate $\cdot$ time.)

107. The volume of the following cone is 80π ft³. Find its height. (*Hint:* Volume $= \dfrac{1}{3}\pi \cdot$ (radius)² $\cdot$ height.)

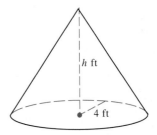

108. The smallest of three consecutive odd integers is n. If seven added to the middle integer is the same as the sum of the first and last integers decreased by 10, what are the three integers?

109. Jake bought t tapes at $7.99 each and spent $63.92. How many tapes did he buy?

110. Robert worked y hours at Wendy's during the week and $y + 7$ hours during the weekend. He gets paid $3.85 per hour during the week and makes $.45 per hour more on the weekends. If he earned $136.05 for the week, how many hours did he work during the week?

Challenge Problems

In Problems 111 and 112, solve the equation. The variable is x and q is a constant.

111. $3(x - q) = q$

112. $\dfrac{x - q}{4} = q$

In Your Own Words

113. Explain the procedure for solving a linear equation.

3.3 Literal Equations and Formulas

TAPE 6

Some equations contain more than one variable. For example,

$$x + y = 1$$

Notice that if we subtract y from both sides, we can write the equation in another form.

$$x + y \;-\; y \;=\; 1 \;-\; y \qquad \text{Subtract } y.$$
$$x = 1 - y$$

We say we have solved the equation for x. Similarly, we can subtract x from both sides of the original equation and solve it for y.

$$x + y \;-\; x \;=\; 1 \;-\; x \qquad \text{Subtract } x.$$
$$y = 1 - x$$

We have solved for y.

We sometimes call equations with more than one variable **literal equations.** When solving a literal equation for one of its variables, it's *important* to use the *same* tools we used in solving linear equations. That is, add the same number to both sides, or multiply both sides by the same nonzero number.

Formulas are literal equations that express the relationship between two or more physical conditions. For example, the area of a circle is given by πr^2, where r is the radius of the circle. We write

$$A = \pi r^2$$

Two other examples are the perimeter of a rectangle and the distance–rate–time formula.

Perimeter: $P = 2l + 2w$ Where l is length and w is width.

Distance: $d = rt$ Distance is rate times time.

Sometimes a formula would be more convenient to use if it were solved for a different variable.

► EXAMPLE 1

Solve each formula for the variable indicated.

(a) $d = rt$ for r (b) $P = 2l + 2w$ for l

Solutions

(a) We are to solve $d = rt$ for r. Notice that if we divide the right side by t we will solve for r. Divide both sides by t.

$$\frac{d}{t} = \frac{rt}{t}$$

$$\frac{d}{t} = r$$

Solution sets are not used when solving literal equations. Rather, the equation (or formula) is usually written with the variable being solved for on the left side.

$$r = \frac{d}{t}$$

(b) Solve $P = 2l + 2w$ for l.

We subtract $2w$ from both sides.

$$P - 2w = 2l + 2w - 2w$$

$$P - 2w = 2l$$

Now we divide both sides by 2.

$$\frac{P - 2w}{2} = \frac{2l}{2}$$

$$\frac{P - 2w}{2} = l$$

We then write the answer.

$$l = \frac{P - 2w}{2} \qquad \blacktriangleleft$$

A Procedure for Solving Literal Equations

1. Clear the equation of fractions, if necessary.
2. Remove all symbols of grouping and combine like terms where possible.
3. Collect all terms containing the variable for which the equation is being solved on one side of the equation and all other terms on the other side of the equation.
4. Combine like terms.
5. Divide both sides of the equation by the coefficient of the variable for which the equation is being solved.

It may be convenient to interchange steps 1 and 2.

▶ **EXAMPLE 2**

The formula for the area of a triangle is one-half the base times the height. Solve this literal equation for height.

Solution

We are to solve $A = \frac{1}{2}bh$ for h.

STEP 1   Clear fractions.

First, we clear fractions by multiplying both sides by 2.

$$2 \cdot A = 2 \cdot \left(\frac{1}{2}bh\right) \qquad \text{Multiply by 2}$$

STEP 2   Remove grouping symbols.

$$2A = \left(2 \cdot \frac{1}{2}\right) \cdot bh \qquad \text{Associative Property}$$

$$2A = 1 \cdot bh$$

$$2A = bh$$

STEPS 3 AND 4. Collect and combine like terms.

These steps are unnecessary in this problem.

STEP 5   Divide both sides by the coefficient of the variable for which we are solving.

Divide both sides by b.

$$\frac{2A}{b} = \frac{bh}{b}$$

$$\frac{2A}{b} = h$$

Thus the formula becomes $h = \dfrac{2A}{b}$ when solved for h.   ◄

Often literal equations are simply equations containing two or more letters.

► **EXAMPLE 3**

Solve the following literal equations for the variable indicated.

(a) $\dfrac{1}{2}x + s = 3t$ for x (b) $4(p + y) = 3$ for y

(c) $\dfrac{q}{3}(w - 11b) = 4$ for w

Solutions

(a) We have a fraction in this equation, which we can clear by multiplying both sides by 2.

$$\frac{1}{2}x + s = 3t$$

$$2 \cdot \left(\frac{1}{2}x + s\right) = 2 \cdot (3t) \qquad \text{Clear fractions}$$

We use the distributive property and the associative property for multiplication to simplify this equation.

$$2 \cdot \left(\frac{1}{2}x\right) + 2 \cdot s = 2 \cdot (3t) \qquad \text{Distributive Property}$$

$$\left(2 \cdot \frac{1}{2}\right)x + 2s = (2 \cdot 3)t \qquad \text{Associative Property}$$

$$(1)x + 2s = (6)t$$

$$x + 2s = 6t$$

Now that we have cleared the fractions, we can subtract $2s$ from both sides to solve for x.

$$x + 2s - 2s = 6t - 2s \qquad \text{Subtract } 2s$$

$$x = 6t - 2s$$

(b) There are no fractions in this equation, but we have a grouping symbol. We use the distributive property to remove the parentheses.

$$4(p + y) = 3$$

$$4p + 4y = 3 \qquad \text{Distributive Property}$$

$$4p + 4y - 4p = 3 - 4p \qquad \text{Subtract } 4p$$

$$4y = 3 - 4p \qquad \text{Combine like terms}$$

$$\frac{4y}{4} = \frac{3 - 4p}{4} \qquad \text{Divide by 4}$$

$$y = \frac{3 - 4p}{4}$$

(c)
$$\frac{q}{3}(w - 11b) = 4$$

$$3 \cdot \left[\frac{q}{3}(w - 11b) \right] = 3 \cdot 4 \qquad \text{Multiply by 3}$$

$$\left(3 \cdot \frac{q}{3} \right)(w - 11b) = 3 \cdot 4 \qquad \text{Associative Property}$$

$$q(w - 11b) = 12$$

$$qw - 11bq = 12 \qquad \text{Distributive Property}$$

$$qw - 11bq + 11bq = 12 + 11bq \qquad \text{Add } 11bq$$

$$qw = 12 + 11bq \qquad \text{Combine like terms}$$

$$w = \frac{12 + 11bq}{q} \qquad \text{Divide by } q \qquad \blacktriangleleft$$

► **EXAMPLE 4**

Solve $-3p = w$ for p.

Solution

We divide both sides by -3.

$$\frac{-3p}{-3} = \frac{w}{-3}$$

$$p = \frac{w}{-3}$$

The fraction $\dfrac{w}{-3}$ can also be written as $\dfrac{-w}{3}$ or as $-\dfrac{w}{3}$. We usually write the negative sign in the numerator or in front of the fraction.

$$p = -\frac{w}{3}$$

◀

Problem Set 3.3

Warm-Ups

In Problems 1 through 7, solve the formula for the variables indicated. See Examples 1 and 2.

1. $A = lw$ (area of rectangle); solve for w.

2. $A = \dfrac{1}{2}bh$ (area of a triangle); solve for b.

3. $C = 2\pi r$ (circumference of a circle); solve for r.

4. $I = prt$ (simple interest); solve for t.

5. $V = \dfrac{1}{3}\pi r^2 h$ (volume of a cone); solve for h.

6. $E = mc^2$ (mass–energy formula); solve for m.

7. $F = \dfrac{9}{5}C + 32$ (Celsius to Fahrenheit); solve for C.

In Problems 8 through 27, solve each for the variable indicated. See Examples 3 and 4.

8. $x + y = 1$ for x

10. $x + y = a$ for x

12. $2x = q$ for x

14. $ax + 2y = k$ for y

16. $ax + by = z$ for y

18. $\dfrac{2x}{b} + C = D$ for x

20. $\dfrac{bz}{C} - s = 14$ for z

22. $2(x + a) = 5$ for x

24. $\dfrac{1}{g}(3 + w) = u$ for w

26. $\dfrac{2}{b}(x + y) = 1$ for y

9. $x + y = 1$ for y

11. $x - y = a$ for x

13. $-3x = p$ for x

15. $2k - 2y = 6k$ for y

17. $3z - by = 4z$ for y

19. $\dfrac{ay}{4} - L = f$ for y

21. $q - \dfrac{aw}{k} = 3q$ for w

23. $3(y - q) = 1$ for y

25. $6 = \dfrac{1}{p}(a + x)$ for x

27. $\dfrac{a}{b}(k + w) = 2L$ for w

Practice Exercises

In Problems 28 through 31, solve the formulas for the variables indicated.

28. $A = lh$ (area of a parallelogram); solve for l.

29. $F = \dfrac{mv^2}{r}$ (centripetal force); solve for m.

30. $V = \dfrac{1}{3}Ah$ (volume of a pyramid); solve for A.

31. $C = \dfrac{5}{9}(F - 32)$ (Fahrenheit to Celsius); solve for F.

In Problems 32 through 71, solve each for the variable indicated.

32. $x + 2z = 5$ for x

34. $x + 3z = k$ for x

36. $3x = p$ for x

33. $2x + z = 5$ for z

35. $x - 3z = k$ for x

37. $-2x = q$ for x

38. $-qx = 4$ for x

39. $px = -7$ for x

40. $kx = a$ for x

41. $kx + a = 0$ for x

42. $3h - y = h$ for y

43. $5x + y = h$ for y

44. $2k - 3y = 5k$ for y

45. $ax + 4y = k$ for y

46. $ax + by = z$ for x

47. $ax - by = z$ for y

48. $\dfrac{x}{5} + \dfrac{b}{5} = \dfrac{1}{5}$ for x

49. $\dfrac{x}{2} - \dfrac{k}{2} = 3$ for x

50. $\dfrac{q}{3} + \dfrac{y}{3} = p$ for y

51. $\dfrac{w}{4} - \dfrac{y}{4} = w$ for y

52. $\dfrac{z}{7} + a = J$ for z

53. $\dfrac{z}{p} - q = T$ for z

54. $\dfrac{3x}{k} + M = L$ for x

55. $\dfrac{qy}{-2} - L = f$ for y

56. $\dfrac{bz}{2v} - s = 11$ for z

57. $t + \dfrac{aw}{k^2} = 0$ for w

58. $3(x + s) = 4$ for x

59. $2(y - p) = 1$ for y

60. $5(y + 2z) = p$ for z

61. $p(x - k) = -2$ for x

62. $j = q(s - 4)$ for s

63. $A(3x - B) = C$ for x

64. $\dfrac{1}{3}(5 + x) = C$ for x

65. $\dfrac{1}{2}(A - s) = A$ for s

66. $k^2 = \dfrac{1}{5}(x - y)$ for x

67. $\dfrac{2}{3}(x + d) = r$ for x

68. $\dfrac{1}{C}(8 + w) = q$ for w

69. $5 = \dfrac{1}{r}(3r - x)$ for x

70. $\dfrac{2}{r}(x^2 + y) = 1$ for y

71. $\dfrac{c}{d}(k + w) = 5L$ for w

72. If C is the cost of an item and M is the markup, the selling price of the item S is given by the formula $S = \dfrac{C}{1 - M}$. Solve for M.

73. Suppose a store sells certain items at discount. The percentage discount D is given by the formula $D = \dfrac{100(P - S)}{P}$, where P is the original price and S is the sale price. Solve for S.

Challenge Problems

Many times in applications, we would like to use the same letter to name different variables. For example, suppose we have two different temperatures. We use subscripts to name the different temperatures. We might call one of them t_1 and the other one t_2. Each of these is a different variable. We read t_1 as "t sub-one."

74. $v = -32t + v_0$ (velocity in free fall); solve for t. (v_0 represents initial velocity.)

75. $\dfrac{P_1 V_1}{T_1} = \dfrac{P_2 V_2}{T_2}$ (ideal gas laws); solve for P_2 (The variables represent different pressures, volumes, and temperatures.)

76. $F = G\dfrac{m_1 m_2}{R^2}$ (gravitational force); solve for m_1. (m_1 and m_2 represent different masses.)

In Your Own Words

77. What is a literal equation?

3.4 Linear Equations as Models

TAPE 6

We do word problems because they are an imitation of life. They are descriptions of the various applications of mathematics in the real world.

To be successful with word problems we must resist the temptation to rush in and puzzle out the answer all at once. Instead, we must organize our efforts to solve the problem one piece at a time. We must also avoid the temptation to solve the problem by trial and error. This only works for the simplest problems and does not provide a method for the more complicated cases.

► EXAMPLE 1

John wishes to cut 17 ft of cable into two pieces such that one piece is 5 ft longer than the other. How long should each piece be?

Solution

First, we read the problem to see its general nature and to find out what it is we want to know. When we reread this problem, we see it is a relationship between the lengths of two pieces of cable and we are to find the *length of each piece.* This is one of the most important steps in solving a word problem. We let some letter (often *x*) stand for one of the things we are looking for. *We make this assignment in writing.*

Let *x* be the length of the shorter piece (in feet).

Next, we write an expression for the length of the other piece *in terms of x.* It is 5 feet longer than the piece whose length is *x* feet, so we can write

x + 5 is the length of the longer piece.

A picture, figure, or table is often useful in seeing the relationships.

$$\left|\!\!\leftarrow\!\!-\!\!-\!\!-\!\!-\!\!-\!\!-17\text{ ft}-\!\!-\!\!-\!\!-\!\!-\!\!\right|$$
$$\left|\!\!\leftarrow\!\!-x\text{ ft}\!\!\rightarrow\!\!\right|\!\!\leftarrow\!\!-\!\!-\!\!-x+5\text{ ft}\!\!-\!\!\right|$$

Now that we can visualize the lengths of the two pieces of cable to work with (they are *x* feet and *x* + 5 feet), we reread the problem to see the relationship between them. From the problem we see that the lengths must add up to 17 ft. The picture also makes this clear. We write this relationship as an equation.

Length of shorter + length of longer = 17

$$x \qquad + \qquad x + 5 \qquad = 17$$

This equation is called a **mathematical model** of the problem. Finding it is usually the hardest part of a word problem. Next, we solve the equation.

$$x + x + 5 = 17$$
$$2x + 5 = 17 \qquad \text{Collect like terms}$$
$$2x + 5 - 5 = 17 - 5 \qquad \text{Subtract 5}$$
$$2x = 12 \qquad \text{Collect like terms}$$
$$x = 6 \qquad \text{Divide by 2}$$

Therefore, the length of the shorter piece is 6 ft. The longer piece is 5 ft more than the shorter piece, so it must be 11 ft long. We were asked a question in writing, so we give a written reply.

The lengths of the pieces are 6 ft and 11 ft.

We should check our solution. We check word problems by taking our written solution (the lengths of the pieces are 6 ft and 11 ft) back to the *original* problem statement and make sure it does two things:

1. Makes sense.

2. Answers the question.

Since $6 + 11$ is 17, our written solution both answers the question and makes sense.   ◄

Notice that we *did not* stop at the point where we wrote

$$x = 6$$

The problem asked a question, and we must provide a written answer. We now give a general procedure for solving word problems.

A Procedure to Solve Word Problems

1. Read the problem and determine what quantities are to be found.

2. Assign a variable, such as x, to represent one of the quantities to be found.

3. Express all other quantities to be found in terms of the chosen variable.

4. Draw a picture or figure to illustrate the problem, if possible. Charts or tables can also be used.

5. Reread the problem and form an equation.

6. Solve the equation.

7. Reread the problem and prepare a *written* answer to the question(s) asked.

8. Check the answer in the original problem statement. It should make sense and answer the question.

► **EXAMPLE 2**

The length of a rectangle is 5 m more than its width. If the perimeter is 78 m, find the dimensions of the rectangle.

Solution

STEP 1   Determine what is to be found.
When we read the problem, we see there are two things to find, the length and the width of the rectangle. However, as the length is 5 m more than the width, we can make the following assignments.

STEP 2   Assign a variable to one quantity to be found.
Let x be the width of the rectangle (meters).

STEP 3   Express other quantities in terms of x.
Then $x + 5$ is the length of the rectangle.

STEP 4   Draw a picture.

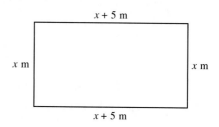

STEP 5   Reread the problem and form a mathematical model.
From the problem statement we see that the perimeter of the rectangle is 78 meters and from the picture we see it is 2 times x, plus 2 times $(x + 5)$. Therefore, the model is

$$2 \cdot x + 2 \cdot (x + 5) = 78$$

STEP 6   Solve the equation.

$$2 \cdot x + 2 \cdot (x + 5) = 78$$

$2x + 2x + 10 = 78$	Distributive Property
$4x + 10 = 78$	Collect like terms
$4x = 68$	Subtract 10
$x = 17$	Divide by 4

Therefore, the width is 17 m (see step 2) and the length is 22 m (step 3).

STEP 7   Answer the question.
The dimensions of the rectangle are 17 m by 22 m.

STEP 8   Check in the *original* problem statement.
The solution certainly makes sense. And because 22 is 5 more than 17 and $2 \cdot 17 + 2 \cdot 22 = 78$, these dimensions answer the question. ◄

► EXAMPLE 3

Ida is four times as old as Kathy. In 5 yr the sum of their ages will be 35 yr. How old is each person now?

Solution

STEP 1   Determine what is to be found.

We are asked to find Ida's age and Kathy's age now.

STEP 2   Assign a variable to one quantity to be found.

Let x be Kathy's age now.

STEP 3   Express other quantities in terms of x.

Then $4x$ is Ida's age now.

STEP 4   Make a table.

When solving age problems, it is best to write down statements giving ages at other times mentioned in the problem. For example, in this problem we are given the sum of their ages 5 years *from now*. A table is convenient for recording this information.

	Now	5 Yr from Now
Kathy's age	x	$x + 5$
Ida's age	$4x$	$4x + 5$

STEP 5   Reread the problem and form a mathematical model.

Now we reread the problem and see that the sum of their ages five years from now is 35 years. Or

$$(x + 5) + (4x + 5) = 35$$

STEP 6   Solve the equation.

$$(x + 5) + (4x + 5) = 35$$
$$x + 5 + 4x + 5 = 35$$

$5x + 10 = 35$ Collect like terms

$5x = 25$ Subtract 10

$x = 5$ Divide by 5

Thus Kathy's age now must be 5 and Ida's is $4 \cdot 5 = 20$.

STEP 7   Answer the question.

Kathy is 5 and Ida is 20.

STEP 8   Check in the *original* problem statement.

This answers the question and, because $(20 + 5) + (5 + 5)$ is 35, it makes sense. ◀

► EXAMPLE 4

Wan has 20 coins in his pocket. They are all nickels and dimes and their total value is $1.35. How many of each type coin does Wan have?

Solution

Determine what is to be found.

This is a problem concerning the value of a collection. We need to find out how many nickels and dimes Wan has in his pocket.

Assign a variable to a quantity to be found.

Let x be the number of nickels in Wan's pocket.

Express other quantities in terms of x.

We also need to find the number of dimes in Wan's pocket. We know that there are a total of 20 coins in his pocket and x of them are nickels. Therefore, 20 less x must be the number of dimes since there is no other type of coins in his pocket.

$20 - x$ is the number of dimes in Wan's pocket.

Make a table.

- The **value** of x nickels is $5x$ cents.
- The **value** of $20 - x$ dimes is $10(20 - x)$ cents.
- The **value** of the collection is $1.35 or 135 cents.

	Nickels	Dimes	Total
Number of coins	x	$20 - x$	20
Their value (in cents)	$5x$	$10(20 - x)$	135

Form a mathematical model.

Value of nickels + value of dimes = total value

$$5x \quad + \quad 10(20 - x) \quad = 135$$

Solve the equation.

$$5x + 200 - 10x = 135$$
$$200 - 5x = 135 \qquad \text{Collect like terms}$$
$$200 - 5x - 200 = 135 - 200 \qquad \text{Subtract 200}$$
$$-5x = -65 \qquad \text{Collect like terms}$$
$$\frac{-5x}{-5} = \frac{-65}{-5} \qquad \text{Divide by } -5$$
$$x = 13$$

There are 13 nickels in Wan's pocket. Thus there must be $20 - 13$, or 7, dimes.

Answer the question.

Wan has 13 nickels and 7 dimes in his pocket.

Check in the original word problem.

Since 13 nickels have a value of 65¢ and 7 dimes are worth 70¢, the value of the nickels plus the value of the dimes is 65¢ plus 70¢, or $1.35. Our answer makes sense and answers the question.   ◄

► EXAMPLE 5

How many gallons of a 60% alcohol solution should be added to 30 gal of a 10% alcohol solution to make a 20% alcohol solution?

Solution

Determine what is to be found.

We are asked for the *number of gallons* of the 60% solution to be added.

Assign a variable.

Let x be the number of gallons of the 60% solution to be added.

Draw a figure.

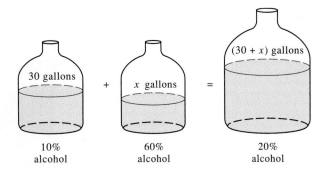

A figure often helps in a mixture problem. We draw three containers: the original 30 gal of 10% solution, the x gallons of 60% solution, and the result of mixing them together.

Form a mathematical model.

We always change percents to decimals or fractions before we do arithmetic. Now, 30 gal of a 10% alcohol solution would contain $30 \cdot \frac{10}{100}$ gallons of alcohol, whereas x gallons of a 60% solution would contain $x \cdot \frac{60}{100}$ gallons of alcohol. After mixing, we have $(30 + x)$ gallons of mixture that is 20% alcohol, or $(30 + x) \cdot \frac{20}{100}$ gallons of alcohol. But the amount of alcohol present before mixing is the same as the amount present after mixing, so we have the model

$$30 \cdot \frac{10}{100} + x \cdot \frac{60}{100} = (30 + x) \cdot \frac{20}{100}$$

Solve the equation.

$$30 \cdot \frac{1}{10} + \frac{3}{5} \cdot x = \frac{1}{5} \cdot (30 + x) \qquad \text{Reduce fractions}$$

$$3 + \frac{3}{5}x = \frac{1}{5} \cdot (30 + x)$$

$$5 \cdot \left(3 + \frac{3}{5}x\right) = 5 \cdot \left[\frac{1}{5}(30 + x)\right] \qquad \begin{array}{l}\text{Multiply by 5 (to clear}\\ \text{fractions)}\end{array}$$

$$15 + 3x = 30 + x \qquad \begin{array}{l}\text{Distributive and}\\ \text{Associative Properties}\end{array}$$

$$2x = 15$$

$$x = \frac{15}{2} = 7.5$$

Answer the question.

We should add 7.5 gal of the 60% solution.

Check in the original problem.

This amount makes sense and answers the question.   ◄

▶ **EXAMPLE 6**

Two trains leave Atlanta at noon, traveling in opposite directions. If one train travels at a rate of 55 mph and the other at a rate of 60 mph, what time will it be when they are 414 mi apart?

Solution

Determine what is to be found.

When we read the problem, we see that we must find the number of hours it will take until the trains are 414 mi apart.

Assign a variable.

Let x be the number of hours it takes until the trains are 414 mi apart.

Draw a diagram and make a table.

Distance–rate–time problems are usually easier with a figure illustrating the problem. When we draw the figure, we *look for two parts.*

```
         ┌──────────────────────────────┐
      ◄──────Train 1──────►|◄──────Train
      ◄─────────────── 414 mi ───────────►
```

The two parts are the train 1 portion and the train 2 portion. In distance–rate–time problems we use the basic formula *distance = rate × time* and the two related formulas we get by solving for rate and time.

$$d = rt \qquad \text{Distance}$$

$$r = \frac{d}{t} \qquad \text{Rate}$$

$$t = \frac{d}{r} \qquad \text{Time}$$

The best approach for these problems is to identify two parts, as shown above, and then find d, r, and t for both parts. One of the three will come from the assignment statement(s) and another one should be found in the problem statement. The third can then be calculated from one of the above formulas.

For example, in this problem each train traveled x hours (time for both parts came from the assignment statement), and the rates for each train are given in the problem statement. We enter this information in a table. Remember, the train left Atlanta at noon.

	Distance	Rate	Time
Train 1		60 mph	x hours
Train 2		55 mph	x hours

We now use one of our formulas to find d for each part. The appropriate formula is $d = rt$. Inserting this calculation, we get

	Distance	Rate	Time
Train 1	$60 \cdot x$	60 mph	x hours
Train 2	$55 \cdot x$	55 mph	x hours

Form an equation.

After completely filling in d, r, and t for both parts, we reread the problem. Clearly, the distance train 1 travels plus the distance train 2 travels must be 414 miles.

$$60 \cdot x + 55 \cdot x = 414$$

Solve the equation and answer the question.

$$115x = 414$$

$$x = \frac{414}{115}$$

$$x = 3.6$$

Because the trains are 414 miles apart after 3.6 h and 0.6 hours is 36 min, we answer the question as follows:

The trains will be 414 miles apart at 3:36 P.M.

This statement makes sense and answers the question.   ◄

Problem Set 3.4

Warm-Ups

The first four word problems deal with relationships between numbers. See Example 1.

1. The sum of 18 and a number is 52. What is the number?

2. Find two consecutive even integers whose sum is 126.

3. Margaret accidentally runs over a 16-ft garden hose while mowing the lawn. If she cuts the hose into two pieces, one of which is 4 ft longer than the other, what are the lengths of the two pieces?

4. Nipa cuts of 42-in. piece of material into two pieces, one of which is twice as long as the other. What are the lengths of the two pieces?

Problems 5 through 8 are concerned with perimeters of geometric figures. See Example 2.

5. One section of a patio deck is rectangular, with a length of 2 m more than its width. If the perimeter is 40 m, what are the dimensions of this section?

6. The width of a rectangle is 5 cm less than its length. What are its dimensions if its perimeter is 98 centimeters?

7. The perimeter of a rectangle is 136 yd. If the width is one-third the length, what are its dimensions?

8. The length of a rectangle is 7 m less than four times the width. If the perimeter is 126 m, what are its dimensions?

Problems 9 through 12 are age problems. See Example 3.

9. Homer is twice as old as Pete. If the sum of their ages is 96 years, how old are Homer and Pete?

10. John is 3 years older than Sam. In ten years, the sum of their ages will be 69 years. How old is each now?

11. Ellen is 4 years younger than Sarah. In 5 years, the sum of their ages will be 74 years. How old is each now?

12. Bill is 30 years older than Traci. In 10 years, he will be twice her age then. How old are Bill and Traci now?

Problems 13 through 16 are value problems. See Example 4.

13. June has $1.70 in nickels and dimes. If she has two more dimes than nickels, how many of each kind of coin does she have?

14. Tommy has 30 bills worth $200. If he has only fives and tens, how many of each kind does he have?

15. A collection of coins consists of 54 nickels and pennies. If the value of the collection is $1.50, how many of each type of coin are there in the collection?

16. Tickets to the ALTA Tennis Championships sold for $5.75 for adults and $2.50 for children under 12. If 450 tickets were sold and $2363.25 was collected, how many adult tickets were sold?

Problems 17 through 20 deal with mixtures. See Example 5.

17. JoAnne needs a 20% brine solution for her pickles. How many liters of a 30% brine solution should she add to 2 L of a 5% brine solution to get a 20% solution?

18. Marion wants to dilute 3 gal of a 10% soap solution to make a 3% solution. How much pure water should she add to the 3 gal of 10% solution to make the 3% solution?

19. Crude oil worth $16 per barrel is to be mixed with crude worth $12 per barrel to make a mixture worth $15 a barrel. How many barrels of $16 crude should be mixed with 210 barrels of $12 crude?

20. Shirley wishes to mix 30 lb of cinnamon coffee worth $5 per pound with plain coffee worth $3 per pound in order to make a mixture worth $3.50 per pound. How many pounds of the plain coffee should she use?

Problems 21 through 24 are distance–rate–time problems. See Example 6.

21. Rowland and Martin are enjoying the lake on a still day. They leave buoy 16 in their boats traveling in opposite directions. Martin's speed is 40 mph and Rowland's is 50 mph. In how many hours will they be 225 mi apart?

22. The slow freight and the Midnight Express are 300 mi apart but are heading toward each other on the same railroad track. If the slow freight is traveling at 40 mph and the Midnight Express at 110 mph, how much time do we have to warn them of their impending collision?

23. Federico leaves Mack's Texaco and drives toward San Diego with an average speed of 40 mph. Augusto leaves Mack's 1 hr later and drives the same route at 60 mph. How long will it take Augusto to overtake Federico?

24. A jet flew at an average speed of 1000 km/h for a period of time and then reduced its speed to 700 km/h. If it flew at 700 km/h for 1 hr longer than it flew at 1000 km/h and it flew a total a distance of 4100 km, how long did it fly at 1000 km/h?

Practice Exercises

Solve the following word problems.

25. Find two consecutive odd integers whose sum is 140.

26. One number is 7 more than another number. The sum of the two numbers is 51. What are the numbers?

27. The length of a rectangle is 3 m more than its width. If its perimeter is 102 m, what are its dimensions?

28. The width of a rectangle is 11 cm less than its length. What are its dimensions if its perimeter is 106 cm?

29. Kitty is half as old as Charlie. If the sum of their ages is 48 years, how old are Kitty and Charlie?

30. Ban is 6 years older than Rieko. Ten years ago he was twice her age then. How old is each now?

31. Betty has $2.75 in nickels and dimes. If she has twice as many dimes as nickels, how many of each type of coin does she have?

32. Sandi has $2.25 in nickels and quarters. If she has ten times as many nickels as quarters, how many of each type of coin does she have?

33. The sum of two consecutive integers is 69. Find the integers.

34. The degree measures of the angles in a triangle are three consecutive even integers. Find the number of degrees in each angle.

35. One of two supplementary angles is four times the other. Find the angles.

36. The perimeter of a triangle is 47 in. The length of one side is 7 in. more than the length of the shortest side, and the length of the other side is twice the length of the shortest side. What are the lengths of the three sides?

37. Tom is four times as old as Kimberly. Four years ago, Tom's age was 10 times Kimberly's age then. How old is each now?

38. Sondra is 3 yr older than Jim. One year ago, Sondra's age was four times Jim's age then. How old is each now?

39. A meter attendant emptied nickels, dimes, and pennies from a parking meter. There were 10 fewer dimes than nickels and 95 more pennies than nickels. If the total collected was $4.75, how many of each type of coin were in the meter?

40. Kaled has $2.25 in nickels, dimes, and quarters. If he has twice as many dimes as quarters and three times as many nickels as dimes, how many of each type of coin does he have?

41. Pecans costing $3.50 per pound are mixed with 20 lb of pecans costing $1.70 per pound to make a mixture costing $2.30 per pound. How many pounds of the $3.50 pecans are required?

42. Kenya coffee beans selling for $8.75 per pound are to be mixed with 10 lb of Blue Mountain beans, which sell for $12.50 per pound, to make a mixture that sells for $10.00 per pound. How many pounds of the Kenya beans are required?

43. A truck and a bus leave Lansing at the same time, both traveling south. If the bus averages 60 mph and the truck averages 45 mph, how long will it be until they are 50 mi apart?

44. Two autos are racing on the Indianapolis 2.5-mi oval track. If they start together and one auto averages 150 mph while the other averages 155 mph, how long will it take for the faster car to lap the slower?

45. Jo needs a 15% brine solution. How many liters of a 20% brine solution should she add to 10 L of a 12% brine solution to get a 15% solution?

46. Joan wants to dilute 6 gal of a 5% soap solution to make a 3% solution. How much pure water should she add to the 6 gal of 5% solution to make the 3% solution?

47. Jack and Jim leave San Antonio in their cars traveling in opposite directions. Jack's speed is 62 mph and Jim's is 55 mph. In how many hours will they be 351 mi apart?

48. At noon two trains are 385 miles apart but are heading toward each other on the same railroad track. If the slow train is traveling at 50 mph and the faster train at 60 mph, at what time will they collide?

49. One of two complementary angles is 6 degrees less than three times the other angle. Find the angles.

50. The supplement of an angle is 20° more than three times its complement. Find the angle.

51. The perimeter of a rectangle is 300 yd. If the width is one-fourth the length, what are its dimensions?

52. The length of a rectangle is 4 m less than three times the width. If the perimeter is 96 m, what are its dimensions?

53. J.R. is 6 yr younger than Kim. Three years ago she was twice J.R.'s age. How old is each now?

54. Sue is 14 yr younger than Jan. Five years ago, Jan's age was three times Sue's age then. How old is each now?

55. Jerry has 40¢ in pennies and nickels. If he has 20 coins, how many of each type does he have?

56. Tickets to the MASDA Charity Cake Walk were $1.25 for adults and 75¢ for children. If $1050 was collected and three times as many children as adults bought tickets, how many adults bought tickets?

57. Fine oil worth $35 per gallon is to be mixed with oil worth $20 per gallon to make a mixture worth $25 a gallon. How many gallons of $35 oil should be mixed with 50 gal of $20 oil?

58. Nguyen wishes to mix 30 lb of nuts worth $4 a pound with nuts worth $2 a pound in order to make a mixture worth $3.50 a pound. How many pounds of the $2 nuts should he use?

59. Barney leaves Minneapolis and drives toward St. Cloud with an average speed of 50 mph. Gerald leaves Minneapolis $\frac{1}{2}$ hr later and drives the same route at 60 mph. How long will it take Gerald to overtake Barney?

60. An airplane flew at an average speed of 600 km/h for a period of time and then reduced its speed to 400 km/h. If it flew at 400 km/h for 2 hr longer than it flew at 600 km/h and it flew a total distance of 2800 km, how long did it fly at 600 km/h?

61. One number is 5 more than twice a certain number. If their sum is 134, what is the certain number?

62. The length of Rika's camper is 8 ft less than twice the length of Kim's camper. If together they are 40 ft long, what is the length of each of their campers?

63. Find the length of the side of a square if its perimeter is 76 mm.

64. The perimeter of a triangle is 68 in. The length of one side is 8 in. more than the length of the shortest side, and the length of another side is twice the length of the shortest side. What are the lengths of the three sides?

65. Louise is three times as old as her daughter. In 7 yr the sum of their ages will be 70. How old is each now?

66. Dan's age is twice Jonathan's age now. Five years ago, Dan's age was seven times Jonathan's age then. How old is each now?

67. A toll box on the George Washington Bridge contains $450 in quarters, dimes, and nickels. If there are twice as many dimes as nickels and five times as many quarters as nickels, how many of each type of coin are in the box?

68. Amanda has $4.50 in nickels and dimes. If she has twice as many dimes as nickels, how many dimes does she have?

69. Cashew clusters costing $4.50 per pound are mixed with 30 lb of peanut clusters costing $2.20 per pound to make a mixture costing $4.00 per pound. How many pounds of the cashew clusters are required?

70. Amaretta coffee beans selling for $6.50 per pound are to be mixed with 10 lb of Kona beans that sell for $11.50 per pound to make a mixture that sells for $8.50 per pound. How many pounds of the $6.50 beans are required?

71. A dogsled and a snowmobile leave Calgary at the same time, both traveling south. If the snowmobile averages 40 mph and the dogsled averages 8 mph, how long will it be until they are 48 mi apart?

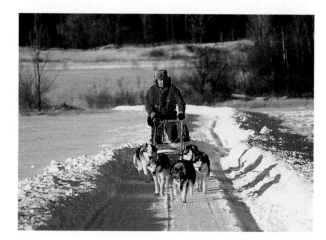

Problem 71

72. Two autos are racing on a 1.5-mi oval track. If they start together and one auto averages 175 mph while the other averages 160 mph, how long will it take for the faster to lap the slower?

Challenge Problems

73. Omar drove from home to a place in the country at an average rate of 50 mph and then returned at an average rate of 60 mph. If the total trip took 2 hr, how far is the place in the country from Omar's home?

74. Lois wants to make 10 L of 29% sugar syrup by mixing syrup containing 50% sugar and syrup containing 20% sugar. How much of each should she mix?

In Your Own Words

75. What is the procedure for solving a word problem?

76. Make up a word problem with an answer of 10 nickels.

3.5   Linear Inequalities

TAPE 7

Just as an equation is a statement that two numbers are equal, an **inequality** is a statement that one number is greater than another. We use the symbols $>$, $<$, $\geq$, and $\leq$.

$$p > q \qquad p \text{ is \textbf{greater than} } q$$

$$p < q \qquad p \text{ is \textbf{less than} } q$$

$$p \geq q \qquad p \text{ is \textbf{greater than or equal to} } q$$

$$p \leq q \qquad p \text{ is \textbf{less than or equal to} } q$$

Some *true* inequalities are

$$10 > 7$$
$$4 > -8$$
$$2 < 17$$
$$-5 < 3$$
$$6 \geq 1$$
$$6 \geq 6$$

Some *false* inequalities are

$$10 < 7$$
$$-7 > -1$$
$$4 \leq 3$$

Like equations, inequalities are interesting when they contain variables. For example, the inequality

$$x + 5 > 7$$

is *true* if x is 11 and is *false* if x is 1. In fact, this statement is *true* if x is *any* number greater than 2 and *false* if x is 2 or *any* number less than 2. The collection of numbers that make the inequality true is called the **solution set.** To **solve** an inequality is to find its solution set.

We cannot write the solution set for an inequality, such as the one just given, by listing its elements. We must use the set-builder notation introduced in Section 1.1. As the solution set is the collection of *all* numbers greater than 2, it can be written

$$\{x \mid x > 2\}$$

Inequalities have a **left side,** an inequality symbol, and a **right side.**

Inequality symbol

$$\downarrow$$

$$x + 5 > 7$$

$$\uparrow \qquad \uparrow$$

Left side   Right side

► **EXAMPLE 1**

For each inequality, determine which of the given numbers are in the solution set.

(a) $4 + 2x < 13$; $-5, 0, 5$ (b) $16 - 5x \le -4$; $-10, 4, 10$

Solutions

(a) If x is -5, then

$$4 + 2x = 4 + 2(-5)$$
$$= -6$$

Since $-6 < 13$ is *true*, -5 *is* in the solution set of $4 + 2x < 13$.
If x is 0, then

$$4 + 2x = 4 + 2(0)$$
$$= 4$$

Since $4 < 13$ is true, 0 *is* in the solution set of $4 + 2x < 13$.
If x is 5, then

$$4 + 2x = 4 + 2(5)$$
$$= 14$$

Because $14 < 13$ is *false*, 5 *is not* in the solution set of $4 + 2x < 13$.

(b) If x is -10, then

$$16 - 5x = 16 - 5(-10)$$
$$= 66$$

Since $66 \le -4$ is *false*, -10 is *not* in the solution set of $16 - 5x \le -4$.
If x is 4, then

$$16 - 5x = 16 - 5(4)$$
$$= -4$$

Because $-4 \le -4$ is *true* (-4 *is* less than or *or equal* to -4), 4 *is* in the solution set of $16 - 5x \le -4$.
If x is 10, then

$$16 - 5x = 16 - 5(10)$$
$$= -34$$

Because $-34 \le -4$ is *true*, 10 *is* in the solution set of $16 - 5x \le -4$.

◄

We can sometimes understand an idea much better if we can draw a picture of it. Such is the case with inequalities. We have seen that the solution set of the inequality $x < 1$ consists of all numbers less than 1 (that is, all numbers that lie to the left of 1 on a number line). We draw a picture of this set of numbers by shading a number line to the left of 1.

1

We indicate that the set continues indefinitely by an arrowhead. We indicate that the number 1 *does not* belong to the solution set with a *parenthesis*. This picture of the solution set is called the **graph** of the inequality.

We can also use an open circle to indicate that a number is not included.

To graph the inequality $x \geq -\frac{1}{2}$, we draw a number line and shade the numbers to the right of $-\frac{1}{2}$. We use a *bracket* to indicate that $-\frac{1}{2}$ *does* belong to the solution set.

We can also use a darkened circle to show that $-\frac{1}{2}$ is included.

► **EXAMPLE 2**

Graph the solution set for each inequality.

(a) $x > 0$ (b) $x \geq -2$ (c) $x < 4$ (d) $3 < x$

Solutions

(a) We shade a number line to the right of 0. The parenthesis indicates that 0 is not included.

$x > 0$:

(b) We shade a number line to the right of -2. The bracket shows that -2 is included.

$x \geq -2$:

(c) We shade a number line to the left of 4. The parenthesis indicates that 4 is not included.

$x < 4$:

(d) Here $3 < x$ means the same as $x > 3$. We shade a number line to the right of 3. The parenthesis shows that 3 is not included.

$3 < x$:

Notice that we wrote the inequality with the variable on the left side. This makes it easier to graph.   ◄

Linear inequalities in one variable contain a variable to the first power only. We can solve linear inequalities in a manner very similar to the method we used to solve linear equations. The idea behind solving linear equations was to use the addition and multiplication tools to change the equation to a form such as

$$x = 4$$

and then write the solution, $\{4\}$. We change an inequality to a form such as

$$x < 4 \quad \text{or} \quad 4 > x$$

and then write the solution set, $\{x \mid x < 4\}$.

Tools for Solving Inequalities

The solution set of an inequality will not change if we perform any of the following:

Addition

Add (or *subtract*) the same number on both sides of the inequality.

Multiplication by Positive Number

Multiply (or *divide*) both sides by the same *positive* number.

Multiplication by Negative Number

Multiply (or *divide*) both sides by the same *negative* number *and reverse* the direction of the inequality symbol.

To help us understand why we need two multiplication tools, consider the following example. The statement $-4 < 8$ is a true statement. What happens if we multiply both sides by -2 and do not reverse the direction of the inequality symbol?

$$-4 < 8 \qquad \text{True statement}$$
$$-4(-2) < 8(-2) \qquad \text{Multiply both sides by } -2.$$
$$8 < -16 \qquad \textit{False statement}$$

So, when we multiply (or divide) both sides of an inequality by a negative number, we must change the direction of the inequality symbol to maintain a true statement.

The inequalities

$$x - 3 < 7$$

and

$$x < 10$$

have the same solution set, since the second one was obtained from the first by adding 3 to both sides.

The inequalities

$$5x \leq 45$$

and

$$x \leq 9$$

have the same solution set, since the second was obtained from the first by dividing both sides by the *positive* number 5.

The inequalities

$$-\frac{1}{2}z > 11$$

and

$$z < -22$$

have the same solution set, since the second was obtained from the first by multiplying both sides by -2 *and reversing the direction of the inequality symbol.*

► **EXAMPLE 3**

Solve each inequality and graph each solution set.

(a) $x - 4 > 2$ (b) $-\frac{1}{3}z \leq 5$ (c) $-x < 2$

Solutions

(a) We add 4 to both sides and simplify.

$$x - 4 > 2$$
$$x - 4 + 4 > 2 + 4$$
$$x > 6$$

The solution set is

$$\{x \mid x > 6\}$$

The graph of this set is

Be Careful!   (b) We multiply both sides by -3. Since -3 is a *negative* number, we must remember to reverse the direction of the inequality symbol.

$$-\frac{1}{3}z \leq 5$$
$$(-3)\left(-\frac{1}{3}z\right) \geq (-3) \cdot 5$$
$$z \geq -15$$

The solution set is

$$\{z \mid z \geq -15\}$$

The graph is

Be Careful!

(c) We need to do something about the negative sign in order to solve this inequality. If we multiply both sides by -1, we will have x alone on the left. Because -1 is negative, *we must reverse the direction of the inequality symbol.*

$$-x < 2$$
$$(-1) \cdot (-x) > (-1) \cdot 2$$
$$x > -2$$
$$\{x \mid x > -2\}$$

► **EXAMPLE 4**

Find the solution set for each inequality and graph each solution set.

(a) $2x - 3 < 0$ (b) $16 - 3y \geq 22$

Solutions

(a) $2x - 3 < 0$

$2x - 3 + 3 < 0 + 3$ Add 3

$2x < 3$ Combine like terms

$\dfrac{2x}{2} < \dfrac{3}{2}$ Divide by 2

$x < \dfrac{3}{2}$

Since 2 is positive, the inequality symbol did not change direction when we divided. We write the solution set:

$$\left\{ x \mid x < \frac{3}{2} \right\}$$

The graph of this set is

(b) $16 - 3y \geq 22$

$16 - 3y - 16 \geq 22 - 16$ Subtract 16

$-3y \geq 6$ Combine like terms

$\dfrac{-3y}{-3} \leq \dfrac{6}{-3}$ Divide by -3

Notice that when we divided by the *negative* number -3 we had to reverse the direction of the inequality symbol. Now we simplify, write the solution set, and draw the graph.

$$y \le -2$$

$$\{y \mid y \le -2\}$$

◀

A Procedure for Solving Linear Inequalities

1. Clear the inequality of fractions (or decimals), if necessary, by multiplying by the least common denominator (LCD) of all the fractions in the equation. Be careful to reverse the inequality symbol when multiplying by a negative number.

2. Simplify each side by removing grouping symbols and combining like terms, where possible.

3. Collect all terms containing the variable on one side of the inequality and all other terms on the other side by using the addition tool.

4. Combine like terms.

5. Divide both sides by the coefficient of the variable. Be careful to reverse the inequality symbol when dividing by a negative number.

6. Write the solution set.

► EXAMPLE 5

Solve each inequality and graph the solution set of each.

(a) $3x + 2 < 2x - 1$ (b) $2(x - 1) \ge 5x$ (c) $\frac{1}{2}x < \frac{2}{3}x + 3$

Solutions

(a) $3x + 2 < 2x - 1$

We first collect the x terms on one side of the inequality. We can do this if we subtract $2x$ from both sides.

$3x + 2 - 2x < 2x - 1 - 2x$ Subtract $2x$

$x + 2 < -1$ Combine like terms

$x + 2 - 2 < -1 - 2$ Subtract 2

$x < -3$ Combine like terms

$\{x \mid x < -3\}$

(b) We remove the parentheses first.

$$2(x - 1) \geq 5x$$

$$2x - 2 \geq 5x \qquad \text{Distributive Property}$$

We can collect the x terms on either side. We subtract $2x$ from both sides to get them on the right side.

$$2x - 2 - 2x \geq 5x - 2x \qquad \text{Subtract } 2x$$

$$-2 \geq 3x \qquad \text{Combine like terms}$$

$$\frac{-2}{3} \geq \frac{3x}{3} \qquad \text{Divide by 3}$$

$$-\frac{2}{3} \geq x$$

This inequality can be written as $x \leq -\frac{2}{3}$.

$$\left\{ x \mid x \leq -\frac{2}{3} \right\}$$

(c) We clear fractions first by multiplying both sides by 6.

$$\frac{1}{2}x < \frac{2}{3}x + 3$$

$$6\left(\frac{1}{2}x\right) < 6\left(\frac{2}{3}x + 3\right) \qquad \text{Multiply by 6}$$

$$6\left(\frac{1}{2}x\right) < 6\left(\frac{2}{3}x\right) + 6(3) \qquad \text{Distributive Property}$$

$$3x < 4x + 18$$

We collect the x terms on the left side.

$$3x - 4x < 4x + 18 - 4x \qquad \text{Subtract } 4x$$

$$-x < 18$$

$$(-1)(-x) > (-1)(18) \qquad \text{Multiply by } -1$$

$$\qquad\qquad\qquad\qquad\qquad \text{(Reverse the inequality symbol)}$$

$$x > -18$$

$$\{x \mid x > -18\}$$

The graphs that we have seen so far consist of numbers to the right or to the left of a particular number. There are many times when we want to graph solution sets that are *between* two numbers. Such a graph would be like this:

This is the graph of the numbers between 2 and 7. We write the inequality that this set represents as $2 < x < 7$. This is read "two is less than x and x is less than seven." It can also be read as "x is greater than two and x is less than seven." This type of inequality is called a **compound inequality.**

▶ EXAMPLE 6

Graph the compound inequalities.

(a) $-3 < x < 0$ (b) $5 \le y < 10$
(c) $-5 < x \le -1$ (d) $-4 \le x \le 4$

Solutions

(a) The graph of $-3 < x < 0$ is all numbers between -3 and 0.

(b) The graph of $5 \le y < 10$ consists of the numbers between 5 and 10 as well as 5. We use a bracket to show that 5 is included.

(c) The graph of $-5 < x \le -1$ is all numbers between -5 and -1 as well as -1. We use a bracket to show that -1 is included.

(d) The graph of $-4 \le x \le 4$ contains all numbers between -4 and 4. We use brackets to include both -4 and 4.

◀

Problem Set 3.5

Warm-Ups

For each of the inequalities in Problems 1 through 5, determine which of the given numbers are in the solution set. See Example 1.

1. $6 + x < 11$; $-8, -2, 0, 2, 8$
3. $2x + 3 > 11$; $-20, 0, 4, 40, 400$
5. $2(x + 3) > 7$; $-6, -5, -4, -3, -2, -1, 0,$
     $1, 2, 3, 4, 5, 6$

2. $22 - x < 10$; $-13, 0, 13, 40$
4. $16 - 3x < 22$; $-10, -2, 0, 2, 10$

In Problems 6 through 9, graph the inequality. See Example 2.

6. $x < 9$

7. $y \le -1$

8. $5 < z$

9. $-3 \ge x$

In Problems 10 through 23, solve the inequalities and graph each solution set. See Example 3.

10. $x - 2 > 11$

11. $x + 6 < 24$

12. $x + 8 \leq 7$

13. $x - 23 \geq -3$

14. $2x < 18$

15. $3x \geq 18$

16. $-6x \geq 24$

17. $-11x \geq -33$

18. $\frac{1}{2}y > 5$

19. $\frac{3}{4}t \leq -2$

20. $-s \leq 14$

21. $-\frac{1}{2}x < 2$

22. $-\frac{2}{3}x > -8$

23. $\frac{6}{5}w \geq -30$

In Problems 24 through 39, find the solution set and graph each inequality. See Example 4.

24. $3 + 2x \leq -17$

25. $6 + \frac{1}{3}x < 9$

26. $7x - 4 \geq 10$

27. $4x + 5 > 1$

28. $7 - 2x > 11$

29. $4 - 5t \leq 14$

30. $3 + 7s \leq 5$

31. $13 - 6k < 11$

32. $2 - 2x > 1$

33. $7 \geq 5 - y$

34. $0 \leq 6x - 18$

35. $2x + 5 < -5$

36. $3 - 3y < 6$

37. $1 - w > 0$

38. $10x + 8 \geq -2$

39. $15 > 6 - 3z$

In Problems 40 through 51, solve each inequality and graph each solution set. See Example 5.

40. $6x < 5x - 2$

41. $11x - 10 > 7x + 2$

42. $3y - 8 \geq 5y + 2$

43. $z - 1 \leq 3z + 12$

44. $\frac{1}{2}x + \frac{2}{3} \leq 1$

45. $\frac{3}{4} - \frac{1}{3}x < \frac{1}{2}x$

46. $\frac{x}{2} > \frac{x}{4} - \frac{1}{3}$

47. $\frac{5}{3} < \frac{x}{2} - 1$

48. $2(x + 3) < 3$

49. $3(4 - x) < -12$

50. $-2(1 - x) \geq 4x$

51. $-3(x - 2) \leq 9x$

In Problems 52 through 55, graph each inequality. See Example 6.

52. $-1 < x < 3$

53. $0 \le y < 4$

54. $\dfrac{1}{2} < x \le 3$

55. $-\dfrac{3}{2} \le z \le \dfrac{5}{3}$

Practice Exercises

In Problems 56 through 85, solve each inequality and graph each solution set.

56. $x + 2 > 11$

57. $x - 6 < 24$

58. $2(x - 12) \le 6$

59. $3(x + 25) \ge -3$

60. $3x \le 18$

61. $2x > 16$

62. $\dfrac{1}{5}t \ge -8$

63. $-\dfrac{1}{4}z < 2$

64. $-x \le -4$

65. $-\dfrac{2}{7}x < 1$

66. $-\dfrac{3}{2}x > 1$

67. $\dfrac{3}{7}s \ge -21$

68. $2 + 3x \le -16$

69. $9 + \dfrac{1}{2}x < 9$

70. $\dfrac{1}{3}(t - 9) < 1$

71. $\dfrac{1}{2}(s + 7) \le 1$

72. $x \ge 1 + x$

73. $y - 3 < 2 + y$

74. $5 - 3x > 11$

75. $5 - 4t \le -5$

76. $7 + 3s \le 5$

77. $23 - 5k < 43$

78. $\dfrac{9 - x}{3} > 1$

79. $1 \ge \dfrac{7 - y}{5}$

80. $\dfrac{2}{3}w + \dfrac{1}{2} > 1$

81. $\dfrac{2}{3} - \dfrac{3}{5}v \ge \dfrac{1}{10}$

82. $3(x + 2) < 10$

83. $18 > -3(4 - x)$

84. $2(w - 3) < 7 + 2w$

85. $3k + 4 \le 3(k - 1)$

In Problems 86 through 89, graph the compound inequality.

86. $0 < x < 1$

87. $-2 \le t \le 2$

88. $-3 < y \le -1$

89. $-7 \le x < -5$

Don't do

Challenge Problems

In Problems 90 through 93, find and graph each solution set.

90. $4 < 2x < 11$

91. $-7 \leq 3g \leq 6$

92. $-3 < y - 2 \leq 0$

93. $3 \leq 2x + 3 < 9$

Solve the following inequalities. The variable is x and p is a constant.

94. $x - p < 3p$

95. $2x > p$

96. $px > 2 \ (p > 0)$

97. $px > 2 \ (p < 0)$

In Your Own Words

98. How is solving an inequality different from solving an equation?

3.6 Linear Inequalities as Models

TAPE 7

Inequalities arise in applications just as equations do. The following phrases are frequently used.

Statement	Symbols
p is more than q	$p > q$
p is less than q	$p < q$
p is not less than q	$p \geq q$
p is not more than q	$p \leq q$
p is at least q	$p \geq q$
p is at most q	$p \leq q$

► **EXAMPLE 1**

Translate each statement into an inequality.

(a) s is at most 6.

(b) -3 is less than z.

(c) Five more than x is at least 10.

(d) Twice n is greater than y.

Solutions

(a) *s is at most* 6

$$s \le 6$$

(b) -3 *is less than z.*

$$-3 < z$$

(c) Five more than *x is at least* 10.

$$x + 5 \ge 10$$

(d) Twice *n is greater than y.*

$$2n > y$$ ◄

► ## EXAMPLE 2

The area of a triangle is to be no more than 48 m². If the base is 8 m, find the largest possible height.

Solution

We let *h* be the height and sketch the triangle.

Since the area of a triangle is one-half the product of the base and the height, the area of this triangle is

$$\frac{1}{2}(8)h = 4h$$

The problem states that the area *is no more than* 48, so we form the inequality

$$4h \le 48$$

$$h \le 12 \qquad \text{Divide by 4}$$

Thus we see the largest possible height is 12 m.   ◄

► ## EXAMPLE 3

Sandra has grades of 68 and 72 on her history tests. What grade must she make on her third test to have an average of at least 75?

Solution

Let t be her third test grade. The sum of her three grades divided by three is her average. This averge is to be *at least* 75. So we have the inequality

$$\frac{68 + 72 + t}{3} \geq 75$$

$$\frac{140 + t}{3} \geq 75$$

$$3 \cdot \frac{140 + t}{3} \geq 3 \cdot 75 \qquad \text{Multiply by 3}$$
$$\qquad\qquad\qquad\qquad \text{(to clear fractions)}$$

$$140 + t \geq 225$$

$$140 + t - 140 \geq 225 - 140 \qquad \text{Subtract 140}$$

$$t \geq 85$$

She must make at least 85 on her third test.   ◄

Problem Set 3.6

Warm-Ups

In Problems 1 through 8, translate each statement into an inequality. See Example 1.

1. y is less than -2.

2. Five more than t is at most 15.

3. The sum s and $s + 1$ is greater than 0.

4. The absolute value of y is at least 5.

5. The product of z and -3 is at most 30.

6. The square of m is greater then 0.

7. Three more than twice x is greater than -5.

8. The absolute value of the opposite of k is greater than 0.

In Problems 9 through 14, solve each problem. In 9 through 11, see Example 2.

9. The perimeter of a square must be greater than or equal to 42 ft. How long must a side be?

10. The length of Lisa's rectangular dining room is 12 ft. If the area of the room is at least 96 ft², what is the smallest width the room could have?

11. Teresa cuts a ribbon into two pieces so that one piece is 6 in. longer than the other. How long can the shorter piece be if the ribbon is no longer than 30 in.?

In Problems 12 through 14, see Example 3.

12. Sally has grades of 82 and 72 in algebra. What grade must she make on her third test to have an average of at least 80?

13. In a gymnastics meet Jennifer has a score of 8.1 on vault, 7.5 on beam, and 7.8 on bars. What score must she make on floor to have an all around score of at least 30?

14. The sum of two consecutive positive integers must be no more than 5. What are the integers?

Practice Exercises

In Problems 15 through 28, translate each statement into an inequality.

15. The product of x and y is greater than or equal to -3.

16. The reciprocal of t is less than 0.

17. The sum of one-third of w and 5 is at least 9.

18. The cube of y is at most 1.

19. The opposite of x is less than 0.

20. The reciprocal of b is greater than b.

21. One-half of p is less than or equal to -4.

22. The square root of z is greater than 5.

23. The square root of p is greater than 7.

24. The opposite of z is less than w.

25. The reciprocal of $\dfrac{3}{5}$ is less than x.

26. Five less than three d is greater than -6.

27. The absolute value of k is at least 4.

28. The product of t and 5 is at most -6.

In Problems 29 through 42, answer the question.

29. Kristen has a score of 7.85 on her first vault. What score must she make on her second vault to have an average score of at least 8.45?

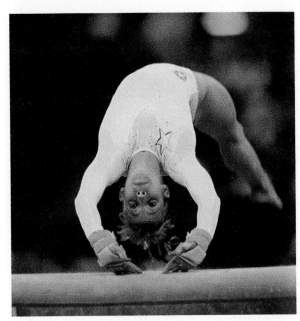

Problem 29

30. The width of a rectangular rug is 4 ft. If the perimeter of the rug must be at most 56 ft, what is the largest possible length it could have?

31. The sum of twice a number and one-half the number is less than 15. Find all possible numbers for which this is true.

32. The sum of two positive consecutive even integers is less than or equal to 6. What are the integers?

33. Jason sells popcorn at the ball game for 75¢ per bag. How many bags must he sell to collect at least $100?

34. Chuck is stacking square boxes in the stock room. Each box is 16 cm high and the room is 300 cm high. What is the largest number of boxes that he can put in a vertical stack?

35. Steve wants to buy as many compact discs as he can afford. Turtle's has them on sale for $11.99 each. If Steve has $140, how many can he buy?

36. Robert makes $3.50 per hour working at Wal-Mart. If he gets a bonus of $25 this week, how many hours must he work to make at least $165?

37. Joe works at B. Dalton bookstore and makes $4.25 per hour. How many hours must he work to make at least $190?

38. The low temperatures for the last two days were 28° and 15°. What must the low temperature for the next day be in order for the average temperature for the 3-day period to be less than 19°?

39. Find all numbers such that twice the number is less than 5 more than the number.

40. Lucy earned $400 and $550 in interest the last 2 yr. How much interest must she earn the next year to earn an average of $600 over the 3 yr?

41. A Duke University team pennant is in the shape of an isoceles triangle with a base of 10 in. If the perimeter of the pennant must be at most 42 in., what is the longest possible length for the other sides?

42. Janice is sending some computer components and supplies to San Francisco via UPS. There is a weight limit of 70 pounds on each package. If Janice's package contains a keyboard weighing 3.2 pounds, a monitor weighing 9.8 pounds, some computer paper weighing 12.5 pounds, and two identical computer carts each weighing the same unknown amount, how much weight is allowed for each cart in order to stay within the UPS limitations?

Challenge Problems

43. Keith has scores of 76, 81, and 85 on three tests in English. What score must he make on his fourth test to have an average of at least 90? (The highest possible score on a test is 100.)

In Your Own Words

44. Make up a word probblem that could be solved using the inequality $\frac{1}{2}x - \frac{1}{3}x \leq 1$.

▶ Chapter Summary

GLOSSARY

Linear equation in one variable   An equation that can be written in the form $Ax + B = 0$ with $A \neq 0$.

Literal equation   Equation with more than one letter.

Inequality   Statement that one number is greater than another.

Solution of an inequality   A number that makes an inequality a true statement.

Solution set of the inequality   The set of all solutions of the inequality.

Linear inequality in one variable   Inequality of the form $Ax + B < 0$, where A is not zero.

Solving Linear Equations

1. Add or subtract the same number to or from both sides.
2. Multiply or divide both sides by the same nonzero number.

Procedure to Solve Word Problems

1. Determine what is to be found.
2. Assign a variable, such as x, to represent one of the quantities to be found.
3. Express all other quantities to be found in terms of x (or the chosen variable).
4. Draw a figure or picture, if possible. Label it.
5. Reread the problem and form an equation.
6. Solve the equation and find the values of all the quantities to be found.
7. Write an answer to the original question.
8. Check the values in the original word problem. They should answer the question and make sense.

Solving Linear Inequalities

1. Add or subtract the same number to or from both sides.
2. Multiply or divide both sides by the same *positive* number.
3. Multiply or divide both sides by the same *negative* number *and* reverse the direction of the inequality symbol.

CHECK UPS

1. Solve the following equations.

   a. $x + 3 = 3x - 5$ Section 3.1; Example 4b

   b. $\dfrac{x}{2} - 1 = \dfrac{2}{5}x + 3$ Section 3.2; Example 3a

2. Solve for y: $4(p + y) = 3$ Section 3.3; Example 3b

3. The length of a rectangle is 5 m more than its width. If the perimeter is 78 m, find the dimensions of the rectangle.   Section 3.4; Example 2

4. Solve the inequality $16 - 3y \geq 22$.   Section 3.5; Example 4b

5. Graph the inequality $x \geq -2$.   Section 3.5; Example 2b

6. The area of a triangle is to be no more than 48 m². If the base is 8 m, find the largest possible height.   Section 3.6; Example 2

► **Review Problems**

In Problems 1 through 18, solve the given equations and inequalities.

1. $x - 6 = 14$

2. $x + 11 = 4$

3. $4 - x = 13$

4. $6 + x = 5$

5. $2x + 13 = 17$

6. $3x - 8 = -2$

7. $x - 6 < 14$

8. $x + 11 \geq 4$

9. $7 - 4x = -9$

10. $2(x - 2) = 12$

11. $4y - 11 > 9$

12. $x + 4 = 3x - 16$

13. $3(t + 2) = 10 - t$

14. $-s > 3$

15. $-x \leq -6$

16. $\frac{1}{2}x - 11 = \frac{1}{2}$

17. $5\left(x + \frac{1}{3}\right) = 3x + \frac{2}{3}$

18. $\frac{2}{3}(3 - y) > 2y$

In Problems 19 through 26, solve for the variable indicated.

19. $s + x = t$ for x

20. $u - v = 2w$ for v

21. $ax - b = c$ for x

22. $a + Bz = c$ for z

23. $2(r - at) = K$ for r

24. $2(r + at) = K$ for t

25. $\frac{a}{3}x - bh = 0$ for x

26. $\frac{abt}{c} = \pi v^2$ for t

27. One number is 3 less than twice another number. What are the numbers if their sum is 126?

28. The width of a rectangle is 3 ft more than half the length. If the perimeter is 78 ft, what are its dimensions?

29. Carol is 4 years younger than Russ. Thirty-four years ago, he was half as old as she is now. How old are Russ and Carol today?

30. A collection of nickels, dimes, and quarters has a value of $3.60. If there are twice as many quarters as dimes and two more nickels than quarters, how many of each kind of coin are there in the collection?

31. How many grams of 50% gold should be added to 30 g of 15% gold to produce a 20% gold alloy?

32. Lee leaves Monterey walking 3 mph and Artie Leaves Carmel Valley walking 2 mph. How many hours will it take for Lee and Artie to meet if they walk toward each other over the 20-mi distance?

33. Twice a number added to one-half the number is at most -3. Find all possible numbers.

34. Jim earns $25 per day giving tours of a park. How many days must he work to earn at least $500?

35. If the perimeter of a square photograph must be at most 60 in.², what is the largest possible size of the photograph?

36. In 1992, Atlanta Braves baseball pitcher Tom Glavine had two less than twice as many wins as losses. If he pitched 31 games, how many wins did he have?

37. In 1991, the Miami airport serviced 3.5 million more passengers than the Newark airport. Together 49.7 million passengers used these two facilities. How many passengers used the Miami airport in 1991?

38. Chuck ran the 10-kilometer Peachtree Road Race in 48 min. What was his speed in km/hr?

39. The Paleozoic Era (known as the Age of Fishes) lasted 215 million years longer than the Mesozoic Era (known as the Age of Reptiles). The Mesozoic and Paleozoic eras together lasted 535 million years. How long did the Mesozoic Era last?

40. Greg and Clayton paddled a canoe 1.5 hours down the White River, which had a current of 5 mph. Their return trip upstream took 3 hours. Find the rate at which Greg and Clayton can paddle in still water.

41. A person would weigh about one-sixth as much on the moon as on Earth. How much would Tunisia weigh on Earth if her weight on the moon is 18 pounds?

Let's Not Forget . . .

In Problems 42 through 47, perform the operations indicated.

42. $(-2x)(3x)$

43. $-2x(3 + x)$

44. $\dfrac{6}{0}$

45. $(x - 9)^2$

46. -6^2

47. $\dfrac{-18z^2}{-6z^3}$

In Problems 48 through 50, simplify each expression. Watch the role of the parentheses.

48. $2 + x(x - 1)$

49. $(2 + x)(x - 1)$

50. $(2 + x)x - 1$

In Problems 51 through 54, determine if x is a factor *or a* term.

51. $x - ab$

52. $3ax$

53. $x(a + b)$

54. $x + b$

In Problems 55 through 57, multiply each expression by $\dfrac{1}{2}$.

55. $-6x$

56. $6(x + 7)$

57. $6x - 4$

In Problems 58 through 63, identify each of the following as an expression, equation, *or* inequality. *Solve the equations and inequalities and simplify the expressions, if possible.*

58. $2x + 3(x - 5) - 4$

59. $3x - 7 > x + 3$

60. $2x - 18 = 5(x - 3)$

61. $5 - 2(x + 6)$

62. $5 - 2x \le 6$

63. $\dfrac{3}{5}x + \dfrac{1}{2} = \dfrac{1}{10}x$

► You Decide

You plan to raise celery in a small commercial garden. You are offered the two rectangular plots of prime bottom land as shown next.

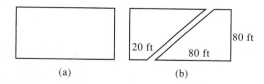

(a)   (b)

Plot (a) is twice as long as it is wide and contains 7200 ft². Plot (b) is divided into two pieces by a straight, little-used public road. The price of each plot is the same. Which plot do you choose and why?

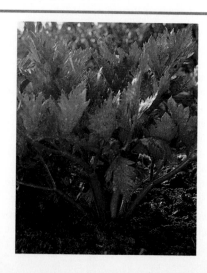

► Chapter 3 Test

In *Problems 1 through 6, choose the correct answer.*

1. The solution set for the equation $6x - 3 = 4 - x$ is (?).

(a) $\left\{\dfrac{5}{7}\right\}$ (b) $\left\{\dfrac{7}{5}\right\}$

(c) $\left\{\dfrac{1}{5}\right\}$ (d) $\{1\}$

2. The solution set for the equation $3x - 5 = 5(x - 1)$ is (?).

(a) $\{-2\}$ (b) $\left\{\dfrac{5}{4}\right\}$

(c) $\{0\}$ (d) $\varnothing$

3. When solved for x, the literal equation $6 + Abx = 2A$ becomes (?).

(a) $x = \dfrac{2A - 6}{Ab}$ (b) $x = \dfrac{-4}{b}$

(c) $x = 2A - 6 - Ab$ (d) $x = 6 - 2A - Ab$

4. The solution set for $4x - 4 > 2x - 8$ is (?).

(a) $\{x \mid x > -2\}$ (b) $\{-2\}$

(c) $\left\{x \mid x > -\dfrac{7}{2}\right\}$ (d) $\left\{-\dfrac{7}{2}\right\}$

5. A board 19 ft long is cut into two pieces such that one is 5 ft longer than the other. If x is the length of the shorter piece, which of the following is true?

(a) $x + 5 = 19$ (b) $x + x + 5 = 19$

(c) $x(x + 5) = 19$ (d) $x + x - 5 = 19$

6. The solution set for the equation $\dfrac{1}{2}x + \dfrac{2}{3} = 1$ is (?).

(a) $\left\{\dfrac{3}{2}\right\}$ (b) $\left\{\dfrac{2}{3}\right\}$

(c) $\{-1\}$ (d) $\left\{\dfrac{4}{3}\right\}$

In *Problems 7 through 10, find the solution set.*

7. $3 - \dfrac{1}{2}x = \dfrac{5}{2}$

8. $2(2 - x) \le 4 - 3x$

9. $5(t - 2) = 2 - t$

10. $\dfrac{1}{2}x > 1 + 2x$

In *Problems 11 and 12, solve for the variable indicated.*

11. $3t - ax = 5t$; solve for x.

12. $V = \dfrac{1}{3}\pi r^2 h$; solve for h.

13. A number is one less than twice another number. If their sum is 50, what are the two numbers?

14. Billy is 5 yr older than his sister. Six years ago he was twice her age then. How old is each now?

15. A collection of 20 coins, consisting of nickels and dimes has a value of $1.35. How many of each type of coin are there in the collection?

4 Factoring

See Problem Set 4.6, Exercise 7.

Connections

Factoring natural numbers has long been of utmost importance to mathematicians. The Fundamental Theorem of Arithmetic says that a natural number can be expressed as a product of prime numbers in only one way, apart from the order in which the factors are written. Because of this result, prime factors are often the key to understanding concepts of arithmetic. Likewise, in algebra factoring is often the key to solving a problem.

A contractor is building a warehouse that will store stacks of 18-in. and 24-in. boxes on the same shelf. To save space, he wants to know the shortest distance to put between shelves. The answer to his question is 72 in. To figure this out, he used the prime factorizations of 24 and 18 to find the least common multiple of 18 and 24.

Often in solving applications one intermediate step is to factor a polynomial. Equations such as $r^2 - 49 = 0$ are very often the mathematical model for solving a problem. Factoring the polynomial $r^2 - 49$ will provide the solutions.

In arithmetic, factoring provides the basis for working with fractions. In Chapter 5 we will see the connection between factoring polynomials and working with algebraic fractions. Factoring is also a very useful tool in solving certain kinds of equations.

4.1   Greatest Common Factor and Grouping

TAPE 7

In this chapter we will study factoring. To **factor** a number means to write the number as a **product.** To factor a polynomial means to write the polynomial as a product of polynomials.

First we will study **factoring out common factors.** This factoring is done by using the distributive property. When we multiplied polynomials, we used the distributive property:

$$p(q + r) = pq + pr$$

To factor, we turn this around and write

$$pq + pr = p(q + r)$$

It is important to notice that the expression $pq + pr$ is a sum of two terms, while $p(q + r)$ is a *product.* That is, $pq + pr$ is not factored, but $p(q + r)$ is factored.

To factor out common factors, we must be able to identify the *factors* in each term. Consider the polynomial $5xy + 2x$. It has terms $5xy$ and $2x$. The factors in the first term are 5, x, and y. In the second term, 2 and x are factors. We can use the distributive property to factor.

$$5xy + 2x = x \cdot 5y + x \cdot 2 \qquad \text{Commutative Property}$$

$$= x(5y + 2) \qquad \text{Distributive Property}$$

Notice that the Distributive Property worked because x was a factor in both terms, $5xy$ and $2x$.

Let's factor $12x + 6$. Using the distributive property, we can write

$$\boxed{1} \quad 12x + 6 = 2(6x + 3)$$

or

$$\boxed{2} \quad 12x + 6 = 3(4x + 2)$$

or

$$\boxed{3} \quad 12x + 6 = 6(2x + 1)$$

Each of these is a correct way to factor $12x + 6$. However, we say that $\boxed{3}$ is factored completely. Notice that in $\boxed{1}$ $6x + 3$ can be factored as $3(2x + 1)$ and in $\boxed{2}$ $4x + 2$ can be factored as $2(2x + 1)$. If no more factoring can be done, we say that a polynomial is **factored completely.** If no factoring can be done, we say that the polynomial is **prime.** We consider monomials to be prime polynomials.

▶ **EXAMPLE 1**

Factor each polynomial completely.

(a) $18y - 12$ (b) $15xy + 10y$

Solutions

(a)$18y - 12 = 6(3y - 2)$
(b)$15xy + 10y = 5y(3x + 2)$ ◀

To factor a polynomial completely involves finding the largest factor common to all the terms in the polynomial. Suppose that we want to find the largest factor that two numbers have in common. Consider the numbers 12 and 18. The *largest* factor of both numbers is 6. If the numbers are so large that we cannot find this greatest common factor by inspection, we must rely on the prime factorization of the numbers. We call such a factor the **greatest common factor.**

> **To Find the Greatest Common Factor (GCF)**
>
> **1.** Factor each number completely.
> **2.** List the prime factors that are common to all the numbers.
> **3.** The GCF is the product of the factors in step 2 with each factor raised to the smallest power of that factor in any number.

▶ **EXAMPLE 2**

Find the GCF for each pair of numbers.

(a) 25 and 60   (b) 32 and 72

Solutions

(a) $25 = 5^2$ Write each number as a product of prime factors
$60 = 2^2 \cdot 3 \cdot 5$

There is only one prime factor that is common to both numbers. It is 5. Since 5 occurs two times in 25 and one time in 60, the GCF is 5^1, or 5.

(b) $32 = 2^5$
$72 = 2^3 \cdot 3^2$

The common prime factor is 2. Since 2 occurs five times in 32 and three times in 72, the GCF is 2^3, or 8. ◀

► ### EXAMPLE 3

Find the GCF of $18x^3y^6$ and $12x^2y^3$.

Solution

$$18x^3y^6 = 2 \cdot 3^2 \cdot x^3y^6$$

$$12x^2y^3 = 2^2 \cdot 3 \cdot x^2y^3$$

The common prime factors are 2, 3, x, and y. The GCF must contain one factor of 2, one factor of 3, two factors of x, and three factors of y. The GCF is $6x^2y^3$. ◀

Be Careful!   Using the Distributive Property to factor a polynomial completely, we factor out the GCF of all the terms in the polynomial. *Always check by multiplying.*

► ### EXAMPLE 4

Factor $36x^3 - 54x^2 + 18x$ completely.

Solution

We must find the GCF for $36x^3$, $54x^2$ and $18x$.

$$36x^3 = 2^2 \cdot 3^2 \cdot x^3$$

$$54x^2 = 2 \cdot 3^3 \cdot x^2$$

$$18x = 2 \cdot 3^2 \cdot x$$

The GCF is $2 \cdot 3^2 \cdot x$, or $18x$.

$$36x^3 - 54x^2 + 18x = \boxed{18x} \cdot 2x^2 - \boxed{18x} \cdot 3x + \boxed{18x} \cdot 1$$

$$= \boxed{18x} \, (2x^2 - 3x + 1)$$

Check by using the Distributive Property. ◀

► ### EXAMPLE 5

Factor $-3y^2 + 6y - 3$ completely.

Solution

If the first term has a negative coefficient, we have a choice of factoring out 3 or -3. If we factor out 3,

$$-3y^2 + 6y - 3 = 3(-y^2 + 2y - 1)$$

Check by using the Distributive Property. If we factor out -3, we must be very careful with signs.

$$-3y^2 + 6y - 3 = -3(y^2 - 2y + 1)$$

Be Careful!   *Be sure to multiply this to check signs.* We usually prefer to factor out the negative number.   ◄

Let's try to factor $a(m + n) + b(m + n)$. There are two terms in this expression, $a(m + n)$ and $b(m + n)$. We see that $(m + n)$ is a factor in both. So it is a common factor.

$$a \;(m + n) \;+ b\; (m + n) \;=\; (m + n)\; (a + b)$$

Notice how we treated $(m + n)$ as one quantity. It represents a number, just as x represents a number.

Also, it is very important to see that the expression $a(m + n) + b(m + n)$ is a *sum*, whereas $(m + n)(a + b)$ is a *product*.

Be Careful!
$$a(m + n) + b(m + n) \text{ is not factored.}$$
$$(m + n)(a + b) \text{ is factored.}$$

► **EXAMPLE 6**

Factor each completely.

(a) $x(a - b) + y(a - b)$ (b) $w^2(2a + c) - z(2a + c)$
(c) $y(c + d) + (c + d)$

Solutions

(a) $(a - b)$ is a factor common to both terms.

$$x\; (a - b) \;+ y\; (a - b) \;=\; (a - b)\; (x + y)$$

(b) $(2a + c)$ is a factor in both terms.

$$w^2\; (2a + c) \;- z\; (2a + c) \;=\; (2a + c)\; (w^2 - z)$$

(c) $(c + d)$ is a factor in both terms.

Be Careful!
$$y\; (c + d) \;+\; (c + d) \;=\; (c + d)\; (y + 1)$$

Notice the 1 in the second factor. Think of $(c + d)$ as $1(c + d)$.   ◄

If we do some grouping, we can factor expressions such as $am + an + bm + bn$. We notice that there are four terms and that there is no common factor. However, if we group the first two terms and last two terms together, we have,

$$am + an + bm + bn = (am + an) + (bm + bn)$$

Now we have a common factor in the first group and a common factor in the second group.

$$(am + an) + (bm + bn) = a(m + n) + b(m + n)$$

We factor out the common factor $(m + n)$.

$$a(m + n) + b(m + n) = (m + n)(a + b)$$

This process is called factoring by **grouping.**

► **EXAMPLE 7**

Factor by grouping.

(a) $xy + xt + ay + at$ (b) $x^2 - xy + ax - ay$

(c) $ar^2 + at^2 - br^2 - bt^2$ (d) $zx - zy + x - y$

Solutions

(a) $xy + xt + ay + at = (xy + xt) + (ay + at)$
$$= x(y + t) + a(y + t)$$
$$= (y + t)(x + a)$$

(b) $x^2 - xy + ax - ay = (x^2 - xy) + (ax - ay)$
$$= x(x - y) + a(x - y)$$
$$= (x - y)(x + a)$$

(c) If we group the terms as they appear, we must be very careful with signs. Watch the negative sign with the parentheses.

$$\downarrow \qquad\qquad\qquad\qquad \downarrow$$

Be Careful!

$$ar^2 + at^2 - br^2 - bt^2 = (ar^2 + at^2) - (br^2 + bt^2)$$
$$= a(r^2 + t^2) - b(r^2 + t^2)$$
$$= (r^2 + t^2)(a - b)$$

(d) $zx - zy + x - y = (zx - zy) + (x - y)$
$$= z(x - y) + (x - y)$$
$$= (x - y)(z + 1)$$ ◄

There may be more than one way to group a polynomial. Let's look at Example 7c and group another way. Suppose we group the first and third terms and the second and fourth terms together.

$$ar^2 + at^2 - br^2 - bt^2 = (ar^2 - br^2) + (at^2 - bt^2)$$
$$= r^2(a - b) + t^2(a - b)$$
$$= (a - b)(r^2 + t^2)$$

Are the two answers the same? Why is $(a - b)(r^2 + t^2)$ the same as $(r^2 + t^2)(a - b)$?

Problem Set 4.1

Warm-Ups

In Problems 1 through 10, factor each completely. See Example 1.

1. $6x + 3$

2. $2x - 4$

3. $xy + xz$

4. $mn - mp$

5. $5x - 15y$

6. $18a + 9b$

7. $9ab - 6ac$

8. $7rs + 14r$

9. $6x + 6$

10. $3z - 3$

In Problems 11 through 20, find the GCF for each set of numbers. See Examples 2 and 3.

11. 12 and 18

12. 27 and 36

13. 54, 81 and 36

14. x^2 and x^3

15. xy^2 and xy

16. ab^2c^2 and a^2b^3c

17. $12x^2$, $18x^2$, and $21x$

18. $27p^2q^3$, $36pq^2$, and $45q^2$

19. $8xy^4$, $12x^2y^3$, and $16xy$

20. $20rst^2$, $30r^2st$, and $45rs^2t$

In Problems 21 through 31, factor each completely. See Example 4.

21. $5x^2 - 15x$

22. $18a^2b + 27ab^2$

23. $25x^2yz^2 - 75x^2yz$

24. $30z^3 + 45z^2$

25. $6r^2s^2t^2 + 3rst$

26. $21y^5 - 14y^7$

27. $26x^6 - 13x^4 + 39x^2$

28. $32y^8 + 36y^7 - 48y^5$

29. $7m^2n - 21m^3n^2 + 7m^2$

30. $a^2b^4c^7 - a^3b^5c^8 - a^5b^3c^5$

31. $56r^4s^2t + 64r^2s^2t^2 - 24rst$

In Problems 32 through 36, factor each completely by factoring out a negative coefficient. See Example 5.

32. $-5y - 10$

33. $-4x + 4$

34. $-2z - 2$

35. $-15y^5 + 10y^4 - 20y^2$

36. $-12a^2b - 15ab^2 + 3abc$

In Problems 37 through 44, factor completely. See Example 6.

37. $r(s + t) + 2(s + t)$

38. $z(a - b) + w(a - b)$

39. $x^2(y - z) + y^2(y - z)$

40. $x(c - d) - y(c - d)$

41. $q(p + r) - (p + r)$

42. $m^2(x + y^2) + (x + y^2)$

43. $3r^3(s^2 - 2) - 2w(s^2 - 2)$

44. $x^4(x^2 + y^2) - z^7(x^2 + y^2)$

In Problems 45 through 56, factor by grouping. See Example 7.

45. $ms + mt + ns + nt$

46. $xz - yz + xw - yw$

47. $r^2y - r^2z + s^2y - s^2z$

48. $ap - ar + bp - br$

49. $rpx + rqz - 3px - 3qz$

50. $x^2s^2 + x^2t^2 - 7s^2 - 7t^2$

51. $2z^5 + 2x - yz^5 - xy$

52. $2x^2 + 4 - wx^2 - 2w$

53. $3xy^2 - 6x - 2wy^2 + 4w$

54. $m^6 - m^3y^2 - m^3z + y^2z$

55. $ax^2 + bx^2 + a + b$

56. $as - at - s + t$

Practice Exercises

In Problems 57 through 100, factor each completely.

57. $5x + 10$

58. $2x - 4$

59. $rw + rz$

60. $st - sp$

61. $6x - 15y$

63. $-9ab + 6ac$

65. $2x^3 + 4x$

67. $3x^2y - 15xy$

69. $15x^3yz^2 - 30x^2yz^2$

71. $-6r^3st^3 + 3r^3st$

73. $6x^2 + y^2$

75. $72r^4s^2 + 48r^3s^2$

77. $14x^6 - 21x^4 + 28x^3$

79. $21a^2b - 14ab^2 + 7ab$

81. $108r^4s^2t + 96r^2s^2t^2 - 84rst$

83. $r(a - b) + 2t(a - b)$

85. $x(x^2 + 1) - y(x^2 + 1)$

87. $2r^2(a^2 - b) - (a^2 - b)$

89. $as - bs + 2at - 2bt$

91. $y^3 + y - xy^2 - x$

93. $pq + pr + 3q + 3r$

95. $2a^2 - 2b - a^2z + bz$

97. $3a + 6 + ay + 2y$ $(A+2)(B+y))$

99. $x^3 + 3x^2 + x + 3$

62. $15c + 5d$

64. $7r^2s + 14rs^2$

66. $4t - 4$

68. $8a^2b^2 + 20a^3b^2$

70. $35z^6 + 42z^4$

72. $16xy^5 - 24x^2y^7$

74. $36p^3q^4 - 48p^3q^5$

76. $x^2y + 5x^2$

78. $18y^6 + 27y^4 - 45y^3$

80. $a^3b^2c^6 - a^2b^4c^7 - a^5b^3c^5$

82. $w(x + y) + (x + y)$

84. $2b(c + d) - 3d(c + d)$

86. $x^2(b - 2d) + 4(b - 2d)$

88. $5(b + 1) - x(b + 1)$

90. $3bc + 3bd - 2cd - 2d^2$

92. $bt^2 - 2dt^2 + 4b - 8d$

94. $2a - 2b + am - bm$

96. $ab + at - b - t$

98. $a^2r^2 - br^2 - a^2z + bz$

100. $a^3 - 2a^2 - 3a + 6$

Challenge Problems

In Problems 101 through 104, factor each completely.

101. $ams + amt + ans + ant$

103. $6bc + 6bd - 6cd - 6d^2$

102. $2xz - 2yz + 2xw - 2yw$

104. $a^4 - 2a^3 - 3a^2 + 6a$

In Your Own Words

105. What is factoring?

106. What do we mean when we say a polynomial is factored completely?

107. What is factoring out the greatest common factor?

108. What is factoring by grouping?

4.2   Factoring Binomials

TAPE 7

We saw that the product $(p + q)(p - q)$ was $p^2 - q^2$. If asked whether $p^2 - q^2$ will factor, we know that it will. This type of factoring is called the **difference of two squares.**

> ### Difference of Two Squares
> $$p^2 - q^2 = (p - q)(p + q)$$

Notice that there are two terms in $p^2 - q^2$ and that each is a square. The factors are the sum and difference of the same two numbers. After we have decided that we are factoring the difference of squares, the factors will always look like

$$(\quad + \quad)(\quad - \quad)$$

► **EXAMPLE 1**

Factor each completely.

(a) $x^2 - 4$ (b) $16x^2 - 25$

Solutions

(a) This is the difference of two terms and both terms are squares. So this is the difference of two squares.

$$x^2 - 4 = x^2 - 2^2$$
$$= (x + 2)(x - 2)$$

(b) $16x^2 - 25 = (4x)^2 - 5^2$
$$= (4x + 5)(4x - 5)$$ ◄

We can combine taking out common factors with the difference of squares.

► **EXAMPLE 2**

Factor each completely.

(a) $x^3 - 4x$ (b) $4x^2 - 4$

Solutions

(a) This is a difference of two terms, but the terms are not squares. However, if we first factor out a common factor of x, it will be the difference of squares.

$$x^3 - 4x = x(x^2 - 4)$$
$$= x(x + 2)(x - 2)$$

(b) This is the difference of squares. What is wrong in factoring it as follows?

$$4x^2 - 4 = (2x - 2)(2x + 2)$$

It is not factored completely. To avoid this, it is best to take out common factors first.

$$4x^2 - 4 = 4(x^2 - 1)$$
$$= 4(x + 1)(x - 1)$$ ◄

Does the sum of squares factor? Can we factor $p^2 + q^2$? The answer is no. It is *not* $(p + q)^2$! We say that $p^2 + q^2$ is *prime*.

We can factor the sum and difference of two cubes, as the following products show.

$$(p + q)(p^2 - pq + q^2) = p^3 - p^2q + pq^2 + p^2q - pq^2 + q^3$$
$$= p^3 + q^3$$
$$(p - q)(p^2 + pq + q^2) = p^3 + p^2q + pq^2 - p^2q - pq^2 - q^3$$
$$= p^3 - q^3$$

Sum and Difference of Cubes

$$p^3 + q^3 = (p + q)(p^2 - pq + q^2)$$
$$p^3 - q^3 = (p - q)(p^2 + pq + q^2)$$

To recognize that we have the sum or difference of cubes, we see that there are two terms and that each term is a cube. Notice the similarity in the two formulas. The second factor is obtained from the first factor. Notice that the middle term is the product of p and q and that the sign is opposite to the sign in the first factor. Let's factor $x^3 + 8$. We see that it is the sum of cubes.

$$x^3 + 8 = x^3 + 2^3$$

We use the pattern and replace p with x and q with 2.

$$p^3 + q^3 = (p + q)(p^2 - pq + q^2)$$
$$x^3 + 8 = x^3 + 2^3 = (x + 2)(x^2 - 2x + 4)$$

To factor $x^3 - 8$, we use the other pattern and replace p with x and q with 2.

$$p^3 - q^3 = (p - q)(p^2 + pq + q^2)$$
$$x^3 - 8 = x^3 - 2^3 = (x - 2)(x^2 + 2x + 4)$$

▶ **EXAMPLE 3**

Factor each.

(a) $w^3 + 27$ (b) $8c^3 - 27$

Solutions

(a) This is the sum of cubes. We use the pattern and replace p with w and q with 3.

$$p^3 + q^3 = (p + q)(p^2 - pq + q^2)$$
$$w^3 + 27 = w^3 + 3^3 = (w + 3)(w^2 - 3w + 9)$$

(b) This is the difference of cubes.

$$p^3 - q^3 = (p - q)(p^2 + pq + q^2)$$
$$8c^3 - 27 = (2c)^3 - 3^3 = (2c - 3)[(2c)^2 + (2c)(3) + 3^2]$$
$$= (2c - 3)(4c^2 + 6c + 9) \quad ◀$$

We can also have common factors along with the sum or difference of cubes. Common factors should be taken out first.

▶ **EXAMPLE 4**

Factor $2x^3 - 16$ completely.

Solution

Notice that there are two terms and that we have a difference. However, 2 and 16 are not cubes. We must first factor out the common factor of 2.

$$2x^3 - 16 = 2(x^3 - 8)$$
$$= 2(x - 2)(x^2 + 2x + 4)$$ ◄

Factoring Binomials

$p^2 - q^2 = (p - q)(p + q)$	Difference of squares
$p^2 + q^2$ is prime	Sum of squares
$p^3 - q^3 = (p - q)(p^2 + pq + q^2)$	Difference of cubes
$p^3 + q^3 = (p + q)(p^2 - pq + q^2)$	Sum of cubes

Problem Set 4.2

Warm-Ups

In Problems 1 through 15, factor each completely if possible. See Example 1.

1. $x^2 - 9$
2. $y^2 - 81$
3. $4x^2 - 9$
4. $x^2 - 49$
5. $16x^2 - 25$
6. $x^2 - y^2$
7. $a^2 - 4b^2$
8. $4b^2 - 1$
9. $a^2b^2 - 36$
10. $4t^2 - x^2y^2$
11. $r^2 - s^2t^2$
12. $x^2 + 4$
13. $4z^2 - 49$
14. $9x^2 - m^2$
15. $100c^2 - 9d^2$

In Problems 16 through 26, factor each completely if possible. See Example 2.

16. $2x^2 - 50$
17. $x^3 - x$
18. $4x^2 - 16$
19. $8x^2 - 18$
20. $a^3 - ab^2$
21. $12b^2 - 3$
22. $5z^2 + 20$
23. $2a^3 - 8a$
24. $3ab^2 - 27a$
25. $4c^3 - 16c$
26. $x^2 + x^4$

In Problems 27 through 38, factor each completely. See Example 3.

27. $y^3 - x^3$
28. $c^3 + d^3$
29. $x^3 + 125$
30. $x^3 + 64$
31. $z^3 - 1$
32. $d^3 - 8c^3$
33. $8x^3 + 27$
34. $27k^3 + 64$
35. $8p^3 - 1$
36. $8x^3 - y^3$
37. $27t^3 + 8s^3$
38. $125z^3 - 27w^3$

In Problems 39 through 47, factor each completely. See Example 4.

39. $2y^3 + 2x^3$

40. $3z^3 - 24$

41. $8p^3q - q$

42. $2x^3 + 128$

43. $8z^3 - 8$

44. $8a^3b - 27b$

45. $64x^3 - 8y^3$

46. $16p^3 + 2q^3$

47. $-5z^3 - 5$

Practice Exercises

In Problems 49 through 92, factor each completely if possible.

48. $x^2 - 25$

49. $y^2 - 64$

50. $9x^2 - 4$

51. $x^2 - 121$

52. $9x^2 - 25$

53. $4x^2 - y^2$

54. $9a^2 - 4b^2$

55. $4b^2 - 81$

56. $9a^2b^2 - 1$

57. $t^2 - 4x^2y^2$

58. $25r^2 - s^2t^2$

59. $4x^2 + 4$

60. $4z^2 - 36$

61. $9x^2 - 9m^2$

62. $64c^2 - 9d^2$

63. $2x^2 - 18$

64. $x^3 - 9x$

65. $64x^2 - 16$

66. $5x^2 - 20$

67. $s^3 - st^2$

68. $6b^2 - 54$

69. $8z^2 + 72$

70. $3a^3 - 3a$

71. $50xy^2 - 2xz^2$

72. $4bc^2d^2 - 36b$

73. $c^3 - d^3$

74. $y^3 + x^3$

75. $x^3 - 64$

76. $z^3 + 1$

77. $d^3 + 8c^3$

78. $8x^3 - 27$

79. $27k^3 - 64$

80. $8p^3 + 1$

81. $8x^3 + y^3$

82. $27t^3 + 8s^3$

83. $125z^3 + 27w^3$

84. $2y^3 + 2x^3$

85. $3z^3 + 24$

86. $8p^3q + q$

87. $2x^3 - 128$

88. $8z^3 + 8$

89. $8a^3b + 27b$

90. $8x^3 + 8y^3$

91. $16p^3 - 2q^3$

92. $-8z^3 - 8$

Challenge Problems

In Problems 93 through 101, factor each completely.

93. $x^6 - 4$

94. $x^8 - 25$

95. $x^4 - 16$

96. $x^9 - 27$

97. $x^9 - 1$

98. $(a + b)^2 - 4$

99. $(a + b)^3 + 8$

100. $(a + b)^3 - c^3$

101. $16 - (a + b)^2$

In Your Own Words

102. What kinds of binomials can we factor?

103. Give the rule for factoring the difference of two squares without using a formula.

104. Give the rule for factoring the sum or difference of two cubes without using a formula.

4.3   Factoring Trinomials

TAPE 8

In this section we examine factoring trinomials. We learned that the product of two binomials is sometimes a trinomial. So we look for two binomial factors.

We begin by looking at how $x^2 + 8x + 15$ can be factored. Notice that the last term, 15, is *positive*. We look for two binomials whose product is the given trinomial. Since the first term is x^2, the first terms in the factors must be x.

$$x^2 + 8x + 15 = \left(x \;\Box \right)\left(x \;\Box \right)$$

The numbers in the boxes must have a product of 15. There are four pairs of integers whose product is 15.

$$(-15)(-1) = 15$$
$$(-5)(-3) = 15$$
$$(15)(1) = 15$$
$$(5)(3) = 15$$

Not only must the product of the numbers be 15, but the sum of the outside and inside products must be $8x$.

$$x^2 + 8x + 15 = \left(x \;\Box \right)\left(x \;\Box \right)$$

Sum must be $8x$

So we look for a pair of integers whose product is 15 and whose sum is 8.

Integers	Product	Sum
$-15, -1$	15	-16
$-5, -3$	15	-8
$15, 1$	15	16
$5, 3$	15	8

Thus

$$x^2 + 8x + 15 = (x + 5)(x + 3)$$

Now let's factor $x^2 - 8x + 15$. Notice again that the last term, 15, is *positive*. The factors must have x as the first term.

$$x^2 - 8x + 15 = \left(x \;\Box \right)\left(x \;\Box \right)$$

The numbers that go in the boxes must have a product of 15. To make a middle term of $-8x$, we must have two numbers whose sum is -8. So we must use -3 and -5. Thus $x^2 - 8x + 15 = (x - 3)(x - 5)$.

Comparing both situations, the factors look alike except for the signs.

$$x^2 + 8x + 15 = (x + 5)(x + 3), \qquad x^2 - 8x + 15 = (x - 5)(x - 3)$$

If the *last* term of a trinomial is *positive*, then the signs in the factors are *alike* and can be determined by the *middle sign*.

► **EXAMPLE 1**

Factor each trinomial.

(a) $x^2 + 4x + 3$ (b) $x^2 - 3x + 2$

(c) $x^2 + 7x + 12$ (d) $x^2 - 9x + 18$

Solutions

(a) $x^2 + 4x + 3$

Since the last term is positive and the middle term is also positive, the factors must look like

$$(\quad + \quad)(\quad + \quad)$$

We are looking for integers with a product of 3 and a sum of 4. They are 1 and 3.

$$x^2 + 4x + 3 = (x + 3)(x + 1)$$

(b) $x^2 - 3x + 2$

Since the last term is positive and the middle term is negative, the factors must look like

$$(\quad - \quad)(\quad - \quad)$$

We are looking for integers with a product of 2 and a sum of 3. They are 2 and 1.

$$x^2 - 3x + 2 = (x - 2)(x - 1)$$

(c) $x^2 + 7x + 12$

The factors must look like

$$(\quad + \quad)(\quad + \quad)$$

We are looking for integers with a sum of 7 and a product of 12. There are several factors of 12.

Product of 12	Sum
12, 1	13
6, 2	8
3, 4	7

$$x^2 + 7x + 12 = (x + 3)(x + 4)$$

(d) $x^2 - 9x + 18$

The factors must look like

$$(\quad - \quad)(\quad - \quad)$$

We look for a product of 18 and a sum of 9.

Product of 18	Sum
18, 1	19
9, 2	11
6, 3	9

$$x^2 - 9x + 18 = (x - 6)(x - 3)$$

◄

Be Careful!   After factoring any polynomial, it is *always* a good idea to *check by multiplying the factors to see if the product is the original polynomial.*

Now let's look at two examples of a second situation, where the last term is negative.

$$x^2 - 2x - 15, \qquad x^2 + 2x - 15$$

The first terms in the factors must be x.

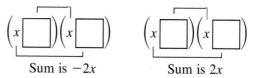

The product of the numbers in the boxes must be -15 and the middle terms must be $-2x$ or $2x$.

$$\left(x\ \Box \right)\left(x\ \Box \right) \qquad \left(x\ \Box \right)\left(x\ \Box \right)$$

Sum is $-2x$ Sum is $2x$

We look for possibilities in the table.

Integers	Product	Sum
$-15, 1$	-15	-14
$-1, 15$	-15	14
$-5, 3$	-15	-2
$-3, 5$	-15	2

Thus we see that, in the first case,

$$x^2 - 2x - 15 = (x - 5)(x + 3)$$

and, in the second case,

$$x^2 + 2x - 15 = (x - 3)(x + 5)$$

Notice that the factors look like

$$(\qquad - \qquad)(\qquad + \qquad)$$

That is, the signs are different in the factors. We made the middle term $2x$ or $-2x$ by where we placed the negative and positive signs.

► **EXAMPLE 2**

Factor each trinomial.

(a) $x^2 - 2x - 3$ (b) $x^2 + 3x - 4$

Solutions

(a) $x^2 - 2x - 3$

Since the last term is negative, the factors must look like

$$(\qquad - \qquad)(\qquad + \qquad)$$

We are looking for integers with a product of 3 and a difference of 2. They must be 1 and 3. We must be careful where we put them. We want the middle term to be $-2x$. So the 3 must be placed with the negative sign.

$$x^2 - 2x - 3 = (x - 3)(x + 1)$$

(b) $x^2 + 3x - 4$

Again, the factors must look like

$$(\quad - \quad)(\quad + \quad)$$

We need a product of 4 and a difference of 3.

Product of 4	Difference
1, 4	3
2, 2	0

We must place the 4 with the positive sign, since the middle term is $+3x$.

$$x^2 + 3x - 4 = (x - 1)(x + 4) \qquad \blacktriangleleft$$

When factoring trinomials that are in standard form with a positive leading coefficient, the following guidelines are useful.

Guidelines for Factoring Trinomials

1. If the last term is *positive,* then the factors will look like

$$(\quad + \quad)(\quad + \quad) \quad \text{or} \quad (\quad - \quad)(\quad - \quad)$$

The $+$ or $-$ sign is determined by the coefficient of the middle term.

2. If the last term is *negative,* then the factors will look like

$$(\quad + \quad)(\quad - \quad) \quad \text{or} \quad (\quad - \quad)(\quad + \quad)$$

These guidelines place the proper signs in the factors for us.

In the discussion of factoring trinomials, we looked only at trinomials that have a leading coefficient of 1. That is, the first term was x^2. Let's consider other possibilities.

▶ **EXAMPLE 3**

Factor each trinomial.

(a) $2x^2 - 5x + 3$ (b) $6x^2 + x - 2$

Solutions

(a) $2x^2 - 5x + 3$

Since the last term is positive and the middle term is negative, the factors must look like

$$(\quad - \quad)(\quad - \quad)$$

The first term must be $2x^2$, which means that we must use $2x$ and x as first terms in the factors.

$$(2x -)(x -)$$

We are looking for a product of 3. The only possibilities are 1 and 3. We must be very careful where we place the 1 and 3. We must have a middle term of $-5x$. There are two choices.

$$(2x - 3)(x - 1) \quad \text{or} \quad (2x - 1)(x - 3)$$

Multiplying tells us that $(2x - 3)(x - 1)$ is correct.

$$2x^2 - 5x + 3 = (2x - 3)(x - 1)$$

(b) $6x^2 + x - 2$

Since the last term is negative, the factors must look like

$$(-)(+)$$

There are many possibilities for this product. The first term of $6x^2$ could give $3x$ and $2x$ or $6x$ and x as first terms in the factors. The last term of 2 gives 2 and 1 for the last terms in the factors. We list all the possible factors.

1	$(6x - 2)(x + 1)$	2	$(6x - 1)(x + 2)$
3	$(6x + 2)(x - 1)$	4	$(6x + 1)(x - 2)$
5	$(3x - 2)(2x + 1)$	6	$(3x - 1)(2x + 2)$
7	$(3x + 1)(2x - 2)$	8	$(3x + 2)(2x - 1)$

This seems to be a difficult job. However, we can eliminate 1, 3, 6, and 7 because the terms in one factor of each have a common factor. These are not the factors because the original trinomial does not have a common factor. Multiplying tells us that 8 is correct.

$$6x^2 + x - 2 = (3x + 2)(2x - 1) \quad \blacktriangleleft$$

Be Careful!   Anytime that we factor we must always look for a common factor first. The next example shows a common factor in a trinomial

► **EXAMPLE 4**

Factor $2a^3 - 4a^2b - 6ab^2$ completely.

Solution

There is a common factor of $2a$. We factor it out first.

$$2a^3 - 4a^2b - 6ab^2 = 2a(a^2 - 2ab - 3b^2)$$
$$= 2a(a + b)(a - 3b) \quad \blacktriangleleft$$

Alternative Approach to Factoring Trinomials: The ac Method

To factor $ax^2 + bx + c$ by the *ac* method, we look for two numbers whose product is *ac* and whose sum is *b*. Then, we factor by grouping. The procedure is illustrated in the next examples.

► **EXAMPLE 5**

Factor $3x^2 + 17x + 10$.

Solution

The product ac is $3(10)$ or 30. Now we look for factors of 30 whose sum is 17.

Factors of 30	Sum of Factors
$1(30)$	$1 + 30 = 31$
$2(15)$	$2 + 15 = 17$
$3(10)$	$3 + 10 = 13$
$5(6)$	$5 + 6 = 11$

Of course, we can stop when we find the correct factors.

The factors we are looking for are 2 and 15. Write $17x$ as $2x + 15x$ and use grouping to factor.

$$3x^2 + \boxed{17x} + 10 = 3x^2 + \boxed{2x + 15x} + 10$$
$$= x(3x + 2) + 5(3x + 2)$$
$$= (3x + 2)(x + 5) \qquad ◄$$

► **EXAMPLE 6**

Factor $6a^2 + 7a - 3$.

Solution

The product of 6 and -3 is -18. We look for factors of -18 whose sum is 7.

Factors of -18	Sum of Factors
$1(-18)$	$1 + (-18) = -17$
$2(-9)$	$2 + (-9) = -7$
$3(-6)$	$3 + (-6) = -3$
$6(-3)$	$6 + (-3) = 3$
$9(-2)$	$9 + (-2) = 7$

The factors we are looking for are 9 and -2. Now we write $7a$ as $9a - 2a$ and use grouping.

$$6a^2 + \boxed{7a} - 3 = 6a^2 + \boxed{9a - 2a} - 3$$
$$= 3a(2a + 3) - (2a + 3)$$
$$= (2a + 3)(3a - 1) \qquad ◄$$

If a trinomial is the square of a binomial, we call the trinomial a **perfect trinomial square.** The special products that we learned in Section 2.6 can be used as factoring formulas. For example, $x^2 + 6x + 9$ is $(x + 3)^2$. Perfect trinomial squares can be factored just like any other trinomial. Recognizing that the trinomial is a perfect square may save some time.

Factoring Perfect Trinomial Squares

$$p^2 + 2pq + q^2 = (p + q)^2$$
$$p^2 - 2pq + q^2 = (p - q)^2$$

Problem Set 4.3

Warm-Ups

In Problems 1 through 15, factor each trinomial completely. See Example 1.

1. $x^2 + 3x + 2$
2. $x^2 + 4x + 3$
3. $x^2 + 5x + 4$
4. $x^2 + 6xy + 5y^2$
5. $x^2 + 7x + 6$
6. $x^2 + 4x + 4$
7. $x^2 + 5x + 6$
8. $x^2 + 6x + 8$
9. $x^2 - 8x + 15$
10. $x^2 - 7x + 12$
11. $x^2 - 3xy + 2y^2$
12. $x^2 - 4x + 3$
13. $x^2 - 8x + 16$
14. $x^2 - 6x + 5$
15. $x^2 - 7x + 6$

In Problems 16 through 25, factor each trinomial completely. See Example 2.

16. $x^2 - x - 2$
17. $a^2 - 2ab - 3b^2$
18. $z^2 + 3z - 4$
19. $x^2 + 4x - 5$
20. $y^2 - 5y - 6$
21. $x^2 - x - 6$
22. $x^2 - 2x - 8$
23. $x^2 - 3x - 10$
24. $t^2 + t - 12$
25. $x^2 - x - 20$

In Problems 26 through 46, factor each trinomial completely. See Example 3 or Examples 5 and 6.

26. $2x^2 + 3x + 1$
27. $3b^2 + 4b + 1$
28. $2x^2 - 5x + 2$
29. $3x^2 - 7xz + 2z^2$
30. $5w^2 + 7w + 2$
31. $3x^2 + 10x + 3$
32. $9x^2 - 12x + 4$
33. $5x^2 - 16x + 3$
34. $2x^2 - x - 1$
35. $3c^2 - 2c - 1$
36. $2x^2 + 3x - 2$
37. $3x^2 - 5x + 2$
38. $6x^2 + 5x + 1$
39. $2x^2 + 7x + 6$
40. $10x^2 - 7x + 1$
41. $3a^2 - 14a + 8$
42. $6x^2 - 7xy - 3y^2$
43. $7x^2 + 33x - 10$
44. $6t^2 + 23t + 20$
45. $8x^2 - 26x + 21$
46. $12p^2 - 5p - 25$

In Problems 47 through 50, factor each trinomial completely. See Example 4.

47. $2x^2 - 2x - 12$

48. $3x^2 - 15x + 12$

49. $4x^2 - 14x - 8$

50. $2ax^2 + 9ax + 9a$

Practice Exercises

In Problems 51 through 100, factor each trinomial completely.

51. $x^2 + 8x + 7$

52. $x^2 + 9x + 8$

53. $x^2 + 10x + 9$

54. $x^2 + 7x + 10$

55. $x^2 + 8x + 12$

56. $x^2 - 8x + 7$

57. $x^2 - 2x + 1$

58. $x^2 - 10x + 9$

59. $x^2 - 9x + 18$

60. $a^2 - 5ab + 6b^2$

61. $y^2 + y - 2$

62. $x^2 + 2x - 3$

63. $x^2 - 3x + 4$

64. $x^2 - 4x - 5$

65. $x^2 + 5x - 6$

66. $x^2 + xy - 6y^2$

67. $a^2 - 2a - 8$

68. $z^2 + 3z - 10$

69. $x^2 - x - 12$

70. $16y^2 + 8y + 1$

71. $5x^2 + 6x + 1$

72. $4x^2 - 6x + 1$

73. $5a^2 - 6ab + b^2$

74. $5t^2 + 8t + 3$

75. $3x^2 + 7x + 2$

76. $3x^2 - 7x + 2$

77. $49b^2 - 28b + 4$

78. $3x^2 - 8x - 3$

79. $2a^2 + ab - b^2$

80. $2w^2 - 3w - 2$

81. $3x^2 + 5x - 2$

82. $4x^2 + 8x + 3$

83. $12x^2 + 11x - 2$

84. $2x^2 - 7x + 6$

85. $6c^2 - 5c + 1$

86. $14x^2 + 15x - 9$

87. $6x^2 + 5x - 6$

88. $4x^2 - 12x + 9$

89. $6x^2 - 23x + 20$

90. $6x^2 - x - 12$

91. $16a^2 - 10ab - 9b^2$

92. $24x^2 - 38x + 15$

93. $24x^2 - 25x + 6$

94. $8t^2 + 22t + 15$

95. $6x^2 - 25x - 25$

96. $2a^2x^2 - ax^2 - 5x^2$

97. $4ax^2 + 16ax + 16a$

98. $10m^4 + 3m^3n - m^2n^2$

99. $a^4 + a^3b - 2a^2b^2$

100. $12p^3 - 9p^2q - 3pq^2$

Challenge Problems

In Problems 101 through 103, factor each trinomial completely.

101. $x^4 - 5x^2 + 6$

102. $x^4 + x^2 - 2$

103. $x^4 - 5x^2 + 4$

In Your Own Words

104. What are the rules that determine the signs in the factors of a trinomial?

105. How do we recognize a perfect square trinomial?

4.4 Summary of Factoring

TAPE 8

After studying several types of factoring, we can summarize what we have learned by formulating a procedure to factor a polynomial completely.

> **Procedure to Factor a Polynomial Completely**
>
> **1.** Factor out the greatest common factor (GCF) if there is one. This should be done *first*.
> **2.** Count the number of terms.
> **3.** If there are two terms, check for a difference of squares or cubes or a sum of cubes. Remember, $p^2 + q^2$ *is prime.*
> **4.** If there are three terms, look for two binomial factors.
> **5.** If there are four or more terms, try grouping.
> **6.** Make sure each factor is prime.
> **7.** Check to see if the product of the factors is the original polynomial.

► **EXAMPLE 1**

Factor $50 - 2x^2$ completely.

Solution

There is a common factor.

$$50 - 2x^2 = 2(25 - x^2)$$

There are two terms in $25 - x^2$. This is the difference of squares.

$$50 - 2x^2 = 2(25 - x^2)$$
$$= 2(5 - x)(5 + x)$$

Check by multiplying. ◄

► **EXAMPLE 2**

Factor $t^3 - 1$ completely.

Solution

There is no common factor. There are two terms, which are the difference of cubes.

$$t^3 - 1 = t^3 - 1^3$$
$$= (t - 1)(t^2 + t + 1)$$

Check by multiplying. ◄

► **EXAMPLE 3**

Factor $5(a - b) - t(a - b)$ completely.

Solution

There are two terms and a common factor of $(a - b)$.

$$5(a - b) - t(a - b) = (a - b)(5 - t) \qquad \blacktriangleleft$$

▶ **EXAMPLE 4**

Factor $2x^2 - x - 1$ completely.

Solution

There is no common factor. There are three terms.

$$2x^2 - x - 1 = (2x + 1)(x - 1)$$

Check by multiplying. $\qquad \blacktriangleleft$

▶ **EXAMPLE 5**

Factor $x^4 + x^3 - 2x^2$ completely.

Solution

There is a common factor.

$$x^4 + x^3 - 2x^2 = x^2(x^2 + x - 2)$$
$$= x^2(x + 2)(x - 1)$$

Check by multiplying. $\qquad \blacktriangleleft$

▶ **EXAMPLE 6**

Factor $6r^2 - 2rt + 3rs - st$ completely.

Solution

There is no common factor and there are four terms. We use grouping.

$$6r^2 - 2rt + 3rs - st = (6r^2 - 2rt) + (3rs - st)$$
$$= 2r(3r - t) + s(3r - t)$$
$$= (3r - t)(2r + s) \qquad \blacktriangleleft$$

Many times we need to determine if a polynomial is factored. Which of the following polynomials is factored?

$$a^3 - b^3, \qquad (a - b)^3$$

The polynomial $a^3 - b^3$ is not factored, whereas the polynomial $(a - b)^3$ is factored. To factor a polynomial means to write it as a **product** of polynomials. Thus $a^3 - b^3$ is a difference, *not* a product, but $(a - b)^3$ *is* a product:

$$(a - b)^3 = (a - b)(a - b)(a - b)$$

Factored	Not Factored
$(a - b)(a + b)$	$a^2 - b^2$
$(a + b)^2$	$a^2 + 2ab + b^2$
$(a - b)^2$	$a^2 - 2ab + b^2$
$(a - b)(a^2 + ab + b^2)$	$a^3 - b^3$
$(a + b)(a^2 - ab + b^2)$	$a^3 + b^3$
$(a + b)(m + n)$	$m(a + b) + n(a + b)$

Problem Set 4.4

Warm-Ups

In Problems 1 through 18, factor each polynomial completely if possible. See Examples 1, 2, and 3.

1. $25x^2 - 25$

2. $8w^3 - 27$

3. $9b^2 - x^2$

4. $225a^5b^2c - 90a^2b^2c$

5. $s^3 + 27$

6. $x^2 - x$

7. $4x^2 + 4$

8. $49 - 4s^2$

9. $y(z - 4) + (z - 4)$

10. $x^2 + 4$

11. $8 - 2b^2$

12. $a^2b + c$

13. $16 - 25x^2$

14. $w^3 + 125x^3$

15. $ab(x - y) + c(x - y)$

16. $z^3 + 1$

17. $2000 - 16y^3$

18. $z^2(1 - y) - 4(1 - y)$

In Problems 19 through 28, factor each polynomial completely if possible. See Examples 4 and 5.

19. $x^2 + x - 2$

20. $x^2 + x + 1$

21. $3x^2 - 30x + 27$

22. $9r^2 - 21r - 8$

23. $x^2 + 6x + 9$

24. $18a^3 + 36b^3c^3 + 27b^2c^2$

25. $21x^2 - 17x + 2$

26. $y^2 - 6y - 7$

27. $x^2 - 5x - 6$

28. $2x^2 + x + 1$

In Problems 29 through 32, factor each polynomial comletely. See Example 6.

29. $a^2z + a^2x - z - x$

30. $x^6 + x^5 + x^4 + x^3$

31. $y^2 + yz + az + ay$

32. $2ab + 2ax + by + xy$

Practice Exercises

In Problems 33 through 68, determine whether or not each polynomial is factored. Factor it completely if possible.

33. $x^2 + 9x + 14$

34. $9 - 16x^2$

35. $135r^3t^5 + 225r^2t^4$

36. $x^2 - 8x + 16$

37. $1 - 27a^3$

38. $x^2 - 64$

39. $ax + ay + x + y$

40. $a^3(x + 1) + b^2(1 + x)$

41. $s^3t(2st + u)$

42. $x^2 + x$

43. $32 - 2x^2$

44. $r^3t(s - 1) + rt(s - 1)$

45. $r^3(t - x) - (t - x)$

46. $6s^2 + st - 15t^2$

47. $z^2 - x^2y^2$

48. $27r^3 + 64$

49. $a^2b + ab^2 + a + b$

50. $x(r + s) - y(s + r)$

51. $x^2 + 3x + 2$

52. $16x^2 - 16$

53. $8x^3 + y^3$

54. $ax - ay - by + bx$

55. $x^2 + 12x + 36$

56. $x^2 - 6x + 5$

57. $3x^2 + x + 1$

58. $x^4 - x^3 + x^2 - x$

59. $8y^3 + z^3$

60. $2 - 2x^2$

61. $x^2 + x - 20$

62. $x^2 + x - 6$

63. $81 - 18x + x^2$

64. $z(a + b)$

65. $a^3 - 125$

66. $(a + b)^2$

67. $a^7b^5c^4 - a^6b^6c^6 + a^5b^6c^3$

68. $4a^2 - 4b^2$

Challenge Problems

In Problems 69 through 74, factor each polynomial completely.

69. $x^4 - x^2 - 6$

70. $x^4 - 1$

71. $5x^4 + 5x^2 - 100$

72. $x^7 - 8x$

73. $x^6 - 9$

74. $x^6 - 1$

In Your Own Words

75. What procedure do we follow in factoring a polynomial?

4.5 Solving Equations by Factoring

TAPE 8

We now look at solving equations by factoring. Remember that an equation is a statement that two numbers are equal. To solve an equation means to find its solution set. The solution set consists of all numbers that will make the equation a true statement.

The two tools that we used to solve first-degree equations, adding the same number to both sides and multiplying both sides by the same nonzero number, are not enough to solve all equations. Our number system has the following property, which provides us with another tool necessary to solve many important kinds of equations.

Property of Zero Products

The statement $AB = 0$ is *true* if either $A = 0$ or $B = 0$ and *false* if neither A nor B is 0.

Let's solve the equation

$$x^2 - x - 6 = 0$$

The left side is a trinomial, which we learned to factor in Section 4.3. So

$$(x + 2)(x - 3) = 0$$

This is a statement that the product of two real numbers is 0. The Property of Zero Products says that this statement will be true if *either* $(x + 2)$ or $(x - 3)$

is zero. That is,

$$x + 2 = 0 \quad \text{or} \quad x - 3 = 0$$

$$x = -2 \quad \text{or} \quad x = 3$$

Thus -2 and 3 are the solutions to the equation $x^2 - x - 6 = 0$. The solution set is $\{-2, 3\}$. We can check -2 and 3 in the original equation.

► **EXAMPLE 1**

Solve $x^2 + x = 2$.

Solution

We write the equation so that one side is 0 and factor:

$$x^2 + x - 2 = 0$$

$$(x + 2)(x - 1) = 0$$

Apply the Property of Zero Products.

$$x + 2 = 0 \quad \text{or} \quad x - 1 = 0$$

$$x = -2 \quad \text{or} \quad x = 1$$

We write the solution set.

$$\{-2, 1\} \qquad \blacktriangleleft$$

Once we have factored, it is easy to write the solution set directly.

Solving Equations by Factoring

1. Write the equation so that one side is 0.

2. Factor completely.

3. Set each factor equal to 0 and solve each equation.

4. Check if required.

5. Write the solution set.

► **EXAMPLE 2**

Solve each equation.

(a) $2y^2 - y = 0$ (b) $x^2 - 9 = 0$

Solutions

(a)
$$2y^2 - y = 0 \qquad \text{The right side is 0}$$

$$y(2y - 1) = 0 \qquad \text{Factor}$$

$$y = 0 \quad \text{or} \quad 2y - 1 = 0 \qquad \text{Set factors equal to 0}$$

$$2y = 1 \qquad \text{Solve}$$

$$y = \frac{1}{2}$$

$$\left\{ 0, \frac{1}{2} \right\} \qquad \text{Write the solution set}$$

Be Careful!

Notice that 0 is a solution just as $\frac{1}{2}$ is. A common mistake is to disregard the 0.

(b)

$$x^2 - 9 = 0 \qquad \text{The right side is 0}$$
$$(x + 3)(x - 3) = 0 \qquad \text{Factor}$$
$$x + 3 = 0 \quad \text{or} \quad x - 3 = 0 \qquad \text{Set factors equal to 0}$$
$$x = -3 \quad \text{or} \quad x = 3 \qquad \text{Solve}$$
$$\left\{-3, 3\right\} \qquad \text{Write the solution set}$$

A common abbreviation for $\{-3, 3\}$ is $\{\pm 3\}$. We read ± 3 as "plus or minus 3." Also, we often write statements such as $x = \pm 3$, which means $x = -3$ *or* $x = 3$. ◀

► **EXAMPLE 3**

Solve each equation.

(a) $-2x^2 - x + 3 = 0$ (b) $t^2 + 4t + 4 = 0$

Solutions

(a) It is much easier to factor if the leading coefficient is positive. So we multiply both sides by -1.

$$-1 \, (-2x^2 - x + 3) = -1 \, (0)$$
$$2x^2 + x - 3 = 0$$
$$(2x + 3)(x - 1) = 0$$
$$2x + 3 = 0 \qquad \text{or} \quad x - 1 = 0$$
$$2x = -3 \quad \text{or} \qquad x = 1$$
$$x = -\frac{3}{2}$$
$$\left\{-\frac{3}{2}, 1\right\}$$

(b) $\quad t^2 + 4t + 4 = 0$
$\quad (t + 2)(t + 2) = 0$
$\qquad t + 2 = 0 \text{ or } t + 2 = 0$
$\qquad\qquad \{-2\}$

The solution set contains just one member because both factors are the same. We call the single solution in such cases a **repeated root** or a **root of multiplicity 2.** ◀

The next example points out how important it is to write the equation so that one side is 0. The Property of Zero Products applies only to a product that is equal to *zero*.

► **EXAMPLE 4**

Solve $(x - 2)(x + 2) = 5$.

Solution

Be Careful!

Be careful! The equation is not in the proper form to use the Property of Zero Products. (Why?) To have one side equal to zero, we multiply $(x - 2)(x + 2)$. So

$$(x - 2)(x + 2) = 5$$
$$x^2 - 4 = 5 \qquad \text{Multiply factors}$$
$$x^2 - 9 = 0 \qquad \text{The right side is } 0$$
$$(x + 3)(x - 3) = 0 \qquad \text{Factor left side}$$
$$\{\pm 3\} \quad \text{or} \quad \{-3, 3\} \qquad \text{Set factors equal to zero} \qquad ◄$$

The Property of Zero Products can be extended to products with more than two factors.

► **EXAMPLE 5**

Solve each equation.

(a) $x(x - 3)(x + 5) = 0$ (b) $y^3 - 25y = 0$

Solutions

(a) $x(x - 3)(x + 5) = 0$
We set each factor equal to 0 and solve.

$$x = 0 \quad \text{or} \quad x - 3 = 0 \quad \text{or} \quad x + 5 = 0$$
$$x = 3 \qquad\qquad x = -5$$
$$\{-5, 0, 3\}$$

(b) $y^3 - 25y = 0$
We factor first.

$$y(y^2 - 25) = 0$$
$$y(y + 5)(y - 5) = 0$$
$$\{\pm 5, 0\} \qquad ◄$$

Problem Set 4.5

Warm-Ups

In Problems 1 through 22, write the solution set.

1. $(x - 9)(x + 7) = 0$

2. $y(y - 6) = 0$

3. $(x + 7)(x - 1) = 0$

4. $(x - 6)(x + 6) = 0$

5. $3t(t - 8) = 0$

6. $(x + 8)(x + 8) = 0$

7. $(2x - 5)(x + 3) = 0$

8. $(3z + 7)(2z - 5) = 0$

9. $(6x + 5)(6x - 5) = 0$

10. $(5y + 3)(2y + 9) = 0$

11. $x(7x - 8)(x + 4) = 0$

12. $(5t + 9)(2t - 1) = 0$

13. $(y - 9)^2 = 0$

14. $x^2 = 0$

15. $(2w + 9)(w + 8) = 0$

16. $(x + 2)(5x - 7)(x - 5) = 0$

17. $(x + 1)(x - 1) = 0$

18. $(2x + 3)(2x + 3) = 0$

19. $(4x + 1)(3x - 5) = 0$

20. $(x - \sqrt{2})(x + \sqrt{2}) = 0$

21. $(x + \sqrt{5})(x - \sqrt{5}) = 0$

22. $(x + \sqrt{7})(x - \sqrt{7}) = 0$

In Problems 23 through 38, find the solution set by factoring.
See Examples 1, 2, and 3.

23. $x^2 = 1$

24. $x^2 = 4$

25. $x^2 - 64 = 0$

26. $4x^2 - 25 = 0$

27. $9y^2 = 4$

28. $100x^2 = 81$

29. $5x = x^2$

30. $x^2 + x = 0$

31. $x^2 + 3x + 2 = 0$

32. $z^2 + 6z + 5 = 0$

33. $x^2 - 5x - 6 = 0$

34. $w^2 = 8w - 15$

35. $x^2 + 8x + 16 = 0$

36. $y^2 - 8y = 9$

37. $2x^2 - 7x + 6 = 0$

38. $x - 10 = -3x^2$

In Problems 39 through 42, solve each equation. See Example 4.

39. $x(x + 3) = 4$

40. $z(z - 2) = -1$

41. $(x - 6)(x - 2) = -3$

42. $(x - 3)(x + 4) = -6$

In Problems 43 through 46, solve each equation. See Example 5.

43. $x^3 - 4x = 0$

44. $x^3 - x^2 - 6x = 0$

45. $(x - 8)(x + 3)(x - 1) = 0$

46. $3y^3 - 27y = 0$

Practice Exercises

In Problems 47 through 74, find the solution set by factoring.

47. $x^2 = 16$

48. $x^2 = 25$

49. $x^3 - 81x = 0$

50. $4x^2 - 81 = 0$

51. $16y^2 = 25$

52. $25x^2 = 16$

53. $x^2 - 4x = 0$

54. $x^2 = 6x$

55. $3x = x^2$

56. $x^3 - x = 0$

57. $(x - 1)(2x + 1)(x + 3) = 0$

58. $z^2 + 11z + 10 = 0$

59. $t^2 + 6t + 8 = 0$

60. $x^2 + 8x + 15 = 0$

61. $x^2 + 5x - 6 = 0$

62. $w^2 = 2w + 15$

63. $x^2 - 6x + 9 = 0$

64. $y^2 - 3y = 4$

65. $x(x + 3) - 2(x + 3) = 0$

66. $2x^2 + 7x + 3 = 0$

67. $2x^2 - x - 15 = 0$

68. $5x + 28 = 3x^2$

69. $6y^2 + y - 2 = 0$

70. $5t^2 = t + 4$

71. $x(x - 7) = -12$

72. $z(z + 1) = 2$

73. $(x - 4)(x + 1) = 14$

74. $(x + 5)(x + 2) = 4$

Challenge Problems

In Problems 75 through 77, solve each equation.

75. $\frac{1}{2}x^2 - 2 = 0$

76. $\frac{1}{4}x^2 = 16$

77. $\frac{1}{3}x^2 - 2x + 3 = 0$

In Your Own Words

78. How do we solve an equation by factoring?

79. Make up an equation that can be solved by factoring.

4.6 Applications

TAPE 8

Word problems can lead to mathematical models involving equations that can be solved by factoring. We follow the procedure developed in Section 3.4 to solve them.

> **A Procedure to Solve Word Problems**
>
> **1.** Read the problem to determine what quantities are to be found.
> **2.** Assign a variable, such as x, to represent one of the quantities to be found.
> **3.** Express all other quantities to be found in terms of x (or the chosen variable).
> **4.** Draw a figure or picture to illustrate the problem if possible. Charts are helpful.
> **5.** Read the problem again and form an equation.
> **6.** Solve the equation.
> **7.** Write an answer to the original question.
> **8.** Check the answer in the original word problem to see if it makes sense and answers the question.

The first type of word problem that we will study deals with *consecutive* integers.

► **EXAMPLE 1**

The product of two consecutive positive integers is 156. Find the integers.

Solution

STEP 2 Assign a variable.
Let n be the first integer.
Then $n + 1$ is the next integer.

STEP 3 Form an equation.
$n(n + 1) = 156$

STEP 6 Solve the equation.

$$n^2 + n = 156$$

$$n^2 + n - 156 = 0 \qquad \text{Make the right side 0}$$

$$(n + 13)(n - 12) = 0$$

$$n = -13 \quad \text{or} \quad n = 12$$

Since the problem asked for *positive* integers, we cannot use -13.

STEP 7 Write an answer to the question
The integers are 12 and 13

STEP 8 Check the answer.
We check by noting that the product of 12 and 13 is 156. ◄

The Pythagorean Theorem is one of the most used results in all of mathematics. It deals with a right triangle (a triangle containing an angle of 90°). The side opposite the right angle is called the **hypotenuse.** The hypotenuse is always the longest side of the triangle. The other two sides are often called **legs.**

The Pythagorean Theorem

In a right triangle, let a and b be the lengths of the legs and c be the length of the hypotenuse.

Then

$$a^2 + b^2 = c^2$$

► **EXAMPLE 2**

The hypotenuse of a right triangle is 13 m. Find the lengths of the other two sides (legs) if one of them is 7 m longer than the other.

Solution

Assign a variable.

Let s be the length of the shorter side (in meters).
Then $s + 7$ is the length of the other side (in meters).

Draw a figure.

s m   13 m

$s + 7$ m

Form an equation.

$$s^2 + (s + 7)^2 = 13^2 \qquad \text{Pythagorean Theorem}$$

Solve the equation.

$$s^2 + s^2 + 14s + 49 = 169$$

$$2s^2 + 14s - 120 = 0$$

$$s^2 + 7s - 60 = 0 \qquad \text{Divide by 2}$$

$$(s + 12)(s - 5) = 0 \qquad \text{Factor}$$

$$s = -12 \quad \text{or} \quad s = 5 \qquad \text{Apply the Property of Zero Products}$$

Since s represents the length of a side, it cannot be negative. So we cannot use -12 as an answer.
Thus $s = 5$ and $s + 7 = 12$.

Answer the question.

The sides are 5 m and 12 m.

Check the answer.

$$5^2 + 12^2 = 25 + 144 = 169$$

$$13^2 = 169$$

So $5^2 + 12^2 = 13^2$. ◄

► **EXAMPLE 3**

The area of a photograph of Garth Brooks is 24 in.^{2} If its length is 2 in. more than its width, find the dimensions of the photo.

Solution

Assign a variable.

Let w be the width of the photo (in inches).
Then $w + 2$ is the length of the photo (in inches).

Draw a picture.

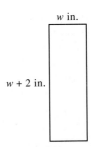

w in.

w + 2 in.

Form a model.

The area of a photograph of Garth Brooks is the product of its length and width. So

$$24 = w(w + 2)$$

Solve the equation.

$$24 = w^2 + 2w$$

$$0 = w^2 + 2w - 24$$

$$0 = (w + 6)(w - 4)$$

$$w = -6 \quad \text{or} \quad w = 4$$

The width of the photo cannot be -6. So $w = 4$ and $w + 2 = 6$.

Answer the question.

The dimensions are 4 in. by 6 in.

Check the answer.

The area of a rectangle with dimensions of 4 in. by 6 in. is 24 in², and we note that 6 is 2 more than 4. ◄

► **EXAMPLE 4**

The area of the triangular sail (the jib) on the sloop Mary K. is 54 square meters. The longest side of this triangular sail is 6 meters less than twice the length of the altitude to that side. Find the length of the longest side.

Solution

Assign a variable.

Let h be the length of the altitude (in meters).

The problem says that the base is 6 less than twice the altitude.

So $2h - 6$ is the length of the base (in meters).

Draw a picture.

h m

2h – 6 m

Form a model.

The formula for the area of a triangle is $A = \frac{1}{2}bh$. So

$$54 = \frac{1}{2}(2h - 6)h$$

Solve the equation.

$$54 = (h - 3)h \qquad \text{Distributive Property}$$

$$54 = h^2 - 3h$$

$$0 = h^2 - 3h - 54$$

$$0 = (h - 9)(h + 6)$$

$$h = 9 \quad \text{or} \quad h = -6$$

We cannot use -6 as a length. So $h = 9$ and $2h - 6 = 12$.

Answer the question.

The length of the longest side is 12 m.

Check the answer.

We check by noting that the base is 6 less than twice the altitude and that the area is $\frac{1}{2} \cdot 12 \cdot 9$ m^2, which is 54 m^2.   ◄

Problem Set 4.6

Warm-Ups

In Problems 1 through 5, see Example 1.

1. Find two consecutive positive integers whose product is 72.

2. One positive number is 4 more than another positive number and the product of the two numbers is 45. Find the numbers.

3. The low temperature was below zero both nights of the biology trip. Mrs. Carroll, the biology teacher, observed that the lows were consecutive odd integers and that the product of the two temperatures was 63. What were the low temperatures?

4. Currin has 5 more aces than double faults in a tennis match. The product of the number of aces and the number of double faults is 24. How many aces does he have?

5. The product of two consecutive positive integers is 5 more than the sum of the two integers. Find the integers.

In Problems 6 through 11, see Example 2.

6. The hypotenuse of a right triangle is 10 in. If the shorter leg is 2 inches less than the longer leg, find the lengths of both legs.

7. A hot air balloon rises 500 feet straight up from a grassy spot in the park. If the wind blows it east for 1200 feet, what is the distance between the grassy spot and the balloon?

8. The bottom of a 17-ft ladder is 8 ft from the base of a house and the top of the ladder is leaning against the house. Find the height to the top of the ladder.

9. If one leg of a right triangle is 8 less than the hypotenuse and the other leg is 1 less than the hypotenuse, find the length of each side of the triangle.

10. The Ace Hardware Store is having a grand opening. John Taylor has several strings of banners that are each 17 feet long. He wants to extend them from the front of the building to the ground. The distance from the building to the point of attachment on the ground must be 7 feet less than the height of the point of attachment on the building. How high up should he attach them to the building?

11. The hypotenuse of a right triangle is 6 more than twice the shorter leg and the longer leg is 4 more than twice the shorter leg. Find the length of each side of the triangle.

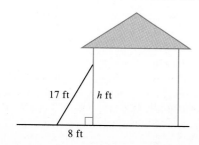

17 ft   *h* ft

8 ft

For Problems 12 through 15, see Example 3.

12. The area of a rectangle is 108 m². If the width is 3 m less than the length, find the dimensions of the rectangle.

13. Sallie Paschal's children helped her make a small

Problem 13

rectangular flower bed. The children decided that the length of the bed should be 1 yard more than twice the width. They also made the flower bed 1 square yard in area, as their mother had requested. What are the dimensions of this flower bed?

14. The width of a rectangle is 10 cm less than its length. If the area of the rectangle is 375 cm², find the dimensions of the rectangle.

15. If the length of a rectangle is three times its width and its area is 48 in.², find the dimensions of the rectangle.

For Problems 16 through 20, see Example 4.

16. The base of a triangle is twice the height. If the area of the triangle is 25 ft², find the lengths of the base and the height.

17. Jean Hess is a famous quilt maker in Louisiana. The latest design that she is using calls for triangles whose heights are four times their bases and whose areas are 32 square inches. What is the length of the base and the length of the height for her triangles?

18. One leg of a right triangle is twice the other leg and its area is 16 in.². Find the length of each leg.

19. If the length of the base of a triangle is 2 m more than twice the length of the height and its area is 42 m², find the lengths of the base and the height.

20. The area of the following figure is 230 square units. Find x.

Practice Exercises

21. Henry and Charlotte have two children born in consecutive years. If the product of the children's ages is 56, how old is each child?

22. One positive number is seven more than another positive number and the product of the two numbers is 228. Find the numbers.

23. Today's weather summary showed that the noon temperature was below zero and was 3° higher than the 6:00 A.M. temperature. If the product of the two temperatures is 88, what was the temperature at noon?

24. The product of two positive integers is 36. If one integer is 5 more than the other, find the integers.

25. The product of two consecutive positive integers is 1 more than the sum of the two integers. Find the integers.

26. If the hypotenuse of a right triangle is 13 in. and one leg is 12 in., find the length of the other leg.

27. Priscilla is making a string-art project. Her masterpiece is rectangular with dimensions of 8 inches by 15 inches. The design requires an expensive gold string to run across the diagonals of the rectangle. How much gold string does Priscilla need?

28. The hypotenuse of a right triangle is 20 in. If the shorter leg is 4 less than the other leg, find the lengths of both legs.

29. If one leg of a right triangle is 3 less than the hypotenuse and the other leg is 6 less than the hypotenuse, find the length of each side of the triangle.

30. The hypotenuse of a right triangle is 5 in. Find the lengths of the legs if one leg is 1 less than the other leg.

31. The hypotenuse of a right triangle is 2 more than twice the shorter leg and the longer leg is 2 less than twice the shorter leg. Find the length of each side of the triangle.

32. The length of the hypotenuse of a right triangle is 5 m and one leg has length 3 m. Find the length of the other leg.

33. A helicopter is searching for wreckage from a plane crash in the Atlantic Ocean. A rectangular area of 112 square miles has been designated as the main target area. If the width of the target area is 6 miles less than the length of the target area, what is the size of the target area?

34. The length of a rectangle is 1 less than three times its width. If the area of the rectangle is 2 yd^2, find the dimensions of the rectangle.

35. The width of a rectangle is 2 cm less than its length. If the area of the rectangle is 195 cm^2, find the dimensions of the rectangle.

36. If the length of a rectangle is five times its width and its area is 125 in.^{2}, find the dimensions of the rectangle.

Challenge Problems

40. Find all numbers such that the cube of the number is equal to four times the number.

41. The Luong children, Son and Tran, were born 5 years apart. Mrs. Luong was born in 1952 and is 2 years

In Your Own Words

42. Make up a word problem that can be solved by factoring.

37. The length of the base of a triangle is four times the length of the height. If the area of the triangle is 200 ft^2, find the lengths of the base and the height.

38. If the length of the base of a triangle is 2 less than four times the length of the height and its area is 45 m^2, find the lengths of the base and the height.

39. The area of the following figure is x^2 square units. Find x.

younger than her husband. The product of their children's ages is 150. How old are Son and Tran?

► Chapter Summary

GLOSSARY

Factor a polynomial   To write the polynomial as a product.

Prime polynomial   A polynomial that has no factors other than itself and 1.

Factor a polynomial completely   To write the polynomial as a product of polynomials that are prime.

Greatest common factor of a set of numbers   The largest number that is a factor of each number in the set.

Perfect trinomial square   A trinomial that will factor into a binomial squared.

Procedure to Factor a Polynomial Completely

1. Factor out the GCF if there is one. Do this *first*.

2. Count the number of terms.

3. If there are two terms, try difference of squares, difference of cubes, or sum of cubes.

Note! $p^2 + q^2$ is prime.

4. If there are three terms, look for two binomial factors.

5. If there are four or more terms, try grouping.

6. Make sure each factor is prime.

7. Check to see if the product of the factors is the original polynomial.

Factoring Rules

1. $p^2 - q^2 = (p - q)(p + q)$ Difference of squares

2. $p^3 - q^3 = (p - q)(p^2 + pq + q^2)$ Difference of cubes

3. $p^3 + q^3 = (p + q)(p^2 - pq + q^2)$ Sum of cubes

4. $p^2 + 2pq + q^2 = (p + q)^2$ Perfect trinomial square

5. $p^2 - 2pq + q^2 = (p - q)^2$ Perfect trinomial square

Procedure to Solve an
Equation by Factoring

1. Write the equation so that one side is 0.

2. Factor completely.

3. Set each factor equal to zero and solve.

4. Check, if required.

5. Write the solution set.

CHECK UPS

1. Factor $18y - 12$ completely.	Section 4.1; Example 1a
2. Factor $x(a - b) + y(a - b)$.	Section 4.1; Example 6a
3. Factor $x^2 - xy + ax - ay$ by grouping.	Section 4.1; Example 7b
4. Factor $16x^2 - 25$.	Section 4.2; Example 1b
5. Factor $8c^3 - 27$.	Section 4.2; Example 3b
6. Factor $x^2 - 9x + 18$.	Section 4.3; Example 1d
7. Factor $2a^3 - 4a^2b - 6ab^2$ completely.	Section 4.3; Example 4
8. Solve $x^2 + x = 2$.	Section 4.5; Example 1
9. Solve $(x - 2)(x + 2) = 5$.	Section 4.5; Example 4
10. The product of two consecutive positive integers is 156. Find the integers.	Section 4.6; Example 1
11. The hypotenuse of a right triangle is 13 m. Find the lengths of the other sides if one of them is 7 m longer than the other.	Section 4.6; Example 2

► Review Problems

In Problems 1 through 30, determine whether each polynomial is factored completely. Factor each polynomial completely, if possible.

1. $z^2 - 16t^2$

2. $s(b + k) - t(b + k)$

3. $r^2 + 16$

4. $3(5x + 10)$

5. $2x + 2t + xt + t^2$

6. $k^2 - 11k + 10$

7. $8u^3 - v^3$

8. $m + n^2$

9. $a^2b^3c^4 - a^3b^2c^3 + a^2b^3c^5$

10. $27 + 8q^3$

11. $12d + 4$

12. $x^2 - 2x - 2$

13. $az^2 + bz^2 - 5a - 5b$

14. $5t^2 + 5t - 30$

15. $(2x - 6)(x + 3)$

16. $100 - d^2$

17. $x^2 + 6x + 9$

18. $27t^2 - 15t - 2$

19. $x^3y^3 - 64$

20. $36x^4 - 27x^6 + 54x^9$

21. $(a + b)^3$

22. $2a(x + y^2) + b(x + y^2)$

23. $x^2 + 9x + 14$

24. $4r^2 - 4s^2$

25. $k^3 - 125j^3$

26. $-3xy^2 + 6y^2$

27. $k^2 - 2k^2n + 4 - 8n$

28. $4x^2 + x - 14$

29. $2b^3 - 5b^2 - 3b$

30. $81x^2 - 64y^2$

In Problems 31 through 40, solve each equation.

31. $(x - 7)(x - 3) = 0$

32. $x(x + 6) = 0$

33. $(z - \sqrt{15})(z + \sqrt{15}) = 0$

34. $w^2 = 144$

35. $x^2 = 3x$

36. $x^2 - 3x - 40 = 0$

37. $x^2 - 3x = 10$

38. $y^2 + 8y + 16 = 0$

39. $3t^2 - 48 = 0$

40. $x^3 - 2x^2 - 8x = 0$

In Problems 41 through 44, solve each word problem.

41. Find two consecutive negative integers whose product is 506.

42. One leg of a right triangle is 3 units less than the length of the hypotenuse and the other leg is 24 units less than the hypotenuse. Find the length of the hypotenuse.

43. The length of a rectanglar mirror is 5 units more than its width. If the area of the mirror is 176 square units, find the dimensions of the mirror.

44. The intersection of DeKalb Avenue and Piedmont Avenue as shown contains two grassy areas, each in the shape of a right triangle. The highway department has covered the two grassy areas with 216 square meters of Bermuda grass sod. Suppose the engineers know that one leg of each of the right triangles is 6 meters less than the other leg. What is the length of the legs?

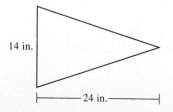

45. A short banner, 24 in. long, is in the shape of an isosceles triangle. What are the lengths of the sides of the banner?

46. Three tubes in the frame of a particular racing bicycle are welded to form a right triangle, as shown. How much tubing is required to form this triangle?

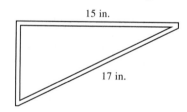

15 in.

17 in.

47. The auditorium in the new Fine Arts Building contains 560 seats. If the number of rows is 5 less than the number of seats in each row, how many rows of seats are there in the auditorium?

48. Sue Henderson received the surprise package shown from Federal Express. The volume of the package is 84 square inches. Find the dimensions of the package.

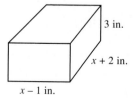

3 in.

$x + 2$ in.

$x - 1$ in.

49. The mainsail of the boat that won the Americo Cup race is in the shape of a right triangle with legs measuring 12 feet and 5 feet. What is the length of the other side of the sail?

50. LaToya rode 24 miles on her bicycle. If her rate of speed is five more than the number of hours she rode, how long did she ride?

51. In the 1988 summer Olympic games, the athletes of Poland won five more medals than did the athletes of Japan. If the product of these two numbers is 176, how many Olympic medals were won by Polish competitors?

Let's Not Forget . . .

In Problems 52 through 57, perform the operations indicated.

52. $(-x)^2$

53. $\dfrac{0}{5}$

54. $\dfrac{6t^3 + t^2}{t^2}$

55. $(y - 1)^2$

56. $(x^2 - 2) - (x^2 - x + 3)$

57. -3^2

In Problems 58 through 60, simplify each expression. Watch the role of the parentheses!

58. $x + 2(x - 1)$

59. $(x + 2)(x - 1)$

60. $(x + 2)x - 1$

In Problems 61 through 63, determine if 5 is a factor or a term.

61. $\sqrt{3} + 5$

62. $5(x + 2)$

63. $\dfrac{1}{x + 5}$

In Problems 64 through 66, multiply each expression by $\dfrac{2}{3}$.

64. $\dfrac{3}{2}x$

65. $6(x + 7)$

66. $9y + 12$

In Problems 67 through 70, label each as an expression, equation, or inequality. Solve the equations and inequalities and simplify the expressions.

67. $(x - 2)(x + 3)$

68. $(x - 2)(x + 3) = 0$

69. $5(x + 3) - 2 < -7$

70. $5(x + 3) - 2$

In Problems 71 through 74, identify the expressions that are factored. Factor those that are not factored if possible.

71. $ay + y$

72. $x^3 + 1$

73. $(x + 1)^3$

74. $4(b + 2)$

► You Decide

You have been offered a very desirable summer job. There are two salary options. Plan A is an hourly wage of $4 plus a $60 bonus if you work a full 6-day week. Plan B is a straight $5 an hour. Let h be the hours worked in a week. Write an expression for the weekly salary under each plan. Choose a plan and give your reasons.

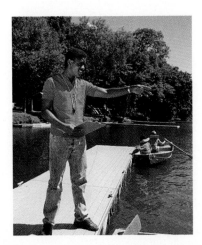

► Chapter 4 Test

In Problems 1 through 5, choose the correct answer.

1. The GCF of $48x^2y$ and $80xy$ is (?).
  (a) 16   (b) $16xy$
  (c) $8xy$ (d) xy

3. One of the factors of $x(a-1) - y(a-1)$ is (?).
  (a) $x + y$ (b) $x - y$
  (c) x (d) $x(a-1)$

5. Which polynomial is factored?
  (a) $x^2 + 2x$ (b) $a(r-s) + b(r-s)$
  (c) $x^3 - 8$ (d) $(x+2)^3$

2. One of the factors of $x^2 - 64$ is (?).
  (a) $x^2 - 8x + 16$ (b) $x^2 + 8x + 16$
  (c) $x + 8$ (d) $x + 4$

4. $x^2 - 12x + 36 = $ (?).
  (a) $(x-9)(x-4)$ (b) $(x-6)(x+6)$
  (c) $(x+6)^2$ (d) $(x-6)^2$

In Problems 6 through 15, factor each polynomial completely.

6. $8a^3 - 27$

8. $ax - bx + 2a - 2b$

10. $a^2 - 2a - 8$

12. $-6c^2d - 3cd^2$

14. $27a^2b^3 - 36a^3b^2 + 45a^4b^4$

7. $2x^2 + 9x - 5$

9. $x(p-q) - y(p-q)$

11. $x^2 + 14x + 49$

13. $4y^2 - 25$

15. $t^2 + 25$

In Problems 16 through 18, solve each equation.

16. $x^2 - 81 = 0$ **17.** $x^2 - 4x = 12$ **18.** $x(x-1)(x+3) = 0$

19. The area of a rectangle is 234 yd². If the width is 5 yd less than the length, find the dimensions of the rectangle.

20. The hypotenuse of a right triangle is 2 m less than three times the shortest side. If the other side is 2 m more than twice the shortest side, what are the lengths of the three sides?

5 Rational Expressions

See Problem Set 5.7, Exercise 18.

Connections

The radio station WKIX often conducts unusual contests offering cash prizes. In one such contest, a secret 7-digit integer is to be guessed by listeners who call in. The contest lasts two weeks, daily guesses are taken, and hints given. The prize for the first correct guess is $1 on the first day, $2 on the second day, and $4 on the third day, and it continues to double each day that the number remains unguessed. The winner is awarded the sum of the daily prizes to the day of his or her correct guess, or $1000 whichever he or she chooses. Suppose you guess correctly on the 10th day of the contest and have 15 seconds to choose between $1000 or the sum of the first 10 days' prizes. What do you choose?

A mathematician would think of this as a polynomial $1 + x + x^2 + x^3 + \cdots + x^9$. This polynomial can be written as the fraction $\dfrac{x^{10} - 1}{x - 1}$. Replace x with 2 and you can see that your prize would be $1023 if you select the sum of the daily prizes.

We studied polynomials in Chapters 2 and 4. In this chapter we study the quotient of two polynomials, which is called a *rational expression*. Rational expressions are algebraic fractions. We see that performing operations with rational expressions is just like working with fractions in arithmetic. Being successful with rational expressions is strongly connected to being successful with factoring polynomials.

5.1   Rational Expressions

TAPE 9

A **rational expression** is a fraction that has a polynomial for its numerator and a polynomial for its denominator. Three examples of rational expressions are

$$\frac{x^3 + 2x^2 - 17}{x^2 + 5x + 1}, \quad \frac{11}{x^3}, \quad \text{and} \quad \frac{16x^7}{3(x + 1)}$$

The value of a rational expression depends on the value of the polynomials.

► **EXAMPLE 1**

Find the value of $\dfrac{x + 3}{2x^2 + 1}$ when:

(a) x is 2

(b) x is -1

Solutions

(a) Replace x with 2.

$$\frac{x + 3}{2x^2 + 1} = \frac{2 + 3}{2(2)^2 + 1}$$

$$= \frac{5}{8 + 1}$$

$$= \frac{5}{9}$$

The value of $\dfrac{x + 3}{2x^2 + 1}$ when x is 2 is $\dfrac{5}{9}$.

(b) Replace x with -1.

$$\frac{x + 3}{2x^2 + 1} = \frac{-1 + 3}{2(-1)^2 + 1}$$

$$= \frac{2}{2 + 1}$$

$$= \frac{2}{3}$$

The value of $\dfrac{x + 3}{2x^2 + 1}$ when x is -1 is $\dfrac{2}{3}$.   ◀

Consider the rational expression $\dfrac{1}{x - 2}$. If x is 4, then its value is $\dfrac{1}{2}$. However, if x is 2, the expression takes the form $\dfrac{1}{0}$. This is *not defined*. We say that the rational expression $\dfrac{1}{x - 2}$ is *undefined* when x is 2. A rational expression is **undefined** when its denominator has the value of 0.

► **EXAMPLE 2**

Find the values of x for which each rational expression is undefined.

(a) $\dfrac{x^2 - 15}{x}$ (b) $\dfrac{x^2}{5}$

Solutions

(a) If x has the value 0, then the rational expression $\dfrac{x^2 - 15}{x}$ becomes $\dfrac{-15}{0}$. So we can say that $\dfrac{x^2 - 5}{x}$ is undefined when x is 0.

(b) The denominator, 5, is *never* zero for any value of x. We say that the rational expression $\dfrac{x^2}{5}$ is defined for all values of x.   ◀

► **EXAMPLE 3**

Find the values of x for which $\dfrac{x + 3}{(x - 1)(x + 2)}$ is undefined.

Solution

If x is 1, the rational expression $\dfrac{x + 3}{(x - 1)(x + 2)}$ becomes $\dfrac{4}{0(3)}$, or $\dfrac{4}{0}$. If x has the value of -2, then the expression becomes $\dfrac{1}{-3(0)}$, or $\dfrac{1}{0}$. Therefore, the rational expression $\dfrac{x + 3}{(x - 1)(x + 2)}$ is undefined when x is 1 *or* x is -2.   ◄

An important property of fractions concerns signs. Notice the signs in the following computations.

$\boxed{1}$ $\dfrac{-6}{2} = -3$ A negative number divided by a positive number.

$\boxed{2}$ $\dfrac{6}{-2} = -3$ A positive number divided by a negative number.

$\boxed{3}$ $-\dfrac{6}{2} = -3$ The opposite of a positive number.

That is, because of the sign rules of arithmetic, we have

$$\frac{-6}{2} = \frac{6}{-2} = -\frac{6}{2}$$

This idea is also true for rational expressions.

The Sign Property of Rational Expressions

If A and B are polynomials and $\dfrac{A}{B}$ is defined, then

$$\frac{-A}{B} = \frac{A}{-B} = -\frac{A}{B}$$

► **EXAMPLE 4**

Use the Sign Property to write $\dfrac{-1}{x}$ two different ways.

Solution

$$\frac{-1}{x} = \frac{1}{-x} = -\frac{1}{x}$$ ◄

► **EXAMPLE 5**

Use the Sign Property to write each rational expression two different ways.

(a) $-\dfrac{x^2}{x+2}$ (b) $\dfrac{-x-1}{x^3}$

Solutions

(a) $-\dfrac{x^2}{x+2} = \dfrac{-x^2}{x+2} = \dfrac{x^2}{-(x+2)}$

Be Careful!   The parentheses in the denominator are *very* important, since $-(x+2)$ is *not* the same as $-x+2$.

(b) We must be very careful here. In using $\dfrac{-A}{B}$ from the Sign Property, the $-A$ in the numerator is the opposite of A. In the numerator of the example, $-x-1$ is *not* the opposite of $x-1$. However, since $-x-1 = -(x+1)$, $-x-1$ *is* the opposite of $(x+1)$.

Be Careful!

$$\frac{-x-1}{x^3} = \frac{-(x+1)}{x^3} = \frac{x+1}{-x^3} = -\frac{x+1}{x^3}$$ ◄

► **EXAMPLE 6**

The airline distances between Atlanta, San Diego, and Seattle are shown below.

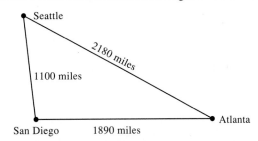

(a) Delta flight 250 flies from San Diego to Atlanta in 3 hours. What is its average ground speed?

(b) Delta flight 835 flies from Atlanta to Seattle in $t-2$ hours. What is its average ground speed?

(c) Northwest flies from Seattle to San Diego with an average ground speed of $r+25$ mph. Write an expression for the time of the flight.

(d) How long did the flight in part (c) take if r is 525 mph?

Solutions

(a) Recall the formula, $d = rt$. For our use in this problem, we use the form $r = \dfrac{d}{t}$. So ground speed is distance divided by time.

$$\frac{1890}{3} = 630$$

The average ground speed is 630 mph.

(b) Ground speed is distance divided by time. The distance from Atlanta to Seattle is 2180 miles and time is $t - 2$ hours. So we have $\dfrac{2180}{t - 2}$. The average ground speed is $\dfrac{2180}{t - 2}$ mph.

(c) Here, we want to find time, so we use the form $t = \dfrac{d}{r}$. Thus, we have

$$\frac{1100}{r + 25}.$$

(d) We must substitute 525 for r.

$$\frac{1100}{525 + 25} = \frac{1100}{550} = 2$$

The flight took 2 hours. ◀

Problem Set 5.1

Warm-Ups

In Problems 1 through 4, evaluate the given rational expressions when x has the value −1. See Example 1.

1. $\dfrac{2}{x - 1}$

2. $\dfrac{x}{x^2 + 1}$

3. $\dfrac{x^2}{x^2 - x + 1}$

4. $\dfrac{2x + 3}{3x - 1}$

In Problems 5 through 12, find the value of x, if any, for which each rational expression is undefined. See Example 2.

5. $\dfrac{5}{x}$

6. $\dfrac{x}{5}$

7. $\dfrac{2x + 7}{x^2}$

8. $\dfrac{-2}{x - 1}$

9. $\dfrac{3}{x - 3}$

10. $\dfrac{22}{5 + x}$

11. $\dfrac{x}{x + 8}$

12. $\dfrac{x + 8}{x}$

In Problems 13 through 18, find the values of x for which each rational expression is undefined. See Example 3.

13. $\dfrac{2x - 5}{(x - 5)(x - 2)}$

14. $\dfrac{x^2}{(x + 1)(x + 9)}$

15. $\dfrac{1}{(4 - x)(11 - x)}$

16. $\dfrac{x}{(x - 1)(x + 1)}$

17. $\dfrac{5x^3 - 2}{x^2 - 9}$

18. $\dfrac{3x}{25 - x^2}$

In Problems 19 through 21, write each of the given rational expressions in two different ways. See Example 4.

19. $\dfrac{-5}{x}$

20. $\dfrac{8}{-x}$

21. $-\dfrac{17}{x^2}$

*In Problems 22 through 27, write each rational expression two different ways. See
Example 5.*

22. $\dfrac{-(x+4)}{x}$

23. $\dfrac{11x^3}{-(2x-3)}$

24. $-\dfrac{x+7}{7-x}$

25. $\dfrac{-3x}{x+4}$

26. $\dfrac{6-5x}{-x^3}$

27. $-\dfrac{x+6}{x-3}$

For Problems 28 and 29, see Example 6.

28. Barney and Joan are driving their camper from Anchorage to Fairbanks, a distance of 360 miles.
   (a) If they make the trip in 9 hours, what is their average speed?
   (b) If they make the trip in $t-1$ hours, what is their average speed?
   (c) What is their average speed in part (b) if t is 9?

29. A top-fuel dragster is judged on the elapsed time it takes to cover one-fourth of a mile from a standing start.
   (a) What is the elapsed time if the driver averages 150 mph?
   (b) Write an expression for the elapsed time if the driver averages $s+100$ mph.
   (c) What is the elapsed time if s is 20?

Problem 29

Practice Exercises

*In Problems 30 through 35, evaluate the given rational expressions when x has the value
1 and y has the value -2.*

30. $\dfrac{3}{x+1}$

31. $\dfrac{2x+y}{5+y}$

32. $\dfrac{16}{2y-x}$

33. $\dfrac{y}{y^2-1}$

34. $\dfrac{x^2y}{y^2+1}$

35. $\dfrac{2xy-3}{y-2}$

*In Problems 36 through 53, find the values of x, if any, for which each rational expression
is undefined.*

36. $\dfrac{7}{x}$

37. $\dfrac{x}{7}$

38. $\dfrac{x-6}{x^3}$

39. $\dfrac{1}{x-2}$

40. $\dfrac{2}{2-x}$

41. $\dfrac{14}{x+5}$

42. $\dfrac{-9}{4+x}$

43. $\dfrac{-3}{x+2}$

44. $\dfrac{4-x}{x}$

45. $\dfrac{x+2}{x-2}$

46. $\dfrac{x-5}{x+5}$

47. $\dfrac{15-x}{16-x}$

48. $\dfrac{x+5}{(x-3)(x-6)}$

49. $\dfrac{x^3+5}{(x+2)(x+4)}$

50. $\dfrac{x}{(1-x)(16-x)}$

51. $\dfrac{1}{(x-2)(x+2)}$

52. $\dfrac{x^2+2}{x^2-16}$

53. $\dfrac{x^2-1}{49-x^2}$

In Problems 54 through 62, write each of the given rational expressions in two different ways.

54. $\dfrac{-2}{x}$

55. $\dfrac{9}{-2x}$

56. $-\dfrac{22}{3x}$

57. $\dfrac{-(x-1)}{x}$

58. $\dfrac{7x^5}{-(3x+2)}$

59. $-\dfrac{x-3}{3+x}$

60. $\dfrac{-4x}{x-6}$

61. $\dfrac{9+8x}{-x}$

62. $-\dfrac{x-2}{x-7}$

63. Suppose that *unit price* is defined as price per ounce.
  (a) Find the unit price of a 32-oz jar of Paul Newman spaghetti sauce that sells for $1.60.
  (b) Find the unit price of a 48-oz jar of Paul Newman spaghetti sauce that sells for $2.52.
  (c) Which size, the 32- or 48-oz jar, is the better buy?
  (d) What is the unit price of a *k*-oz jar of a product that sells for *x* dollars?

64. Suppose Terry Pendleton has 28 hits in his first 100 "at bats," giving him a batting average average of 0.280.
  (a) If Terry gets 17 hits in his next 50 "at bats," what does his batting average become?
  (b) If Terry gets *h* hits in his next *b* "at bats," what does his batting average become?

65. A space shuttle is flying at a speed of 15,000 miles per hour.
  (a) How many minutes will it take the shuttle to cross the United States? The distance across the United States is 2500 miles.
  (b) If the speed is increased by 5000 miles per hour, how many minutes will it take the shuttle to cross the United States?
  (c) If the speed is increased by *r* miles per hour, how many minutes will it take the shuttle to cross the United States?

66. If Maria rides her bicycle 5 miles per hour faster than Walt in a 26-mile race, write a rational expression for:
  (a) Maria's time.

  (b) Walt's time.

Challenge Problems

Find the value of x, if any, for which each rational expression is undefined.

67. $\dfrac{6}{2x-1}$

68. $\dfrac{x}{3x-1}$

69. $\dfrac{x+1}{2x+1}$

70. $\dfrac{x^2-2x-17}{3x+1}$

71. $\dfrac{7}{3x-4}$

72. $\dfrac{x}{5x+19}$

73. $\dfrac{x}{x^2+1}$

74. $\dfrac{3x+1}{x^2-2}$

75. $\dfrac{5}{x^2-5}$

In Your Own Words

76. What does it mean to say a rational expression is undefined?

5.2 Fundamental Principle of Rational Expressions

TAPE 9

We know that $\frac{1}{2}, \frac{2}{4}$, and $\frac{10}{20}$ are just different ways of writing the same number. They illustrate an important basic principle of fractions.

The Fundamental Principle of Rational Expressions

If A, B, and C are polynomials, then

$$\frac{AC}{BC} = \frac{A}{B}$$

where B and C do not have the value 0.

The Fundamental Principle is the most important tool we have for the algebra of fractions. The Fundamental Principle allows us to divide out **factors** that are common to both the numerator and the denominator. This is how we reduce fractions to their simplest form.

$$\frac{10}{20} = \frac{1(10)}{2(10)} = \frac{1}{2}$$

Since 10 was a common *factor* of both the numerator and the denominator, the fraction was reduced by using the Fundamental Principle.

A fraction is reduced to **lowest terms** if there are no common factors in the numerator and denominator.

We also use the Fundamental Principle to build up fractions. For example, suppose we wish to find a fraction equal in value to $\frac{3}{4}$ but with 28 as its denominator. We use the Fundamental Principle to rewrite such a fraction.

$$\frac{3}{4} = \frac{3(7)}{4(7)} = \frac{21}{28}$$

In this case, we used the Fundamental Principle to multiply *both* the numerator and the denominator by the same (nonzero) number, 7.

We can reduce the rational expression $\frac{5x}{7x}$ by using the Fundamental Principle because x is a *factor* of both the numerator and the denominator of the fraction.

$$\frac{5x}{7x} = \frac{5}{7} \qquad \text{Fundamental Principle}$$

The rational expression $\frac{5 + x}{7 + x}$ *cannot* be reduced, since the Fundamental Principle applies only to *factors*.

Be Careful!

$$\frac{5 + x}{7 + x} \qquad \text{Does } not \text{ reduce.}$$

We *must* master this idea before we continue.

► **EXAMPLE 1**

Use the Fundamental Principle to reduce each of the following rational expressions to lowest terms.

(a) $\dfrac{-6x^2}{10x^3}$ (b) $\dfrac{x^2 + 2y}{x^2 + 4y}$

Solutions

(a) We notice that $2x^2$ is a common *factor* of both the numerator and denominator of this fraction. We can use the Fundamental Principle.

$$\frac{-6x^2}{10x^3} = \frac{-3 \cdot 2x^2}{5x \cdot 2x^2} = \frac{-3}{5x}$$

Be Careful!

(b) We *must* resist the temptation to reduce this fraction until we factor it. However, neither the numerator nor the denominator will factor, so the fraction *cannot be reduced*.

$$\frac{x^2 + 2y}{x^2 + 4y} \qquad \text{Does not reduce} \qquad ◄$$

► **EXAMPLE 2**

Use the Fundamental Principle to reduce each of the following rational expressions to lowest terms.

(a) $\dfrac{3x - 3y}{x - y}$ (b) $\dfrac{3x + 9}{x^2 - 9}$

Solutions

(a) Because the Fundamental Principle applies only to *factors*, we must factor the numerator and denominator completely before applying it.

$$\frac{3x - 3y}{x - y} = \frac{3(x - y)}{x - y} \qquad \text{Factor the numerator}$$

$$= \frac{3\,(x - y)}{1\,(x - y)}$$

$$= \frac{3}{1} = 3 \qquad \begin{array}{l}\text{Reduce by using the}\\ \text{Fundamental Principle}\end{array}$$

(b) Again, the Fundamental Principle *cannot* be used until numerator and denominator are *factored*.

$$\frac{3x + 9}{x^2 - 9} = \frac{3\,(x + 3)}{(x - 3)\,(x + 3)} \qquad \text{Factor}$$

$$= \frac{3}{x - 3} \qquad \begin{array}{l}\text{Reduce by using the}\\ \text{Fundamental Principle} \quad ◄\end{array}$$

► **EXAMPLE 3**

Use the Fundamental Principle to reduce $\dfrac{4x^2 - 9}{2x^2 + x - 6}$ to lowest terms.

Solution

The numerator and denominator must be factored. Notice that the numerator is the difference of two squares and the denominator is a trinomial.

$$\frac{4x^2 - 9}{2x^2 + x - 6} = \frac{(2x + 3)\,(2x - 3)}{(x + 2)\,(2x - 3)} \qquad \text{Factor}$$

$$= \frac{2x + 3}{x + 2} \qquad \begin{array}{l}\text{Reduce by using the}\\ \text{Fundamental Principle} \blacktriangleleft\end{array}$$

► EXAMPLE 4

Use the Fundamental Principle to reduce $\dfrac{x^2 - 1}{1 - x}$ to lowest terms.

Solution

$$\frac{x^2 - 1}{1 - x} = \frac{(x + 1)(x - 1)}{1 - x} \qquad \text{Factor the numerator}$$

Be Careful! There would be a common factor and we could reduce this fraction if the denominator were $x - 1$ instead of $1 - x$. Is there some relationship between them? Recall that $(x - 1)$ and $(1 - x)$ are opposites. That is,

$$(1 - x) = -(x - 1)$$

Notice what happens if we *replace* $1 - x$ in the denominator with $-(x - 1)$.

$$\frac{(x + 1)(x - 1)}{1 - x} = \frac{(x + 1)\,(x - 1)}{-\,(x - 1)}$$

Now there is a common factor of $(x - 1)$ and the fraction can be reduced.

$$\frac{(x + 1)(x - 1)}{-(x - 1)} = \frac{x + 1}{-1} = -(x + 1) \qquad \blacktriangleleft$$

Opposites often come up in rational expressions. It is important to recognize opposites.

Some Opposites
$x - 2$ and $2 - x$
$6 - t$ and $t - 6$
$3y - 8$ and $8 - 3y$
$x + 1$ and $-x - 1$

Don't become confused and think that $x + 2$ and $2 + x$ are opposites. The Commutative Property for Addition says that $x + 2$ and $2 + x$ are *equal*.

Reducing Opposites

$$\frac{p - q}{q - p} = \frac{p - q}{-(p - q)} = -1$$

► **EXAMPLE 5**

Use the Fundamental Principle to reduce each of the following rational expressions to lowest terms.

(a) $\dfrac{x^2 - 4}{8 + x^3}$ (b) $\dfrac{ax - bx + 2ay - 2by}{3x + 6y}$

Solutions

(a) We *must* factor before we give any thought to reducing this expression. We notice that the numerator is the difference of two squares and the denominator is the sum of two cubes.

$$\frac{x^2 - 4}{8 + x^3} = \frac{(x + 2)\,(x - 2)}{(2 + x)\,(4 - 2x + x^2)} \qquad \text{Factor}$$

Now that it is factored, note the common factor, $x + 2$.

$$\frac{x^2 - 4}{8 + x^3} = \frac{x - 2}{x^2 - 2x + 4}$$

(b) The first two terms of the *numerator* have x as a common factor, whereas the last two terms have $2y$ as a common factor.

$$\frac{ax - bx + 2ay - 2by}{3x + 6y} = \frac{x(a - b) + 2y(a - b)}{3(x + 2y)}$$

$$= \frac{(a - b)\,(x + 2y)}{3\,(x + 2y)}$$

$$= \frac{a - b}{3} \qquad \blacktriangleleft$$

When working with rational expressions, we often wish to change the denominator. If we consider the Fundamental Principle from the other direction, we see how to do this.

$$\frac{A}{B} = \frac{AC}{BC}$$

Notice that we used the Fundamental Principle to *multiply* the numerator and the denominator of the fraction $\dfrac{A}{B}$ by the *same* number, C.

► **EXAMPLE 6**

Write $\dfrac{5}{x}$ with a denominator of $2x^3$.

Solution

We need to find a number such that,

$$\frac{5}{x} = \frac{5 \cdot ?}{x \cdot ?} = \frac{\square}{2x^3}$$

If we multiply x by $2x^2$, we will have $2x^3$, the desired denominator. So we use the Fundamental Principle to multiply both the numerator and denominator by $2x^2$.

$$\frac{5}{x} = \frac{5}{x} \frac{(2x^2)}{(2x^2)} = \frac{10x^2}{2x^3}$$ ◄

► **EXAMPLE 7**

Write each given rational expression as a rational expression with the indicated denominator.

(a) $\dfrac{x+1}{x-1}$; $x^2 - 1$ (b) $\dfrac{2}{x+3}$; $x^2 - x - 12$

Solutions

(a) Since the denominator we *want*, $x^2 - 1$, can be factored as $(x-1)(x+1)$, we must multiply the denominator by $(x+1)$. Of course, we also must multiply the numerator by $(x+1)$.

$$\frac{x+1}{x-1} = \frac{(x+1)}{(x-1)} \frac{(x+1)}{(x+1)} = \frac{x^2+2x+1}{x^2-1}$$

(b) If we factor the desired denominator, we see that

$$x^2 - x - 12 = (x+3)(x-4)$$

We need to multiply the numerator and the denominator by $(x-4)$.

$$\frac{2}{x+3} = \frac{2}{(x+3)} \frac{(x-4)}{(x-4)} = \frac{2x-8}{x^2-x-12}$$ ◄

Problem Set 5.2

Warm-Ups

In Problems 1 through 15, use the Fundamental Principle, when it applies, to reduce the rational expressions to lowest terms. See Example 1.

1. $\dfrac{4x}{6y}$

2. $\dfrac{6x^2}{9x^3}$

3. $\dfrac{14x^3y^2}{7x^2y^2}$

4. $\dfrac{2y^4}{3y^2}$

5. $\dfrac{3x(x+2)}{6xy}$

6. $\dfrac{-2x^2y}{xy^2}$

7. $\dfrac{(-2x)^2y}{xy^2}$

8. $\dfrac{-n^4}{(-n)^3(3+n)}$

9. $\dfrac{(x+1)^2}{(1+x)^3}$

10. $\dfrac{(y-5)^5}{(y-5)^4}$

11. $\dfrac{xy}{x+y}$

12. $\dfrac{5(2x+3y)}{15y}$

13. $\dfrac{xy^2}{xy}$

14. $\dfrac{x+y^2}{xy}$

15. $\dfrac{xy+y^2}{xy}$

In Problems 16 through 27, reduce to lowest terms. See Example 2.

16. $\dfrac{(x-7)(x-1)}{(x+1)(x-7)}$

17. $\dfrac{8(x-3)}{6x-18}$

18. $\dfrac{11-33y}{22(1-3y)}$

19. $\dfrac{7x^3(a+b)}{-14x^2(a+b)^4}$

20. $\dfrac{5x^2(x+2)}{20x+40}$

21. $\dfrac{x^2+3x}{6+2x}$

22. $\dfrac{3xy+9x^2y}{x^2z+3x^3z}$

23. $\dfrac{2(x+4)}{x^2-16}$

24. $\dfrac{1-x^2}{2-2x}$

25. $\dfrac{8x^2z^2-2x^2}{16z+8}$

26. $\dfrac{axyz+txyz}{a^2-t^2}$

27. $\dfrac{x^2-9}{(x-3)^2}$

In Problems 28 through 33, reduce to lowest terms. See Example 3.

28. $\dfrac{x^2+3x+2}{3x+6}$

29. $\dfrac{2x^2-6x}{x^2-2x-3}$

30. $\dfrac{x^2-x-2}{x^2-1}$

31. $\dfrac{27x^2-3y^2}{9x^2+6xy+y^2}$

32. $\dfrac{x^2+3x+2}{x^2-x-2}$

33. $\dfrac{x^2-x-6}{x^2+x-2}$

In Problems 34 through 39, reduce to lowest terms. See Example 4.

34. $\dfrac{5-7t}{7t-5}$

35. $\dfrac{3t-1}{1-3t}$

36. $\dfrac{2x-6}{3-x}$

37. $\dfrac{10x-6}{3-5x}$

38. $\dfrac{3-2x}{4x^2-9}$

39. $\dfrac{1-x}{x^2-3x+2}$

In Problems 40 through 45, use the Fundamental Principle to reduce the rational expressions to lowest terms. See Example 5.

40. $\dfrac{x+1}{x^3+1}$

41. $\dfrac{3-y}{27-y^3}$

42. $\dfrac{ax+ay+bx+by}{5a+5b}$

43. $\dfrac{(x-y)^3}{x^3-y^3}$

44. $\dfrac{3x-y}{3x+3x^2-xy-y}$

45. $\dfrac{1-8x^3}{1-2x-2y+4xy}$

In Problems 46 through 48, write each rational expression as a fraction with the indicated denominator. See Example 6.

46. $\dfrac{1}{x}$; $3x^2$

47. $\dfrac{7}{2x}$; $6x^3$

48. $\dfrac{2y}{3ax}$; $12a^2x$

In Problems 49 through 52, write each rational expression as a fraction with the indicated denominator. See Example 7.

49. $\dfrac{2}{x+1}$; $5(x+1)$

50. $\dfrac{-6}{x-9}$; x^2-81

51. $\dfrac{x+1}{x-2}$; x^2-4x+4

52. $\dfrac{-16}{2x-7}$; $6x^2-7x-49$

Practice Exercises

In Problems 53 through 94, use the Fundamental Principle, when it applies, to reduce the rational expressions to lowest terms.

53. $\dfrac{8x}{6y}$

54. $\dfrac{6x^3}{8x^4}$

55. $\dfrac{16x^5t^3}{8x^3t^4}$

56. $\dfrac{2t^6}{3t^3}$

57. $\dfrac{2x(x - 3)}{6xy}$

58. $\dfrac{12x^3y}{xy^4}$

59. $\dfrac{-2x^2y}{x^3y}$

60. $\dfrac{(-n)^4}{-n^3(3 + n)}$

61. $\dfrac{(s + 5)^3}{(s + 5)^5}$

62. $\dfrac{(y - 5)^4}{(y - 5)^2}$

63. $\dfrac{st^2}{t - s}$

64. $\dfrac{6(3x + 2y)}{15y}$

65. $\dfrac{(x + 7)(x + 1)}{(x + 1)(x - 7)}$

66. $\dfrac{6(x + 2)}{8x + 16}$

67. $\dfrac{13 + 52y}{26(1 + 4y)}$

68. $\dfrac{6x^2(a + b)}{-42x^3(a + b)^2}$

69. $\dfrac{7t^2(t - 3)}{21t - 63}$

70. $\dfrac{z^3 - 3z}{2z^2 - 6}$

71. $\dfrac{2st - 8s^2t}{s^2t - 4s^3t}$

72. $\dfrac{3(x - 4)}{x^2 - 16}$

73. $\dfrac{1 - x^2}{3 + 3x}$

74. $\dfrac{8x^2z^2 - 2x^2}{8z - 4}$

75. $\dfrac{aklm - tklm}{a^2 - t^2}$

76. $\dfrac{x^2 - 4}{(x + 2)^2}$

77. $\dfrac{x^2 - 3x + 2}{3x - 6}$

78. $\dfrac{2x^2 + 6x}{x^2 + 2x - 3}$

79. $\dfrac{x^2 + x - 2}{x^2 - 1}$

80. $\dfrac{16x^2 - 4y^2}{4x^2 - 4xy + y^2}$

81. $\dfrac{x^2 + 3x + 2}{x^2 + x - 2}$

82. $\dfrac{3x - 6}{2 - x}$

83. $\dfrac{x^2 + x - 6}{x^2 - x - 2}$

84. $\dfrac{2 - 3y}{9y^2 - 4}$

85. $\dfrac{2 - t}{t^2 - 3t + 2}$

86. $\dfrac{6x^2 + 13x - 15}{2x^2 + 5x - 3}$

87. $\dfrac{7 - 3t}{3t - 7}$

88. $\dfrac{6x^3 + 7x^2y - 20xy^2}{6x^2y + 23xy^2 + 20y^3}$

89. $\dfrac{s - 1}{s^3 - 1}$

90. $\dfrac{4 + x}{64 + x^3}$

91. $\dfrac{ax - ay + bx - by}{7a + 7b}$

92. $\dfrac{(x + y)^2}{x^3 + y^3}$

93. $\dfrac{25x^2 - y^2}{5x + 5x^2 - xy - y}$

94. $\dfrac{1 - 4x^2}{1 + 2x + 2y + 4xy}$

In Problems 95 through 103, write each rational expression as a fraction with the indicated denominator.

95. $\dfrac{2}{y}$; $5y^2$

96. $\dfrac{16}{3x}$; $6x^3$

97. $\dfrac{2b}{3ax}$; $21a^2x^2$

98. $\dfrac{2}{x - 2}$; $5(x - 2)$

99. $\dfrac{-x}{x - 7}$; $x^2 - 49$

100. $\dfrac{x + 3}{x - 3}$; $x^2 - 9$

101. $\dfrac{x - 1}{x + 2}$; $x^2 + 4x + 4$

102. $\dfrac{x - 5}{2x + 7}$; $2x^2 - 3x - 35$

103. $\dfrac{-10}{6x + 5}$; $6x^2 - 13x - 15$

Challenge Problems

In Problems 104 through 111, use the Fundamental Principle to reduce the rational expressions to lowest terms.

104. $\dfrac{x^4 - 1}{x^2 + 1}$

105. $\dfrac{x^4 - 16}{x + 2}$

106. $\dfrac{2t^3 - 1}{4t^6 - 1}$

107. $\dfrac{1 - x^4}{x - 1}$

108. $\dfrac{s^6 - 16}{4 - s^3}$

109. $\dfrac{6y^4 - y^2 - 35}{6y^5 + 11y^3 - 7y}$

110. $\dfrac{ax + bx + ay + by}{2ax - ay + 2bx - by}$

111. $\dfrac{pr + qs - qr - ps}{pr - qs - qr + ps}$

In Your Own Words

112. What does the Fundamental Principle of Rational Expressions allow us to do with a fraction?

113. Explain the steps in reducing $\dfrac{x^3 + 4x^2}{x^2 - 16}$ to lowest terms.

5.3   Multiplication and Division

TAPE 9

Because rational expressions are fractions, we multiply and divide them exactly as we multiply and divide fractions.

Let's multiply $\dfrac{3}{5}$ times $\dfrac{5}{9}$ to review the procedure for multiplication of fractions.

$$\frac{3}{5} \cdot \frac{5}{9} = \frac{3(5)}{5(9)} \qquad \text{Multiply numerators and}$$
$$\text{multiply denominators}$$

Instead of actually multiplying, we look for common factors in both the numerator and denominator so that we can reduce. 5 is a factor of both the numerator and denominator.

$$\frac{3(5)}{5(9)} = \frac{1}{3} \qquad \text{Reduce}$$

Multiplication of Rational Expressions

If A, B, C, and D are polynomials and B and D do not have value 0.

$$\frac{A}{B} \cdot \frac{C}{D} = \frac{A \cdot C}{B \cdot D}$$

► **EXAMPLE 1**

Perform the operations indicated.

(a) $\dfrac{2x}{3y} \cdot \dfrac{5x}{7y^2}$ (b) $\dfrac{3x^2}{10y} \cdot \dfrac{5xy^2}{6}$

Solutions

(a) $\dfrac{2x}{3y} \cdot \dfrac{5x}{7y^2} = \dfrac{2x(5x)}{3y(7y^2)} = \dfrac{10x^2}{21y^3}$

(b) $\dfrac{3x^2}{10y} \cdot \dfrac{5xy^2}{6} = \dfrac{3x^2(5xy^2)}{10y(6)} = \dfrac{15x^3y^2}{60y}$

We notice that this fraction will reduce. Very often, multiplication of fractions results in a fraction that will reduce. We *must* always look for common factors in the numerator and denominator.

$$\frac{15x^3y^2}{60y} = \frac{15y(x^3y)}{15y(4)} = \frac{x^3y}{4}$$

◄

► **EXAMPLE 2**

Perform the operations indicated.

(a) $\dfrac{5x - 10}{x + 3} \cdot \dfrac{2x + 6}{3x - 6}$

(b) $\dfrac{x^2 - 1}{x^2 - 4} \cdot \dfrac{x + 2}{x - 1}$

Solutions

Be Careful!

(a) $\dfrac{5x - 10}{x + 3} \cdot \dfrac{2x + 6}{3x - 6} = \dfrac{(5x - 10)(2x + 6)}{(x + 3)(3x - 6)}$

Notice the parentheses! If we forget the parentheses in this step, we will surely make a mistake in this problem. Now we wish to simplify the resulting fraction, if possible. The first step is to factor.

$$\dfrac{5\ (x - 2)\ \cdot 2\ (x + 3)}{(x + 3)\ \cdot 3\ (x - 2)} = \dfrac{10}{3} \qquad \text{Reduce}$$

(b) $\dfrac{x^2 - 1}{x^2 - 4} \cdot \dfrac{x + 2}{x - 1} = \dfrac{(x^2 - 1)(x + 2)}{(x^2 - 4)(x - 1)}$

Again note the parentheses. Now we factor to see if there is a common factor, and, if so, we simplify.

$$\dfrac{(x^2 - 1)(x + 2)}{(x^2 - 4)(x - 1)} = \dfrac{(x + 1)\ (x - 1)(x + 2)}{(x + 2)\ (x - 2)\ (x - 1)} = \dfrac{x + 1}{x - 2} \qquad ◄$$

Procedure for Multiplying Rational Expressions

1. Multiply the numerators to get the new numerator and the denominators to get the new denominator.
2. Reduce the resulting fraction, if possible.

► **EXAMPLE 3**

Perform the operation indicated.

$$\dfrac{x^2 - 2x + 1}{x^2 - x - 6} \cdot \dfrac{x^2 - 4}{x^2 - 3x + 2}$$

Solution

$$\dfrac{x^2 - 2x + 1}{x^2 - x - 6} \cdot \dfrac{x^2 - 4}{x^2 - 3x + 2} = \dfrac{(x^2 - 2x + 1)(x^2 - 4)}{(x^2 - x - 6)(x^2 - 3x + 2)}$$

$$= \dfrac{(x - 1)^2(x + 2)(x - 2)}{(x - 3)(x + 2)(x - 1)(x - 2)}$$

$$= \dfrac{x - 1}{x - 3} \qquad ◄$$

► **EXAMPLE 4**

Perform the operation indicated.

$$\frac{3 + x}{3 - x} \cdot \frac{x^2 - 9}{x + 2}$$

Solution

$$\frac{3 + x}{3 - x} \cdot \frac{x^2 - 9}{x + 2} = \frac{(3 + x)(x^2 - 9)}{(3 - x)(x + 2)}$$

$$= \frac{(3 + x)(x + 3)(x - 3)}{(3 - x)(x + 2)}$$

Be Careful!

Notice that $x - 3$ and $3 - x$ are opposites.

$$= \frac{-(3 + x)(x + 3)}{(x + 2)}$$

Since $(3 + x)$ *does* equal $(x + 3)$ (by the Commutative Property for Addition), we would write the answer

$$= \frac{-(x + 3)^2}{x + 2}$$

$$= -\frac{(x + 3)^2}{x + 2} \qquad ◄$$

To divide fractions, multiply by the reciprocal of the divisor. For example, let's divide $\frac{4}{7}$ by $\frac{2}{21}$.

$$\frac{4}{7} \div \frac{2}{21} = \frac{4}{7} \cdot \frac{21}{2} \qquad \text{Multiply by reciprocal}$$

$$= \frac{4(21)}{7(2)} \qquad \begin{array}{l}\text{Multiply numerators and}\\ \text{multiply denominators}\end{array}$$

$$= 6 \qquad \text{Simplify}$$

Division of Rational Expressions

If A, B, C, and D are polynomials and B, C, and D do not have value 0,

$$\frac{A}{B} \div \frac{C}{D} = \frac{A}{B} \cdot \frac{D}{C} = \frac{A \cdot D}{B \cdot C}$$

► **EXAMPLE 5**

Perform the division indicated.

$$\frac{x + 1}{x - 1} \div \frac{x + 1}{x - 2}$$

Solution

Multiply by the reciprocal of the divisor.

$$\frac{x+1}{x-1} \div \frac{x+1}{x-2} = \frac{x+1}{x-1} \cdot \frac{x-2}{x+1}$$

$$= \frac{(x+1)(x-2)}{(x-1)(x+1)}$$

$$= \frac{x-2}{x-1} \qquad \text{Reduce} \qquad \blacktriangleleft$$

Procedure for Dividing Rational Expressions

1. Change the problem to multiplication by multiplying by the reciprocal of the divisor.
2. Multiply the fractions.
3. Reduce the resulting fraction, if possible.

► **EXAMPLE 6**

Perform the operations indicated.

(a) $\dfrac{xy}{x^2-4} \div \dfrac{yz}{x^2+4x+4}$ (b) $\dfrac{3x}{5} \div (3x-6)$

Solutions

(a) $\dfrac{xy}{x^2-4} \div \dfrac{yz}{x^2+4x+4} = \dfrac{xy}{x^2-4} \cdot \dfrac{x^2+4x+4}{yz}$

$$= \frac{xy(x^2+4x+4)}{(x^2-4)yz}$$

Be Careful!

Important! Note the use of parentheses. Now, factor and reduce.

$$= \frac{xy(x+2)^2}{(x+2)(x-2)yz}$$

$$= \frac{x(x+2)}{(x-2)z}$$

(b) The reciprocal of $(3x-6)$ is $\dfrac{1}{3x-6}$, so

$$\frac{3x}{5} \div (3x-6) = \frac{3x}{5} \cdot \frac{1}{3x-6}$$

$$= \frac{3x}{5(3x-6)}$$

$$= \frac{3x}{(5)(3)(x-2)} \qquad \text{Factor}$$

$$= \frac{x}{5(x-2)} \qquad \text{Reduce} \qquad \blacktriangleleft$$

► **EXAMPLE 7**

Perform the operation indicated.

$$\frac{6x^3}{2x - 3} \div \frac{9x}{3 - 2x}$$

Solution

$$\frac{6x^3}{2x - 3} \div \frac{9x}{3 - 2x} = \frac{6x^3}{2x - 3} \cdot \frac{3 - 2x}{9x}$$

$$= \frac{6x^3(3 - 2x)}{(2x - 3)9x}$$

Be Careful!   Notice that $3 - 2x$ and $2x - 3$ are opposites.

$$= \frac{-2x^2}{3}$$ ◄

Problem Set 5.3

Warm-Ups

In Problems 1 through 6, perform the operations indicated. See Example 1.

1. $\dfrac{2x}{3y} \cdot \dfrac{4x}{7y}$

2. $\dfrac{3a}{5b^2} \cdot \dfrac{3x}{10b^2}$

3. $\dfrac{11m^3}{7n} \cdot \dfrac{2m}{13n^2}$

4. $\dfrac{2x}{5z} \cdot \dfrac{3z^2}{8}$

5. $\dfrac{-2x^2y}{7m^2n^3} \cdot \dfrac{21mn^4}{2xy^2}$

6. $\dfrac{18}{5x^3} \cdot 25x^4$

In Problems 7 through 14, perform the operations indicated. See Example 2.

7. $\dfrac{x + 1}{-2x} \cdot \dfrac{6x^2}{x + 1}$

8. $\dfrac{-4x^3}{y^2 + 2} \cdot \dfrac{y^2 + 2}{12x}$

9. $\dfrac{x + 5}{x - 5} \cdot \dfrac{x - 5}{x + 6}$

10. $\dfrac{4x - 2}{x + 1} \cdot \dfrac{x - 1}{2x - 1}$

11. $\dfrac{3k^2}{2x^3 - 6x} \cdot \dfrac{5x^4 - 15x^2}{9k}$

12. $\dfrac{x^2 - 1}{3x + 1} \cdot \dfrac{3x + 1}{x - 1}$

13. $\dfrac{2x - 3}{2x + 3} \cdot \dfrac{4x^2 - 9}{4x - 6}$

14. $\dfrac{2 + 6x^2}{3 - 6x} \cdot \dfrac{5 - 10x}{6 + 18x^2}$

In Problems 15 through 19, perform the operations indicated. See Example 3.

15. $\dfrac{x^2 + 2x + 1}{x - 1} \cdot \dfrac{2x + 1}{x^2 - 1}$

16. $\dfrac{x^2 + 4x + 4}{4x + 2} \cdot \dfrac{4x^2 - 1}{x^2 - 4}$

17. $\dfrac{s^2 + s - 6}{s^2 - 4} \cdot \dfrac{2 + s}{s^2 - s - 6}$

18. $\dfrac{t^2 - 3t - 4}{t + 4} \cdot \dfrac{t^2 + 3t - 4}{t - 4}$

19. $\dfrac{2x^2 + x - 1}{x^2 + 2x - 3} \cdot \dfrac{x^2 + 7x + 12}{2x^2 - 9x + 4}$

In Problems 20 and 21, perform the operations indicated. See Example 4.

20. $\dfrac{x}{x-y} \cdot \dfrac{y-x}{y}$

21. $\dfrac{2-3w}{5} \cdot \dfrac{10}{3w-2}$

In Problems 22 through 27, perform the operations indicated. See Example 5.

22. $\dfrac{2x}{3y} \div \dfrac{3y}{2x}$

23. $\dfrac{3a}{5b^2} \div \dfrac{4x}{7a^2}$

24. $\dfrac{12m^2}{5n^2} \div \dfrac{2n}{-5m}$

25. $\dfrac{2x}{5y} \div \dfrac{3x}{7y}$

26. $\dfrac{5x^2y^3}{6mn^2} \div \dfrac{10x^3y^2}{11mn^3}$

27. $\dfrac{-5x^3}{8y^2} \div 25x^2$

In Problems 28 through 40, perform the operations indicated. See Example 6.

28. $\dfrac{3x^2-2x}{13y^3} \div \dfrac{6x-4}{-39xy}$

29. $\dfrac{12y^2}{7y^2+21} \div \dfrac{30y}{5y^3+15y}$

30. $\dfrac{3-9x^2}{4x+18x^3} \div \dfrac{4x-12x^3}{10+45x^2}$

31. $\dfrac{x^2+2x^3}{-7} \div \dfrac{x^2}{7}$

32. $\dfrac{3x}{4y} \div (3y-6)$

33. $(x^2+1) \div \dfrac{x^2+1}{x-1}$

34. $\dfrac{(x+1)^2}{x-1} \div (2x+2)$

35. $(3x-9x^2) \div \dfrac{4-12x}{3}$

36. $\dfrac{x^2-4}{2x-1} \div \dfrac{2+x}{x-2}$

37. $\dfrac{x^2-2x+1}{x+1} \div \dfrac{x^2-1}{2x+1}$

38. $\dfrac{x^2-6x+9}{3x+9} \div \dfrac{x^2-9}{2x+6}$

39. $\dfrac{y^2-2y-8}{y+1} \div \dfrac{y^2+y-20}{1+y}$

40. $\dfrac{6x^2-x-2}{x^2+4x-5} \div \dfrac{3x^2+x-2}{x^2+3x-10}$

In Problems 41 and 42, perform the operations indicated. See Example 7.

41. $\dfrac{x^2+1}{2x+3} \div \dfrac{1-x}{4x+6}$

42. $\dfrac{q^2-9}{5-q} \div \dfrac{q^2-2q-15}{2q-10}$

Practice Exercises

In Problems 43 through 90, perform the operations indicated.

43. $\dfrac{3x}{4y} \cdot \dfrac{5x}{7y}$

44. $\dfrac{2b}{5a^2} \cdot \dfrac{3y}{15a^3}$

45. $\dfrac{9m^2}{5n} \cdot \dfrac{3m}{11n^3}$

46. $\dfrac{3x}{4y} \cdot \dfrac{3y^3}{2}$

47. $\dfrac{2xy^2}{5m^3n^2} \cdot \dfrac{20mn^4}{2x^2y}$

48. $\dfrac{12x^3y^3}{65x^2} \cdot 65x^5$

49. $\dfrac{-3x}{2y} \div \dfrac{2y}{3x}$

50. $\dfrac{2a}{3b^2} \div \dfrac{5x}{7a}$

51. $\dfrac{18x^2}{5y^3} \div 6x$

52. $\dfrac{3x}{4y} \div \dfrac{5x}{7y}$

53. $\dfrac{6x^3y^2}{5mn^3} \div \dfrac{-12x^2y^3}{13mn^3}$

54. $-24x^4y^2 \div \dfrac{16x^2y^2}{13z}$

55. $\dfrac{y-1}{3y} \cdot \dfrac{6y^2}{y-1}$

56. $\dfrac{3x^2}{x^2-5} \cdot \dfrac{x^2-5}{12x}$

57. $\dfrac{x-5}{x-6} \cdot \dfrac{x+5}{x-5}$

58. $\dfrac{6x - 3}{x - 1} \cdot \dfrac{x + 1}{2x - 1}$

59. $\dfrac{2k^2}{9x - 3x^3} \cdot \dfrac{7x^5 - 21x^3}{8k}$

60. $\dfrac{3 - 6x^2}{2 + 6x} \cdot \dfrac{5 + 15x}{6 - 12x^2}$

61. $\dfrac{3x - 2}{4y} \div \dfrac{3x - 2}{y^3}$

62. $\dfrac{-5m}{2mn - 3} \div \dfrac{3mn}{2mn - 3}$

63. $\dfrac{3x + 2}{3x - 2} \div \dfrac{2 + 3x}{2x + 1}$

64. $\dfrac{2s^2 - 3s}{11t^3} \div \dfrac{4s - 6}{33st}$

65. $\dfrac{16w^2}{5w^2 - 20} \div \dfrac{28w}{3w^3 - 12w}$

66. $\dfrac{2 + 8x^2}{3x + 18x^3} \div \dfrac{6x + 24x^3}{15 + 90x^2}$

67. $\dfrac{-2x}{5z} \cdot 4z^3$

68. $3(x + 4) \cdot \dfrac{-11}{x + 4}$

69. $\dfrac{x - 2}{2 + x} \cdot (x + 2)$

70. $2(y + 3y^3) \cdot \dfrac{2}{x + 3xy^2}$

71. $\dfrac{2x}{5y} \div (2y - 2)$

72. $(x^2 - 4) \div \dfrac{x^2 - 4}{x - 2}$

73. $\dfrac{(r - 2)^2}{r + 2} \div (4 - 2r)$

74. $(2x + 8x^2) \div \dfrac{4 + 16x}{3}$

75. $\dfrac{x^2 - 9}{2x - 1} \cdot \dfrac{2x - 1}{x + 3}$

76. $\dfrac{3x + 2}{2x - 3} \cdot \dfrac{4x^2 - 9}{6x + 4}$

77. $\dfrac{x^2 - 1}{2x + 1} \div \dfrac{x + 1}{x - 1}$

78. $\dfrac{2x - 3}{2x + 3} \div \dfrac{4x^2 - 9}{4x + 6}$

79. $\dfrac{x^2 - 2x + 1}{x + 1} \cdot \dfrac{2x + 1}{x^2 - 1}$

80. $\dfrac{x^2 - 4x + 4}{4x - 2} \cdot \dfrac{4x^2 - 1}{x^2 - 4}$

81. $\dfrac{x^2 + 2x + 1}{x - 1} \div \dfrac{x^2 - 1}{2x - 1}$

82. $\dfrac{x^2 + 6x + 9}{3x - 9} \div \dfrac{x^2 - 9}{2x - 6}$

83. $\dfrac{y^2 - y - 6}{y^2 - 9} \cdot \dfrac{y + 3}{y^2 + y - 6}$

84. $\dfrac{s^2 - 4s - 5}{s - 1} \cdot \dfrac{s^2 + 4s - 5}{s + 1}$

85. $\dfrac{x^2 - 4}{4 - x} \div \dfrac{x^2 - 2x - 8}{3x - 12}$

86. $\dfrac{w + 3}{w^2 - 25} \div \dfrac{w^2 + 6w + 9}{w - 5}$

87. $\dfrac{3x^2 + x - 2}{x^2 + 2x - 3} \cdot \dfrac{x^2 + 8x + 15}{3x^2 + 13x - 10}$

88. $\dfrac{u^2}{u - 2v} \cdot \dfrac{2v - u}{v^2}$

89. $\dfrac{6x^2 + x - 2}{x^2 - 4x - 5} \div \dfrac{3x^2 - x - 2}{x^2 - 3x - 10}$

90. $\dfrac{x^2 - 9}{2x + 4} \div \dfrac{3 - x}{4x + 8}$

Challenge Problems

In Problems 91 through 94, perform the operations indicated.

91. $\dfrac{x^2 - 5x - 14}{3x^2 - 2x - 1} \div \dfrac{x^2 - 9x + 14}{1 + 2x - 3x^2}$

92. $\dfrac{16 - x^2}{x^2 + 3x + 2} \cdot \dfrac{x^2 + x - 2}{x^2 - 5x + 4}$

93. $\dfrac{3x^2 - 2x - 8}{2x^2 - 3x - 5} \cdot \dfrac{2x^2 - 11x + 15}{3x^2 + 10x + 8} \cdot \dfrac{x^2 - 6x - 7}{4x^2 - 7x - 2}$

94. $\dfrac{6x^2 - 5x + 1}{x^2 + x - 2} \cdot \dfrac{3x^2 + 8x + 4}{4x^2 + 8x - 5} \div \dfrac{3x^2 - 7x - 6}{3x^2 - 4x + 1}$

In Your Own Words

95. Explain how to multiply two rational expressions.

96. Explain how to divide two rational expressions.

5.4 Addition and Subtraction

TAPE 9

We add and subtract rational expressions exactly as we add and subtract fractions. If necessary, we use the Fundamental Principle of Fractions to adjust the expressions so that they have the same denominator; then we use that denominator and add or subtract the numerators.

Let's add $\frac{2}{7}$ to $\frac{3}{7}$ to review the procedure of adding fractions with the same denominator.

$$\frac{2}{7} + \frac{3}{7} = \frac{2 + 3}{7}$$

$$= \frac{5}{7}$$

Addition and Subtraction of Rational Expressions

If A, B, and C, are polynomials and if C does not have value 0,

$$\frac{A}{C} + \frac{B}{C} = \frac{A + B}{C} \quad \text{and} \quad \frac{A}{C} - \frac{B}{C} = \frac{A - B}{C}$$

► **EXAMPLE 1**

Perform the operations indicated.

(a) $\dfrac{5}{x + y} + \dfrac{3}{x + y}$ (b) $\dfrac{x + 1}{x - 3} - \dfrac{x - 1}{x - 3}$

Solutions

(a) $\dfrac{5}{x + y} + \dfrac{3}{x + y} = \dfrac{5 + 3}{x + y}$

$\qquad\qquad\qquad = \dfrac{8}{x + y}$

Be Careful!

(b) It is *important* to understand the use of parentheses in the following example!

$$\frac{x + 1}{x - 3} - \frac{x - 1}{x - 3} = \frac{(x + 1) - (x - 1)}{x - 3}$$

We wish to subtract the *entire numerator*, $x - 1$, not just x, from $x + 1$.

$$\frac{(x + 1) - (x - 1)}{x - 3} = \frac{x + 1 - x + 1}{x - 3} \qquad \text{Note signs}$$

$$= \frac{2}{x - 3} \qquad ◄$$

When adding and subtracting rational expressions, we usually reduce answers to lowest terms.

► **EXAMPLE 2**

Perform the operations indicated.

(a) $\dfrac{5}{12xy} + \dfrac{1}{12xy}$ (b) $\dfrac{x^2}{(x + 1)(x - 3)} - \dfrac{x + 6}{(x + 1)(x - 3)}$

Solutions

(a) $\dfrac{5}{12xy} + \dfrac{1}{12xy} = \dfrac{5 + 1}{12xy}$

$= \dfrac{6}{12xy}$

$= \dfrac{1}{2xy}$ Reduce

(b) Notice the parentheses in this example.

Be Careful!

$\dfrac{x^2}{(x + 1)(x - 3)} - \dfrac{x + 6}{(x + 1)(x - 3)} = \dfrac{x^2 - (x + 6)}{(x + 1)(x - 3)}$

Be careful with signs!

$= \dfrac{x^2 - x - 6}{(x + 1)(x - 3)}$

If the numerator is factored, this can be reduced.

$= \dfrac{(x - 3)(x + 2)}{(x + 1)(x - 3)}$

Now we see there is a common factor in the numerator and denominator.

$= \dfrac{x + 2}{x + 1}$ ◄

Often we wish to add (or subtract) rational expressions that have different denominators. In this case, we must decide on an appropriate denominator and rewrite each fraction with that denominator. We use the Fundamental Principle of Fractions to rewrite the fractions.

Suppose we were to add the fractions $\dfrac{1}{12}$ and $\dfrac{1}{18}$. We need a common denominator. So we inspect 12 and 18 to find the smallest number that is a multiple of each of them. We select 36 because 36 is $2 \cdot 18$ and $3 \cdot 12$. There is no smaller integer that is a multiple of both. Now we can add.

$\dfrac{1}{12} + \dfrac{1}{18} = \dfrac{1(3)}{12(3)} + \dfrac{1(2)}{18(2)}$ Fundamental Principle of Fractions

$= \dfrac{3}{36} + \dfrac{2}{36} = \dfrac{5}{36}$ Addition of Fractions

How do we find the number 36? First, we factor 12 and 18 completely.

$12 = 2 \cdot 2 \cdot 3$

$18 = 2 \cdot 3 \cdot 3$

The factors of 36 are 2 · 2 · 3 · 3. Notice that 2 occurs twice in the factors of the least common denominator and 2 occurs twice in 12. Notice that 3 occurs twice in the least common denominator and 3 occurs twice in 18. This is the key. The least common denominator is made up of the factors of the denominators, each taken the greatest number of times it occurs *in any one denominator.*

Procedure for Finding a Least Common Denominator (LCD)

1. Factor each denominator completely.
2. List all the *different prime factors.*
3. For each factor in step 2, find the *greatest* number of times it occurs in any *one* denominator.
4. The LCD is the product of the factors found in step 2, each taken the number of times found in step 3.

► **EXAMPLE 3**

Assume that each number or expression is a denominator in a fraction. Find the LCD for the given denominators.

(a) 72 and 48   (b) x^2y^3z and x^4y^2 (c) $7y$ and y

(d) $x(x + 2)$ and x (e) x and $x + 3$ (f) $x + 4$ and $(x + 4)^2$

Solutions

(a) 72 and 48

STEP 1   Factor each denominator completely.
$$72 = 2^3 \cdot 3^2, \qquad 48 = 2^4 \cdot 3$$

STEP 2   List all the different prime factors.
The different prime factors are 2 and 3.

STEP 3   Determine the greatest number of times each prime factor occurs in any one denominator.
The factor 2 occurs three times in 72 and four times in 48. The greatest number of times that it occurs in either denominator is *four*.
The factor 3 occurs two times in 72 and one time in 48. The greatest number of times that it occurs in either denominator is *two*.

STEP 4   Write the LCD.
The LCD contains *four* factors of 2 and *two* factors of 3. The LCD is $2^4 \cdot 3^2$, which is 144.

(b) x^2y^3z and x^4y^2
Factor each denominator.
The denominators are factored.
List the different prime factors.
The different prime factors are x, y, and z.

Count the number of times each prime factor occurs in each denominator. x occurs as a factor twice in x^2y^3z and four times in x^4y^2. Thus we need four factors of x.

y occurs three times in x^2y^3z and two times in x^4y^2. So we need three factors of y.

z occurs one time in x^2y^3z and no times x^4y^2. Thus we need one factor of z. The LCD is x^4y^3z.

(c) $7y$ and y

Both denominators are factored.

The different prime factors are 7 and y.

7 occurs one time in $7y$ and no times in y.

y occurs one time in $7y$ and one time in y.

Thus we need one 7 and one y.

The LCD is $7y$.

(d) $x(x + 2)$ and x

Both are factored.

The different prime factors are x and $x + 2$.

x occurs in $x(x + 2)$ one time and one time in x. We need one x in the LCD.

$x + 2$ occurs one time in $x(x + 2)$ and no times in x. We need one $x + 2$ in the LCD.

The LCD is $x(x + 2)$.

(e) x and $x + 3$

Both denominators are prime.

The different prime factors are x and $x + 3$.

x occurs one time in x and *no* times in $x + 3$. (The x in $x + 3$ is not a factor.)

$x + 3$ occurs no times in x and one time in $x + 3$.

The LCD is $x(x + 3)$.

(f) $x + 4$ and $(x + 4)^2$

Both denominators are factored.

The prime factor is $x + 4$.

$x + 4$ occurs one time in $x + 4$ and two times in $(x + 4)^2$. So we need two factors of $x + 4$.

The LCD is $(x + 4)^2$. ◀

► **EXAMPLE 4**

Assume that each polynomial is a denominator of a fraction. Find the LCD for the given denominators.

(a) $x - y$ and $x^2 - y^2$ (b) $x^2 - 4x + 4$ and $x^2 - 4$

Solutions

(a) $x - y$ and $x^2 - y^2$. We factor the two denominators completely.

$$x - y = x - y \qquad \text{It is prime}$$
$$x^2 - y^2 = (x + y)(x - y) \qquad ◀$$

The different factors that occur are $x - y$ and $x + y$. The greatest number of times that either one occurs *in any one denominator* is once. We need one of each.

The LCD is $(x + y)(x - y)$.

(b) $x^2 - 4x + 4$ and $x^2 - 4$

Factor each expression completely.

$$x^2 - 4x + 4 = (x - 2)^2$$
$$x^2 - 4 = (x - 2)(x + 2)$$

The different factors are $(x - 2)$ and $(x + 2)$; we need two factors of $(x - 2)$ and one of $(x + 2)$.
The LCD is $(x - 2)^2(x + 2)$.   ◄

To add or subtract rational expressions with different denominators, we must first find the least common denominator. We then use the Fundamental Principle of Rational Expressions to write each fraction with the common denominator.

► **EXAMPLE 5**

Perform the operations indicated.

(a) $\dfrac{x}{5} - \dfrac{1}{x}$ (b) $\dfrac{1}{x} + \dfrac{1}{x + 3}$

Solutions

(a) The LCD is $5x$.

$$\frac{x}{5} - \frac{1}{x} = \frac{x \cdot x}{5 \cdot x} - \frac{1 \cdot 5}{x \cdot 5} \qquad \text{Fundamental Principle}$$

$$= \frac{x^2}{5x} - \frac{5}{5x}$$

$$= \frac{x^2 - 5}{5x}$$

Notice that this cannot be reduced.

(b) The denominators are x and $x + 3$. Neither factors, and the LCD is $x(x + 3)$. We use the Fundamental Principle to write each of the fractions with a denominator of $x(x + 3)$.

$$\frac{1}{x} + \frac{1}{x + 3} = \frac{1 \cdot (x + 3)}{x \cdot (x + 3)} + \frac{1 \cdot x}{(x + 3) \cdot x}$$

$$= \frac{x + 3}{x(x + 3)} + \frac{x}{x(x + 3)}$$

Now that we have a common denominator, we can add the fractions.

$$\frac{x + 3}{x(x + 3)} + \frac{x}{x(x + 3)} = \frac{x + 3 + x}{x(x + 3)}$$

$$= \frac{2x + 3}{x(x + 3)} \qquad ◄$$

► **EXAMPLE 6**

Perform the operations indicated.

(a) $\dfrac{3}{2x^2} + \dfrac{4}{x^3}$ (b) $\dfrac{1}{12x^2y} - \dfrac{1}{18xy^2}$

Solutions

(a) The LCD is $2x^3$.

$$\dfrac{3}{2x^2} + \dfrac{4}{x^3} = \dfrac{3 \cdot x}{2x^2 \cdot x} + \dfrac{4 \cdot 2}{x^3 \cdot 2} \qquad \text{Fundamental Principle}$$

$$= \dfrac{3x}{2x^3} + \dfrac{8}{2x^3}$$

$$= \dfrac{3x + 8}{2x^3}$$

(b) We factor the two denominators.

$$12x^2y = 2 \cdot 2 \cdot 3 \cdot x^2y$$

$$18xy^2 = 2 \cdot 3 \cdot 3 \cdot xy^2$$

The LCD is $2 \cdot 2 \cdot 3 \cdot 3 \cdot x^2y^2$, or $36x^2y^2$.

$$\dfrac{1}{12x^2y} - \dfrac{1}{18xy^2} = \dfrac{1 \cdot 3y}{12x^2y \cdot 3y} - \dfrac{1 \cdot 2x}{18xy^2 \cdot 2x}$$

$$= \dfrac{3y}{36x^2y^2} - \dfrac{2x}{36x^2y^2}$$

$$= \dfrac{3y - 2x}{36x^2y^2} \qquad ◄$$

► **EXAMPLE 7**

Perform the operations indicated.

(a) $\dfrac{x}{x + 2} + \dfrac{4x}{x^2 - 4}$ (b) $\dfrac{1}{x^2 + 2x + 1} - \dfrac{1}{x^2 - 1}$

Solutions

(a) The denominators are $x + 2$ and $x^2 - 4$. We must factor $x^2 - 4$ to find the LCD.

$$x + 2 \qquad \text{Prime}$$

$$x^2 - 4 = (x + 2)(x - 2)$$

The factors are $x + 2$ and $x - 2$, and the most either one occurs *in any one denominator* is once. Therefore, the LCD is $(x + 2)(x - 2)$.

$$\frac{x}{x + 2} + \frac{4x}{x^2 - 4} = \frac{x}{x + 2} + \frac{4x}{(x + 2)(x - 2)}$$

$$= \frac{x \cdot (x - 2)}{(x + 2)\,(x - 2)} + \frac{4x}{(x + 2)(x - 2)}$$

$$= \frac{x^2 - 2x + 4x}{(x + 2)(x - 2)}$$

$$= \frac{x^2 + 2x}{(x + 2)(x - 2)}$$

$$= \frac{x(x + 2)}{(x + 2)(x - 2)} \qquad \text{Factor the numerator}$$

$$= \frac{x}{x - 2} \qquad \begin{array}{l}\text{The common factor}\\ \text{is } x + 2\end{array}$$

(b) We factor the two denominators.

$$x^2 + 2x + 1 = (x + 1)^2$$

$$x^2 - 1 = (x + 1)(x - 1)$$

The LCD is $(x + 1)^2(x - 1)$.

$$\frac{1}{x^2 + 2x + 1} - \frac{1}{x^2 - 1} = \frac{1}{(x + 1)^2} - \frac{1}{(x + 1)(x - 1)}$$

$$= \frac{1 \cdot (x - 1)}{(x + 1)^2\,(x - 1)} - \frac{1 \cdot (x + 1)}{(x + 1)^2\,(x - 1)}$$

$$= \frac{(x - 1) - (x + 1)}{(x + 1)^2(x - 1)} \qquad \text{Note parentheses}$$

$$= \frac{x - 1 - x - 1}{(x + 1)^2(x - 1)} \qquad \text{Note signs}$$

$$= \frac{-2}{(x + 1)^2(x - 1)} \qquad \blacktriangleleft$$

► **EXAMPLE 8**

Perform the operations indicated.

(a) $3y^3 + \dfrac{2}{5y}$ (b) $\dfrac{2}{3x - 1} - 2$

Solutions

(a) Think of $3y^3$ as $\dfrac{3y^3}{1}$. The LCD is $5y$.

$$3y^3 + \frac{2}{5y} = \frac{3y^3 \ (5y)}{1 \ (5y)} + \frac{2}{5y} \qquad \text{Fundamental Principle}$$

$$= \frac{15y^4}{5y} + \frac{2}{5y}$$

$$= \frac{15y^4 + 2}{5y}$$

(b) If we think of the number 2 as the fraction $\dfrac{2}{1}$, we see that the LCD is $3x - 1$.

$$\frac{2}{3x - 1} - 2 = \frac{2}{3x - 1} - \frac{2}{1}$$

$$= \frac{2}{3x - 1} - \frac{2 \ (3x - 1)}{1 \ (3x - 1)}$$

$$= \frac{2 - 2(3x - 1)}{3x - 1}$$

$$= \frac{2 - 6x + 2}{3x - 1}$$

$$= \frac{4 - 6x}{3x - 1} \quad \text{or} \quad \frac{2(2 - 3x)}{3x - 1}$$

The numerator will factor, but the fraction cannot be reduced. So it is not necessary to write the answer in factored form. ◀

► EXAMPLE 9

Perform the operations indicated.

(a) $\dfrac{x}{2x - 3} + \dfrac{6}{3 - 2x}$ (b) $\dfrac{5}{x - 7} - \dfrac{3}{7 - x}$

Solutions

(a) The denominators are $2x - 3$ and $3 - 2x$. We should *never forget* that $2x - 3$ and $3 - 2x$ are *opposites* of each other. That is,

$$(3 - 2x) = -(2x - 3)$$

We replace $3 - 2x$ in the second fraction with $-(2x - 3)$ and simplify.

$$\frac{x}{2x - 3} + \frac{6}{3 - 2x} = \frac{x}{2x - 3} + \frac{6}{-(2x - 3)}$$

Be Careful!

$$= \frac{x}{2x - 3} + \frac{-6}{2x - 3} \qquad \begin{array}{l}\text{Sign Property of}\\ \text{Rational Expressions}\end{array}$$

$$= \frac{x - 6}{2x - 3}$$

(b) The denominators are opposites. Replace $7 - x$ with $-(x - 7)$.

$$\frac{5}{x - 7} - \frac{3}{7 - x} = \frac{5}{x - 7} - \frac{3}{-(x - 7)}$$

$$= \frac{5}{x - 7} + \frac{3}{x - 7} \qquad \text{Sign Property of Rational Expressions}$$

$$= \frac{5 + 3}{x - 7}$$

$$= \frac{8}{x - 7} \qquad \blacktriangleleft$$

In adding and subtracting rational expressions, we must be very careful when simplifying. Consider this example.

Be Careful!

$$\frac{2}{a + b} + \frac{3}{a - b} = \frac{2(a - b)}{(a + b)(a - b)} + \frac{3(a + b)}{(a + b)(a - b)}$$

Why do we not reduce these fractions at this point? Notice that if we did reduce, we would end up where we started. Continuing,

$$\frac{2(a - b)}{(a + b)(a - b)} + \frac{3(a + b)}{(a + b)(a - b)} = \frac{2(a - b) + 3(a + b)}{(a + b)(a - b)}$$

Can we reduce this fraction now? Are there common factors in both the numerator and denominator? $(a - b)$ and $(a + b)$ appear in the numerator and denominator. However, the numerator is *not* factored. Thus neither $(a - b)$ nor $(a + b)$ is a *factor* of the numerator. We *cannot* simplify this fraction. We continue by using the distributive property.

$$\frac{2(a - b) + 3(a + b)}{(a + b)(a - b)} = \frac{2a - 2b + 3a + 3b}{(a + b)(a - b)}$$

$$= \frac{5a + b}{(a + b)(a - b)}$$

Problem Set 5.4

Warm-Ups

In Problems 1 through 10, perform the operations indicated and reduce answers to lowest terms. See Examples 1 and 2.

1. $\dfrac{5}{x} + \dfrac{8}{x}$

2. $\dfrac{3}{x + 2} + \dfrac{2x}{2 + x}$

3. $\dfrac{x}{x - 1} + \dfrac{x - 2}{x - 1}$

4. $\dfrac{3t}{2 - t} + \dfrac{t - 1}{2 - t}$

5. $\dfrac{x}{x - 1} - \dfrac{x + 1}{x - 1}$

6. $\dfrac{3t}{2 - t} - \dfrac{t - 1}{2 - t}$

7. $\dfrac{x^2}{x - 2} - \dfrac{4}{x - 2}$

8. $\dfrac{2x + 3}{x - 1} - \dfrac{x + 1}{x - 1}$

9. $\dfrac{x^2 + 3}{5 + x} + \dfrac{3x - 13}{x + 5}$

10. $\dfrac{x^2}{x^2 - x - 2} + \dfrac{2x + 1}{x^2 - x - 2}$

In Problems 11 through 22, assume that each number or expression is a denominator of a fraction. Find the LCD of the given denominators. See Examples 3 and 4.

11. 5 and 7

12. x and $x + 2$

13. 3 and 27

14. x and x^3

15. 60 and 90

16. $6x$ and $9x$

17. x^2 and $x(x - 2)$

18. x^7y^4 and x^2y^5z

19. $5x$ and $x + 5$

20. $x^2 + 4x + 4$ and $x^2 - 4$

21. $x^2 - x - 6$ and $3x + 6$

22. $18x^3y^4$ and $27xy^3$

In Problems 23 through 32, perform the operation indicated and reduce answers to lowest terms. See Example 5.

23. $\dfrac{5}{x} + \dfrac{2}{3y}$

24. $\dfrac{x}{7} + \dfrac{3}{2x}$

25. $\dfrac{5}{x} - \dfrac{3}{2}$

26. $\dfrac{3}{2x} - \dfrac{x}{3}$

27. $\dfrac{3}{x + 7} - \dfrac{1}{x}$

28. $\dfrac{3}{x - 1} + \dfrac{1}{x}$

29. $\dfrac{2}{r - 2} - \dfrac{1}{r + 2}$

30. $\dfrac{3}{3z - 2} - \dfrac{1}{z + 1}$

31. $\dfrac{t}{t + 1} + \dfrac{t}{t - 1}$

32. $\dfrac{1}{x + 1} + \dfrac{2}{2x - 3}$

In Problems 33 through 38, perform the operations indicated. See Example 6.

33. $\dfrac{1}{x^2} + \dfrac{1}{2x}$

34. $\dfrac{1}{3x^2} + \dfrac{2}{x^3}$

35. $\dfrac{3}{4a^2} - \dfrac{5}{2a}$

36. $\dfrac{7}{6xy} - \dfrac{1}{12y}$

37. $\dfrac{4}{15x^2y} - \dfrac{3}{10xy^2}$

38. $\dfrac{1}{3x^2} + \dfrac{5}{2xy}$

In Problems 39 through 50, perform the operations indicated. See Example 7.

39. $\dfrac{3x}{1 - 5x} + \dfrac{3x}{4(1 - 5x)}$

40. $\dfrac{2z}{z(z + 9)} - \dfrac{11}{z + 9}$

41. $\dfrac{5}{4(k - 6)} - \dfrac{1 - k}{8(k - 6)}$

42. $\dfrac{2x}{(x - 7)^2} - \dfrac{1}{x - 7}$

43. $\dfrac{2y}{(y + 4)^2} - \dfrac{1}{y + 4}$

44. $\dfrac{2}{k - 7} + \dfrac{3}{2k - 14}$

45. $\dfrac{3d}{5d + 10} - \dfrac{2}{6 + 3d}$

46. $\dfrac{x}{x^2 - 1} + \dfrac{1}{1 + x}$

47. $\dfrac{3x}{x^2 - 2x + 1} - \dfrac{1}{x - 1}$

48. $\dfrac{2}{x^2 + 4x + 4} - \dfrac{3}{x^2 - 4}$

49. $\dfrac{c}{c^2 + c - 12} + \dfrac{c}{c^2 - 2c - 3}$

50. $\dfrac{t + 5}{t^2 - 2t - 15} + \dfrac{t}{t^2 - 6t + 5}$

For Problems 51 through 60, perform the operations indicated. See Example 8.

51. $1 + \dfrac{x}{7}$

52. $x + \dfrac{1}{3}$

53. $y - \dfrac{2}{y}$

54. $3x - \dfrac{1}{x}$

55. $1 + \dfrac{2}{x + 3}$

56. $\dfrac{1}{z + 2} + 1$

57. $\dfrac{x}{1 + 2x} - 2$

58. $\dfrac{5}{t - 1} - 3$

59. $\dfrac{x + 1}{x - 1} + 1$

60. $2 - \dfrac{x - 1}{x + 1}$

For Problems 61 through 64, perform the operations indicated. See Example 9.

61. $\dfrac{x}{2x - 1} + \dfrac{1}{1 - 2x}$

62. $\dfrac{3}{t - 14} - \dfrac{3}{14 - t}$

63. $\dfrac{a}{b - 1} + \dfrac{1}{1 - b}$

64. $\dfrac{x}{x - y} + \dfrac{y}{y - x}$

Practice Exercises

In Problems 65 through 73, find the LCD of the given denominators.

65. 8 and 3

66. y and $y - 3$

67. 8 and 2

68. t and t^4

69. 126 and 84

70. $4x$ and $6x$

71. x^2 and $x(x + 3)$

72. $x^2 - 4$ and $x - 2$

73. $2x^2 - 5x - 3$ and $2x^2 + 7x + 3$

In Problems 74 through 103, perform the operation indicated and reduce answers to lowest terms.

74. $\dfrac{r^2}{r - 3} + \dfrac{r - 12}{r - 3}$

75. $\dfrac{2y}{5 - y} + \dfrac{5 - 3y}{5 - y}$

76. $\dfrac{r}{r - 3} - \dfrac{1 + r}{r - 3}$

77. $\dfrac{2y}{5 - y} - \dfrac{1 - y}{5 - y}$

78. $\dfrac{x^2 - 2}{x + 1} - \dfrac{2x + 1}{1 + x}$

79. $\dfrac{2t}{2 - 3t} + \dfrac{2t}{3(2 - 3t)}$

80. $\dfrac{7}{8(x - 5)} - \dfrac{1 + x}{6(x - 5)}$

81. $\dfrac{2}{3x^3} + \dfrac{1}{2xy}$

82. $\dfrac{3}{16x^2y} - \dfrac{5}{18xy^2}$

83. $\dfrac{3}{x - 4} - \dfrac{1}{x}$

84. $\dfrac{1}{t - 1} - \dfrac{1}{t + 1}$

85. $\dfrac{2}{r + 2} + \dfrac{r}{r - 2}$

86. $\dfrac{3}{m - 2} + \dfrac{3}{m + 2}$

87. $\dfrac{a}{a - 3} - \dfrac{a}{a + 2}$

88. $\dfrac{x}{7} - \dfrac{2}{28x}$

89. $\dfrac{1}{6t^2} - \dfrac{3}{4t^3}$

90. $\dfrac{5}{x - 7} - \dfrac{7x}{(x - 7)^2}$

91. $\dfrac{1}{y + 5} + \dfrac{y}{(y + 5)^2}$

92. $\dfrac{3}{s - 5} - \dfrac{2}{3s - 15}$

93. $\dfrac{7}{4d + 12} - \dfrac{5}{3d + 9}$

94. $\dfrac{1}{x^2 - 1} - \dfrac{x}{x + 1}$

95. $\dfrac{2x}{x^2 + 2x + 1} - \dfrac{5}{x + 1}$

96. $\dfrac{2}{x^2 - 6x + 9} - \dfrac{1}{x^2 - 9}$

97. $\dfrac{c}{c^2 - c - 12} - \dfrac{c}{c^2 + 2c - 3}$

98. $1 - \dfrac{1}{x - 1}$

99. $\dfrac{2}{x - 2} + 1$

100. $\dfrac{9}{x} - 3$

101. $\dfrac{3}{x} - 4$

102. $\dfrac{x}{3x - 2} - \dfrac{1}{2 - 3x}$

103. $\dfrac{2}{s - 11} + \dfrac{1}{11 - s}$

104. Find a single rational expression to represent the sum of the reciprocals of two consecutive even integers.

105. An American Airlines 737 flies from Kansas City to Indianapolis at a rate of $r + 100$ miles per hour and then reduces its speed by 100 mph from Indianapolis to St. Louis.

(a) Write a rational expression to represent the flying time from Kansas City to Indianapolis.
(b) Write a rational expression to represent the flying time from Indianapolis to St. Louis.
(c) Write the total flying time as the sum of two rational expressions.
(d) Add the rational expressions in part (c).
(e) If r is 125, find the total flying time.

106. One number is 2 less than another number. Write a rational expression to represent the sum of the reciprocals of the two numbers.

107. Four minutes separated the first- and second-place winners in the 26-kilometer marathon.
(a) Write a rational expression to represent the speed of the first-place winner in kilometers per minute.
(b) Write a rational expression to represent the speed of the first-place winner in kilometers per hour.
(c) Write a rational expression to represent the speed of the second-place winner in kilometers per minute.
(d) Write a rational expression to represent the speed of the second-place winner in kilometers per hour.

Problem 107

Challenge Problems

Perform the operations indicated and simplify.

108. $\dfrac{16}{s^2 - 10s + 21} - \dfrac{40}{s^2 - 4s - 21} + \dfrac{3(s + 1)}{s^2 - 9}$

109. $\dfrac{6}{25 + 5x + x^2} + \dfrac{1}{x - 5} + \dfrac{15x}{125 - x^3}$

110. $\dfrac{1}{t - 5} - \dfrac{1}{t^2 - 10t + 25} + \dfrac{1}{5 - t} - \dfrac{1}{25 - t^2}$

In Your Own Words

111. Explain how to find the least common denominator for 108 and 48.

5.5 Complex Fractions

TAPE 10

A fraction that contains a fraction is called a **complex fraction.** For example,

$$\frac{\dfrac{1}{x+1}}{x+3}, \qquad \frac{4x^2}{\dfrac{1}{x}-\dfrac{2}{y}}, \qquad \text{and} \qquad \frac{\dfrac{1}{x}+1}{5-\dfrac{1}{x}}$$

are complex fractions. Two methods are generally used for simplifying complex fractions. The first is to think of the complex fraction as a division problem and then use the rules for dividing fractions. The second method employs the Fundamental Principle of Fractions.

► **EXAMPLE 1**

Simplify $\dfrac{\dfrac{1}{x+1}}{\dfrac{2}{(x+1)^2}}$ using both methods.

Solution

Method 1 | We think of the example as a division problem.

$$\frac{\dfrac{1}{x+1}}{\dfrac{2}{(x+1)^2}} = \frac{1}{x+1} \div \frac{2}{(x+1)^2}$$

$$= \frac{1}{x+1} \cdot \frac{(x+1)^2}{2}$$

$$= \frac{1 \cdot (x+1)^2}{(x+1) \cdot 2}$$

which simplifies to

$$\frac{x+1}{2}$$

Method 2 | In Method 2 we use the Fundamental Principle of Fractions and multiply both the numerator and denominator of the original fraction by the LCD of the fractions *within* the numerator and denominator. This will result in a single fraction. The LCD of $x+1$ and $(x+1)^2$ is $(x+1)^2$.

$$\frac{\dfrac{1}{x+1}}{\dfrac{2}{(x+1)^2}} = \frac{\dfrac{1}{x+1} \cdot (x+1)^2}{\dfrac{2}{(x+1)^2} \cdot (x+1)^2}$$

$$= \frac{\dfrac{1 \cdot (x+1)^2}{x+1}}{\dfrac{2 \cdot (x+1)^2}{(x+1)^2}}$$

Both the numerator fraction and the denominator fraction will simplify, yielding

$$\frac{1 \cdot (x + 1)}{2}$$

This is a single fraction, which we write as

$$\frac{x + 1}{2}$$ ◄

Procedure for Simplifying Complex Fractions

Method 1

1. Write the fraction as a division problem.
2. Perform the division.
3. Reduce the result to lowest terms.

Method 2

1. Find the LCD of all denominators.
2. Multiply the numerator and denominator by the LCD.
3. Reduce the result to lowest terms.

► EXAMPLE 2

Simplify $\dfrac{\dfrac{1}{x} + 2}{\dfrac{3}{x^2} - 1}$ using both methods.

Solution

Method 1 To divide as we did in Example 1, we must first have a single fraction in the numerator and denominator. The *numerator* becomes

$$\frac{1}{x} + 2 = \frac{1}{x} + \frac{2x}{x}$$

$$= \frac{1 + 2x}{x}$$

The *denominator* becomes

$$\frac{3}{x^2} - 1 = \frac{3}{x^2} - \frac{x^2}{x^2}$$

$$= \frac{3 - x^2}{x^2}$$

Therefore, the *fraction* becomes

$$\frac{\dfrac{1}{x} + 2}{\dfrac{3}{x^2} - 1} = \frac{\dfrac{1 + 2x}{x}}{\dfrac{3 - x^2}{x^2}}$$

Notice that we have a single fraction divided by a single fraction. That is, there is only *one* fraction in the numerator and *one* fraction in the denominator. The complex fraction *must* be in that form to use Method 1. Now we divide.

$$\frac{\dfrac{1 + 2x}{x}}{\dfrac{3 - x^2}{x^2}} = \frac{1 + 2x}{x} \div \frac{3 - x^2}{x^2}$$

$$= \frac{1 + 2x}{x} \cdot \frac{x^2}{3 - x^2}$$

$$= \frac{(1 + 2x)x^2}{x(3 - x^2)} \qquad \text{Note parentheses}$$

x is a common factor of the numerator and the denominator.

$$\frac{(1 + 2x)x^2}{x(3 - x^2)} = \frac{(1 + 2x)x}{(3 - x^2)}$$

Method 2 The LCD is x^2. We multiply the numerator and denominator by x^2.

$$\frac{\dfrac{1}{x} + 2}{\dfrac{3}{x^2} - 1} = \frac{\left(\dfrac{1}{x} + 2\right)x^2}{\left(\dfrac{3}{x^2} - 1\right)x^2} \qquad \text{Note parentheses}$$

We use the distributive property to multiply in the numerator and denominator.

$$\frac{\left(\dfrac{1}{x} + 2\right)x^2}{\left(\dfrac{3}{x^2} - 1\right)x^2} = \frac{\left(\dfrac{1}{x}\right)x^2 + 2 \cdot x^2}{\left(\dfrac{3}{x^2}\right)x^2 - 1 \cdot x^2}$$

$$= \frac{\dfrac{x^2}{x} + 2x^2}{\dfrac{3x^2}{x^2} - x^2} = \frac{x + 2x^2}{3 - x^2} \qquad \blacktriangleleft$$

A complex fraction may be simplified by either method. However, as Examples 1 and 2 indicated, one method may be better than another for a particular problem.

In general, Method 1 is best if the numerator and denominator each contain a *single* term. Method 2 is preferred if *either* the numerator or the denominator has more than one term.

▶ **EXAMPLE 3**

Simplify $\dfrac{\dfrac{1}{x-1}+5}{\dfrac{1}{x-1}-1}$

Solution

Since there is more than one term in the numerator (and in the denominator), we choose Method 2. The LCD is $(x-1)$.

$$\frac{\dfrac{1}{x-1}+5}{\dfrac{1}{x-1}-1}=\frac{\left(\dfrac{1}{x-1}+5\right)(x-1)}{\left(\dfrac{1}{x-1}-1\right)(x-1)}$$

$$=\frac{\dfrac{1}{x-1}\cdot(x-1)+5\cdot(x-1)}{\dfrac{1}{x-1}\cdot(x-1)-1\cdot(x-1)}\qquad\text{Distributive Property}$$

$$=\frac{1+5x-5}{1-x+1}$$

$$=\frac{5x-4}{2-x}\qquad\qquad◀$$

▶ **EXAMPLE 4**

Simplify $\dfrac{\dfrac{1}{x}+\dfrac{1}{x+1}}{\dfrac{1}{x}-\dfrac{1}{x+1}}$

Solution

There is more than one term in the numerator, so we choose Method 2. The LCD is $x(x+1)$.

$$\frac{\dfrac{1}{x}+\dfrac{1}{x+1}}{\dfrac{1}{x}-\dfrac{1}{x+1}}=\frac{\left(\dfrac{1}{x}+\dfrac{1}{x+1}\right)\cdot x(x+1)}{\left(\dfrac{1}{x}-\dfrac{1}{x+1}\right)\cdot x(x+1)}$$

$$=\frac{\dfrac{1}{x}\cdot x(x+1)+\dfrac{1}{x+1}\cdot x(x+1)}{\dfrac{1}{x}\cdot x(x+1)-\dfrac{1}{x+1}\cdot x(x+1)}\qquad\text{Distributive Property}$$

$$=\frac{(x+1)+x}{(x+1)-x}$$

$$=\frac{2x+1}{1}=2x+1\qquad\qquad◀$$

Problem Set **5.5**

Warm-Ups

For Problems 1 through 15, simplify each complex fraction. Write answers in lowest terms. See Example 1.

1. $\dfrac{\dfrac{1}{2}}{\dfrac{3}{2}}$

2. $\dfrac{\dfrac{4}{5}}{\dfrac{4}{3}}$

3. $\dfrac{\dfrac{2}{3}}{\dfrac{4}{9}}$

4. $\dfrac{\dfrac{1}{x}}{\dfrac{1}{x^2}}$

5. $\dfrac{\dfrac{2}{y^2}}{\dfrac{6}{y}}$

6. $\dfrac{\dfrac{2x^2}{3}}{\dfrac{4x}{12}}$

7. $\dfrac{\dfrac{4}{x+1}}{\dfrac{18}{x+1}}$

8. $\dfrac{\dfrac{xy}{x+y}}{\dfrac{x^2}{x+y}}$

9. $\dfrac{\dfrac{6w^2}{xyz}}{\dfrac{8w^3}{xyz}}$

10. $\dfrac{\dfrac{x+1}{x}}{\dfrac{x-1}{x}}$

11. $\dfrac{\dfrac{3x^2}{y^2-3}}{\dfrac{12x}{y^2-3}}$

12. $\dfrac{\dfrac{2-3x}{2x}}{\dfrac{3-4x}{3x}}$

13. $\dfrac{\dfrac{35x^2}{24}}{\dfrac{49x}{12}}$

14. $\dfrac{\dfrac{1}{6z^3}}{\dfrac{z+1}{9z^2}}$

15. $\dfrac{\dfrac{2}{x-1}}{\dfrac{1}{x+1}}$

In Problems 16 through 21, simplify each complex fraction. See Example 2.

16. $\dfrac{\dfrac{1}{2}+\dfrac{5}{2}}{\dfrac{3}{2}+\dfrac{7}{2}}$

17. $\dfrac{\dfrac{2}{3}-\dfrac{1}{2}}{\dfrac{4}{3}+\dfrac{1}{2}}$

18. $\dfrac{\dfrac{3}{4}+1}{\dfrac{5}{8}-2}$

19. $\dfrac{\dfrac{1}{x}+1}{\dfrac{1}{x}-1}$

20. $\dfrac{\dfrac{2}{y}-3}{\dfrac{4}{y}+5}$

21. $\dfrac{2+\dfrac{3}{z}}{4-\dfrac{5}{z}}$

In Problems 22 through 24, simplify each complex fraction. See Example 3.

22. $\dfrac{\dfrac{1}{z-3}+1}{\dfrac{2}{z-3}}$

23. $\dfrac{\dfrac{2}{s+1}}{\dfrac{s}{s+1}+1}$

24. $\dfrac{\dfrac{5}{x+2}-1}{\dfrac{6}{x+2}}$

In Problems 25 through 27, simplify each complex fraction. See Example 4.

25. $\dfrac{\dfrac{1}{x}+1}{\dfrac{1}{y}+1}$

26. $\dfrac{\dfrac{1}{x}+\dfrac{1}{y}}{\dfrac{1}{x}-\dfrac{1}{y}}$

27. $\dfrac{\dfrac{3}{x-1}-\dfrac{2}{x}}{\dfrac{2}{x-1}+\dfrac{3}{x}}$

Practice Exercises

In Problems 28 through 54, simplify each complex fraction. Reduce answers.

28. $\dfrac{\dfrac{1}{3}}{\dfrac{2}{3}}$

29. $\dfrac{\dfrac{3}{4}}{\dfrac{3}{5}}$

30. $\dfrac{\dfrac{3}{4}}{\dfrac{9}{16}}$

31. $\dfrac{\dfrac{1}{3} + \dfrac{4}{3}}{\dfrac{4}{3} + \dfrac{7}{3}}$

32. $\dfrac{\dfrac{3}{5} - \dfrac{1}{4}}{\dfrac{4}{5} + \dfrac{1}{4}}$

33. $\dfrac{\dfrac{2}{3} - 1}{\dfrac{5}{6} + 2}$

34. $\dfrac{\dfrac{1}{x^2}}{\dfrac{1}{x}}$

35. $\dfrac{\dfrac{6}{y^2}}{\dfrac{8}{y^3}}$

36. $\dfrac{\dfrac{5s}{16}}{\dfrac{s^3}{12}}$

37. $\dfrac{\dfrac{6}{x-1}}{\dfrac{27}{x-1}}$

38. $\dfrac{\dfrac{2xy^2}{x-y}}{\dfrac{x^2}{x-y}}$

39. $\dfrac{\dfrac{6w^2}{xyz}}{\dfrac{15w}{xyz}}$

40. $\dfrac{\dfrac{y-2}{xy}}{\dfrac{y+3}{xy}}$

41. $\dfrac{\dfrac{4x^3}{y^2+1}}{\dfrac{12x}{y^2+1}}$

42. $\dfrac{\dfrac{2t+5}{5t}}{\dfrac{3t+7}{6t}}$

43. $\dfrac{\dfrac{35x^3}{54}}{\dfrac{25x}{27}}$

44. $\dfrac{\dfrac{s-1}{12s^2}}{\dfrac{3}{18s^4}}$

45. $\dfrac{\dfrac{x}{x-2}}{\dfrac{x}{x+2}}$

46. $\dfrac{\dfrac{1}{x} - 2}{\dfrac{1}{x} + 3}$

47. $\dfrac{\dfrac{3}{y} - 4}{\dfrac{5}{y} + 6}$

48. $\dfrac{3 + \dfrac{4}{z}}{5 - \dfrac{6}{z}}$

49. $\dfrac{\dfrac{2}{z-4} - 1}{\dfrac{1}{z-4}}$

50. $\dfrac{\dfrac{t}{t-1}}{\dfrac{t^2}{t-1} + t}$

51. $\dfrac{\dfrac{2}{k+3} - 4}{\dfrac{4}{k+3}}$

52. $\dfrac{\dfrac{1}{x} - 1}{\dfrac{1}{x^2} - \dfrac{1}{x}}$

53. $\dfrac{\dfrac{1}{xy} + \dfrac{1}{xy^2}}{\dfrac{1}{x} - \dfrac{1}{xy^2}}$

54. $\dfrac{\dfrac{2}{x+1} - \dfrac{3}{x}}{\dfrac{3}{x+1} + \dfrac{2}{x}}$

Challenge Problems

55. Is $\dfrac{\dfrac{a}{b}}{c}$ the same as $\dfrac{a}{\dfrac{b}{c}}$?

In Your Own Words

56. Discuss the two methods for simplifying a complex
fraction.

5.6 Fractional Equations

TAPE 10

In Section 3.2 we solved linear equations that contained fractional coefficients. In this section we look at solving equations that contain variables in their denominators. We call such equations **fractional equations.**

Let's solve

$$\frac{x}{x-1} - \frac{2}{x-1} = 3$$

Notice that there are variables in the denominators. Anytime we have a variable in the denominator of a fraction, we must be *very* careful. For example, if x has the value 1, then the first fraction in the equation is of the form

$$\frac{1}{1-1} \quad \text{or} \quad \frac{1}{0}$$

However, in Section 5.1 we learned that a fraction with a denominator that is 0 is undefined. Thus 1 cannot be in the solution set of this equation.

So how do we solve equations of this type? We solve them by multiplying both sides of the equation by the least common denominator. However, before writing the solution set, we check to make sure the possible solution does not make any denominator in the equation have the value of zero.

We multiply both sides by the least common denominator, $(x - 1)$, to clear fractions.

$$(x-1) \cdot \left(\frac{x}{x-1} - \frac{2}{x-1} \right) = (x-1) \cdot 3$$

$$(x-1) \cdot \frac{x}{x-1} - (x-1) \cdot \frac{2}{x-1} = (x-1) \cdot 3 \qquad \text{Distributive Property}$$

$$\frac{(x-1)x}{x-1} - \frac{(x-1)2}{x-1} = (x-1)3$$

$$x - 2 = (x-1)3$$

$$x - 2 = 3x - 3$$

$$-2 = 2x - 3$$

$$1 = 2x$$

$$\frac{1}{2} = x$$

Replacing x with $\frac{1}{2}$ does *not* make any denominator in the original equation have the value of zero, so we write the solution set:

$$\left\{\frac{1}{2}\right\}$$

A Procedure for Solving Fractional Equations

1. Clear the equation of fractions by multiplying both sides by the least common denominator of all denominators in the equation.

2. Solve the resulting equation for possible solutions.

3. Check to see if any of the possible solutions make a denominator of the original equation have a value of zero. If so, we *cannot* include it in the solution set.

► **EXAMPLE 1**

Solve $\dfrac{1}{x} + \dfrac{2}{3} = \dfrac{3}{5}$.

Solution

STEP 1   Clear the equation of fractions and solve the resulting equation. The least common denominator of all the fractions in the equation is $15x$. Multiply both sides by $15x$.

$$15x \cdot \left(\frac{1}{x} + \frac{2}{3}\right) = 15x \cdot \frac{3}{5}$$

$$15x \cdot \frac{1}{x} + 15x \cdot \frac{2}{3} = 15x \cdot \frac{3}{5} \qquad \text{Distributive Property}$$

$$\frac{15x}{x} + \frac{30x}{3} = \frac{45x}{5}$$

$$15 + 10x = 9x$$

$$15 + 10x - 9x = 9x - 9x$$

$$15 + x = 0$$

$$x = -15 \qquad \text{Possible solution}$$

STEP 2   Check that -15 does not make a denominator have value 0. Replacing x by -15 does not make any denominator in the original equation equal 0, so we may put -15 in the solution set.

STEP 3   Write the solution set.

$$\{-15\}$$

◄

► **EXAMPLE 2**

Solve $\dfrac{x + 2}{x + 5} - \dfrac{2x - 1}{x + 5} = 1$.

Solution

Multiply both sides by $(x + 5)$.

$$(x + 5) \left(\dfrac{x + 2}{x + 5} - \dfrac{2x - 1}{x + 5} \right) = (x + 5) \, 1$$

$$(x + 5) \cdot \dfrac{x + 2}{x + 5} - (x + 5) \cdot \dfrac{2x - 1}{x + 5} = x + 5 \qquad \text{Distributive Property}$$

$$x + 2 - (2x - 1) = x + 5 \qquad \text{Note parentheses}$$

$$x + 2 - 2x + 1 = x + 5$$

$$-x + 3 = x + 5$$

$$3 = 2x + 5$$

$$-2 = 2x$$

$$-1 = x \qquad \text{Possible solution}$$

Since -1 does not make a denominator have value zero, the solution set is

$$\{-1\} \qquad \qquad ◄$$

► **EXAMPLE 3**

Solve $\dfrac{x}{x + 1} + 2 = -\dfrac{1}{x + 1}$.

Solution

We multiply both sides by $(x + 1)$ and simplify.

$$(x + 1) \cdot \left(\dfrac{x}{x + 1} + 2 \right) = (x + 1) \cdot \left(-\dfrac{1}{x + 1} \right)$$

$$(x + 1) \cdot \dfrac{x}{x + 1} + (x + 1) \cdot 2 = (x + 1) \cdot \left(-\dfrac{1}{x + 1} \right)$$

$$\text{Distributive Property}$$

$$\dfrac{(x + 1)x}{x + 1} + (x + 1)2 = -\dfrac{x + 1}{x + 1}$$

$$x + 2x + 2 = -1$$

$$3x + 2 = -1$$

$$3x + 2 - 2 = -1 - 2$$

$$3x = -3$$

$$x = -1 \qquad \text{Possible solution}$$

Be Careful!

Before we write the solution set, we must be sure the possible solution does not make any denominator have the value zero. However, when we inspect the original equation, we find that if x has the value -1, two denominators are 0. The number -1 is not a solution. So, because -1 was our *only* possible solution, there are no numbers in the solution set. Therefore, the solution set is the empty set, which we write

$$\{ \ \} \quad \text{or} \quad \varnothing \qquad \blacktriangleleft$$

The least common denominator is used in working several different kinds of problems in this chapter. In Section 5.4 we used the LCD to add and subtract fractions; in Section 5.5 the LCD was used to simplify complex fractions; in this section we used the LCD to clear an equation of fractions.

▶ **EXAMPLE 4**

Follow the directions in each part and notice how the LCD is used.

(a) Perform the operation indicated: $\dfrac{3}{x} + \dfrac{1}{7}$.

(b) Solve $\dfrac{3}{x} + \dfrac{1}{7} = \dfrac{22}{7}$.

(c) Simplify $\dfrac{\dfrac{3}{x}}{\dfrac{1}{7}}$.

Solutions

(a) This is addition of fractions. The LCD is $7x$.

$$\frac{3}{x} + \frac{1}{7} = \frac{3\,(7)}{x\,(7)} + \frac{1\,(x)}{7\,(x)} \qquad \text{Fundamental Principle}$$

$$= \frac{21}{7x} + \frac{x}{7x}$$

$$= \frac{21 + x}{7x}$$

(b) This is a fractional equation. The LCD is $7x$. We clear fractions by multiplying by $7x$.

$$\frac{3}{x} + \frac{1}{7} = \frac{22}{7}$$

$$7x\left(\frac{3}{x} + \frac{1}{7}\right) = 7x \cdot \frac{22}{7} \qquad \text{Multiply by } 7x$$

$$7x \cdot \frac{3}{x} + 7x \cdot \frac{1}{7} = 7x \cdot \frac{22}{7} \qquad \text{Distributive Property}$$

$$21 + x = 22x$$

$$1 = x$$

$$\{1\}$$

(c) This is a complex fraction. The LCD is $7x$. We multiply the numerator and denominator by $7x$.

$$\frac{\dfrac{3}{x}}{\dfrac{1}{7}} = \frac{\dfrac{3}{x} \cdot \boxed{7x}}{\dfrac{1}{7} \cdot \boxed{7x}}$$ Multiply by $7x$

$$= \frac{21}{x}$$ ◄

Fractional equations can lead to equations other than linear equations.

► **EXAMPLE 5**

Solve $\dfrac{x}{2} - \dfrac{4}{x} = 1$.

Solution

We clear fractions by multiplying by $2x$.

$$\boxed{2x} \left(\frac{x}{2} - \frac{4}{x} \right) = \boxed{2x}\,(1)$$

$$x^2 - 8 = 2x$$ Distributive Property

$$x^2 - 2x - 8 = 0$$ Make one side 0

$$(x - 4)(x + 2) = 0$$ Factor

$$x = 4 \quad \text{or} \quad x = -2$$ Apply Property of Zero Products

Neither -2 nor 4 make a denominator in the original equation have a value of 0.

$$\{-2, 4\}$$ ◄

Problem Set 5.6

Warm-Ups

In Problems 1 through 12, solve the equations. See Example 1.

1. $\dfrac{3}{y} - \dfrac{4}{y} = 1$

2. $\dfrac{x + 2}{x} + \dfrac{x - 1}{x} = 3$

3. $\dfrac{2p - 1}{p} - \dfrac{p + 1}{p} = 3$

4. $\dfrac{3t - 1}{2t} - \dfrac{2 - t}{2t} = 1$

5. $\dfrac{12}{x} = \dfrac{6}{5}$

6. $\dfrac{3}{2y} = \dfrac{1}{2}$

7. $\dfrac{1}{x} - \dfrac{1}{2} = -\dfrac{1}{6}$

8. $\dfrac{2}{5} - \dfrac{1}{x} = \dfrac{1}{5}$

9. $\dfrac{x - 4}{x} + \dfrac{1}{7} = \dfrac{4}{7}$

10. $\dfrac{2}{3} - \dfrac{3}{x} = 1$

11. $\dfrac{4}{3x} + \dfrac{1}{x} = \dfrac{1}{2}$

12. $\dfrac{7}{2y} - \dfrac{1}{y} = \dfrac{5}{4}$

In Problems 13 through 18, solve each equation. See Example 2.

13. $\dfrac{3}{x+1} - \dfrac{x-2}{x+1} = 1$

14. $\dfrac{x+1}{x-1} - \dfrac{2x-6}{x-1} = 2$

15. $\dfrac{1}{w+3} + \dfrac{1}{2} = \dfrac{3}{w+3}$

16. $\dfrac{3}{v-1} + \dfrac{2}{3} = 1$

17. $\dfrac{2}{x+3} + \dfrac{10}{7} = 2$

18. $\dfrac{2x-1}{x-1} - \dfrac{3}{2} = 1$

In Problems 19 through 24, solve each equation. See Example 3.

19. $\dfrac{x-3}{x-2} + \dfrac{1}{x-2} = 2$

20. $\dfrac{x}{x+2} + \dfrac{3}{x+2} = \dfrac{1}{x+2}$

21. $\dfrac{3}{y+3} - 2 = \dfrac{-y}{y+3}$

22. $\dfrac{2}{t-2} = \dfrac{t}{t-2} - 2$

23. $\dfrac{2k+3}{k} - \dfrac{3k+3}{k} = 1$

24. $\dfrac{x+3}{x} - \dfrac{3-x}{x} = 3$

In Problems 25 through 28, perform any operations indicated and simplify any complex fractions. See Example 4.

25. $\dfrac{\dfrac{x}{x+2}}{\dfrac{3}{x+2}}$

26. $\dfrac{7}{2y} - \dfrac{1}{y}$

27. $\dfrac{3}{x+1} - \dfrac{x+2}{x+1}$

28. $\dfrac{1}{w+3} + \dfrac{1}{2}$

In Problems 29 through 32, solve each equation. See Example 5.

29. $x + \dfrac{3}{x} = -4$

30. $x - \dfrac{4}{x} = 3$

31. $\dfrac{x}{2} + \dfrac{5}{2} + \dfrac{3}{x} = 0$

32. $\dfrac{x}{4} - \dfrac{1}{4} = \dfrac{3}{x}$

Practice Exercises

Solve any equations, perform any additions or subtractions, or simplify any complex fractions.

33. $\dfrac{5}{y} = \dfrac{1}{2}$

34. $\dfrac{3}{4} + \dfrac{1}{x} = \dfrac{5}{4}$

35. $\dfrac{4}{w} = \dfrac{2}{w} + 4$

36. $\dfrac{r+3}{r} + \dfrac{2r+1}{r} = 2$

37. $\dfrac{1}{x} + \dfrac{1}{3} = \dfrac{5}{6}$

38. $\dfrac{1}{x} - \dfrac{2}{7} = \dfrac{1}{21}$

39. $\dfrac{5-z}{z} - \dfrac{3-2z}{z} = -1$

40. $\dfrac{2t-3}{t} - \dfrac{5t+7}{t} = 2$

41. $\dfrac{6x-5}{x} - \dfrac{7-x}{x} = 3$

42. $\dfrac{x+7}{x} = \dfrac{2}{3}$

43. $\dfrac{2b-5}{b} = \dfrac{5}{3}$

44. $\dfrac{7}{5y} = \dfrac{2}{y}$

45. $\dfrac{3}{8} - \dfrac{1}{x} = \dfrac{1}{4}$

46. $\dfrac{5}{12} + \dfrac{3}{x} = \dfrac{2}{3}$

47. $\dfrac{7}{2} - \dfrac{x+3}{x} = 1$

48. $\dfrac{4}{5} - \dfrac{3}{x} = \dfrac{x-4}{5}$

49. $\dfrac{1}{x} - \dfrac{1}{3} + \dfrac{1}{6}$

50. $\dfrac{1}{x} + \dfrac{2}{7}$

51. $\dfrac{2}{x+1} + \dfrac{x}{x+1} = 2$

52. $\dfrac{\dfrac{x+1}{x-1}}{\dfrac{5}{x-1}}$

53. $\dfrac{x}{x+2} - \dfrac{1}{2} = \dfrac{1}{x+2}$

54. $\dfrac{y+9}{y-3} + 1 = \dfrac{4y}{y-3}$

55. $x + 2 = \dfrac{3}{x}$

56. $\dfrac{x-3}{x-2} - \dfrac{1}{x-2} = 2$

57. $\dfrac{x-16}{x+4} - 2 = \dfrac{5x}{x+4}$

58. $\dfrac{3x-1}{x-1} - \dfrac{3}{2} = 2$

59. $\dfrac{3}{x+2} = \dfrac{3}{2}$

60. $x - \dfrac{10}{x} = 3$

61. $\dfrac{5}{k+5} - 4 = \dfrac{-k}{k+5}$

62. $\dfrac{-2}{x-2} + 2 = \dfrac{-x}{x-2}$

Challenge Problems

In the Problems 63 through 70, Q, R , and S are constants, with S ≠ 0. Find the solution set for each.

63. $\dfrac{x}{S} + \dfrac{1}{S} = \dfrac{3}{S}$

64. $\dfrac{x}{3} + \dfrac{Q}{3} = Q$

65. $\dfrac{Q}{5} - \dfrac{x}{5} = \dfrac{R}{5}$

66. $\dfrac{Q}{S} + \dfrac{x}{S} = \dfrac{R}{S}$

67. $\dfrac{1}{2}(x + R) = R$

68. $\dfrac{2(x - Q)}{3} = 2Q$

69. $\dfrac{3x}{2S} - \dfrac{x}{S} = \dfrac{5}{S}$

70. $\dfrac{4}{3S} + \dfrac{1}{2S}x = 1$

In Your Own Words

71. Explain the procedure for solving a fractional equation.

5.7   Ratio and Proportion, Similar Triangles, and Other Applications

TAPE 10

Ratios and proportions are commonly used to express the relationship of one quantity to another. Both are used in many fields, and using them leads to fractional equations.

A church choir has 5 men and 13 women. We say the ratio of men to women in the choir is 5 to 13, which is often written $5 : 13$. We also write ratios as fractions. The ratio of men to women in the choir is $\dfrac{5}{13}$. Because ratios are fractions, they may often be reduced. A bin of oranges at the Farmers Market contains 500 navel oranges and 20 tangerines. The ratio of navel oranges to tangerines in the bin is $500 : 20$. Since this can be written as the fraction $\dfrac{500}{20}$, which can be reduced to $\dfrac{25}{1}$, we say the ratio is $25 : 1$.

Ratio

If p and q are numbers with $q \neq 0$, the **ratio** of the number p to the number q is written

$$p : q \quad \text{or} \quad \frac{p}{q}$$

► **EXAMPLE 1**

Write a ratio for each.

(a) The ratio of 3 males to 5 females.   (b) The ratio of 7 cats to 11 dogs.

Solutions

(a) $\dfrac{\text{Males}}{\text{Females}} = \dfrac{3}{5}$ (b) $\dfrac{\text{Cats}}{\text{Dogs}} = \dfrac{7}{11}$ ◄

Application problems can often be solved by finding two ratios that are equal.

Proportion

A **proportion** is a statement that two ratios are equal.

We read the proportion $p:q = s:t$ as "p is to q as s is to t." The "inside" numbers in this proportion, q and s, are called the **means,** and the "outside" numbers, p and t, are called the **extremes.** This proportion can also be written

$$\frac{p}{q} = \frac{s}{t}$$

In this equation, if we multiply both sides by qt, we see that

$$pt = qs$$

That is, in a proportion, the *product* of the *means* equals the *product* of the *extremes.*

► **EXAMPLE 2**

Find x in the proportion $7:30 = x:5$.

Solution

Writing $7:30 = x:5$ is the same as writing $\dfrac{7}{30} = \dfrac{x}{5}$. To find x, we must solve this equation.

$$\frac{7}{30} = \frac{x}{5}$$

$$5 \cdot \frac{7}{30} = 5 \cdot \frac{x}{5} \qquad \text{Multiply by 5}$$

$$\frac{7}{6} = x$$

In the proportion $7:30 = x:5$, x has the value $\dfrac{7}{6}$.   ◄

Proportion problems are often stated in words.

► **EXAMPLE 3**

A shirt manufacturer uses 3 lb of cotton for every 2 lb of rayon in its fall hunting shirt. How many pounds of rayon should the manufacturer buy to use with 840 lb of cotton to make these shirts?

Solution

We assume the ratio of cotton to rayon stays the same in the manufacture of these shirts. We are given the ratio

$$\frac{\text{Number of pounds of cotton}}{\text{Number of pounds of rayon}} = \frac{3}{2}$$

If we let x be the number of pounds of rayon needed to use with 840 lb of cotton, then we form the proportion

$$\frac{3}{2} = \frac{840}{x}$$

$$2x \cdot \frac{3}{2} = 2x \cdot \frac{840}{x} \qquad \text{Multiply by } 2x$$

$$3x = 1680$$

$$x = \frac{1680}{3}$$

$$x = 560$$

The manufacturer should buy 560 lb of rayon to use with 840 lb of cotton. ◄

Similar triangles are triangles that are the same shape. In plane geometry we find that corresponding sides of similar triangles are proportional.

► **EXAMPLE 4**

If a triangle has sides of lengths 18 ft, 24 ft, and 40 ft, what is the length of the longest side of the *similar* triangle whose shortest side is 63 ft?

Solution

If x is the length of the longest side of the unknown triangle, then we form the proportion

$$\frac{40}{x} = \frac{18}{63} = \frac{2}{7} \qquad \text{Reduce}$$

$$7x \cdot \frac{40}{x} = 7x \cdot \frac{2}{7} \qquad \text{Multiply by } 7x$$

$$280 = 2x$$

$$140 = x \qquad \text{Divide by 2}$$

The length of the longest side is 140 ft.   ◄

► **EXAMPLE 5**

The ratio of the number of peppermints to the number of butterscotch pieces in a can of party candy is $5:8$. If there are 26 pieces of candy in a can, how many of them are peppermints?

Solution

Let x be the number of peppermints. Since there are 26 pieces in all, we have $26 - x$ pieces of butterscotch.
We are given the ratio

$$\frac{\text{Peppermint}}{\text{Butterscotch}} = \frac{5}{8}$$

We form a proportion

$$\frac{5}{8} = \frac{x}{26 - x}$$

We multiply by the LCD, $8(26 - x)$.

$$8(26 - x) \cdot \frac{5}{8} = 8(26 - x) \cdot \frac{x}{26 - x}$$

$$(26 - x)5 = 8x$$

$$130 - 5x = 8x \qquad \text{Distributive Property}$$

$$130 = 13x$$

$$10 = x$$

There are 10 peppermints in the can.   ◄

Other applications of fractional equations include number problems and distance problems.

► **EXAMPLE 6**

One number is 5 times another number. If the sum of their reciprocals is $\frac{3}{5}$, what are the numbers?

Solution

Let n be the smaller number.
Then $5n$ is the other number.
We form an equation

$$\frac{1}{n} + \frac{1}{5n} = \frac{3}{5}$$

$$5n\left(\frac{1}{n} + \frac{1}{5n}\right) = 5n \cdot \frac{3}{5} \qquad \text{Multiply by } 5n$$

$$5 + 1 = 3n \qquad \text{Distributive Property}$$

$$6 = 3n$$

$$2 = n$$

If n is 2, then $5n$ is 10. The numbers are 2 and 10.   ◄

► **EXAMPLE 7**

A Delta 737 flies 50 mph faster than a British Airways DC 9. The 737 travels 500 mi in the same amount of time that the DC 9 travels 400 mi. Find the speeds of the planes.

Solution

Let s be the speed of the DC 9 in miles per hour. Then $s + 50$ is the speed of the 737 in miles per hour.
We make a table.

	d	r	t
737	500	$s + 50$	
DC 9	400	s	

To find the time for each plane, we use the formula $d = rt$, or $t = \dfrac{d}{r}$.

	d	r	t
737	500	$s + 50$	$\dfrac{500}{(s + 50)}$
DC 9	400	s	$\dfrac{400}{s}$

Since the time each plane flies is the same, we form the equation

$$\frac{500}{s + 50} = \frac{400}{s}$$

We multiply both sides by $s(s + 50)$.

$$s(s + 50)\frac{500}{s + 50} = s(s + 50)\frac{400}{s}$$

$$500s = (s + 50)400$$

$$500s = 400s + 20{,}000$$

$$100s = 20{,}000$$

$$s = 200$$

If s is 200, $s + 50$ is 250.
The speed of the DC 9 is 200 mph and the speed of the 737 is 250 mph.   ◄

Problem Set 5.7

Warm-Ups

For Problem 1, see Example 1.

1. This fall 28 students have enrolled in Ms. Gavant's beginning algebra class, 16 of whom are females. There are 8 students from out of state and 7 from out of the country. Write the following ratios.

   (a) Females to males
   (b) Males to females
   (c) Out-of-state students to out-of-country
   (d) Females to out-of-state
   (e) Males to total number of students

In Problems 2 through 5, find the value of x that makes the proportion correct. See Example 2.

2. $x:5 = 3:1$

3. $8:x = 1:7$

4. $2:3 = x:12$

5. $3:5 = 15:x$

For Problems 6 through 11, see Example 3.

6. If a person 6 ft tall casts a shadow 4 ft long, how tall is a flagpole that casts a 60-ft shadow?

7. If there are 2.54 cm in 1 in., how many inches are in 508 cm?

8. If 5 lb of chicken will serve 18 people, how many pounds will it take to serve 360 people?

9. If it takes 2 hr to plant 30 tomato plants, how many tomato plants can be planted in 20 min?

10. Seven drops of special dye 3 are needed to color 400 L of oil. How many drops should be added to color 20,000 L?

11. Gloria's car averages 22 mi/gal. How many gallons of gasoline did she use during the 209,000-mi lifetime of her car?

In Problems 12 and 13, the triangles are similar. Find x in each. See Example 4.

12.

13.

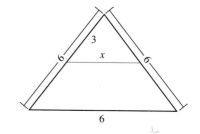

For Problems 14 and 15, see Example 5.

14. The ratio of woodpeckers to mockingbirds in a forest is $2:7$. If a total of 27 of these two kinds of birds was counted one day, how many of them were woodpeckers?

15. A cable 30 m long is divided into two parts so that the ratio of the length of the shorter piece to that of the longer piece is $2:3$. Find the length of each piece.

For Problems 16 and 17, see Example 6.

16. If twice the reciprocal of a number is 6, find the number.

17. One number is four times another number. If the sum of their reciprocals is $\frac{5}{8}$, what are the numbers?

For Problems 18 through 20, see Example 7.

18. Jody and Meredith live 2 and 3 mi, respectively, from the Joffrey Ballet Studio in New York. They walk to the studio from their homes each day to take lessons. The walk takes Jody 15 min less than Meredith. If they walk at the same rate, how long does it take Meredith to walk to the studio?

19. Frank and Doug ran races at a field day. Frank ran 10 km and Doug ran 12 km in the same amount of time.

If Doug's rate was 2 km/hr faster than Frank's rate, how fast did Frank run?

20. Two trains leave Grand Central Station at 5:00 P.M. One train travels 400 mi in the same amount of time that the other travels 600 mi. If one of the trains traveled 10 mph faster than the other, how fast did the slower train travel?

Practice Exercises

21. The triangles pictured are similar. Find x.

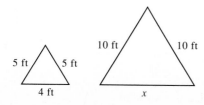

22. There are 1.6 km per mile. How many miles are there in 48.8 km?

23. A box of coffee beans weighing 40 lb is to be divided into two parts so that the ratio of the weight of the ligher part to the weight of the heavier part is $3:5$. How many pounds are there in both parts?

24. If it takes 2 hr to work 30 problems, how many problems can be worked in 12 min?

25. An investment in 100 shares of a certain stock earned a dividend of $23. An investment in how many shares would have received a dividend of $29.90?

26. A sugar solution is made by dissolving 2 lb of sugar in 5 gal of water. How many pounds of sugar are needed for 30 gal of water to make a solution of the same strength?

27. The reciprocal of 2 more than a number is twice the reciprocal of the number. Find the number.

28. One number is twice another number. If the sum of their reciprocals is $\frac{1}{2}$, find the numbers.

29. A train travels 10 mph faster than a boat. The train goes 250 mi in the same amount of time that the boat goes 200 mi. What is the speed of each?

30. It takes Gail 20 min longer to walk 3 mi than it takes Carol to walk 2 mi. If they walk at the same rate, how long does it take Carol to walk the 2 mi?

31. Lois thinks that a 5 by 7 rectangle is the most pleasing to the eye. Therefore, she wants the front of her new home to be in that proportion. If her new house is to be 35 feet high, how wide should the front of the house be to meet her specification?

Challenge Problems

32. Prove that the product of the means is equal to the product of the extremes.

33. The sum of the reciprocals of two consecutive even integers is 0. Find the integers.

In Your Own Words

34. Make up a word problem that uses a ratio of $2 : 3$.

 Chapter Summary

GLOSSARY

Rational expression   The quotient of two polynomials. The expression is **undefined** when the polynomial in the denominator has the value of zero.

Ratio   A quotient of two numbers, written as a fraction.

Reduce a rational expression to lowest terms   To write the expression so that there are no common factors in the numerator and the denominator.

Proportion   A statement that two ratios are equal.

Fundamental Principle of Fractions

$$\frac{AC}{BC} = \frac{A}{B},$$

where B and C do not have a value of 0

We use the Fundamental Principle to reduce and rewrite fractions. B, C, D do not have value 0.

Multiplication and Division of Rational Expressions

Multiply

$$\frac{A}{B} \cdot \frac{C}{D} = \frac{A \cdot C}{B \cdot D} = \frac{AC}{BD}$$

Divide

$$\frac{A}{B} \div \frac{C}{D} = \frac{A}{B} \cdot \frac{D}{C} = \frac{AD}{BC}$$

Addition and
Subtraction of
Rational Expressions

Add

$$\frac{A}{B} + \frac{C}{B} = \frac{A + C}{B}$$

Subtract

$$\frac{A}{B} - \frac{C}{B} = \frac{A - C}{B}$$

Simplifying Complex
Fractions

If there is only one term in the numerator and denominator, we use Method 1.
Method 1: Treat the fraction as a division problem.
If either the numerator or the denominator has more than one term, we use Method 2.
Method 2: Multiply the numerator and denominator by the LCD.

Fractional Equations

Fractional equations can be solved by clearing the equation of fractions (multiply both sides by the LCD) and then solving the resulting equation for possible solutions. Possible solutions that make any denominator in the original equation have a value of zero *cannot* be included in the solution set.

Proportion

The proportion $p : q = r : s$ can be written as $\dfrac{p}{q} = \dfrac{r}{s}$.

CHECK UPS

1. The rational expression $\dfrac{x + 3}{(x - 1)(x + 2)}$ is undefined when x has the value of (?).   Section 5.1; Example 3

2. The rational expression $\dfrac{3x + 9}{x^2 - 9}$ reduces to (?).   Section 5.2; Example 2b

3. $\dfrac{5x - 10}{x + 3} \cdot \dfrac{2x + 6}{3x - 6} = (?)$ Section 5.3; Example 2a

4. $\dfrac{x + 1}{x - 1} \div \dfrac{x + 1}{x - 2} = (?)$ Section 5.3; Example 5

5. The LCD for $x^2 y^3 z$ and $x^4 y^2$ is (?).   Section 5.4; Example 3b

6. $\dfrac{x}{x + 2} + \dfrac{4x}{x^2 - 4} = (?)$.   Section 5.4; Example 7a

7. $\dfrac{1}{x^2 + 2x + 1} - \dfrac{1}{x^2 - 1} = (?)$ Section 5.4; Example 7b

8. $\dfrac{\dfrac{1}{x - 1} + 5}{\dfrac{1}{x - 1} - 1} = (?)$.   Section 5.5; Example 3

9. Solve $\dfrac{1}{x} + \dfrac{2}{3} = \dfrac{3}{5}$.   Section 5.6; Example 1

10. Find x in the proportion $7 : 30 = x : 5$.   Section 5.7; Example 2

11. One number is 5 times another number. If the sum of their reciprocals is $\dfrac{3}{5}$, what are the numbers?   Section 5.7; Example 6

► Review Problems

In Problems 1 through 4, find the values of x, if any, for which the rational expression is undefined.

1. $\dfrac{11}{2x}$

2. $\dfrac{x}{x-2}$

3. $\dfrac{x+1}{(x+2)(x-3)}$

4. $\dfrac{x^2-9}{x^2-3x-10}$

In Problems 5 through 8, express each rational expression as an expression of equal value with the indicated denominator.

5. $\dfrac{21}{x^2}; 4x^3$

6. $\dfrac{3}{x-1}; 4(x-1)$

7. $\dfrac{x-1}{x+2}; x^2+x-2$

8. $\dfrac{4}{x+4}; x^2-16$

In Problems 9 through 12, reduce each expression to lowest terms.

9. $\dfrac{3t^3}{5t^2}$

10. $\dfrac{7-21x}{2(1-3x)}$

11. $\dfrac{x^2-2x}{x^2-4}$

12. $\dfrac{3x-6}{x^2-x-2}$

In Problems 13 through 32, perform the operations indicated. Express each number in lowest terms.

13. $\dfrac{2x}{x-1} \cdot \dfrac{x-1}{2x+6}$

14. $\dfrac{x^2-4}{xy^3} \cdot \dfrac{x^2-2x}{x+2}$

15. $\dfrac{x^3y^2}{6-2x} \div \dfrac{xy^3}{3x-9}$

16. $\dfrac{x^2-x-6}{x^2-25} \div \dfrac{x^2-9}{x^2-2x-15}$

17. $\dfrac{x+2}{3x-2} + \dfrac{x-3}{6x-4}$

18. $\dfrac{4}{x^2-1} + \dfrac{2x}{x+1}$

19. $\dfrac{1}{4x+6} - \dfrac{1}{6x+9}$

20. $\dfrac{3}{2x^2+3x-27} - \dfrac{1}{2x^2+13x+18}$

21. $\dfrac{4t^2-t}{-t^3} \div (16t^2-1)$

22. $\dfrac{1}{2-5y} + \dfrac{1}{5y-2}$

23. $x - \dfrac{3}{2x}$

24. $\dfrac{2-s-s^2}{s+1} \cdot \dfrac{s}{s^2-1}$

25. $\dfrac{1}{w+1} + \dfrac{3}{w^2-w-2}$

26. $\dfrac{ux+3u-2x-6}{u^2x} \cdot \dfrac{ux^2}{u^2+u-6}$

27. $\dfrac{x^2-16}{x^2+8x+16} \div \dfrac{4-x}{4+x}$

28. $\dfrac{3}{2x-1} - \dfrac{5}{3-6x}$

29. $\dfrac{\dfrac{1}{2}-\dfrac{1}{3}}{\dfrac{1}{2}+\dfrac{1}{3}}$

30. $\dfrac{\dfrac{1}{x}+\dfrac{1}{x^2}}{\dfrac{1}{x^2}}$

31. $\dfrac{1+\dfrac{1}{y}}{1-\dfrac{1}{y}}$

32. $\dfrac{\dfrac{3}{t-1}-1}{\dfrac{2}{t-1}}$

In Problems 33 through 40, find the solution set.

33. $\dfrac{3}{x} - 1 = 0$

34. $\dfrac{1}{2x} + \dfrac{2}{x} = \dfrac{5}{2}$

35. $\dfrac{x-3}{x} + \dfrac{1}{4} = \dfrac{7}{x}$

36. $\dfrac{3x+4}{x} - \dfrac{2x+3}{x} = 2$

37. $\dfrac{7}{k+7} - 6 = \dfrac{-k}{k+7}$

38. $\dfrac{4}{x-3} + 1 = \dfrac{5}{x-3}$

39. $x - \dfrac{3}{x} = 2$

40. $\dfrac{x}{2} = \dfrac{8}{x}$

41. The ratio of foxes to wolves in an area is $3 : 8$. If 15 foxes are in the area, how many wolves are in the area?

42. If 1 gal is 3.8 L, how many liters are in 3.5 gal?

43. The sum of the reciprocals of two consecutive odd integers is 0. What are the numbers?

44. Jean Marie and Paul drive different routes from their house to the mall. Jean Marie takes the scenic route, which is 6 mi, and Paul takes the interstate highway, which is 15 mi. It takes the same amount of time, but Paul travels 45 mph faster than Jean Marie. How fast does Jean Marie drive?

45. Camilla bought a leather bookbag in London for 35 pounds. If the exchange rate for British and U.S. currency is 1 pound = \$1.75, what is the equivalent cost of the bookbag in U.S. dollars?

46. Bryan set up a training schedule to run and bike 40 miles a day. On Tuesday he ran 10 miles. His running rate was 5 mph slower than his biking rate. If he trained for 3 hours, find his running rate.

47. Jeremy sent a photograph of his Black Labradors to Clark Photo to be copied poster size. The poster will be proportional in size to the original photograph, which

Problem 47

was 3 in. wide and 5 in. high. If the poster's width is to be 24 in., what will its height be?

48. Lakeesha drove her mother's Maserati the 8 miles from home to school in half the time she can drive her father's car the same route. She drove her mother's car 20 miles per hour faster than her father's car. How long did it take her to get to school driving the Maserati?

Let's Not Forget . . .

In Problems 49 through 54, perform the operations indicated.

49. -5^2

50. $(x - 5)^2$

51. $(-5)^2$

52. $\dfrac{5}{0}$

53. $(x + 2) \cdot \dfrac{x}{x + 2}$

54. $\dfrac{4}{x-1} - \dfrac{3}{1-x}$

In Problems 55 through 60, simplify each expression. Watch the role of the parentheses.

55. $x - 2(3x - 7)$

56. $(x - 2)3x - 7$

57. $(x - 2)(3x - 7)$

58. $x - 2(3x) - 7$

59. $\dfrac{x+1}{y} - \dfrac{x-1}{y}$

60. $\dfrac{x+1}{(x-1)y} - \dfrac{1}{y}$

In Problems 61 through 64, determine if 2 is a factor *or a* term.

61. $\dfrac{3x+2}{5x}$

62. $\dfrac{7x^2 - x + 1}{y + 2}$

63. $\dfrac{7x^3 - 1}{2x(3x + 1)}$

64. $\dfrac{5(x-1) + (1+x)}{2y}$

In Problems 65 through 67, multiply each expression by $\dfrac{2}{x}$.

65. $\dfrac{x}{2}$

66. $x^2(x + 7)$

67. $2x^2 + 3x$

In Problems 68 through 71, label each as an expression, equation, *or* inequality. *Solve the equations and inequalities and simplify the expressions.*

68. $\dfrac{x^2 - 9}{3 - x}$

69. $\dfrac{2}{x} + \dfrac{3}{4x} = \dfrac{11}{4}$

70. $5(x - 1) - 3 < -7$

71. $\dfrac{2}{x} + \dfrac{3}{4x}$

In Problems 72 through 77, identify the expressions that are factored. Factor those that are not factored, if possible.

72. $(x - 3)^2$

73. $x^2 - 9$

74. $x^2 - 6x + 9$

75. $x^3 + 27$

76. $(x + 3)^3$

77. $3x^3 - 81$

► You Decide

Draw two rectangles, one 4 in. by 5 in. and the other 1 in. by 5 in. Most people feel that the 1 by 5 is too long to be pleasing to the eye and the 4 by 5 too tall. Draw several more rectangles, modifying the dimensions until you have what is the most pleasing to *your* eye. Carefully cut a rectangle from cardboard that is $8\frac{1}{2}$ in. on the longest side, with its dimensions in proportion to your favorite rectangle.

The ancient Greeks (and many modern architects) think that a rectangle whose dimensions are in the ratio $\dfrac{\sqrt{5} - 1}{2} \approx 0.618$ is the most pleasing to the eye. This ratio is often called the **golden ratio.** Cut a rectangle with longest side $8\frac{1}{2}$ in. and dimensions in the golden

ratio from cardboard and compare it with the one with your favorite dimensions. Decide which of the two is most pleasing to your eye now.

► Chapter 5 Test

In Problems 1 through 8, choose the correct answer.

1. The rational expression $\dfrac{x - 1}{x(x - 2)}$ is undefined if x has the value of (?).

(a) 1

(b) −1

(c) 0 or 2

(d) −2

2. If the rational expression $\dfrac{3x}{x - 5}$ is written with a denominator of $x^2 - 25$, it becomes (?).

(a) $\dfrac{3x}{x^2 - 25}$

(b) $\dfrac{3x^2 + 15x}{x^2 - 25}$

(c) $\dfrac{9x^2}{x^2 - 25}$

(d) $\dfrac{3x^2 - 15x}{x^2 - 25}$

3. When reduced to lowest terms, the rational expression $\dfrac{2x^2 - 6x}{2x^2 - 5x - 3}$ becomes (?)

(a) $\dfrac{3}{4}$

(b) $\dfrac{6x}{5x + 3}$

(c) $\dfrac{2x}{2x + 1}$

(d) None of the above

4. $\dfrac{12x^2}{7x^2 - 21} \cdot \dfrac{5x^3 - 15x}{9x} = (?).$

(a) $\dfrac{10x^3}{7}$

(b) $\dfrac{20x^4 - 60x^2}{21x^2 - 63}$

(c) $60x(x^3 - x)$

(d) $\dfrac{20x^2}{21}$

5. $\dfrac{6x - 4}{x^2(x + 1)} \div \dfrac{x(3x - 2)}{2x + 2} = (?).$

(a) $\dfrac{4}{x^3}$

(b) $\dfrac{4}{x}$

(c) $\dfrac{1}{x}$

(d) None of the above

6. $\dfrac{x}{x - 2} - \dfrac{2x - 3}{2x - 4} = (?).$

(a) $\dfrac{-3}{2(x - 2)}$

(b) $\left\{\dfrac{1}{2}\right\}$

(c) $\dfrac{-5x + 3}{(x - 2)(2x - 4)}$

(d) $\dfrac{3}{2(x - 2)}$

7. $\dfrac{1}{(x - 1)^2} + \dfrac{1}{x^2 - 1} = (?)$

(a) $\dfrac{2}{(x - 1)^2}$

(b) $\dfrac{2x}{(x - 1)^2(x + 1)}$

(c) $\dfrac{2}{x^2 - 1}$

(d) $\dfrac{x^2 - (x - 1)^2}{(x - 1)^2(x^2 - 1)}$

8. When simplified, the complex fraction $\dfrac{x + \dfrac{1}{x}}{1 - \dfrac{2}{x}}$ becomes (?).

(a) $\dfrac{x^2 + 1}{x - 2}$

(b) $\dfrac{2x}{1 - 2x}$

(c) $\dfrac{x + 1}{-2}$

(d) $\dfrac{(x^2 + 1)(x - 2)}{x^2}$

9. Evaluate the rational expression $\dfrac{3x - y^2}{2x + y^3}$ when x is 3 and y is -2.

In Problems 10 through 14, perform the operations indicated. Write answers in lowest terms.

10. $\dfrac{1}{x - x^2} + \dfrac{x}{1 - x}$

11. $\dfrac{(t + 2)^2}{t^2 + 6t + 8} \cdot \dfrac{t^2 - 2t - 8}{(t + 2)^3}$

12. $\dfrac{x^3 - 3x^2}{x^3 - 6x^2 + 9x} \div \dfrac{x}{x^2 - 9}$

13. $\dfrac{x}{x - 2} - \dfrac{10}{x^2 + x - 6}$

14. $\dfrac{s^2 - s - 6}{s^2 + s - 2} \cdot \dfrac{s^2 + 3s - 4}{s^2 - 7s + 12}$

In Problems 15 through 17, find the solution set.

15. $\dfrac{1}{x} + \dfrac{1}{2x} = \dfrac{3}{2}$

16. $\dfrac{2}{x + 2} - \dfrac{1}{x} = \dfrac{-4}{x(x + 2)}$

17. $x - \dfrac{7}{x} = 6$

18. There are 1500 registered voters in a district, 900 of whom are Republicans; the remainder are Democrats. Express each of the following as a ratio:

(a) The number of Republicans to the number of Democrats

(b) The number of Democrats to the number of voters

19. On one map, 2 in. represents 75 mi. If two cities are 5 in. apart on the map, find the number of miles between the two cities.

20. One number is four times another number. If the sum of their reciprocals is $\dfrac{5}{12}$, find the two numbers.

6 Linear Equations and Inequalities in Two Variables

See Problem Set 6.2, Exercise 65.

▶ Connections

The inclination of the roof on a home in Vermont determines how much snow it can safely hold. A particularly steep ski slope might be developed specifically for advanced skiers. A road with a long, steep grade sometimes has a runaway ramp for 18-wheelers that lose their brakes driving down the incline. These are all applications of the idea of slope of a line, which we will explore in this chapter.

In 1638, the French mathematician René Descartes published the *Discourse on Method*. In one of the appendexes, Descartes describes a method of drawing pictures to represent abstract mathematical ideas. These pictures, called graphs, are the basis of what we today call Analytic Geometry. Analytic Geometry shows the relationship between algebra and geometry.

In this chapter, we introduce the rectangular coordinate system popularized by Descartes and link it to the graphs of points and lines. Lines and their equations and slopes will be emphasized. Linear inequalities in two variables will be discussed and graphed. Functions, a central concept of calculus, will also be introduced.

6.1   The Graph of *Ax + By = C*

TAPE 10

In Chapter 1 we used the number line to help us visualize the set of real numbers. Each point on the number line represents a real number, and each real number can be represented by a point on the number line. To help us visualize another set, we place two copies of the number line, one horizontally and one vertically, so that they intersect at the zero point of each number line.

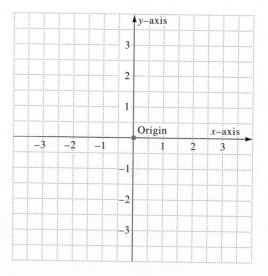

We call this a **rectangular coordinate system,** or a **Cartesian coordinate system,** named for the French mathematician René Descartes (1596–1650). We call the horizontal number line the **x-axis** and the vertical number line the **y-axis.** The point where the *x*-axis and *y*-axis intersect is called the **origin.**

With this coordinate system, we can associate each point in the plane with two numbers, called **coordinates.** Suppose we take two numbers, say 1 and 2, and write them in the form

$$(1, 2)$$

We call this form an **ordered pair.** We associate a point with the ordered pair (1, 2) by first locating 1 on the *x*-axis and then locating 2 on the *y*-axis. The point is determined by the point of intersection of the vertical line drawn through 1 and the horizontal line drawn through 2.

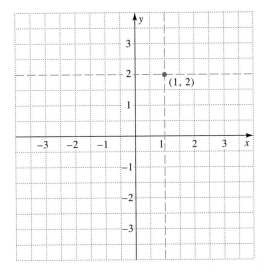

This unique point corresponds to the ordered pair (1, 2) and is called its **graph.** We call the first number in the ordered pair the **x-coordinate,** or **abscissa,** and the second number the **y-coordinate,** or **ordinate.** To **plot** an ordered pair means to graph its ordered pair.

Notice that the graphs of the ordered pairs (1, 2) and (2, 1) are different points. The order in which the numbers are written *is important.*

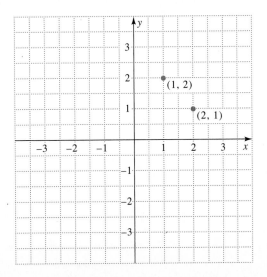

► **EXAMPLE 1**

Graph the points $(0, 2)$, $(-1, 0)$, $(-1, 2)$, $(1, -2)$, and $(-1, -2)$.

Solution

The five points are graphed in the following graph. Notice how the point $(-1, 0)$ lies on the *x*-axis and the point $(0, 2)$ lies on the *y*-axis.   ◄

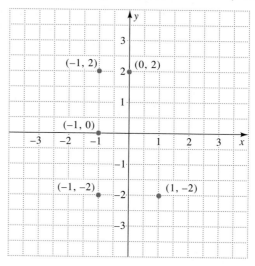

The plane in which the coordinate system lies is divided into four parts called **quadrants.** They are named with the Roman numerals I, II, III, and IV in a counterclockwise direction.

► **EXAMPLE 2**

Identify the quadrants in which the following points lie.

(a) $(2, 3)$ (b) $(-3, 1)$ (c) $(2, -3)$

(d) $(-2, -1)$ (e) $(2, 0)$

Solutions

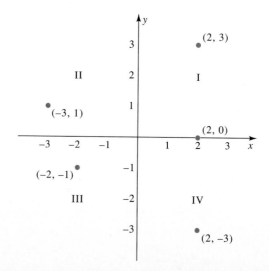

(a) (2, 3) lies in quadrant I.   (b) (−3, 1) lies in quadrant II.

(c) (2, −3) lies in quadrant IV.   (d) (−2, −1) lies in quadrant III.

(e) A point on an axis does not lie in any quadrant.   ◄

Let's consider solving the equation

$$2x - 3y = 6$$

Since there are two variables, a solution will consist of a number x and a number y that make the equation a *true* statement.

We can see that, if x is 3 and y is 0, we have

$$2(3) - 3(0) = 6$$

For these values the equation is a true statement. We can write this solution as an ordered pair, writing the x-value first and the y-value second:

$$(3, 0)$$

Also, if x is 0 and y is −2, we have

$$2(0) - 3(-2) = 6$$

This is also a true statement. Thus we write the ordered pair

$$(0, -2)$$

Likewise, if x is 6 and y is 2, the equation is a true statement:

$$2(6) - 3(2) = 6$$

So we write the ordered pair

$$(6, 2)$$

Still another solution is the ordered pair $(-3, -4)$.

We have found four solutions, which we write as ordered pairs. They are

$$(3, 0), \qquad (0, -2), \qquad (6, 2), \qquad (-3, -4)$$

There are an infinite number of solutions. So we cannot list them all. However, if we plot the ordered pairs that we have found to be in the solution set, we can make an interesting observation.

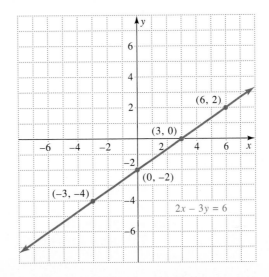

The points appear to lie on a line. In fact, every member of the solution set of the equation $2x - 3y = 6$ is an ordered pair whose graph lies on this line. Furthermore, any point on this line represents an ordered pair that belongs to the solution set. So the line is a picture of the solution set of the equation $2x - 3y = 6$. We call this line the **graph** of the equation.

The Graph of Linear Equations in Two Variables

If A, B, and C are constants with A and B not *both* zero, the graph of

$$Ax + By = C$$

is a line.

$Ax + By = C$ is called the **standard form** of the equation.

Some examples of linear equations in two variables written in standard form are as follows:

$$3x + 7y = 5 \qquad A \text{ is 3, } B \text{ is 7, and } C \text{ is 5.}$$

$$x - 2y = 0 \qquad A \text{ is 1, } B \text{ is } -2, \text{ and } C \text{ is 0.}$$

$$x = 3 \qquad A \text{ is 1, } B \text{ is 0, and } C \text{ is 3.}$$

► **EXAMPLE 3**

Determine if the given ordered pair is in the solution set of the given equation.

(a) $5x + 3y = 8$; $(-1, 3)$ (b) $x - 2y = 0$; $(0, 0)$
(c) $x = y$; $(-1, 1)$ (d) $x = 2$; $(0, 4)$
(e) $y + 3 = 0$; $(-1, -3)$

Solutions

(a) $5x + 3y = 8$; $(-1, 3)$
     Replacing x with -1 and y with 3 gives

$$5(-1) + 3(3) = -5 + 9$$
$$= 4$$

Since 4 is not equal to 8, $(-1, 3)$ is not in the solution set.
(b) $x - 2y = 0$; $(0, 0)$
     Substituting 0 for x and y gives

$$0 - 2(0) = 0$$

Because this makes a true statement, $(0, 0)$ is in the solution set.
(c) $x = y$; $(-1, 1)$
     Since -1 is not equal to 1, $(-1, 1)$ is not in the solution set.
(d) $x = 2$; $(0, 4)$
     This equation is the same as

$$x + (0)y = 2$$

If we replace x with 0 and y with 4, we have

$$0 + 0(4) = 0$$

Since 0 is not equal to 2, $(0, 4)$ is not in the solution set.

(e) $y + 3 = 0$; $(-1, -3)$

If we subtract 3 from both sides, this becomes

$$y = -3$$

This is the same as

$$(0)x + y = -3$$

We replace x with -1 and y with -3.

$$0(-1) + (-3) = -3$$

This makes a true statement, so $(-1, -3)$ is in the solution set.   ◄

► EXAMPLE 4

Find four points that are on the graph of the equation $2x + y = 4$.

Solution

We choose a number for x, say 0, and then determine what number y must be to make the equation a true statement. If we replace x with 0, the equation becomes

$$2(0) + y = 4$$

Solving for y gives us

$$y = 4$$

Thus the point $(0, 4)$ is on the graph.

It is convenient to list this information in a table with the x-coordinate on the left and the y-coordinate on the right.

x	y
0	4

Let's choose another number, this time for y. Suppose we let y be 0 and solve for x.

$$2x + 0 = 4$$
$$2x = 4$$
$$x = 2$$

x	y
0	4
2	0

So the point $(2, 0)$ is also on the graph. Let's choose 3 for x. We replace x by 3:

$$2(3) + y = 4$$
$$6 + y = 4$$
$$y = -2$$

x	y
0	4
2	0
3	-2

The point $(3, -2)$ is on the graph.

To find a fourth point, let y be -3.

$$2x + (-3) = 4$$
$$2x = 7$$
$$x = \frac{7}{2}$$

x	y
0	4
2	0
3	-2
$\frac{7}{2}$	-3

This gives the point $\left(\dfrac{7}{2}, -3\right)$.

Four points on the graph of $2x + y = 4$ are

$$(0, 4), \quad (2, 0), \quad (3, -2), \quad \text{and} \quad \left(\frac{7}{2}, -3\right) \qquad \blacktriangleleft$$

A common problem is to graph a linear equation. Since the graph of a linear equation is a line, two distinct points are enough to determine the line. If we look at Example 4, we can see that the easiest points to find are those where x is 0 and where y is 0. So, to graph a linear equation, we will find these two points. A third point may be found to serve as a check.

▶ ## EXAMPLE 5

Draw the graph of $x + 2y = 2$.

Solution

We let x be 0 and then solve for y.

$$x + 2y = 2$$
$$0 + 2y = 2$$
$$2y = 2$$
$$y = 1$$

x	y
0	1

We let y be 0 and then solve for x.

$$x + 2(0) = 2$$
$$x + 0 = 2$$
$$x = 2$$

x	y
0	1
2	0

To get a third point, we let y be 2 and solve for x.

$$x + 2(2) = 2$$
$$x + 4 = 2$$
$$x = -2$$

x	y
0	1
2	0
-2	2

We plot these points and use a straightedge to draw the line.   ◀

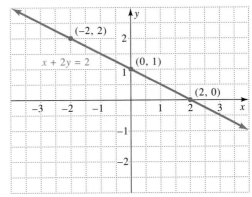

Notice that when we let x have the value 0 and find the y-value that makes the equation true we have found where the line crosses the y-axis. We call this y-value the **y-intercept.** Likewise, the **x-intercept** is the x-value where the line crosses the x-axis. In Example 5, the x-intercept is 2 and the y-intercept is 1.

► **EXAMPLE 6**

Find the intercepts of the equation $2x - 4y = 4$ and draw its graph.

Solution

To find the x-intercept, we replace y with 0 and solve for x. To find the y-intercept, we replace x with 0 and solve for y.

x	y
2	0
0	−1

To graph the line, we plot the points $(0, -1)$ and $(2, 0)$.

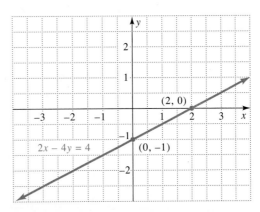

To Find the Intercepts of a Line

1. To find the x-intercept, replace y with 0 and solve for x.

2. To find the y-intercept, replace x with 0 and solve for y.

► **EXAMPLE 7**

Draw the graph of $x + y = 0$.

Solution

To find the *x*-intercept, we replace *y* with 0 and solve for *x*. We see that *y* is also 0. This gives us the point (0, 0). When we find the *y*-intercept, we get (0, 0) again. Both the *x*-intercept and the *y*-intercept are at the origin.

We need to find another point. If we let *y* be 1 and solve for *x*, we get −1 for *x*. Thus (−1, 1) is a second point on the graph.

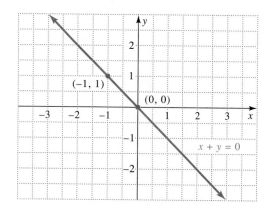

x	y
0	0
−1	1

Notice that the line in Example 7 goes through the origin. If *C* is 0 in the standard form $Ax + By = C$, then the line will go through the origin. The *x*-intercept as well as the *y*-intercept are 0.

Does every line have both an *x*-intercept and a *y*-intercept?

► **EXAMPLE 8**

Draw the graph of $x = 3$.

Solution

If we write this equation as

$$x + (0)y = 3$$

we see that *y* can be any value and *x* is always 3. So (3, 1) and (3, −2) are two points on the graph. We plot these points and draw the line.

x	y
3	1
3	−2

The line in Example 8 is called a *vertical line*.

> **Vertical Lines**
>
> The graph of the equation $x = p$ is a **vertical** line with an x-intercept of p.

► **EXAMPLE 9**

Draw the graph of $y = 2$.

Solution

If we write this equation as

$$(0)x + y = 2$$

we can see that x can be any value and y is always 2.
The points $(1, 2)$ and $(-3, 2)$ are on the graph. We plot these points and draw the line.

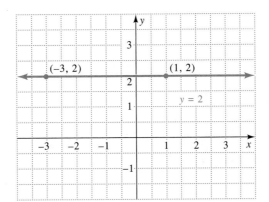

The line in Example 9 is called a *horizontal line*.

> **Horizontal Lines**
>
> The graph of the equation $y = q$ is a **horizontal** line with a y-intercept of q.

Problem Set 6.1

Warm-Ups

In Problems 1 through 4, plot the following ordered pairs. See Example 1.

1. $(2, 3)$ **2.** $(1, -3)$ **3.** $(-2, 2)$ **4.** $(-2, -3)$

In Problems 5 through 8, give the quadrant in which each point lies. See Example 2.

5. $(1, 9)$ **6.** $(-2, 8)$ **7.** $\left(\dfrac{1}{2}, -6\right)$ **8.** $\left(-\dfrac{4}{3}, -\dfrac{9}{5}\right)$

In Problems 9 through 12, determine which of the given ordered pairs are in the solution set of the given equation. See Example 3.

9. $2x - y = 4$; $(2, 0), (0, 2), (1, 3)$ **10.** $x + 3y = 6$; $(0, 2), (-1, 2), (1, 3)$

11. $x + y = 0$; $(0, 0), (1, 1), (1, -1)$ **12.** $x = 2y$; $(1, 2), (2, 1), (-2, -1)$

In Problems 13 through 20, find three points on the graph of each equation. See Example 4.

13. $x + y = 4$ **14.** $x = 3y$

15. $\dfrac{1}{2}x - y = 0$ **16.** $x - \dfrac{1}{2}y = 3$

17. $y = 4$ **18.** $2x - 3y = 6$

19. $x = \dfrac{7}{3}$ **20.** $\dfrac{3}{2}x + y = 5$

In Problems 21 through 28, find the intercepts of each equation and graph each. See Examples 5 and 6.

21. $x + 3y = 6$ **22.** $x - y = 3$ **23.** $x + 3y = 0$ **24.** $2x - 5y = 10$

25. $\dfrac{1}{5}x - 1 = 0$ **26.** $\dfrac{1}{2}y + 1 = 0$ **27.** $x = -1$ **28.** $y = \dfrac{1}{2}$

Practice Exercises

In Problems 29 through 38, plot the following ordered pairs.

29. $(3, 2)$ **30.** $(-3, -2)$ **31.** $(-3, 3)$ **32.** $(2, -2)$

33. $(0, 2)$ **34.** $(0, -1)$ **35.** $\left(-\dfrac{7}{3}, \dfrac{2}{3}\right)$ **36.** $\left(\dfrac{1}{2}, 0\right)$

37. $(0, -3)$

38. $(-3, 0)$

In Problems 39 through 46, give the quadrant in which each point lies.

39. $(1, -9)$

40. $(-2, -8)$

41. $(-4, 6)$

42. $(6, 0)$

In Problems 43 though 50, determine which of the given ordered pairs are in the solution set of the given equation.

43. $3x - y = 4$; $(2, 0), (0, -4), (1, 3)$

44. $x + 2y = 5$; $(0, 5), (-1, -2), (3, 1)$

45. $x - y = 0$; $(0, 0), (1, 1), (1, -1)$

46. $x = -2y$; $(1, 2), (2, 1), (2, -1)$

47. $5x - y = 3$; $(0, -3), (0, -1), (-1, -8)$

48. $x - 4y = 5$; $(3, -1), (5, 0), (0, 5)$

49. $x = 7$; $(0, 7), (7, 7), (7, 3)$

50. $y = 5$; $(0, 5), (-3, 5), (0, 0)$

In Problems 51 through 58, find three points on the graph of each equation.

51. $x + y = 3$

52. $x = -3y$

53. $x - 4y = 0$

54. $3x - y = 4$

55. $y = -3$

56. $2x - 4y = 12$

57. $x = 8$

58. $5x + 2y = 10$

In Problems 59 through 82, find the intercepts of each equation and graph each.

59. $x + 4y = 4$

60. $x - y = 2$

61. $x = -2$

62. $y = \dfrac{5}{2}$

63. $2x - 3y = 12$

64. $3x = y$

65. $\dfrac{1}{3}x - y = 3$

66. $4x + y = 4$

67. $x = \dfrac{-3}{2}$

68. $x = y$

69. $\dfrac{1}{3}x + \dfrac{1}{5}y = 1$

70. $\dfrac{2}{3}x - \dfrac{3}{2}y = 0$

71. $x + 5y = 0$

72. $3x - 5y = 15$

73. $x - 1 = 0$

74. $y + 3 = 0$

75. $2x - 3y = -6$

76. $4x = y$

77. $x - 4y = 4$

78. $3x + y = 3$

79. $x = \dfrac{-3}{4}$

80. $2x = y$

81. $5x + 4y = 20$

82. $x - 5y = 0$

83. A 400-gallon, partly empty, tropical fish tank can be filled according to the equation $25x - y = -100$, where y is the number of gallons in the tank and x is the number of minutes water has been going into the tank.
   (a) Sketch the graph of this equation. (Use different scales on the x- and y-axes.)

Problem 83

(b) What is the y-intercept? How much water was in the tank before any was added?
(c) On the x-axis, positive numbers represent future time and negative numbers represent past time. What is the x-intercept? How many minutes were required to put in the first 100 gallons?

84. Kim's automobile uses a full tank of gas according to the equation $x + 10y = 140$, where y is the number of gallons of gasoline in the tank and x is the miles driven.
   (a) Sketch the graph of this equation.
   (b) What is the y-intercept? What is the capacity of Kim's fuel tank?
   (c) What is the x-intercept? After how many miles will the tank be empty?

Challenge Problems

85. Plot five points with the same x- and y-coordinates on the same set of axes. For example, $(2, 2)$ and $(-3, -3)$. Connect these points. What seems to be true? Write an equation for a line whose points have the same x and y coordinates.

86. Write an equation for a line whose points have coordinates that are opposites.

87. Write an equation for a line on which each y-coordinate is twice the x-coordinate.

88. The Meyer Cookie Company determined that the relationship between the number of chocolate chips in a cookie and the calorie count of those chocolate chips is given by the equation $C - 7N = 0$, where C is the calorie count for the chocolate chips in a cookie and N is the number of chocolate chips in a cookie.
   (a) Graph this equation, letting the horizontal axis be the N-axis.
   (b) What will be the calorie count of the chocolate chips in a cookie with 9 chocolate chips? How is this shown on the graph?
   (c) What are the intercepts? What do they mean?

In Your Own Words

89. What is a rectangular coordinate system?

90. What are the intercepts of a line and how are they found?

6.2 Slope

How are the following two lines different?

Moving along each line from left to right, we could say that line A is going up, whereas line B is going down. The rate at which a line goes up or down is the most important property of a line. We call this property the **slope** of the line. To determine the slope of a line, we start with any two points on the line, say P_1 with coordinates (x_1, y_1) and P_2 with coordinates (x_2, y_2).

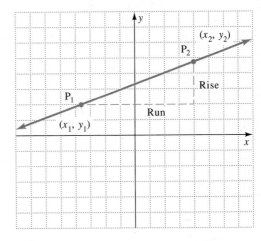

As we move from P_1 to P_2, the vertical change is called the **rise,** and the horizontal change is called the **run.** We define the slope of a line as the ratio of the rise to the run.

$$\text{Slope} = \frac{\text{rise}}{\text{run}}$$

To find the rise, we calculate the difference in the y-coordinates. The y-coordinate of P_1 is y_1 and that of P_2 is y_2.

$$\text{Rise} = y_2 - y_1$$

To find the run, we find the difference in the x-coordinates. The x-coordinate of P_1 is x_1 and that of P_2 is x_2. So

$$\text{Run} = x_2 - x_1$$

We use the letter m to designate the slope. Thus the slope of the line containing the points (x_1, y_1) and (x_2, y_2) is given by the following formula.

> **The Slope Formula**
>
> The slope of the line through the points (x_1, y_1) and (x_2, y_2) is given by
>
> $$m = \frac{y_2 - y_1}{x_2 - x_1}, \qquad x_1 \neq x_2$$

▶ **EXAMPLE 1**

Find the slope of the line that contains the points $(1, 5)$ and $(4, 7)$.

Solution

If we let P_1 be the point $(1, 5)$ and P_2 be the point $(4, 7)$, then we can see that x_1 is 1, y_1 is 5, x_2 is 4, and y_2 is 7. Substituting these values into the formula gives

$$m = \frac{y_2 - y_1}{x_2 - x_1}$$

$$= \frac{7 - 5}{4 - 1}$$

$$= \frac{2}{3}$$

The slope is $\frac{2}{3}$.

The graph of the line is as follows. Notice that the line is going up from left to right.

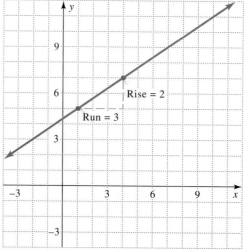

What would happen in Example 1 if we let P_1 be the point $(4, 7)$ and P_2 be the point $(1, 5)$? It would mean that x_1 is 4, y_1 is 7, x_2 is 1, and y_2 is 5. Substituting these values into the slope formula gives

$$m = \frac{5 - 7}{1 - 4}$$

$$= \frac{-2}{-3}$$

$$= \frac{2}{3}$$

Notice that the slope is the same in both cases. It doesn't make any difference which point we call P_1 or P_2. However, it *is* important that the coordinates be subtracted in the same order. The order in which the *x*-values are subtracted *must be the same* as the order in which the *y*-values are subtracted.

► **EXAMPLE 2**

Find the slope of the line that contains the points (3, 9) and (8, 5).

Solution

If we let P_1 be (3, 9) and P_2 be (8, 5), we have $x_1 = 3$, $y_1 = 9$, $x_2 = 8$, and $y_2 = 5$. We substitute these into the slope formula.

$$m = \frac{y_2 - y_1}{x_2 - x_1}$$

$$= \frac{5 - 9}{8 - 3}$$

$$= \frac{-4}{5}$$

$$= -\frac{4}{5}$$

The slope is $-\dfrac{4}{5}$.

The graph of the line is shown next. Notice that it goes down from left to right.

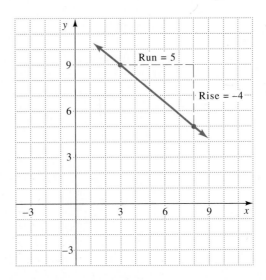

The next examples show that we must be careful with signs.

► **EXAMPLE 3**

Find the slope of the line containing the points (−2, 4) and (3, −1).

Solution

If P_1 is (−2, 4) and P_2 (3, −1), we have x_1 is −2, y_1 is 4, x_2 is 3, and y_2 is −1.

$$m = \frac{y_2 - y_1}{x_2 - x_1}$$

$$= \frac{-1 - 4}{3 - (-2)}$$

$$= \frac{-5}{5}$$

$$= -1$$

The slope is -1.

► EXAMPLE 4

Find the slope of the line that goes through the points $(-3, 0)$ and $(-1, 2)$.

Solution

If we let P_1 be $(-3, 0)$ and P_2 be $(-1, 2)$, we have

$$m = \frac{y_2 - y_1}{x_2 - x_1}$$

$$= \frac{2 - 0}{-1 - (-3)}$$

$$= \frac{2}{2}$$

$$= 1$$

The slope is 1.

The graphs of the lines in Examples 3 and 4 are shown next. Notice that lines going up have a *positive slope,* whereas lines going down have a *negative slope.*

In the next example, the line determined by the given points is a horizontal line. Look carefully at the calculation of its slope.

► **EXAMPLE 5**

Find the slope of the line that contains the points $(1, 3)$ and $(-2, 3)$.

Solution

If we let P_1 be $(1, 3)$ and P_2 be $(-2, 3)$, the slope formula gives

$$m = \frac{y_2 - y_1}{x_2 - x_1} = \frac{3 - 3}{-2 - 1}$$

$$= \frac{0}{-3}$$

$$= 0$$

The slope is 0.   ◄

The y-coordinates of all points on a horizontal line are the same, so the rise is 0. Thus the slope is 0. The graph of the line in Example 5 is shown next.

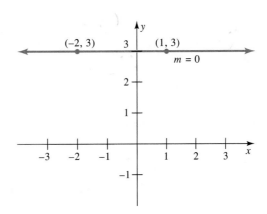

The next example illustrates a vertical line.

► **EXAMPLE 6**

Find the slope of the line through the points $(4, -1)$ and $(4, 1)$.

Solution

If P_1 is $(4, -1)$ and P_2 is $(4, 1)$, we have

$$m = \frac{y_2 - y_1}{x_2 - x_1} = \frac{1 - (-1)}{4 - 4}$$

$$= \frac{2}{0}$$

Since division by 0 is not allowed, the slope is undefined. We say a vertical line has no slope.   ◄

The x-coordinates of all points on a vertical line are the same. The run will be 0. Thus a vertical line has no defined slope. The graph of the line in Example 6 is shown next.

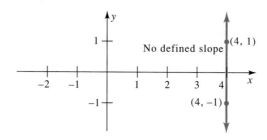

Horizontal and Vertical Lines

1. *Horizontal lines* have equations of the form

$$y = q$$

and have a slope of 0.

2. *Vertical lines* have equations of the form

$$x = p$$

and have *no* defined slope.

If we are given the equation of a line, we can find its slope by finding two points on the line and using the slope formula.

▶ **EXAMPLE 7**

Find the slope of the graph of $x + 2y = 4$.

Solution

The graph is a line. We find two points on the line. The intercepts are convenient to use.

$$x + 2(0) = 4$$
$$x = 4$$
$$0 + 2y = 4$$
$$y = 2$$

x	y
4	0
0	2

We use the points $(4, 0)$ and $(0, 2)$ as P_1 and P_2 in the slope formula.

$$m = \frac{y_2 - y_1}{x_2 - x_1} = \frac{2 - 0}{0 - 4}$$

$$= \frac{2}{-4}$$

$$= -\frac{1}{2}$$

The slope is $-\dfrac{1}{2}$.   ◀

► **EXAMPLE 8**

Find the slope of the graph of $x - 2y = 0$.

Solution

We have a line that goes through the origin. So we need one other point. If y has a value of 1, then

$$x - 2(1) = 0$$

$$x = 2$$

This gives the point $(2, 1)$. We use the points $(0, 0)$ and $(2, 1)$ in the slope formula:

$$m = \frac{y_2 - y_1}{x_2 - x_1} = \frac{1 - 0}{2 - 0}$$

$$= \frac{1}{2}$$

The slope is $\frac{1}{2}$. ◄

► **EXAMPLE 9**

Find the slope of the graphs of each equation.

(a) $x = 14$ (b) $y = -8$

Solutions

(a) The graph is a vertical line. It has no defined slope.

(b) The graph is a horizontal line. Its slope is 0. ◄

Parallel and Perpendicular Lines

l_1 is a line with slope m_1 and l_2 is a line with slope m_2. Assume that neither line is vertical.

1. If l_1 and l_2 are **parallel** lines, then their slopes are the same; that is,

$$m_1 = m_2$$

2. If l_1 and l_2 are **perpendicular** lines, then the product of their slopes is -1; that is,

$$m_1 m_2 = -1 \quad \text{or} \quad m_1 = -\frac{1}{m_2}$$

► **EXAMPLE 10**

If the slope of l_1 is $\frac{3}{4}$, find the slope of a line that is:

(a) Parallel to l_1 (b) Perpendicular to l_1

Solutions

(a) The slope of any line parallel to l_1 is $\frac{3}{4}$.

(b) The slope of any line perpendicular to l_1 is $-\frac{4}{3}$, since $\frac{3}{4} \cdot \left(-\frac{4}{3}\right) = -1$. ◄

► **EXAMPLE 11**

Find the slope of a line parallel to the line containing the points $(-2, 5)$ and $(3, -1)$.

Solution

We first find the slope of the line containing the points $(-2, 5)$ and $(3, -1)$. The slope formula gives

$$m = \frac{y_2 - y_1}{x_2 - x_1}$$

$$= \frac{5 - (-1)}{-2 - 3}$$

$$= \frac{6}{-5}$$

$$= -\frac{6}{5}$$

The slope of a line parallel to this line is the same. So the slope of the line is $-\dfrac{6}{5}$.

◄

► **EXAMPLE 12**

Find the slope of a line that is perpendicular to the line through the points $(-3, 4)$ and $(-1, -5)$.

Solution

We use the slope formula to find the slope of the line through the points $(-3, 4)$ and $(-1, -5)$

$$m = \frac{y_2 - y_1}{x_2 - x_1}$$

$$= \frac{-5 - 4}{-1 - (-3)}$$

$$= \frac{-9}{2}$$

$$= -\frac{9}{2}$$

The slope of a line perpendicular to this line has slope $\dfrac{2}{9}$ because $-\dfrac{9}{2} \cdot \dfrac{2}{9} = -1$.

◄

► **EXAMPLE 13**

Find the slope of a line that is parallel to the graph of $x + 2y = -2$ and find the slope of a line that is perpendicular to the graph of $x + 2y = -2$.

Solution

We can find the slope of the graph of $x + 2y = -2$ by finding two points on the line and then using the slope formula. We see that $(0, -1)$ and $(-2, 0)$ are on the line.

$$m = \frac{0 - (-1)}{-2 - 0}$$

$$= -\frac{1}{2}$$

The slope of the graph of $x + 2y = -2$ is $-\frac{1}{2}$. The slope of a line parallel to this line will also be $-\frac{1}{2}$. The slope of a line perpendicular to this line is 2 because $-\frac{1}{2}(2) = -1$.   ◄

Problem Set 6.2

Warm-Ups

In Problems 1 through 15, find the slope of the line, if it exists, containing the given points. In Problems 1 through 6, see Examples 1 and 2.

1. $(2, 1), (3, 2)$ **2.** $(4, 1), (2, 5)$ **3.** $(0, 2), (4, 3)$
4. $(3, 4), (1, 1)$ **5.** $(1, 1), (2, 2)$ **6.** $(0, 2), (2, 0)$

For Problems 7 through 12, see Examples 3 and 4.

7. $(2, 1), (3, -6)$ **8.** $(-1, 2), (1, 3)$ **9.** $(3, -1), (-2, 2)$
10. $(2, -4), (-3, -5)$ **11.** $(-2, -2), (-3, 0)$ **12.** $(-5, -4), (-1, -3)$

For Problems 13 through 15, see Examples 5 and 6.

13. $(-1, 2), (3, 2)$ **14.** $(3, -2), (3, 4)$ **15.** $(-1, 0), (-1, 4)$

In Problems 16 through 21, find the slope of the graph of each equation. See Examples 7 through 9.

16. $x + 2y = 2$ **17.** $3x - y = 6$ **18.** $x - 4y = 0$
19. $2x + 3y = 0$ **20.** $x = 15$ **21.** $y = -5$

In Problems 22 through 25, the two given points determine a line. For each line, find (a) the slope of a line parallel to the given line, and (b) the slope of a line perpendicular to the given line. See Examples 11 and 12.

22. $(2, 3)$ and $(-1, 2)$ **23.** $(1, 4)$ and $(2, 0)$
24. $(2, 4)$ and $(2, 3)$ **25.** $(-1, -3)$ and $(-4, 2)$

In Problems 26 through 28, the equation of a line is given. For each equation, find (a) the slope of a line parallel to the given line, and (b) the slope of a line perpendicular to the given line. See Example 13.

26. $x + y = 3$ 27. $x - y = 2$ **28.** $2x - y = 2$

Practice Exercises

In Problems 29 through 43, find the slope of the line determined by the two given points, if the slope is defined.

29. (3, 2), (4, 3)　　　　　　**30.** (5, 6), (3, 2)　　　　　　**31.** (0, 3), (5, 4)
32. (−1, 2), (2, 0)　　　　　　**33.** (3, −1), (−2, 2)　　　　**34.** (3, 3), (−1, 9)
35. (4, 4), (2, 2)　　　　　　**36.** (−2, −2), (3, 3)　　　　**37.** (0, 3), (3, 0)
38. (−4, −3), (−2, 3)　　　　**39.** (3, 3), (4, 3)　　　　　**40.** (0, 3), (0, 0)
41. (4, 2), (4, 7)　　　　　　**42.** (−2, 3), (2, 4)　　　　　**43.** (4, −2), (3, 3)

In Problems 44 through 52, find the slope of the graph of each equation.

44. $x + y = 3$　　　　　　　**45.** $x - y = 2$　　　　　　　**46.** $2x + y = 6$
47. $x + 2y = 4$　　　　　　**48.** $3x - y = 6$　　　　　　**49.** $x - 3y = 8$
50. $x = -3$　　　　　　　　**51.** $2x + 4y = 8$　　　　　　**52.** $y = 7$

In Problems 53 through 58, the two given points determine a line. For each line, find (a) the slope of a line parallel to the given line, and (b) the slope of a line perpendicular to the given line.

53. (1, 3) and (−2, 2)　　　　**54.** (2, 5) and (1, 0)　　　　**55.** (1, 4) and (3, 5)
56. (2, −3) and (1, 2)　　　　**57.** (3, −3) and (2, −3)　　**58.** (−1, −3) and (0, 0)

In Problems 59 through 64, the equation of a line is given. For each equation, find (a) the slope of a line parallel to the given line, and (b) the slope of a line perpendicular to the given line.

59. $x + y = 1$　　　　　　　**60.** $x - y = 5$　　　　　　　**61.** $2x - y = -2$
62. $x - 3y = 8$　　　　　　**63.** $3x - 5y = 15$　　　　　**64.** $x - y = 0$

65. An airplane flying from San Francisco to Boston is approaching Logan International Airport for a landing. If the airplane is 32 mi from Logan at an altitude of 2 mi, what should be the slope of its glide path to make a successful landing?

66. The pitch, or slope, of a roof gives it the ability to shed water or snow. Find the pitch of the roof shown.

67. A ski slope descends 2200 ft in a straight-line run of 16,000 feet. What is its slope?

Problem 67

68. An airplane at an altitude of 10,000 ft is going to land at an airport 20 mi distant. What should the slope of its glide path be? (1 mi = 5280 ft).

69. What is the pitch (slope) of the roof shown?

Challenge Problems

70. Find the slope of a line perpendicular to the graph of $x = 0$.

71. Find the slope of a line that is parallel to the graph of $x = 0$.

72. Find the slope of a line that is perpendicular to the graph of $y = 0$.

73. Find the slope of a line that is parallel to the graph of $y = 0$.

In Your Own Words

74. What do we mean by the slope of a line?

75. How is the slope of a line calculated?

6.3   Equations of Lines

TAPE 11

We have been graphing linear equations written in standard form by finding the intercepts. We found the slope of the lines by using the slope formula. Consider the graph of $2x + 3y = 6$. We see that the x-intercept is 3 and the y-intercept is 2, which gives the following graph.

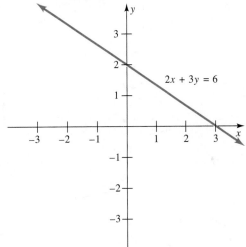

We can calculate the slope of this line by using the two points $(0, 2)$ and $(3, 0)$ in the slope formula.

$$m = \frac{2 - 0}{0 - 3}$$

$$= -\frac{2}{3}$$

The slope of the line is $-\frac{2}{3}$.

The equation $2x + 3y = 6$ can be written in a form other than the standard form. By solving for y, we can rewrite the equation as follows:

$$2x + 3y = 6$$

$$3y = -2x + 6 \qquad \text{Subtract } 2x$$

$$y = -\frac{2}{3}x + 2 \qquad \text{Divide by 3}$$

This form is called the **slope–intercept form.** Looking closely at this, we can make some interesting observations. The slope of the line, $-\frac{2}{3}$, and the y-intercept, 2, are very visible. This is not an accident. The equation of any nonvertical line can be written in this form.

Slope–Intercept Form

If a linear equation can be written in the form

$$y = mx + b$$

then the graph has slope m and y-intercept b.

Any nonvertical line can be written in this very useful form by solving for y.

► EXAMPLE 1

Find the slope and y-intercept of the graph of each equation.

(a) $3x + y = 6$ (b) $5x - 2y = 4$

Solutions

(a) We rewrite the equation in slope–intercept form by solving for y.

$$3x + y = 6$$

$$y = -3x + 6 \qquad \text{Subtract } 3x$$

The slope is -3 and the y-intercept is 6.

(b) Writing the equation in slope–intercept form, we have

$$5x - 2y = 4$$

$$-2y = -5x + 4 \qquad \text{Subtract } 6x$$

$$y = \frac{5}{2}x - 2 \qquad \text{Divide by } -2$$

The slope is $\frac{5}{2}$ and the y-intercept is -2.   ◄

It is often convenient to use the slope–intercept form when drawing the graph of a linear equation. The y-intercept gives us a point on the line, and we use the slope to find a second point.

► EXAMPLE 2

Sketch the graph of the equation $y = 2x + 1$.

Solution

As the equation is in slope–intercept form, we see that the slope is 2 and the y-intercept is 1. To draw the line, we locate the y-intercept. This gives us one point on the line. To find a second point, we use the slope, which is 2.

$$m = 2$$

or
$$m = \frac{2}{1} = \frac{\text{rise}}{\text{run}}$$

The rise is 2 and the run is 1. We start at the y-intercept and move *right* 1 unit and then *up* 2 units. This gives us a second point on the line.   ◄

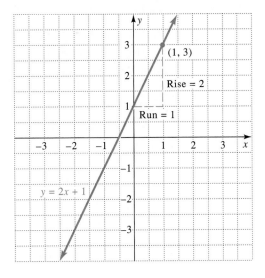

► **EXAMPLE 3**

Sketch the graph of $y = -\frac{2}{3}x - 1$.

Solution

The slope is $-\frac{2}{3}$ and the y-intercept is -1.

We locate the y-intercept and use the slope to find a second point.

$$m = -\frac{2}{3}$$

or

$$m = \frac{-2}{3}$$

The rise is -2 and the run is 3. From the y-intercept, we go *right* 3 units and *down* 2 units.

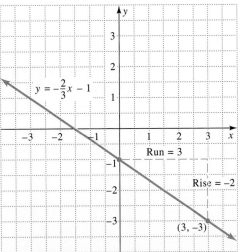

Notice that the run is always from left to right and that the rise is *up* if the slope is *positive* and *down* if the slope is *negative*.

Often we are given a line and asked to find an equation of that line. We need two pieces of information about the line:

1. The slope of the line.

2. The coordinates of any point on the line.

Suppose a line has slope 2 and contains the point $(2, 4)$. Then we know that $m = 2$ and the point $(2, 4)$ is on the line. Let (x, y) be *any* other point *on the line*.

If we substitute this information into the slope formula, we can get an equation of the line. Let P_1 be $(2, 4)$ and P_2 be (x, y). Then

$$m = \frac{y_2 - y_1}{x_2 - x_1}$$

$$\text{Slope} \quad 2 = \frac{y - 4}{x - 2} \qquad \text{From the point } (2, 4)$$

$$2(x - 2) = y - 4 \qquad \text{Multiply by } (x - 2)$$

$$2x - 4 = y - 4 \qquad \text{Distributive Property}$$

$$2x - y = 0$$

Notice that we have found an equation for a line with slope 2 that contains the point $(2, 4)$.

When using a given slope of m, a given point of (x_1, y_1), and a general point of (x, y), the slope formula becomes

$$m = \frac{y - y_1}{x - x_1}$$

If we multiply both sides by $x - x_1$, we have

$$m(x - x_1) = y - y_1$$

We usually turn this around and write,

$$y - y_1 = m(x - x_1)$$

This is called the **point–slope** form of the equation of a line.

Linear equations are often given in *standard* form. We have seen that one very useful form is the *slope–intercept* form. The *point–slope* form is convenient for finding an equation of a line given a point on the line and its slope.

Forms of the Linear Equation in Two Variables

Standard Form

$$Ax + By = C, \qquad A \text{ and } B \text{ not both } 0$$

Slope–Intercept Form

$$y = mx + b$$

Point–Slope Form

$$y - y_1 = m(x - x_1)$$

► **EXAMPLE 4**

Find an equation of the line with slope $\dfrac{1}{2}$ that contains the point $(3, 1)$.

Solution

We use the point–slope form, where m is $\dfrac{1}{2}$ and (x_1, y_1) is $(3, 1)$. Substituting into the point–slope formula gives

$$y - y_1 = m(x - x_1)$$

$$y - 1 = \frac{1}{2}(x - 3)$$

$$2y - 2 = x - 3 \qquad \text{Multiply by 2}$$

$$1 = x - 2y$$

$$x - 2y = 1$$

The desired equation is $x - 2y = 1$.   ◄

► **EXAMPLE 5**

Find an equation of the line with slope -4 that goes through the point $(-3, -2)$.

Solution

In the point–slope formula, we substitute -4 for m, -3 for x_1, and -2 for y_1.

$$y - y_1 = m(x - x_1)$$

$$y - (-2) = -4[x - (-3)]$$

$$y + 2 = -4(x + 3)$$

$$y + 2 = -4x - 12$$

$$4x + y = -14$$ ◄

► **EXAMPLE 6**

Find an equation of the line containing the points $(-1, 2)$ and $(3, -4)$.

Solution

To write an equation of a line, we need to know the *slope* of the line and the *coordinates* of any *point* on the line. We have the coordinates of two points on the line, so we need the slope. We can find it with the slope formula.

$$m = \frac{y_2 - y_1}{x_2 - x_1} = \frac{2 - (-4)}{-1 - 3}$$

$$= \frac{6}{-4}$$

$$= -\frac{3}{2}$$

Now we pick *either* of the two points to be (x_1, y_1). Let's use $(-1, 2)$ and substitute into the point–slope form.

$$y - y_1 = m(x - x_1)$$

$$y - 2 = -\frac{3}{2}[x - (-1)]$$

$$2y - 4 = -3(x + 1) \qquad \text{Multiply by 2}$$

$$2y - 4 = -3x - 3$$

$$3x + 2y = 1$$

An equation of the line is $3x + 2y = 1$.   ◄

If a line has a defined slope, we can use the point–slope form to write an equation for it.

To Write an Equation of a Line That Has Slope

1. Find m, the slope of the line.
2. Find (x_1, y_1), the coordinates of any point on the line.
3. Substitute into the point–slope form:

$$y - y_1 = m(x - x_1)$$

4. Write the equation in whichever form is preferred.

A special case of this occurs if we are given the slope and y-intercept of a line.

► **EXAMPLE 7**

Write an equation of the line with slope of 2 and y-intercept of 3.

Solution

Since we know the slope and the y-intercept, we can use the *slope–intercept* form and write an equation of the line directly.

$$y = mx + b$$

$$y = 2x + 3$$

This is an equation of the line in slope–intercept form.   ◄

Since a vertical line has no defined slope, we must treat it in a different way. It is usually best to treat horizontal lines in the same manner.

Equations of Horizontal and Vertical Lines

Horizontal Lines

A line with slope 0 is *horizontal* and has its equation

$$y = q$$

where q is the y-coordinate of any point on the line.

Vertical Lines

A line with no defined slope is *vertical* and has its equation

$$x = p$$

where p is the x-coordinate of any point on the line.

► **EXAMPLE 8**

Find an equation of the line through the points $(1, 2)$ and $(3, 2)$.

Solution

First we find the slope of the line.

$$m = \frac{2 - 2}{1 - 3} = \frac{0}{-2} = 0$$

Since the slope is 0, the line is horizontal and we can write an equation directly:

$$y = 2$$ ◄

► **EXAMPLE 9**

Find an equation of the vertical line through the point $(3, -2)$.

Solution

We cannot use the point–slope form because a vertical line has no defined slope. We remember that an equation of a vertical line must be of the form

$$x = p$$

An equation of the line is $x = 3$.   ◄

Problem Set 6.3

Warm-Ups

In Problems 1 through 6, write each equation in slope–intercept form. Find the slope and y-intercept of each graph and draw the graph of each. See Examples 1 through 3.

1. $2x + y = 0$

2. $2x + 3y = 6$

3. $2x - 3y = 9$

4. $x + 2y = 2$

5. $3x - y = 1$

6. $4x - 2y = 4$

In Problems 7 through 20, write an equation of the line satisfying the given conditions. For Problems 7 through 9, see Examples 4 and 5.

7. Through the point $(2, -3)$ with slope $\dfrac{1}{2}$.

9. Through the origin with slope of 1.

8. Through the point $(-1, 4)$ with slope $\dfrac{-2}{3}$.

For Problems 10 through 13, see Example 6.

10. Containing the points $(1, 3)$ and $(-2, 1)$.

11. Containing the points $(7, -2)$ and $(3, -5)$.

12. Containing the points $(1, 3)$ and $(0, 0)$.

13. Containing the points $(0, 3)$ and $(-2, 0)$.

For Problems 14 and 15, see Example 7.

14. With y-intercept of 2 and slope of 1.

15. With y-intercept of -1 and slope of $\frac{2}{3}$.

For Problems 16 through 20, see Examples 8 and 9.

16. Containing the points $(2, 4)$ and $(2, 7)$.

17. Through $(1, 2)$ with no defined slope.

18. Through $(1, 2)$ with 0 slope.

19. Horizontal line through the point $(1, -2)$.

20. Vertical line through the point $(1, -2)$.

Practice Exercises

In Problems 21 through 29, write each equation in slope–intercept form. Find the slope and y-intercept of each graph and draw the graph of each.

21. $2x + y = 2$

22. $4x + y = 1$

23. $x - y = 3$

24. $5x + y = 0$

25. $2x - 3y = -6$

26. $2x + 4y = 8$

27. $x + 3y = 3$

28. $2x - y = 2$

29. $3x - 4y = 4$

In Problems 30 through 48, write an equation of the line satisfying the given conditions.

30. Through the point $(4, 3)$ with slope 5.

31. Through the point $(-2, 0)$ with slope -1.

32. Through the point $(3, -3)$ with slope $\dfrac{3}{2}$.

33. Through the point $(-1, 3)$ with slope $\dfrac{-3}{2}$.

34. Through the origin with slope of -1.

35. With y-intercept of -2 and slope of 2.

36. With y-intercept of 4 and slope of $\dfrac{2}{5}$.

37. Through $(-1, 2)$ with no defined slope.

38. Through $(-1, 2)$ with 0 slope.

39. Vertical line through the point $(2, -2)$.

40. Horizontal line through the point $(2, -2)$.

41. Containing the points $(5, 3)$ and $(-2, 4)$.

42. Containing the points $(-2, 5)$ and $(0, 7)$.

43. Containing the points $(4, -6)$ and $(7, -5)$.

44. Containing the points $(2, 2)$ and $(-1, 1)$.

45. Containing the points $(1, 4)$ and $(1, 7)$.

46. Containing the points $(-1, 3)$ and $(0, 0)$.

47. Containing the points $(0, -3)$ and $(2, 0)$.

48. x-intercept of -4 and y-intercept of 1.

Challenge Problems

49. Find an equation of the line through the origin with no defined slope.

50. Find an equation of the line through the origin with 0 slope.

51. Find an equation of the line through the origin parallel to the graph of $y = 2x + 4$.

52. Find an equation of the line through the origin perpendicular to the graph of $y = \dfrac{2}{3}x - 1$.

53. Find an equation of the line through the point $(1, 3)$ and parallel to the line through the points $(1, 4)$ and $(-1, 3)$.

54. Find an equation of the line through the point $(2, 1)$ and perpendicular to the line through the points $(2, -5)$ and $(-3, 7)$.

55. Find an equation of the line through the point $(2,2)$ and parallel to the x-axis.

56. Find an equation of the line through the point $(-1, -3)$ and perpendicular to the y-axis.

57. Northeast Luggage Co. determined that cost of production can be expressed as a linear equation. If the cost is $4150 when 250 suitcases are produced and $400 when 0 suitcases are produced, find the cost-of-production equation.

58. The Great American Frisbee Corporation uses a linear equation to determine the number of vacation days allowed for clerks who have been employed for less than 9 years. If James was given 14 vacation days for working 6 years, and Lief Ann was given 17 days vacation for working 8 years, find the equation that determines the number of days of vacation based on a person's number of years with the corporation.

In Your Own Words

59. Compare the *standard form*, the *slope–intercept form*, and the *point–slope form* of an equation of a line.

60. Describe the general procedure for finding an equation of a line.

6.4   Inequalities in Two Variables

TAPE 11

In this section we examine linear inequalities in two variables. Consider the inequality

$$x + y \leq 2$$

The solution set is the set of all ordered pairs that make the inequality a true statement. Because there are an infinite number of such ordered pairs, we don't try to list them. We draw a picture of the solution set and call it the **graph** of the inequality.

We begin by noting that the inequality $x + y \leq 2$ means

$$x + y = 2 \quad \text{or} \quad x + y < 2$$

We can draw the graph of $x + y = 2$ as follows:

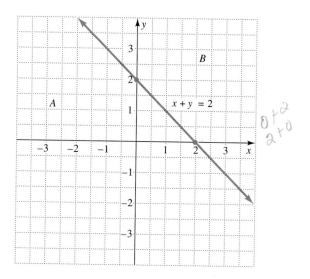

Every point on the line will make $x + y = 2$ a true statement. This line is called a **boundary line** because it divides the plane into two regions, A and B.

Let's pick some points in each region and see if they are in the solution set of $x + y < 2$, as shown in the following table:

Region	Test Point in Region	Statement $x + y < 2$	Truth of Statement	Region in Solution Set?
A	$(0, 0)$	$0 + 0 < 2$	True	Yes
A	$(-1, 0)$	$-1 + 0 < 2$	True	Yes
A	$(0, -1)$	$0 + (-1) < 2$	True	Yes
A	$(-1, -1)$	$-1 + -1 < 2$	True	Yes
B	$(2, 1)$	$2 + 1 < 2$	False	No
B	$(3, 0)$	$3 + 0 < 2$	False	No
B	$(0, 3)$	$0 + 3 < 2$	False	No
B	$(4, 2)$	$4 + 2 < 2$	False	No

It appears that every point in region *A* is in the solution set, whereas no point in region *B* is in the solution set. This is correct. To graph this solution set, we shade region *A*.

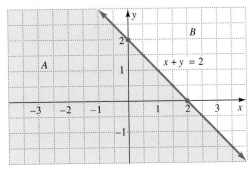

To graph a linear inequality in two variables, we first draw the boundary line. This line divides the plane into two regions. Pick one point in each region and test it in the original inequality to determine if the point is in the solution set. If one point from a region is in the solution set, then the entire region is in the solution set. If any point from a region is not in the solution set, then the entire region is not in the solution set.

If the boundary line itself is in the solution set, it is drawn as a *solid* line. If the boundary line is not in the solution set, it is drawn as a *dashed* line.

► **EXAMPLE 1**

Graph the solution set for $x - y < 0$.

Solution

The boundary line is $x - y = 0$. We graph it as a *dashed* line because equality is not included in $x - y < 0$. We pick a point from region *A* and one from region *B*.

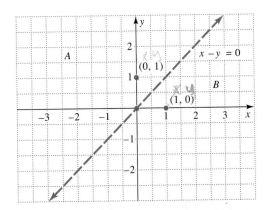

Region	Test Point in Region	Statement $x - y < 0$	Truth of Statement	Region in Solution Set?
A	(0, 1)	0 − 1 < 0	True	Yes
B	(1, 0)	1 − 0 < 0	False	No

We shade region *A*.

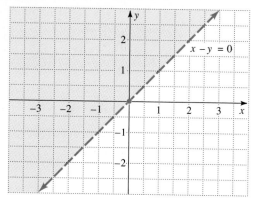

Procedure for Graphing Linear Inequalities in Two Variables

1. Graph the boundary line.

(a) Draw a solid line if equality is included ($\leq$ or $\geq$).
(b) Draw a dashed line if equality is not included ($<$ or $>$).

2. Determine which region formed by the line makes the inequality true by testing with one point from each region.

3. Shade the region(s) that make(s) the inequality true.

▶ **EXAMPLE 2**

Graph the solution set for $x + 2y \geq 2$.

Solution

The boundary line is $x + 2y = 2$.

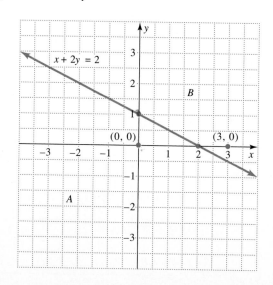

We draw a *solid* line because equality is included.

Region	Test Point in Region	Statement $x + 2y > 2$	Truth of Statement	Region in Solution Set?
A	(0, 0)	$0 + 0 > 2$	False	No
B	(3, 0)	$3 + 0 > 2$	True	Yes

Notice that the origin, (0, 0), is always a good test point to use *if* it lies in one of the regions.   ◄

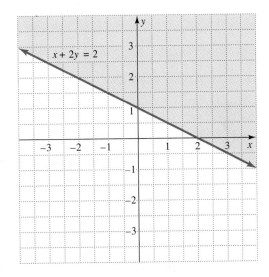

► EXAMPLE **3**

Graph the solution set for $y \leq 1$ on the *xy-plane*.

Solution

The boundary line is $y = 1$.
We draw a solid line because equality is included.

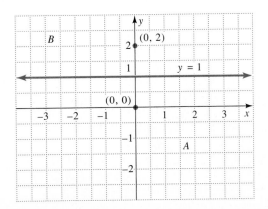

Region	Test Point in Region	Statement $y \leq 1$	Truth of Statement	Region in Solution Set?
A	(0, 0)	$0 \leq 1$	True	Yes
B	(0, 2)	$2 \leq 1$	False	No

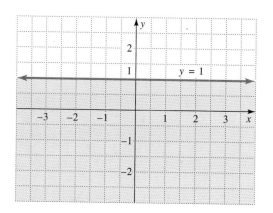

Problem Set 6.4

Warm-Ups

In Problems 1 through 6, graph the solution set for each inequality on the xy-plane. For Problems 1 through 3, see Examples 1 and 2.

1. $x + y \leq 1$

2. $3x + 2y \geq 6$

3. $2x + y > 4$

For Problems 4 through 6, see Example 3.

4. $x \leq 1$

5. $y > -2$

6. $2x \geq 7$

Practice Exercises

In Problems 7 through 18, graph the solution set for each inequality on the xy-plane.

7. $x + y \leq 2$

8. $3x + y > 1$

9. $x - y \geq -1$

10. $x \leq 7$

11. $3x + 2y < -6$

12. $4x - y > 1$

13. $4x - y \geq 0$

14. $y > -5$

15. $2x + 5y < 10$

16. $2x - 5y \geq 10$

17. $x > -1$

18. $y \leq \dfrac{3}{2}$

19. Jose wishes to purchase some computer disks that cost $.95 each and some notebooks that cost $1.39 each. He cannot spend more than $10. Let c be the number of computer disks and n be the number of notebooks.
  (a) Write an inequality to describe this information.
  (b) If Jose buys 5 disks, how many notebooks can he buy?

20. Toyota wants to transport a fleet of at least 120 cars from Chamblee, Georgia, to Charlotte, North Carolina. Two sizes of transit trucks are available. The larger size can transport 12 cars at a time and the smaller size can move 8 automobiles at a time. The manager needs to determine how many trips will be needed by the small and large piggyback carriers. Let s be the number of trips made by the smaller carrier and l be the number of trips made by the larger carrier.
  (a) Write a linear inequality that describes the transport of the cars.
  (b) Graph this linear inequality using the horizontal axis as the s-axis.

Problem 20

  (c) Is (7, 10) a solution of the inequality? Explain why this is true.
  (d) Explain what the ordered pair (5, 9) represents.

Challenge Problems

In Problems 21 through 28, graph the solution set for each inequality on the xy-plane.

21. $x \geq 0$

22. $x > 0$

23. $x \leq 0$

24. $x < 0$ 　　　　　　**25.** $y > 0$ 　　　　　　**26.** $y \geq 0$

27. $y < 0$ 　　　　　　**28.** $y \leq 0$

In Problems 29 through 31, assume that a is a positive real number and that b is a negative real number. Graph the solution set for each inequality on the xy-plane.

29. $x > a$ 　　　　　　**30.** $y \leq b$ 　　　　　　**31.** $x \leq b$

In Your Own Words

32. Describe how to graph a linear inequality in two variables.

6.5   Functions (Optional)

TAPE 11

So far, this chapter has concerned itself with pairs of numbers as coordinates of points on a plane and as solutions of equations and inequalities in two variables. It is very common to have numbers paired together in other situations. For example:

The *number* of hours traveled at 60 mph	⟷	The *number* of miles traveled
The *number* of dollars invested at 14%	⟷	The *number* of dollars of interest received
The *number* of therms used by a home in a month	⟷	The *number* of dollars on the monthly gas bill

Such relationships lead us to an important mathematical concept.

> **Function**
>
> A *function* is a rule that assigns to each number in one set (the domain) exactly one number in another set (the range).

To illustrate the key ideas in this definition, we examine some functions that have domains containing just a few numbers.

Suppose we have a function named f with domain {2, 4, 5}. Furthermore, suppose that f assigns the number 10 to 2, 20 to 4, and 25 to 5. This is conveniently illustrated by the following diagram.

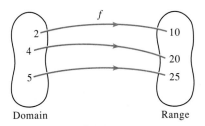

These are sometimes called "peanut drawings."

► EXAMPLE 1

Consider the assignment made by each picture. Is each a function?

(a)

(b)

(c)

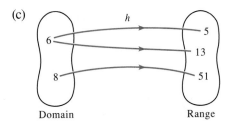

Solutions

(a) Yes, f is a function. It assigns exactly one number in the range to each number in the domain.

(b) Yes, g is also a function. Although g assigns the same number to both 7 and 11, it assigns *exactly one* number to each.

(c) No, h is *not* a function. It assigns *two* numbers from the range to the number 6. ◄

There are several ways to illustrate a function. Using the peanut drawings of Example 1 is one way. Sometimes the assignments are given as ordered pairs or in tables.

► **EXAMPLE 2**

Suppose the rule for the function F is "add 5 to the number" and the domain is the set $\{1, 2, 3, 4\}$.

(a) Illustrate F with a drawing, as in Example 1.

(b) Illustrate F with a set of ordered pairs.

(c) What is the range of F?

Solutions

(a)

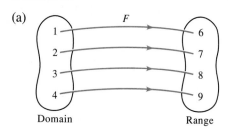

Domain   Range

(b) $F = \{(1, 6), (2, 7), (3, 8), (4, 9)\}$

(c) $\{6, 7, 8, 9\}$ ◄

In mathematical applications of functions, the rule is often given using **functional notation.** The rule for the function F in Example 2 was "add 5 to the number." In functional notation, this is written as

$$F(x) = x + 5$$

The symbol $F(x)$ looks like F times x, but that is *not* what it means. We read $F(x)$ as "F of x" or "F evaluated at x."

Notice what happens if we replace x with a domain element in the functional notation for F.

$$F(1) = 1 + 5 = 6$$
$$F(2) = 2 + 5 = 7$$
$$F(3) = 3 + 5 = 8$$
$$F(4) = 4 + 5 = 9$$

If x is a number in the domain of F, then $F(x)$ is the corresponding number in the range.

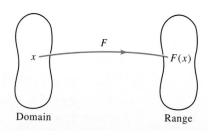

Domain   Range

To define a function, a rule and a domain must be specified. The rule is usually specified with functional notation.

► **EXAMPLE 3**

If $g(x) = 2x - 7$ with domain all real numbers, find each value.

(a) $g(2)$ (b) $g(0)$ (c) $g(-3)$ (d) $g(\pi)$

Solutions

(a) $g(2) = 2(2) - 7 = -3$ (b) $g(0) = 2(0) - 7 = -7$

(c) $g(-3) = 2(-3) - 7 = -13$ (d) $g(\pi) = 2\pi - 7$ ◄

Although to define a function it is necessary to specify both a rule and a domain, it is common practice to state the rule with functional notation and omit the domain. *When this occurs, we assume the domain to be the* natural domain *of the function.* The **natural domain** is the largest subset of the real numbers for which the rule has meaning.

► **EXAMPLE 4**

Give the natural domain of the following rules.

(a) $f(x) = 13x + x^2$ (b) $g(x) = \dfrac{1}{x - 6}$ (c) $h(x) = 7 + \sqrt{x}$

Solutions

(a) We can multiply any number by 13 and we can square any number, so the natural domain of f is all real numbers.

(b) We *cannot* divide by zero, so this rule does not have meaning when x is 6. The natural domain is all real numbers *except* 6.

(c) $\sqrt{x}$ is a real number only if x is greater than or equal to 0. The natural domain of h is

$$\{x \mid x \geq 0\}$$ ◄

The graph of a function is a useful tool for visualizing its properties. To graph the function f, we graph the equation $y = f(x)$ for every x in the domain of f.

► **EXAMPLE 5**

Graph the function $h(x) = 2x + 1$.

Solution

We graph the equation $y = h(x)$. However, $h(x) = 2x + 1$, so we graph $y = 2x + 1$. The domain was not specified, so we use the natural domain, which is all real numbers.

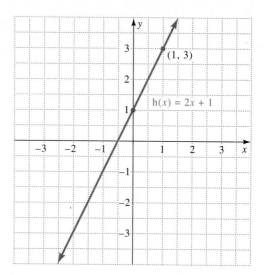

We can examine a graph to determine whether it can be the graph of a function or not. Consider the circle graphed next. Can it be the graph of a function?

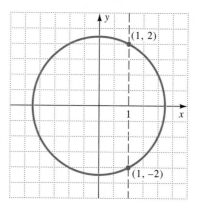

The graph contains the points (1, 2) and (1, −2). This means the domain element 1 must be assigned to *both* 2 and −2. This cannot be the graph of a function. Notice that both of the points are on the same *vertical* line.

Vertical Line Test

If any vertical line intersects a graph in more than one point, the graph is not the graph of a function.

► **EXAMPLE 6**

Is each the graph of a function?

(a)

(b)

(c)

(d)

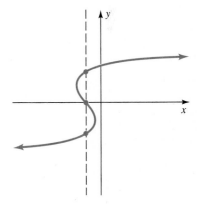

Solutions

(a) Any vertical line strikes the graph at most once. This is the graph of a function.

(b) The dashed vertical line intersects the graph in two places. This graph cannot be the graph of a function.

(c) This graph passes the vertical line test and is the graph of a function.

(d) This graph fails the vertical line test and cannot be the graph of a function.

◄

Notice that the *x*-coordinate of any point on the graph of a function is an element of the domain of the function and the *y*-coordinate is an element of the range. Therefore, we can estimate the domain and range of a function by examining its graph.

► **EXAMPLE 7**

Determine the domain and range of each function graphed next.

(a)

(b)

(c)

(d)

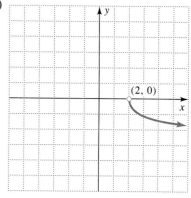

Solutions

(a) Domain: $\{x \mid -1 \le x \le 1\}$
    Range: $\{y \mid 0 \le y \le 1\}$

(b) Domain: {all real numbers}
    Range: $\{y \mid y \ge 1\}$

(c) Domain: $\{x \mid x \ne 1\}$
    Range: $\{y \mid y \ne 2\}$

(d) Domain: $\{x \mid x > 2\}$
    Range: $\{y \mid y < 0\}$

◄

Problem Set **6.5**

Warm-Ups

In Problems 1 through 4, which assignments represent functions? See Example 1.

1.

2.

3.

Domain   Range

4.

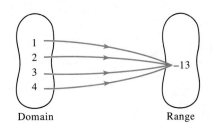

Domain   Range

In Problems 5 and 6, illustrate each function with a set of ordered pairs, and find the range of the function. See Example 2.

5. The rule for f is to square the number and add 2. The domain is $\{-1, 0, 1\}$.

6. The rule for g is to assign 0 if the number is even and 1 if it is odd. The domain is $\{1, 2, 3, 4\}$.

In Problems 7 through 9, assume the domain of each function is the set of real numbers. See Example 3.

7. If $f(x) = x^2 - x$, find $f(-1), f(0), f(1),$ and $f(7)$.

8. If $g(x) = \dfrac{x + 1}{x^2 + 1}$, find $g(0), g(1), g(-1),$ and $g(7)$.

9. If $h(x) = x^2 - 3x + 1$, find $h(0), h(3), h(-1), h(-3)$.

In Problems 10 through 13, write the natural domain of each rule. See Example 4.

10. $f(x) = x^3 + x^2 + x + 1$

11. $g(x) = \dfrac{x - 1}{(x - 2)(x - 3)}$

12. $h(x) = \dfrac{1 - x^2}{1 + x^2}$

13. $F(x) = \sqrt{x - 5}$

In Problems 14 and 15, graph the given function. See Example 5.

14. $f(x) = 4x - 2$

15. $g(x) = -\dfrac{2}{3}x + 1$

In Problems 16 and 17, decide if the graphs could be graphs of functions. See Example 6.

16.

In Problems 18 through 21, determine the domain and range of each function by examining its graph. See Example 7.

18.

19.

20.

21.

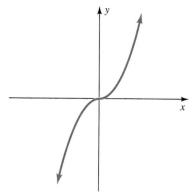

Practice Exercises

In Problems 22 through 25, illustrate each function with a set of ordered pairs, and find the range of the function.

22. The role for f is to add twice the number to -5. The domain is $\{-2, -1, 0\}$.

23. The rule for g is to divide 12 by the number. The domain is $\{1, 2, 3, 4, 5\}$.

24. The rule for F is to divide the number by one more than the number. The domain is $\{0, 1, 4\}$.

25. The rule for G is to square the number and divide by 2. The domain is $\{-1, 0, 1\}$.

In Problems 26 through 29, assume the domain of each function is the set of real numbers.

26. If $H(x) = x^3 + 2x^2 + 3x + 4$, find $H(-2)$, $H(-1)$, $H(0)$, $H(1)$, and $H(2)$.

27. If $f(x) = x^2 - 2x$, find $f(-2)$, $f(-1)$, $f(0)$, $f(1)$, and $f(2)$.

28. If $g(x) = \dfrac{x}{x^2 + x + 1}$, find $g(-5)$, $g(-1)$, $g(0)$, $g(1)$, and $g(5)$.

29. If $h(x) = \sqrt{2x^2 + x + 1}$, find $h(-3)$, $h(0)$, and $h(1)$.

In Problems 30 through 34, write the natural domain of each rule.

30. $f(x) = 5x^3 - 3x^2 + 2x - 1$

31. $g(x) = \dfrac{x - 1}{x^2 + 23}$

32. $h(x) = \dfrac{1 + x^2}{1 - x^2}$ **33.** $F(t) = \dfrac{5 - t}{4 + t}$

34. $G(s) = \sqrt{5 - 2s}$

In Problems 35 through 40, graph the given function.

35. $f(x) = -2x + 1$ **36.** $g(x) = 2x - 1$ **37.** $h(x) = \dfrac{3}{2}x + 2$

38. $F(x) = -\dfrac{2}{3}x - 2$ **39.** $G(x) = 4 - \dfrac{1}{3}x$ **40.** $H(x) = -\dfrac{1}{5}x$

In Problems 41 through 46, determine if the graphs could be graphs of functions.

41.

42.

43.

44.

45.

46.

In Problems 47 through 54, determine the domain and range of each function by examining its graph.

47.

48.

49.

50.

51.

52.

53.

54.

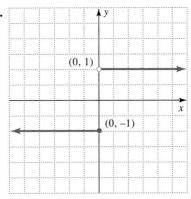

Challenge Problems

55. The following graph shows the relationship between outside temperature in degrees Fahrenheit (T) and time in hours (t) for a 24-hr period last winter.

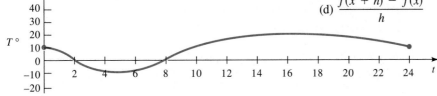

(a) Is this graph the graph of a function?

(b) Estimate the domain and range for this particular period.

(c) What was the temperature at 2 A.M.? 4 P.M.?

56. If $f(x) = x^2 + 2x - 3$, find each value.
 (a) $f(a)$
 (b) $f(a + h)$
 (c) $f(x + h) - f(x)$
 (d) $\dfrac{f(x + h) - f(x)}{h}$

In Your Own Words

57. What is a function?

58. What is the domain of a function?

59. What is the range of a function?

► Chapter Summary

GLOSSARY

Rectangular coordinate system   Two number lines joined at their origins, perpendicular to each other.

x-axis   The horizontal number line of a rectangular coordinate system.

y-axis   The vertical number line of a rectangular coordinate system.

Ordered pair   Two numbers, written in the form (a, b), where the order in which they are written is important.

x-intercept   The x-coordinate of the point on the x-axis at which a graph crosses the axis.

y-intercept   The y-coordinate of the point on the y-axis at which a graph crosses the axis.

Boundary line   A line separating two regions of a linear inequality.

Optional Section

Function A rule that assigns to each number in one set (the domain) exactly one number in another set (the range).

Linear Equations in Two Variables

1. The graph of a linear equation in two variables is a line.
2. Three common forms of the linear equation in two variables.
   (a) Standard form

$$Ax + By = C, \qquad A \text{ and } B \text{ not both } 0$$

   (b) Slope–intercept form

$$y = mx + b, \qquad \text{where } m \text{ is the slope and } b \text{ is the } y\text{-intercept}$$

   (c) Point–slope form

$$y - y_1 = m(x - x_1) \qquad \text{where } m \text{ is the slope and } (x_1, y_1) \text{ is a point on the line}$$

Slope

1. The slope of a line is given by the slope formula:

$$m = \frac{y_2 - y_1}{x_2 - x_1}, \qquad \text{where } (x_1, y_1) \text{ and } (x_2, y_2) \text{ are two points on the line}$$

2. Parallel lines have the same slope.
3. Perpendicular lines have slopes whose product is -1.

Horizontal and Vertical Lines

1. Horizontal lines have a slope of 0 and equations of the form

$$y = q$$

where q is the y-coordinate of any point on the line.
2. Vertical lines have no defined slope and equations of the form

$$x = p$$

where p is the x-coordinate of any point on the line.

Linear Inequalities in Two Variables

To graph the solution set of a linear inequality, follow these steps.

1. Graph the boundary line.
   (a) Draw a solid line if equality is included.
   (b) Draw a dashed line if equality is not included.
2. Test the regions formed by the boundary line in the original inequality.
3. Shade the regions that tested true.

Vertical Line Test (Optional Section)

If any vertical line intersects a graph in more than one point, the graph is not the graph of a function.

CHECK UPS

1. Graph the points $(0, 2)$, $(-1, 0)$, $(-1, 2)$, $(1, -2)$ and $(-1, -2)$.   Section 6.1; Example 1
2. Draw the graph of $x + 2y = 2$.   Section 6.1; Example 5

3. Draw the graph of $x = 3$.   Section 6.1; Example 8

4. Find the slope of the line containing the points   Section 6.2; Example 3
   $(-2, 4)$ and $(3, -1)$.

5. Find the slope of the graph of $x + 2y = 4$.   Section 6.2; Example 7

6. Find an equation of the line containing the   Section 6.3; Example 6
   points $(-1, 2)$ and $(3, -4)$.

7. Find an equation of the vertical line through   Section 6.3; Example 9
   the point $(3, -2)$.

8. Graph the solution set for $x + 2y \geq 2$.   Section 6.4; Example 2

9. If $g(x) = 2x - 7$ with domain all real num-   Section 6.5; Example 3
   bers, find each value.   (Optional)
   (a) $g(2)$ (b) $g(0)$ (c) $g(-3)$ (d) $g(\pi)$

10. Give the natural domain of the rule $h(x) =$ Section 6.5; Example 4c
    $7 + \sqrt{x}$.   (Optional)

► Review Problems

In Problems 1 through 6, give the quadrant in which each point lies.

1. $(2, 5)$ **2.** $(-1, 3)$ **3.** $(-2, -4)$ **4.** $(1, -3)$

In Problems 5 through 10, determine if the point is in the solution set of the equation or inequality.

5. $x + y = 0$; $(-1, 1)$ **6.** $x + y \leq 0$; $(2, 1)$ **7.** $y = -2x$; $(1, -2)$

8. $x - y \geq 1$; $(2, -1)$ **9.** $2x - 3y = 5$; $(1, -1)$ **10.** $4x + 5y = -7$; $(1, -5)$

In Problems 11 through 16, find the intercepts of each graph.

11. $x + y = -2$ **12.** $2x + y = 6$ **13.** $3x - y = 3$

14. $2x + y = 0$ **15.** $y = -4$ **16.** $x = 6$

In Problems 17 through 24, sketch the graph of each equation.

17. $x + y = 4$ **18.** $y = 2x - 1$ **19.** $2x - y = 2$

20. $y = -\dfrac{2}{3}x + 3$ **21.** $x = 5$ **22.** $y = \dfrac{1}{2}x$

23. $x - 2y = 0$ **24.** $y = 7$

In Problems 25 through 30, find the slope of the line containing the two points.

25. $(-2, 7); (4, 5)$ **26.** $(0, -3); (5, 0)$ **27.** $(3, -1); (3, 4)$
28. $(2, -1); (-5, 2)$ **29.** $(-4, 1); (2, 1)$ **30.** $(0, 0); (5, -4)$

*In Problems 31 through 36, write each equation in slope–intercept form. Give the slope-
and y-intercept for each graph.*

31. $2x + y = 4$ **32.** $x - y = 3$
33. $2x + 3y = 6$ **34.** $3x - 4y = 8$
35. $2x - 3y = 0$ **36.** $x - 4y = -4$

In Problems 37 through 44, write an equation of the line that satisfies the given conditions.

37. Through the point $(2, 3)$ with slope 4.   **41.** Horizontal line through the point $(4, -3)$.
38. Through the points $(-1, 3)$ and $(0, -2)$.   **42.** y-intercept of 3 with slope -2.
39. Through the points $(1, 1)$ and $(0, 3)$.   **43.** Through the point $(-1, 4)$ with no slope.
40. Vertical line through the point $(-2, 3)$.   **44.** Through the point $(4, -2)$ with slope 0.

In Problems 45 through 50, graph the solution set for each inequality on the xy-plane.

45. $x + y \geq 4$ **46.** $y < -2$ **47.** $2x - y > 2$

48. $x \geq 4$ **49.** $3x - 4y \leq 12$ **50.** $3x + 2y < -12$

In Problems 51 through 53, find $f(2)$ and $f(-1)$ (optional).

51. $f(x) = x^3 - 2x^2 + 3x - 4$ **52.** $f(x) = \sqrt{x + 2}$ **53.** $f(x) = \dfrac{x + 1}{x - 1}$

Let's Not Forget . . .

In Problems 54 through 61, perform the operations indicated.

54. -2^2 **55.** $(-3)^2$ **56.** $x - (5x + 3)$
57. $\dfrac{a^2 - b^2}{b - a}$ **58.** $\dfrac{0}{7}$ **59.** $\dfrac{x^2}{x^2 - 4} \cdot (x - 2)$
60. $(x + 2y)^2$ **61.** $(a + 4)(a - 4)$

In Problems 62 through 64, simplify each expression. Watch the role of parentheses!

62. $(x + 2)x - 2$ **63.** $(x + 2)(x - 2)$ **64.** $\dfrac{1}{x + 1} - \dfrac{1 - x}{x + 1}$

In Problems 65 through 68, determine whether x^2 is a factor *or a* term.

65. $x^2 + 2$

66. $3(x^2 + 2)$

67. $\dfrac{1}{x^2(x + 2)}$

68. $x^2(y + 3)$

In Problems 69 through 71, multiply each expression by $\dfrac{1}{x + 1}$.

69. $x + 1$

70. $2 + 5(x + 1)$

71. $(x + 1)(x^2 - 1)$

In Problems 72 through 77, identify which expressions are factored. Factor those that are not factored, if possible.

72. $4y + y^2$

73. $x^2 - y^2$

74. $(x - 2)^3$

75. $2(ab + c)$

76. $x^2 + 2xy + y^2$

77. $x^3 - 8$

In Problems 78 through 82, identify whether each problem is an expression, equation, *or* inequality. *Solve the equations and inequalities and perform the operations indicated with the expressions.*

78. $\dfrac{1}{x + 4} - 4$

79. $\dfrac{1}{x + 4} - 4 = \dfrac{x}{x + 4}$

80. $-\dfrac{1}{3}x \le 5$

81. $\dfrac{\dfrac{2}{x} + x}{\dfrac{1}{x} + 3}$

82. $2(x - 3) - (1 - x)$

► You Decide

Your brother who lives in Des Moines, 1300 miles away, is getting married in three weeks. The wedding is to be at 11:00 Saturday morning. However, you cannot leave until the Friday before the ceremony. The following travel options are available from your location:

(a) A direct, nonstop bus that leaves daily at 8:00 A.M. and averages 50 mph.

(b) A flight that leaves everyday at noon, averages 400 mph, and lands at Kansas City, a distance of 1100 miles. A commuter plane connection goes the rest of the way to Des Moines. The commuter, a 12-passenger propeller-driven airplane, leaves once an hour, on the half-hour, and takes one hour for the trip.

(c) The red-eye special that leaves every night at 11:30, flies at an average of 500 mph, and lands at Chicago, 750 miles distant. After a 3-hour lay-over, a direct 1-hour flight to Des Moines is available.

Which option do you choose? When do you arrive? Why did you choose this plan?

▶ Chapter 6 Test

In Problems 1 through 7, choose the letter of the correct answer.

1. In which quadrant does the point $(-3, -4)$ lie?
  (a) I   (c) III
  (b) II   (d) IV

2. Which ordered pair is in the solution set of $x - y = 3$?
  (a) $(0, 3)$ (c) $(-3, 0)$
  (b) $(1, -2)$ (d) $(1, 3)$

3. The slope of the line parallel to the line with equation $2x - 5y = 10$ is (?).
  (a) $\dfrac{2}{5}$ (c) 2

  (b) $-\dfrac{2}{5}$ (d) -2

4. Which of the following is an equation of the line through the point $(1, -5)$ with slope -2?
  (a) $2x + y = -3$ (c) $2x + y = -6$
  (b) $2x + y = 7$ (d) $2x + y = 4$

5. The slope of the graph of $x = 6$ is (?).
  (a) 1 (c) 6
  (b) 0 (d) Not defined

6. The y-intercept of the graph of $2x + 3y = 9$ is (?).
  (a) 9 (c) $\dfrac{9}{2}$
  (b) 3 (d) $-\dfrac{2}{3}$

7. The slope of the line containing the points $(2, -3)$ and $(-1, 5)$ is (?)
  (a) $-\dfrac{8}{3}$ (c) $\dfrac{8}{3}$
  (b) $-\dfrac{3}{8}$ (d) $\dfrac{3}{8}$

In Problems 8 through 11, sketch the graph of each equation.

8. $2x + 3y = 6$

9. $x + 2y = 0$

10. $x = -\dfrac{3}{2}$

11. $y = -\dfrac{2}{3}x + 2$

In Problems 12 and 13, graph the solution set.

12. $x - 3y \le 6$

13. $2x + y > 0$

14. Write an equation of the line through the points $(1, -2)$ and $(3, 4)$.

15. Write an equation of the horizontal line through the point $(2, -3)$.

Optional Section

16. If $f(x) = 3 - 2x + 3x^2 - x^3$, find $f(-2)$.

17. Which of the following graphs are graphs of functions?

(a)

(b)

(c)

(d)

(e)

(f)

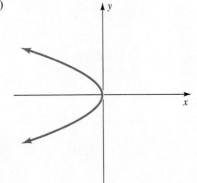

7 Systems of Linear Equations and Inequalities

See Problem Set 7.4, Exercise 21.

▶ **Connections**

Suppose an engineer wishes to cut a 14-foot truss into two sections, ending up with one piece 4 feet longer than the other. The engineer could solve this relatively simple problem by using trial and error or by using his or her mathematical skills. However, engineers routinely deal with much more complicated problems for which trial and error will not work, but the mathematical approach will. Such problems often lead to equations in two variables. Engineers are not the only people who use systems of linear equations, for these ideas can be applied to many complex mathematical concepts.

In this chapter we will learn three methods for solving systems of equations and examine applications of systems. We will also introduce systems of two inequalities in two variables.

7.1 **Graphical Method**

TAPE 12

Suppose we are looking for two numbers whose sum is 6. If we let x and y be the two numbers, then we are looking for solutions to the equation

$$x + y = 6$$

There are an infinite number of such solutions. We learned in the last chapter that a picture of the solution set of this equation is the line shown next.

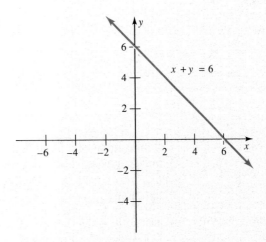

Now suppose that we also know that the difference between these two numbers is 2. That is,

$$x - y = 2$$

All solutions to this equation would lie on the following line.

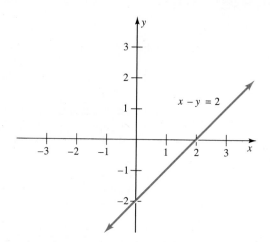

We are looking for ordered pairs that make *both* $x + y = 6$ and $x - y = 2$ true statements. If we draw the graph of each equation on the same coordinate system, we have the following graph.

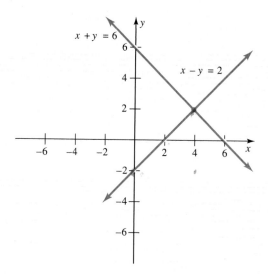

The graphs intersect in a single point. This ordered pair is in the solution set of *both* equations. Thus the coordinates of this point are the numbers for which we are looking. From the graph, we guess this ordered pair to be (4, 2).

To see that (4, 2) is correct, we replace x with 4 and y with 2 in both equations.

$$x + y = 6 \qquad 4 + 2 = 6 \qquad \text{True statement}$$

$$x - y = 2 \qquad 4 - 2 = 2 \qquad \text{True statement}$$

We see that (4, 2) is in the solution set of both equations. Two numbers whose sum is 6 and difference is 2 are 4 and 2.

Situations such as this arise very often in the real world. To indicate that x represents the same number in both equations and that y represents the same number in both equations, we write the system in the following manner:

$$\begin{cases} x + y = 6 \\ x - y = 2 \end{cases}$$

We call this a **system of two equations in two variables.** The set of all ordered pairs that make *both* equations true statements is called the **solution set.** To **solve** a system means to find the solution set.

Determining the solution set from a graph is called the **graphical method** of solving a system of equations. It requires judgment and estimation and is thus not a very satisfactory technique to use. We will learn two other techniques in Sections 7.2 and 7.3.

► **EXAMPLE 1**

Using the graphical method, find the solution of the system of equations.

$$\begin{cases} x + y = 2 \\ 2x - y = 1 \end{cases}$$

Solution

We draw the graph of each equation on the same coordinate system.

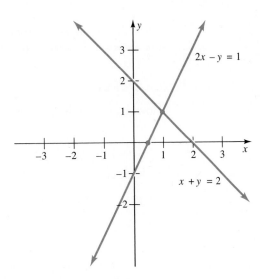

From the graph we guess the solution to be $(1, 1)$. If we replace x with 1 and y with 1 in both equations,

$$x + y = 2 \qquad 2x - y = 1$$
$$1 + 1 = 2 \qquad 2(1) - 1 = 1$$

we see that both are true statements. Thus $(1, 1)$ is the solution.   ◄

The graphing method is not very satisfactory for finding the solution set, but it is very helpful in describing the solution set of a system of equations.

► **EXAMPLE 2**

How many solutions are in the solution set for the system

$$\begin{cases} x - y = 1 \\ x + y = 2 \end{cases}$$

Solution

We draw the graph of each equation.

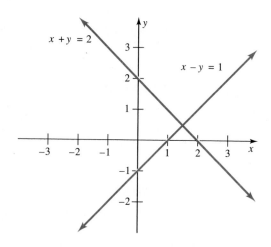

From the graph we see that the lines intersect in one point. So there is exactly one solution to the system. Such a system is called an **independent** system.

◄

► **EXAMPLE 3**

How many solutions are in the solution set of the system

$$\begin{cases} x - y = 1 \\ 2x - 2y = 4 \end{cases}$$

Solution

We graph each equation.

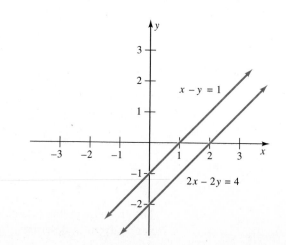

From the graph we see that the lines appear to be parallel. In fact, since each line has a slope of 1, they *are* parallel. Thus the lines do not intersect. This means that there is *no* ordered pair that is in the solution set of both equations. So there is no solution to the system. We call such a system **inconsistent.**   ◄

► **EXAMPLE 4**

How many solutions are in the solution set for the system

$$\begin{cases} x + y = 2 \\ 2x + 2y = 4 \end{cases}$$

Solution

We graph each line.

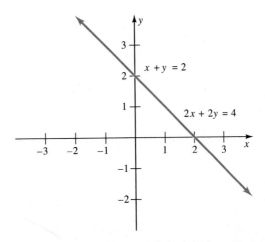

The graph of both equations is the same line. This means that every point on the line is in the solution set for *both* equations. There are an infinite number of points on this line, so there are an infinite number of solutions to the system. We call such a system **dependent.**   ◄

As Examples 2, 3, and 4 suggest, three situations can occur when solving systems of two linear equations in two variables.

Independent

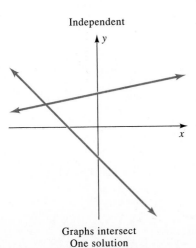

Graphs intersect
One solution

Inconsistent

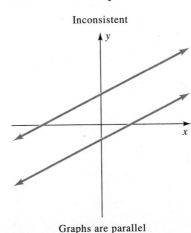

Graphs are parallel
No solution

Dependent

Graphs are same
Infinite number
of solutions

Problem Set 7.1

Warm-Ups

In Problems 1 through 9, find the solution set for each system by the graphical method. See Example 1.

1. $\begin{cases} x - y = 1 \\ x + y = 3 \end{cases}$

2. $\begin{cases} 2x + y = -1 \\ x + y = 1 \end{cases}$

3. $\begin{cases} 2x - y = 4 \\ x + y = -1 \end{cases}$

4. $\begin{cases} x - y = 1 \\ x + y = 5 \end{cases}$

5. $\begin{cases} 2x + y = -3 \\ x + y = -2 \end{cases}$

6. $\begin{cases} 2x - y = 1 \\ x + y = 2 \end{cases}$

7. $\begin{cases} x = 3 \\ x + y = 5 \end{cases}$

8. $\begin{cases} x - y = 3 \\ y = -2 \end{cases}$

9. $\begin{cases} x + 2y = 1 \\ x = -1 \end{cases}$

In Problems 10 through 18, determine the number of solutions in the solution set for each system. Classify each system as independent, inconsistent, or dependent. See Examples 2, 3, and 4.

10. $\begin{cases} x - y = 2 \\ x + y = 2 \end{cases}$

11. $\begin{cases} 2x + 2y = 4 \\ x + y = 0 \end{cases}$

12. $\begin{cases} 2x - y = -1 \\ x - y = 1 \end{cases}$

13. $\begin{cases} x - y = -2 \\ x + y = 6 \end{cases}$

14. $\begin{cases} x + y = 1 \\ 2x + 2y = 2 \end{cases}$

15. $\begin{cases} 2x - y = 1 \\ x + y = -1 \end{cases}$

16. $\begin{cases} x - y = 1 \\ 2x - 2y = 3 \end{cases}$

17. $\begin{cases} 2x + 4y = 2 \\ x + 2y = 1 \end{cases}$

18. $\begin{cases} 2x - y = 3 \\ x - y = 2 \end{cases}$

Practice Exercises

In Problems 19 through 33, find the solution set for each system by the graphical method.

19. $\begin{cases} x - y = 0 \\ 3x + y = 4 \end{cases}$

20. $\begin{cases} 2x + y = 3 \\ x + y = 0 \end{cases}$

21. $\begin{cases} 2x - y = 0 \\ x + y = 0 \end{cases}$

22. $\begin{cases} x - y = 3 \\ x + y = 1 \end{cases}$

23. $\begin{cases} 2x + y = 4 \\ x + y = 2 \end{cases}$

24. $\begin{cases} 2x - y = -2 \\ x + y = -4 \end{cases}$

25. $\begin{cases} x - y = -1 \\ x + y = 3 \end{cases}$

26. $\begin{cases} 2x + y = -1 \\ x + y = -1 \end{cases}$

27. $\begin{cases} 2x + y = 0 \\ 2x + y = -2x + 1 \end{cases}$

28. $\begin{cases} x - y = -1 \\ x + y = -3 \end{cases}$

29. $\begin{cases} 2x + y = 1 \\ x - y = 2 \end{cases}$

30. $\begin{cases} 2x - y = 2 \\ x + y = -2 \end{cases}$

31. $\begin{cases} x = -2 \\ x - y = -5 \end{cases}$

32. $\begin{cases} x - y = -1 \\ y = 2 \end{cases}$

33. $\begin{cases} x - 3y = 3 \\ 2x - 6y = 0 \end{cases}$

In Problems 34 through 42, determine the number of solutions in the solution set for each system. Classify each system as independent, inconsistent, or dependent.

34. $\begin{cases} x - y = 0 \\ x + y = 0 \end{cases}$

35. $\begin{cases} 2x - y = 6 \\ x - y = 3 \end{cases}$

36. $\begin{cases} 2x - 2y = 2 \\ x - y = 1 \end{cases}$

37. $\begin{cases} x - y = 4 \\ x + y = 2 \end{cases}$

38. $\begin{cases} x - y = 1 \\ 4x - 4y = 4 \end{cases}$

39. $\begin{cases} 2x - y = -3 \\ x + y = -6 \end{cases}$

40. $\begin{cases} x - y = 1 \\ 3x - 3y = 1 \end{cases}$

41. $\begin{cases} 2x - 12y = 12 \\ x - 6y = 6 \end{cases}$

42. $\begin{cases} 2x - y = 2 \\ x - y = 2 \end{cases}$

Challenge Problems

In Problems 43 through 45, find the solution set for each system graphed.

43.

44.

45.
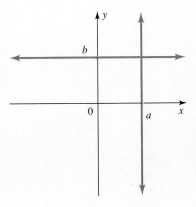

In Your Own Words

46. What is a system of two equations in two variables?

47. Describe the graphical method for solving a system of equations.

7.2   Substitution

TAPE 12

Let's consider how we can solve the system

$$\begin{cases} x + y = 1 \\ \quad y = x + 3 \end{cases}$$

by a method other than the graphical method. Notice that the second equation is solved for y. If y is $x + 3$ in the second equation, it must be $x + 3$ in the first equation, since y represents the same number in both equations. So we replace y by $x + 3$ in the first equation.

$$x + \boxed{(x + 3)} = 1$$

Now we have an equation in one variable, which we solve.

$$2x + 3 = 1$$
$$2x = -2$$
$$x = -1$$

This gives us the x value for the solution. To find the y value, we replace x with -1 in the second equation.

$$y = \boxed{-1} + 3$$
$$y = 2$$

So the solution set is $\{(-1, 2)\}$.

This method is called **substitution.**

Method of Substitution

1. Solve one of the equations for one of the variables.

2. Substitute this expression for the same variable in the other equation.

3. Solve the resulting equation.

4. Substitute the result in the equation formed in step 1 to find the other variable.

5. Check both values in both equations, if required.

6. Write the solution set.

► **EXAMPLE 1**

Solve the system by substitution.

$$\begin{cases} x + y = 1 \\ \quad y = x - 1 \end{cases}$$

Solution

STEP 1   Solve one equation for one variable.

The second equation is already solved for y.

STEP 2   Substitute into the other equation.

We replace y in *the first equation* with $x - 1$.

$$x + \boxed{(x - 1)} = 1$$

STEP 3   Solve the resulting equation.

$$x + (x - 1) = 1$$
$$2x - 1 = 1$$
$$2x = 2$$
$$x = 1$$

We have found the x value of the solution.

STEP 4   Substitute the result into the equation of step 1.

To find the y value, we replace x in the second equation with 1.

$$y = \boxed{1} - 1$$
$$y = 0$$

STEP 5   Check the result in both equations.

We are checking the ordered pair $(1, 0)$. The left side of the first equation has the value

$$x + y = 1 + 0 = \boxed{1}$$

which agrees with the right side. Checking $(1, 0)$ in the second equation gives a value of 0 for each side.

STEP 6   Write the solution set.

$$\{(1, 0)\}$$

◄

► **EXAMPLE 2**

Solve the system by substitution.

$$\begin{cases} x + y = 5 \\ x - y = -1 \end{cases}$$

Solution

STEP 1   Solve one equation for one variable.

Let's solve the first equation for x.

$$x = 5 - y$$

STEP 2   Substitute into the other equation.

We substitute $5 - y$ for x in the *second* equation.

$$\boxed{(5 - y)} - y = -1$$

STEP 3 Solve the resulting equation.

$$(5 - y) - y = -1$$
$$5 - 2y = -1$$
$$-2y = -6$$
$$y = 3$$

STEP 4   Substitute the result into the equation of step 1.

We substitute 3 for y in step 1.

$$x = 5 - \boxed{3}$$
$$x = 2$$

STEP 5   Check in both equations.

We are checking the ordered pair (2, 3). In the first equation the left side has the value

$$x + y = 2 + 3 = \boxed{5}$$

which agrees with the right side. The left side of the second,

$$x - y = 2 - 3 = \boxed{-1}$$

also agrees with the right side. The solution checks.

STEP 6   Write the solution set.

$$\{(2, 3)\}$$

◀

► **EXAMPLE 3**

Solve the system by substitution.

$$\begin{cases} 2x + \ y = 4 \\ 3x + 2y = 7 \end{cases}$$

Solution

Solve one equation for one variable.

We can avoid fractions by solving the first equation for y.

$$y = 4 - 2x$$

Substitute into the other equation.

Substitute $4 - 2x$ for y in the second equation.

$$3x + 2 \boxed{(4 - 2x)} = 7$$

Solve the resulting equation.

$$3x + 2(4 - 2x) = 7$$
$$3x + 8 - 4x = 7$$
$$-x + 8 = 7$$
$$-x = -1$$
$$x = 1$$

Substitute the result into the equation of step 1.

We substitute 1 for x in step 1.

$$y = 4 - 2 \boxed{(1)}$$
$$y = 4 - 2$$
$$y = 2$$

Check in both equations.

In the first equation,

$$2x + y = 2(1) + 2 = \boxed{4}$$

In the second equation,

$$3x + 2y = 3(1) + 2(2) = \boxed{7}$$

Write the solution set.

$$\{(1, 2)\}$$

◀

► **EXAMPLE 4**

Solve the system by substitution.

$$\begin{cases} 2x - y = 5 \\ 3x + 2y = 11 \end{cases}$$

Solution

Solve one equation for one variable.

We solve the first equation for y.

$$2x - y = 5$$
$$-y = 5 - 2x$$
$$y = -5 + 2x \quad \text{or} \quad y = 2x - 5$$

Substitute into the other equation and solve.
Substitute $2x - 5$ for y in the second equation.

$$3x + 2 \boxed{(2x - 5)} = 11$$
$$3x + 4x - 10 = 11$$
$$7x - 10 = 11$$
$$7x = 21$$
$$x = 3$$

Substitute into the equation of step 1.

Substitute 3 for x in $y = 2x - 5$.

$$y = 2 \; (3) \; - 5$$
$$y = 1$$

Check.

(3, 1) checks in both equations.

Write the solution set.

$$\{(3, 1)\}$$

◄

► **EXAMPLE 5**

Solve the system by substitution.

$$\begin{cases} 3x - 2y = 4 \\ 2x + 4y = 8 \end{cases}$$

Solution

In deciding for which variable to solve, we notice that to solve the first equation for x or y would introduce fractions. However, if we solve the second equation for x, we can avoid fractions.

$$2x + 4y = 8$$
$$2x = 8 - 4y$$
$$x = 4 - 2y$$

Substitute $4 - 2y$ for x in the first equation and solve.

$$3 \; (4 - 2y) \; - 2y = 4$$
$$12 - 6y - 2y = 4$$
$$12 - 8y = 4$$
$$-8y = -8$$
$$y = 1$$

We replace y with 1 in $x = 4 - 2y$.

$$x = 4 - 2 \; (1)$$
$$x = 2$$

As (2, 1) checks in both equations, we write the solution set.

$$\{(2, 1)\}$$

◄

► **EXAMPLE 6**

Solve the system by substitution.

$$\begin{cases} x - 2y = 0 \\ 2x - 4y = 3 \end{cases}$$

Solution

We solve the first equation for x.

$$x = 2y$$

We substitute $2y$ for x in the second equation.

$$2\,(2y)\;-4y = 3$$

$$4y - 4y = 3$$

$$0 = 3$$

Notice that both variables have dropped out of the equation. If we draw the graphs of the original two equations, we see what the problem is.

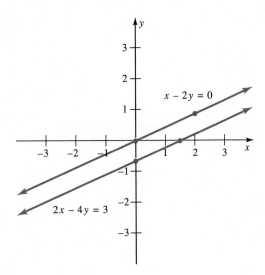

The lines appear to be parallel. In fact, since they have the same slope, they are parallel. This means that the system has no solution. Thus the solution set is the empty set, which we write as

$$\varnothing$$

Notice that when the variable disappeared we were left with the equation $0 = 3$. This is a false statement, which tells us the system is *inconsistent* and that the solution set is the empty set.   ◄

► **EXAMPLE 7**

Solve the system $\begin{cases} x - 2y = 1 \\ 3x - 6y = 3 \end{cases}$ by substitution.

Solution

We solve the first equation for x.

$$x = 1 + 2y$$

We substitute $1 + 2y$ for x in the second equation.

$$3(1 + 2y) - 6y = 3$$
$$3 + 6y - 6y = 3$$
$$3 = 3$$

Again the variables have all dropped out of the equation. Notice, however, that we are left with a *true* statement. If we draw the graph of each equation, we will see that the graph of both equations is the same line. This means that the system is *dependent* and that the solution set has an infinite number of elements. At this stage, we will not write the solution set for dependent systems but will simply mark them *dependent*. ◄

Problem Set 7.2

Warm-Ups

In Problems 1 through 18, solve each system by substitution. If the system is dependent, write dependent. In Problems 1 through 6, see Example 1.

1. $\begin{cases} x + y = 2 \\ y = x \end{cases}$

2. $\begin{cases} x - y = 4 \\ x - y = 3x \end{cases}$

3. $\begin{cases} x + 2y = 0 \\ x + 2y = 4x \end{cases}$

4. $\begin{cases} x - 2y = -5 \\ x - 2y = 5 - 2x \end{cases}$

5. $\begin{cases} 2x + y = 3 \\ 2x + x = 4 - y \end{cases}$

6. $\begin{cases} 3x + 2y = -2 \\ 3x + 2x = 4 - 3y \end{cases}$

For Problems 7 through 12, see Examples 2, 3, and 4.

7. $\begin{cases} x - 2y = 5 \\ x - 2x = 3 \end{cases}$

8. $\begin{cases} x + y = 1 \\ x - y = 3 \end{cases}$

9. $\begin{cases} x + y = 0 \\ 2x + y = 1 \end{cases}$

10. $\begin{cases} 2x - 2y = 14 \\ x - 2y = 10 \end{cases}$

11. $\begin{cases} 5x - 3y = 25 \\ 4x - y = 13 \end{cases}$

12. $\begin{cases} 5x - 2y = -4 \\ 3x + 2y = -5 \end{cases}$

For Problems 13 through 15, see Example 5.

13. $\begin{cases} 2x + 3y = -4 \\ 3x + 6y = -9 \end{cases}$

14. $\begin{cases} 3x - 16y = 0 \\ 6x - 16y = -4 \end{cases}$

15. $\begin{cases} 3x + 5y = -23 \\ 2x - 4y = -8 \end{cases}$

For Problems 16 through 18, see Examples 6 and 7.

16. $\begin{cases} x + y = 5 \\ x + y = 3 \end{cases}$

17. $\begin{cases} 2x + 2y = 2 \\ 2x + 2y = 4 \end{cases}$

18. $\begin{cases} 3x - 16y = -2 \\ 6x - 12y = 4 \end{cases}$

Practice Exercises

In Problems 19 through 43, solve each system by substitution. If the system is dependent, write dependent.

19. $\begin{cases} x + y = 4 \\ x + y = 3x \end{cases}$

20. $\begin{cases} x - y = 6 \\ x - x = 7y \end{cases}$

21. $\begin{cases} 4x + y = 0 \\ 4x + y = -4x \end{cases}$

22. $\begin{cases} x - 3y = -17 \\ x - 3y = -1 - 3x \end{cases}$

23. $\begin{cases} 3x + y = 4 \\ 3x + x = 6 - y \end{cases}$

24. $\begin{cases} 3x - 2y = 8 \\ 3x - 2x = 3y + 12 \end{cases}$

25. $\begin{cases} x + y = 7 \\ x - y = 3 \end{cases}$

26. $\begin{cases} 2x + y = 0 \\ 2x - y = -9 \end{cases}$

27. $\begin{cases} x - 2y = 7 \\ x - 2y = 5 \end{cases}$

28. $\begin{cases} 3x - 3y = 2 \\ 3x - 3y = 1 \end{cases}$ **29.** $\begin{cases} 2x - 2y = 3 \\ 2x - 4y = 6 \end{cases}$ **30.** $\begin{cases} 3x - 5y = -1 \\ 3x - 5y = -4 \end{cases}$

31. $\begin{cases} 3x - 2y = 0 \\ 6x + y = 5 \end{cases}$ **32.** $\begin{cases} x + 2y = -6 \\ 3x - 4y = 2 \end{cases}$ **33.** $\begin{cases} 4x - y = 17 \\ 2x - 3y = 1 \end{cases}$

34. $\begin{cases} 2x + 4y = 20 \\ 3x + 5y = 24 \end{cases}$ **35.** $\begin{cases} 3x - 2y = 2 \\ 5x - 10y = 50 \end{cases}$ **36.** $\begin{cases} 4x + 2y = -18 \\ 3x - 7y = -22 \end{cases}$

37. $\begin{cases} 5x + 6y = -2 \\ 3x - 3y = 12 \end{cases}$ **38.** $\begin{cases} 2x - 5y = 7 \\ 4x - 10y = 14 \end{cases}$ **39.** $\begin{cases} 6x + 8y = 6 \\ 2x + 4y = 2 \end{cases}$

40. $\begin{cases} x + 2y = 3 \\ x = 3 \end{cases}$ **41.** $\begin{cases} 2x + 3y = 0 \\ y = 3 \end{cases}$ **42.** $\begin{cases} 3x - 4y = 9 \\ x = -1 \end{cases}$

Challenge Problems

43. Solve the system by substitution.
$\begin{cases} x + y = 4 \\ y = k \end{cases}$

44. Solve the system by substitution.
$\begin{cases} x - y = a \\ x + y = b \end{cases}$

45. Solve the system by substitution.
$\begin{cases} x^2 + y^2 = 10 \\ y = 3x \end{cases}$

In Your Own Words

46. Describe the substitution method for solving two equations with two variables.

47. How do we detect inconsistent and dependent systems?

7.3   Elimination

TAPE 12

The next method for solving systems of equations is called the method of **elimination.** Because adding the same number to both sides of an equation does not change its solution set, we can add two equations together. This procedure is used in the following examples.

► **EXAMPLE 1**

Solve $\begin{cases} x + y = 1 \\ x - y = 3 \end{cases}$ by elimination.

Solution

We add the two equations together: $\begin{cases} x + y = 1 \\ \underline{x - y = 3} \end{cases}$
$$2x = 4$$

Then we solve the resulting equation:

$$x = 2$$

We have found the x value of the solution. To find the y value of the solution, we replace x with 2 in either of the original equations and solve for y. Suppose we use the first equation.

$$x + y = 1$$

$$\boxed{2} + y = 1$$

$$y = -1$$

The solution set is $\{(2, -1)\}$.   ◄

► **EXAMPLE 2**

Solve $\begin{cases} 2x + 3y = 1 \\ 3x + y = -2 \end{cases}$ by elimination.

Solution

If we add the equations as they are, we get another equation containing x and y. This will not help us solve the system. However, if we multiply the second equation by -3, the solution set for the system does not change. Notice what happens.

$$\begin{cases} 2x + 3y = 1 & \text{First equation unchanged.} \\ \boxed{-9}\,x \boxed{-3}\,y = \boxed{6} & \text{Multiply the second equation by } -3. \end{cases}$$

The solution set for this system is the same as the solution set for the original system. But, when we add the equations, the variable y will be eliminated.

$$\begin{cases} 2x + 3y = 1 \\ \underline{-9x - 3y = 6} \\ -7x = 7 \end{cases}$$

$$x = -1$$

We replace x with -1 in either of the original equations and solve for y. Using the second, we have

$$3x + y = -2$$

$$3\boxed{(-1)} + y = -2$$

$$-3 + y = -2$$

$$y = 1$$

The ordered pair $(-1, 1)$ checks in both of the original equations. Thus the solution set is

$$\{(-1, 1)\}$$ ◄

Method of Elimination

1. Write the equations in the form $Ax + By = C$.

2. If necessary, multiply both sides of each equation by a suitable number so that one of the variables will be eliminated by addition of the equations.

3. Add the equations together and solve the resulting equation.

4. Substitute the value found in step 2 into one of the original equations and solve this equation.

5. Check, if required.

6. Write the solution set.

► **EXAMPLE 3**

Solve $\begin{cases} 2x + 5y = 6 \\ \quad\quad 3x = 2y - 10 \end{cases}$ by elimination.

Solution

STEP 1 Write each equation in the form $Ax + By = C$.

$$\begin{cases} 2x + 5y = 6 \\ \quad\quad 3x = 2y - 10 \end{cases} \qquad \begin{cases} 2x + 5y = 6 \\ 3x - 2y = -10 \end{cases}$$

STEP 2 Multiply each equation by a suitable number so that addition will eliminate a variable.

If we multiply the first equation by -3 and the second equation by 2, the system becomes

$$\begin{cases} 2x + 5y = 6 \\ 3x - 2y = -10 \end{cases} \xrightarrow[\text{Multiply by } 2]{\text{Multiply by } -3} \begin{cases} -6x - 15y = -18 \\ 6x - 4y = -20 \end{cases}$$

STEP 3 Add the equations together and solve.

$$\begin{cases} -6x - 15y = -18 \\ \underline{6x - 4y = -20} \\ -19y = -38 \\ y = 2 \end{cases}$$

STEP 4 Substitute this value into one of the original equations and solve.

We replace y with 2 in the first equation of the original system and solve for x.

$$2x + 5y = 6$$
$$2x + 5(\boxed{2}) = 6$$
$$2x + 10 = 6$$
$$2x = -4$$
$$x = -2$$

STEP 5 Check.

Since we found x with the first of the original equations, we need check only the second.

$$3x = 3(-2) = \boxed{-6}$$
$$2y - 10 = 2(2) - 10 = 4 - 10 = \boxed{-6}$$

The ordered pair $(-2, 2)$ checks.

STEP 6 Write the solution set.

$$\{(-2, 2)\}$$ ◄

► **EXAMPLE 4**

Solve $\begin{cases} x - 3y = 0 \\ 2x - 6y = 5 \end{cases}$ by elimination.

Solution

STEP 1   Write each equation in the form $Ax + By = C$.

Both equations are already in the proper form.

STEP 2   Multiply each equation by a suitable number so that addition will eliminate a variable.

We multiply the first equation by -2.

$$\begin{cases} x - 3y = 0 \\ 2x - 6y = 5 \end{cases} \xrightarrow{\text{Multiply by } -2} \begin{cases} -2x + 6y = 0 \\ 2x - 6y = 5 \end{cases}$$

STEP 3   Add the equations together and solve.

$$\begin{cases} -2x + 6y = 0 \\ \underline{2x - 6y = 5} \\ 0 = 5 \end{cases}$$

As before, a false statement indicates that the system is *inconsistent* and the solution set is the empty set.

STEP 6   Write the solution set.

$$\varnothing$$

◄

► **EXAMPLE 5**

Solve $\begin{cases} 9x - 6y = 18 \\ 3x - 2y = 6 \end{cases}$ by elimination.

Solution

The equations are in the proper form.

STEP 2   Multiply each equation by a suitable number so that addition will eliminate a variable.

We multiply the second equation by -3.

$$\begin{cases} 9x - 6y = 18 \\ 3x - 2y = 6 \end{cases} \xrightarrow{\text{Multiply by } -3} \begin{cases} 9x - 6y = 18 \\ -9x + 6y = -18 \end{cases}$$

STEP 3   Add the equations together and solve.

$$\begin{cases} 9x - 6y = 18 \\ \underline{-9x + 6y = -18} \\ 0 = 0 \end{cases}$$

This is a true statement, so the system is *dependent*.   ◄

Problem Set 7.3

Warm-Ups

In Problems 1 through 12, solve the following systems by elimination. If the system is dependent, write dependent. For Problems 1 through 3, see Example 1.

1. $\begin{cases} x + y = 4 \\ x - y = 4 \end{cases}$
2. $\begin{cases} 2x - y = 3 \\ x + y = 6 \end{cases}$
3. $\begin{cases} x - y = 0 \\ 4x + y = 10 \end{cases}$

For Problems 4 through 9, see Example 2.

4. $\begin{cases} x + y = 4 \\ x + 2y = 5 \end{cases}$
5. $\begin{cases} x - y = 3 \\ x - 2y = 4 \end{cases}$
6. $\begin{cases} 3x + y = -2 \\ 2x + y = -1 \end{cases}$

7. $\begin{cases} 2x - y = 8 \\ 3x + 2y = 5 \end{cases}$
8. $\begin{cases} 3x - y = 1 \\ x + 2y = 12 \end{cases}$
9. $\begin{cases} 4x + 3y = 9 \\ 2x - y = 7 \end{cases}$

For Problems 10 through 12, see Examples 4 and 5.

10. $\begin{cases} 2x + 3y = 1 \\ 2x + 3y = 0 \end{cases}$
11. $\begin{cases} 3x - 2y = 4 \\ 9x - 6y = 12 \end{cases}$
12. $\begin{cases} 6x - 4y = -24 \\ 3x - 2y = 10 \end{cases}$

Practice Exercises

In Problems 13 through 34, solve the following systems by elimination. If the system is dependent, write dependent.

13. $\begin{cases} x + y = 3 \\ x - y = 3 \end{cases}$
14. $\begin{cases} 3x - y = 11 \\ x + y = 5 \end{cases}$

15. $\begin{cases} x - y = 0 \\ 3x + y = 12 \end{cases}$
16. $\begin{cases} x + y = 1 \\ x + 2y = 2 \end{cases}$

17. $\begin{cases} x - y = 8 \\ x + 2y = -7 \end{cases}$
18. $\begin{cases} 3x - y = 2 \\ 2x - y = -4 \end{cases}$

19. $\begin{cases} 2x - 3y = 4 \\ 2x - 3y = 2 \end{cases}$
20. $\begin{cases} x - 3y = -3 \\ 2x - y = 4 \end{cases}$

21. $\begin{cases} 3x - y = 22 \\ x - 3y = 2 \end{cases}$
22. $\begin{cases} 2x - y = 8 \\ 4x + 2y = 0 \end{cases}$

23. $\begin{cases} 9x + 6y = 4 \\ 3x + 2y = 1 \end{cases}$
24. $\begin{cases} -6x - 3y = 9 \\ 2x + y = 4 \end{cases}$

25. $\begin{cases} 7x + 2y = 7 \\ 3x + 4y = -19 \end{cases}$
26. $\begin{cases} 2x - 5y = 8 \\ 4x - 10y = -10 \end{cases}$

27. $\begin{cases} 2x + 3y = 2 \\ 5x + 4y = -9 \end{cases}$
28. $\begin{cases} 5x - 3y = 21 \\ 6x - 5y = 35 \end{cases}$

29. $\begin{cases} 6x + 4y = -26 \\ 3x + 2y = -13 \end{cases}$
30. $\begin{cases} 6x - 9y = 12 \\ 2x - 3y = 4 \end{cases}$

31. $\begin{cases} 3x - 2y = 0 \\ 4x + 5y = -23 \end{cases}$
32. $\begin{cases} 6x + 5y = 1 \\ 5x - 6y = 11 \end{cases}$

33. $\begin{cases} 2x - 4y = -3 \\ 3x + 5y = 1 \end{cases}$
34. $\begin{cases} 6x + 2y = 1 \\ 4x - 3y = 5 \end{cases}$

Challenge Problems

In Problems 35 through 37, solve the following systems by elimination.

35. $\begin{cases} x - y = a \\ x + y = a \end{cases}$

36. $\begin{cases} x + y = b \\ x - y = a \end{cases}$

37. $\begin{cases} x + y = a \\ x - 2y = b \end{cases}$

In Your Own Words

38. Why is the procedure outlined in this section called the method of *elimination*?

39. Describe the method of elimination.

7.4   Applications

TAPE 12

Often word problems are easier to solve if we use more than one variable. In this section we examine solving word problems by using two variables. We follow a procedure similar to that given in Section 3.4. When we use two variables, the mathematical model should be a system of two equations.

> ### A Procedure for Solving Word Problems with Two Variables
>
> 1. Read the problem and determine the two quantities that are to be found.
> 2. Assign two variables, such as x and y, to represent the quantities to be found.
> 3. Draw a figure or picture if possible.
> 4. Reread the problem and form a system of two equations.
> 5. Solve the system of equations by substitution or elimination.
> 6. Translate the solution of the system to answer the question in the word problem and check this answer in the original problem. It should make sense and answer the question.
> 7. Answer the question asked in the word problem.

We will look at some examples that illustrate how this procedure works.

► **EXAMPLE 1**

If the smaller of two numbers is subtracted from the larger, the result is 8. If the larger number is added to twice the smaller number, the result is 53. Find the two numbers.

Solution

STEP 1   Read the problem and determine the two quantities to be found.

We are to find two numbers.

STEP 2   Assign a variable to each number to be found.

Let L be the larger number and S be the smaller.

STEP 4   Form a system of two equations.

Since L is the larger number, the smaller subtracted from the larger is $L - S$. One equation is

$$L - S = 8$$

Twice the smaller number is $2S$. The larger number added to twice the smaller is $L + 2S$. So a second equation is

$$L + 2S = 53$$

Thus we have the system

$$\begin{cases} L - S = 8 \\ L + 2S = 53 \end{cases}$$

STEP 5   Solve the system.

Elimination is a good method to use if both equations are linear. We multiply the first equation by 2 and add the two equations. Thus

$$\begin{cases} L - S = 8 \\ L + 2S = 53 \end{cases} \xrightarrow{\text{Multiply by 2}} \begin{cases} 2L - 2S = 16 \\ L + 2S = 53 \end{cases}$$

$$\begin{cases} 2L - 2S = 16 \\ \underline{L + 2S = 53} \end{cases} \quad \text{Add the equations.}$$

$$3L = 69 \quad \text{Solve.}$$

$$L = 23$$

We replace L with 23 in the first equation and get

$$L - S = 8$$
$$23 - S = 8$$
$$-S = -15$$
$$S = 15$$

STEP 6   Translate into an answer and check.

We can check these two numbers in the original word problem: $23 - 15$ is 8, and $2(15) + 23$ is 53.

STEP 7   Answer the question.

The numbers are 23 and 15.   ◄

The next example deals with the perimeter of a rectangle. Recall the formula $P = 2L + 2W$.

► **EXAMPLE 2**

The crowd barrier around an outdoor stage is in the shape of a rectangle with a perimeter of 128 ft. If the length is three times the width, find the dimensions of the rectangle.

Solution

STEP 1   Read the problem and determine the two quantities to be found.

We are to find the length and width of a rectangle.

STEP 2   Assign a variable to each number to be found.
Let L be the length and W be the width in feet.

STEP 3   Draw a picture.

STEP 4   Form a system of two equations.

Using the formula for perimeter, we form the equation

$$2L + 2W = 128$$

We also know that the length is three times the width, so

$$L = 3W$$

This gives us the system

$$\begin{cases} 2L + 2W = 128 \\ \qquad\quad L = 3W \end{cases}$$

STEP 5   Solve the system.

If one of the equations of the system is already solved for one of the variables, such as the second equation of this system, substitution is a convenient method to use.
Replace L in the first equation with $3W$.

$$2L + 2W = 128$$
$$2(3W) + 2W = 128$$
$$8W = 128$$
$$W = 16$$

Since $L = 3W$, we see that L is 48.

STEP 6   Check.

We notice that 48 is three times 16, and if the rectangle has dimensions of 16 ft and 48 ft, its perimeter is $2(16) + 2(48)$, or 128 ft.

STEP 7   Answer the question.

The rectangle is 16 ft by 48 ft.

◄

► **EXAMPLE 3**

Tickets to a concert are $5 for balcony and $10 for orchestra. If attendance at one show was 800 and if total receipts were $7000, how many people bought orchestra seats?

Solution

Read the problem and determine the two quantities to be found.

We are asked for the number of *orchestra* seats sold. However, in solving the problem, we will also find the number of balcony seats sold.

Assign a variable to each quantity to be found.

Let x be the number of balcony tickets sold.
Let y be the number of orchestra tickets sold.

Form a system of two equations.

As a total of 800 tickets were sold, one equation is

$$x + y = 800$$

The total receipts are found by finding the value of x balcony tickets plus the value of y orchestra tickets. The value of the balcony tickets is $5x$ dollars and the value of the orchestra tickets is $10y$ dollars. So the total receipts are $5x + 10y$ dollars. This gives the equation

$$5x + 10y = 7000$$

We have the following system.

$$\begin{cases} x + y = 800 \\ 5x + 10y = 7000 \end{cases}$$

Solve the system.

We choose the method of elimination. If we multiply the first equation by -5 and add the two equations, we will eliminate x.

Multiply by -5

$$\begin{cases} x + y = 800 \\ 5x + 10y = 7000 \end{cases} \longrightarrow \begin{cases} -5x - 5y = -4000 \\ 5x + 10y = 7000 \end{cases}$$

$$\begin{cases} -5x - 5y = -4000 \\ 5x + 10y = 7000 \end{cases} \quad \text{Add the equations}$$

$$5y = 3000$$
$$y = 600 \qquad \text{Solve}$$

It is not necessary to find x to answer the question. However, to check, we need its value. Replacing y with 600 in the first equation gives a value of 200 for x.

Check.

Since $200 + 600$ is 800 and $5(200) + 10(600)$ is 7000, the solution checks.

Answer the question.
There were 600 orchestra tickets sold. ◄

The next example uses the formula for simple interest:

$$\text{Interest} = \text{principal} \cdot \text{rate} \cdot \text{time}$$
$$I = PRT$$

► **EXAMPLE 4**

Mrs. Bernardo invested $150,000 for one year in two different accounts, which pay 5% and 6.5% simple interest. If she earned a total of $7,950 in interest for the year, how much did she invest at each rate?

Solution

Determine the two quantities to be found.

We are to find the amount invested at 5% and the amount invested at 6.5%.

Assign variables.

Let x be the number of dollars invested at 5%.
Let y be the number of dollars invested at 6.5%

Find two equations.

One equation is $x + y = 150,000$.
A table will help find another equation.

	Principal *P*	Rate *R*	Time *T*	Interest *I = PRT*
5% investment	x	5%	1 year	$x(0.05)(1)$
6.5% investment	y	6.5%	1 year	$y(0.065)(1)$

From the table we see that the interest earned by the 5% investment is $0.05x$, dollars, and the interest earned by the 6.5% investment is $.065y$, dollars. But we know she earned $7,950 on these two investments. That provides the second equation.

$$0.05x + 0.065y = 7950$$

Our system of equations is

$$\begin{cases} x + y = 150,000 \\ 0.05x + 0.065y = 7,950 \end{cases}$$

Solve the system.

First, we clear decimals in the second equation.

$$\begin{cases} x + y = 150,000 \\ 0.05x + 0.065y = 15,600 \end{cases} \xrightarrow{\text{Multiply by 1000}} \begin{cases} x + y = 150,000 \\ 50x + 65y = 7,950,000 \end{cases}$$

We solve by elimination.

Multiply by -50

$$\begin{cases} x + y = 150,000 \\ 50x + 65y = 7,950,000 \end{cases} \longrightarrow \begin{cases} -50x - 50y = -7,500,000 \\ 50x + 65y = 7,950,000 \end{cases}$$

$$\begin{cases} -50x - 50y = -7,500,000 \\ \underline{50x + 65y = 7,950,000} \end{cases} \quad \text{Add the equations}$$

$$15y = 450,000 \qquad \text{Solve}$$
$$y = 30,000$$

Putting this into the first equation gives

$$x + y = 150,000$$
$$x + 30,000 = 150,000$$
$$x = 120,000$$

Check.

Since $120,000 + 30,000 = 150,000$, the total investment is $150,000. Also, 5% of $120,000 is $6000 and 6.5% of $30,000 is $1950, so the total interest is $7,950. The solution checks.

Answer the question.

Mrs. Bernardo invested $120,000 at 5% and $30,000 at 6.5%.

The next example uses the formula for distance, rate, and time: distance = rate · time. ◄

► EXAMPLE 5

Lee's bass boat takes 2 hours, traveling at a constant speed, to go from her dock on the Apple River to marker 8, a distance of 24 miles with the current. However, the trip back, against the current, takes her 3 hours. What is the average speed of the current and the average speed of Lee's boat in still water?

Solution

Let x be the average speed of Lee's boat in still water mph.
Let y be the average speed of the current in the river mph.
Therefore, her speed *with the current* is $x + y$ and her speed *against the current* is $x - y$. Since **distance** equals **rate** times **time,** we can form two equations.

$$24 = (x + y) \cdot 2 \qquad \text{With the current}$$
$$24 = (x - y) \cdot 3 \qquad \text{Against the current}$$

If we divide the first equation by 2 and the second equation by 3, we get the system

$$\begin{cases} x + y = 12 \\ x - y = 8 \end{cases}$$

Choosing the method of elimination, we eliminate y from the system by adding the two equations together.

$$\begin{cases} x + y = 12 \\ \underline{x - y = 8} \end{cases}$$
$$2x = 20$$
$$x = 10$$

We can use the first equation to find y.

$$10 + y = 12$$
$$y = 2$$

Now we can answer the question. The average speed of the current is 2 mph and the average speed of Lee's boat in still water is 10 mph.   ◄

Problem Set 7.4

Warm-Ups

Use two variables to solve the following problems. For Problems 1 through 4, see Example 1.

1. The sum of two numbers is 48. One of the numbers is twice the other number. Find the two numbers.

2. If twice a number is added to another number, the result is 16. If the sum of the numbers is 10, find the numbers.

3. There are 6000 fans at the basketball game. If the home team fans outnumber the visiting team fans by 540, how many fans are for the home team?

4. Beverly made 78 red and blue ribbons. If she made twice as many blue as red, how many of each did she make?

For Problems 5 through 7, see Example 2.

5. The parking lot of Freeman's Hardware Store is a rectangle with a perimeter of 72 m. If the length is 6 m more than the width, find the dimensions of the parking lot.

6. The perimeter of a triangle is 46 ft. The shortest side is 8 ft. The longest side is 1 ft less than twice the middle side. Find the lengths of the longest and middle sides.

7. The perimeter of a rectangle is 200 yd. If the length is three times the width, find the dimensions of the rectangle.

For Problems 8 through 11, see Example 3.

8. Jane has $625 in $5 and $10 bills. If she has a total of 105 bills, how many of each kind does she have?

9. Tickets to the senior play were $6 on Friday and $8 on Saturday. The attendance for both performances was 2850. If the total receipts were $20,300, how many attended the senior play each night?

10. Sam bought $8.44 worth of 29¢ and 19¢ stamps. If he bought 4 more 19¢ stamps than 29¢ stamps, how many of each kind did he buy?

11. Two cans of Wilson tennis balls and three cans of Spalding tennis balls cost a total of $10.55, whereas one can of Wilson and two cans of Spalding balls cost $6.37. How much is a can of Spalding balls?

For Problems 12 through 14, see Example 4.

12. Annie Brown invested a total of $50,000, some at 8% and the rest at 11%. If she earned a total of $4900 in 1 yr, how much did she invest at each rate?

13. Glen Cunningham borrowed a total of $15,000 for 1 yr. On one part he pays 10% interest and on the rest he pays 7% interest. If the total interest he paid in the year was $1320, how much did he borrow at each rate?

14. Sarah borrowed $6000, some at 8% interest and the rest at 10% interest, for 1 yr. If she paid a total of $550 in interest, how much did she borrow at each rate?

For Problems 15 and 16, see Example 5.

15. A private airplane, traveling with the wind, takes 3 hours to fly from El Paso to San Antonio, a distance of 600 miles. The return trip takes 4 hours, flying against the wind. Find the airspeed of the plane and the wind speed assuming that both are constant.

16. A weather service Hurricane Hunter airplane flies a carefully measured distance of 340 miles in one direction at constant speed and then flies the same distance in the opposite direction. If the trip takes 1 hour with the wind and 2 hours against it, what is the average wind speed?

Problem 16

Practice Exercises

In Problems 17 through 45, use two variables to solve the following problems.

17. The sum of two numbers is 6 and one number is 18 more than the other number. Find the two numbers.

18. There are 48 elm and oak trees on the north campus. The number of oaks subtracted from three times the number of elms is 12. Find the number of elms and the number of oaks.

19. A crystal salt and pepper set sells for $35. If the salt shaker sells for $5 more than the pepper shaker, how much does each cost alone?

20. The sum of two numbers is 9. If the larger is subtracted from twice the smaller, the result is 3. Find the numbers.

21. There are a total of 2000 runners in the Los Angeles marathon. If the male runners outnumber the female runners by 400, how many female runners are in the race?

22. One number is 7 more than another number. If the smaller number is subtracted from twice the larger number, the result is 25. Find the two numbers.

23. Two dozen doughnuts and 6 dozen sweet rolls cost $16.52, whereas 3 dozen doughnuts and 1 dozen sweet rolls cost $8.06. How much is a dozen doughnuts?

24. The sum of two numbers is 1. If the sum of one of the numbers and twice the other number is 0, find the numbers.

25. Maria bought 2 acorn squash and 1 spaghetti squash for $2.61. Jose bought 3 acorn and 2 spaghetti for $4.41. How much is a spaghetti squash?

26. Three times the smaller of two numbers is one more than the larger number. If they differ by 37, find the numbers.

27. Tanya served 84 Coca-Colas, some Classic and some caffeine-free. If she served 10 fewer Classic than caffeine-free, how many of each kind did she serve?

28. Walt and John sold 15 boxes of Christmas fruit cake to raise funds for their fraternity. Walt sold twice as many boxes as John. How many boxes did each sell?

29. The length of a rectangle is 6 cm longer than twice the width. The perimeter is 54 cm. Find the dimensions of the rectangle.

30. The width of a rectangle is 4 m less than the length. The perimeter is 40 m. Find the dimensions of the rectangle.

31. A box contains 120 black and white marbles. There are twice as many black as white. How many of each kind are in the box?

32. If four times a number is added to twice another number, the result is 30. The sum of the numbers is 6. Find the numbers.

33. The perimeter of a triangle is 38 ft. The longest side is 18 ft. The shortest side is 4 ft less than the middle side. Find the lengths of the middle and short sides.

34. Odette has $1500 in $50 and $20 bills. If she has a total of 45 bills, how many of each kind does she have?

35. Ricardo bought 4 lb of mixed peppermints and chocolates for $13.16. If peppermints are $1.19 a pound and chocolates are $3.99 a pound, how many pounds of each kind did he buy?

36. The sum of two numbers is 0. The sum of three times the first number and the second number is 6. Find the numbers.

37. Sharon bought 3 pens and 2 pencils for $2.95, and Linda bought 2 pens and 3 pencils for $2.45. What is the price of a pencil?

38. If 1 tuna and 2 chicken sandwiches cost $3.27, but 3 tuna and 1 chicken cost $3.86, how much is a tuna sandwich?

39. If 3 yd of wool and 2 yd of cotton cost $29.92, but 1 yd of wool and 1 yd of cotton cost $10.97, find the price per yard of wool and cotton.

40. Jim LeVert has $60,000 invested at simple interest in two accounts, which pay 11% and 9% interest. If the total annual interest is $5900, how much has Jim invested in each account?

41. Ms. Gould borrowed three times as much money at 7% interest as she did at 9% interest. If she paid $60 in interest in one year, how much money did she borrow at each rate?

42. Joan Baxter invested a total of $1000, some at 10% interest and the rest at 6% interest. If her total annual interest was $90, how much did she invest at each rate?

43. The crew on the school rowing team can row upstream at 4 miles per hour and downstream at 7 miles per hour. Find the crew's rate in still water and the rate of the current.

44. Dr. White and Dr. Hitchcock teach at universities that are 345 miles apart and have been appointed to the nominating committee for their state teacher organization. They agree to meet at a state park that will require 3

Problem 43

hours of driving time for each. Dr. White drives on an interstate highway and can travel 15 mph faster than Dr. Hitchcock. How fast must each drive to arrive at the same time?

45. A boat can travel upstream 30 miles in 5 hours and 35 miles downstream in the same time. What is the speed of the current and the speed of the boat in still water?

Challenge Problems

46. The length of a rectangular field is twice the width. The length is 75 m more than the width. If fencing cost $10.60 per meter, find the cost to fence the field.

47. Work Problems 6, 11, and 35 using the one-variable technique of Section 3.4. Which method is easier?

In Your Own Words

48. What is the two-variable procedure for solving word problems?

49. What is the difference between the two-variable procedure and the one-variable procedure?

50. In what ways are the two-variable procedure and the one-variable procedure the same?

7.5 Systems of Linear Inequalities in Two Variables

TAPE 13

So far in this chapter we have studied systems of linear *equations* in two variables. If we consider two linear inequalities together, we have a **system of linear inequalities.** To solve such a system means to find all ordered pairs that make both of the inequalities true. Since there is usually an infinite number of solutions, we don't try to list them. Instead, we graph the solution set.

► EXAMPLE 1

Graph the solution set of the system of linear inequalities.

$$\begin{cases} x - y < 1 \\ x + y \geq 1 \end{cases}$$

Solution

We graph the solution set of the inequality $x - y < 1$. As we did in Section 6.4, we draw the boundary line $x - y = 1$. We use a dashed line since equality is not included.

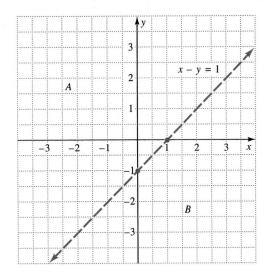

The boundary line divides the plane into two regions, A and B. We test each region.

Region	Test Point in Region	Statement $x - y < 1$	Truth of Statement	Region in Solution Set?
A	$(0, 0)$	$0 - 0 < 1$	True	Yes
B	$(2, 0)$	$2 - 0 < 1$	False	No

We shade the solution set.

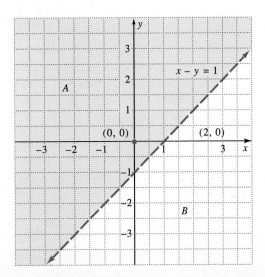

Next we draw the solution set of the second inequality $x + y \geq 1$. We draw the boundary line $x + y = 1$ as a solid line because equality *is* included and test the two regions.

Region	Test Point in Region	Statement $x + y \geq 1$	Truth of Statement	Region in Solution Set?
A	(0, 0)	$0 + 0 \geq 1$	False	No
B	(2, 0)	$2 + 0 \geq 1$	True	Yes

We shade the solution set.

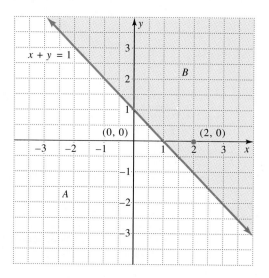

The solution set for the system consists of all ordered pairs that make *both* inequalities true statements. If we graph the solution set for both inequalities on the same coordinate system, we see that both inequalities are true where the shading overlaps.

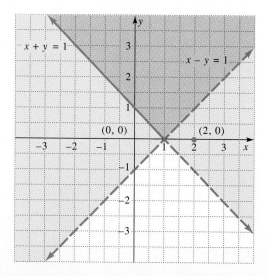

Notice the line $x + y = 1$ in the final graph of the solution set. We have changed the portion of that line *not in the solution set* from solid to dashed.   ◀

> ## A Procedure for Graphing the Solution Set of a System of Linear Inequalities
>
> 1. Graph each inequality on the same coordinate system, shading each solution set lightly.
> 2. Darken where the lightly shaded regions overlap. Change any portions of solid lines that are not in the solution set to dashed lines.

► **EXAMPLE 2**

Graph the solution set of the following system of inequalities.

$$\begin{cases} 2x + y < 2 \\ \qquad x \geq 1 \end{cases}$$

Solution

We graph the solution set for $2x + y < 2$. The boundary line $2x + y = 2$ is dashed because equality is not included. We test each region.

Region	Test Point in Region	Statement $2x + y < 2$	Truth of Statement	Region in Solution Set?
A	(0, 0)	0 + 0 < 2	True	Yes
B	(1, 1)	2 + 1 < 2	False	No

We lightly shade the solution set.

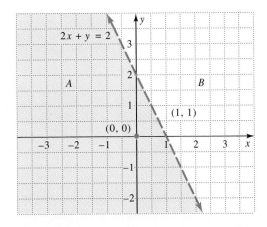

Next we graph the solution set for $x \geq 1$. We draw the boundary $x = 1$ as a solid line because equality is included and test each region.

Region	Test Point in Region	Statement $x \geq 1$	Truth of Statement	Region in Solution Set?
A	(0, 0)	$0 \geq 1$	False	No
B	(2, 0)	$2 \geq 1$	True	Yes

We lightly shade the solution set.

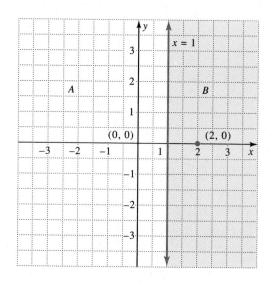

If we graph the two solutions on the same coordinate system and darken the area where the lightly shaded regions overlapped, we have the following graph of the solution set of the system.

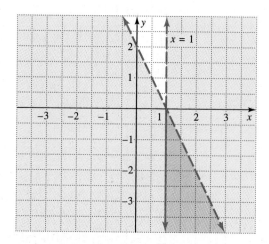

Notice that we changed the portion of the line $x = 1$ that is *above* the x-axis from solid to dashed.   ◄

Problem Set 7.5

Warm-Ups

In Problems 1 through 12, graph the solution set of each system of inequalities. See Examples 1 and 2.

1. $\begin{cases} x + y \geq 3 \\ x - y < 2 \end{cases}$

2. $\begin{cases} x + y \leq 2 \\ x + 2y \geq 2 \end{cases}$

3. $\begin{cases} 2x + y < 2 \\ x - 3y > 3 \end{cases}$

4. $\begin{cases} x - 3y \leq 3 \\ 2x - y \leq 2 \end{cases}$

5. $\begin{cases} x - 3y < 3 \\ 4x + y < 2 \end{cases}$

6. $\begin{cases} 3x - y \geq 3 \\ x - 2y \geq 2 \end{cases}$

7. $\begin{cases} x - y < 0 \\ x + y > 0 \end{cases}$

8. $\begin{cases} 4x - y \geq 2 \\ x + 5y < 2 \end{cases}$

9. $\begin{cases} x - y < 3 \\ x + 2y \geq 0 \end{cases}$

10. $\begin{cases} x + y < 2 \\ x < 2 \end{cases}$

11. $\begin{cases} y \geq 2 \\ 2x - y \leq 2 \end{cases}$

12. $\begin{cases} x \leq 2 \\ x + y > 0 \end{cases}$

Practice Exercises

Graph the solution set of each system of inequalities.

13. $\begin{cases} x - y \geq 1 \\ x + y < 1 \end{cases}$

14. $\begin{cases} x - y \leq 2 \\ x - 2y \geq 1 \end{cases}$

15. $\begin{cases} 2x - y < 2 \\ x + 3y > 3 \end{cases}$

16. $\begin{cases} 4x - 3y > 6 \\ 2x + 3y \leq -6 \end{cases}$

17. $\begin{cases} 2x - 3y \geq 6 \\ 3x + 4y < 4 \end{cases}$

18. $\begin{cases} 4x - 7y < 7 \\ 7x + 4y \geq 4 \end{cases}$

19. $\begin{cases} 2x + 3y \leq 6 \\ 2x + y \leq 2 \end{cases}$

20. $\begin{cases} x + 3y < 3 \\ 4x - y < 2 \end{cases}$

21. $\begin{cases} 3x - y \geq 3 \\ x - 2y \geq 2 \end{cases}$

22. $\begin{cases} x - y \geq 0 \\ x + y \geq 0 \end{cases}$

23. $\begin{cases} 4x + y \geq 4 \\ x - 5y < 5 \end{cases}$

24. $\begin{cases} x + y \leq 3 \\ x - 3y \geq 0 \end{cases}$

25. $\begin{cases} x + 2y > 2 \\ 2x - 3y < 6 \end{cases}$

26. $\begin{cases} 3x + 2y \leq 12 \\ x + 3y \geq -6 \end{cases}$

27. $\begin{cases} 2x - 5y < 10 \\ x + 5y < -5 \end{cases}$

28. $\begin{cases} 2x - 3y > 6 \\ 3x + 5y < 15 \end{cases}$

29. $\begin{cases} 4x + 3y \geq 12 \\ 2x - 7y \leq 14 \end{cases}$

30. $\begin{cases} 3x + 4y < 12 \\ 2x - 5y \geq 10 \end{cases}$

31. $\begin{cases} 4x + 3y > 12 \\ 5x - 3y \leq 15 \end{cases}$

32. $\begin{cases} 2x + 3y \geq -6 \\ 3x - 4y < 12 \end{cases}$

33. $\begin{cases} 4x + 7y < 28 \\ 7x - 4y \geq 28 \end{cases}$

34. $\begin{cases} x - y < 3 \\ x \leq -2 \end{cases}$

35. $\begin{cases} y > 3 \\ 2x + y \leq 4 \end{cases}$

36. $\begin{cases} x \leq 1 \\ x + y \geq 0 \end{cases}$

Challenge Problems

In Problems 37 through 39, graph the solution set.

37. $\begin{cases} y > 0 \\ x < 0 \end{cases}$

38. $\begin{cases} y < 0 \\ x \geq 0 \end{cases}$

39. $\begin{cases} x \geq 0 \\ y \leq 0 \end{cases}$

In Your Own Words

40. Describe how to graph a system of two linear inequalities in two variables.

► Chapter Summary

GLOSSARY

System of two equations in two variables   Two equations linked so that a solution of the system is a solution of *both* equations.

Independent system of linear equations   A system of linear equations with exactly one solution.

Dependent system   A system of equations with an infinite number of solutions.

Inconsistent system   A system of equations with no solutions.

Method of Graphing for Solving Systems of Equations	Graph both equations on the same coordinate system and estimate their point of intersection, if any.

Substitution Method for Solving Systems of Equations

1. Solve one of the equations for one of the variables.
2. Substitute this expression for the same variable in the other equation.
3. Solve the resulting equation.
4. Substitute the solution of step 3 into the equation formed in step 1 to find the value of the other variable.
5. Write the solution set.

Elimination Method for Solving Systems of Equations

1. Multiply both sides of each equation by a suitable number so that adding the equations eliminates a variable.
2. Add the equations and solve the resulting equation.
3. Substitute this value into one of the original equations to find the value of the other variable.
4. Write the solution set.

Graphing Systems of Linear Inequalities

1. Graph each inequality on the same coordinate system. Lightly shade the solution set of each.
2. Darken where the lightly shaded portions overlap. Change any portions of solid lines that are not in the solution set to dashed lines.

CHECK UPS

1. How many solutions are in the solution set for the following system?

$$\begin{cases} x + y = 2 \\ 2x + 2y = 4 \end{cases}$$

Section 7.1; Example 4

2. Solve the system by substitution.

$$\begin{cases} x + y = 5 \\ x - y = -1 \end{cases}$$

Section 7.2; Example 2

3. Solve $\begin{cases} 2x + 3y = 1 \\ 3x + y = -2 \end{cases}$ by elimination.

Section 7.3; Example 2

4. If the smaller of two numbers is subtracted from the larger, the result is 8. If the larger number is added to twice the smaller number, the result is 53. Find the two numbers.

Section 7.4; Example 1

5. Graph the solution set of the system of inequalities.

$$\begin{cases} x - y < 1 \\ x + y \geq 1 \end{cases}$$

Section 7.5; Example 1

► Review Problems

In Problems 1 through 6, use the graphical method to classify each system as independent, inconsistent, or dependent.

1. $\begin{cases} x - 2y = 4 \\ 2x - y = 2 \end{cases}$

2. $\begin{cases} 3x - y = 3 \\ 3x + y = 3 \end{cases}$

3. $\begin{cases} x - y = 0 \\ 2x - 2y = 0 \end{cases}$

4. $\begin{cases} 4x + 2y = 4 \\ 2x + y = 0 \end{cases}$

5. $\begin{cases} x + y = 3 \\ y = 2x \end{cases}$

6. $\begin{cases} x - y = 0 \\ 2x + 2y = 4 \end{cases}$

In Problems 7 through 18, use either substitution or elimination to solve each system. If the system is dependent, write dependent.

7. $\begin{cases} x + y = 3 \\ y = x \end{cases}$

8. $\begin{cases} x + 2y = 4 \\ x - 2y = 2 \end{cases}$

9. $\begin{cases} x - 3y = 5 \\ x + y = -3 \end{cases}$

10. $\begin{cases} y = x + 1 \\ 2x - y = 1 \end{cases}$

11. $\begin{cases} 2x + 3y = 1 \\ x + y = 0 \end{cases}$

12. $\begin{cases} 2x - 4y = 3 \\ x = 2y \end{cases}$

13. $\begin{cases} 2x + 5y = 1 \\ 5x + 2y = -8 \end{cases}$

14. $\begin{cases} 3x - 2y = 4 \\ 6x - 4y = 8 \end{cases}$

15. $\begin{cases} 2x + 3y = 5 \\ y = 2x - 1 \end{cases}$

16. $\begin{cases} 3x - 4y = 18 \\ 2x + 3y = -5 \end{cases}$

17. $\begin{cases} 2x + 7y = 0 \\ 5x - 2y = 0 \end{cases}$

18. $\begin{cases} 3x - 5y = 6 \\ 2x - 3y = 4 \end{cases}$

In Problems 19 through 24, graph the solution set for each system of inequalities.

19. $\begin{cases} x - 2y < 0 \\ x + y \geq 3 \end{cases}$

20. $\begin{cases} 2x + y \geq 2 \\ x + 2y \geq -2 \end{cases}$

21. $\begin{cases} x + 3y < 0 \\ 2x - y > 0 \end{cases}$

22. $\begin{cases} 2x - y \leq 4 \\ x + y > 5 \end{cases}$

23. $\begin{cases} 2x + 3y > 6 \\ x \geq -1 \end{cases}$

24. $\begin{cases} x + 2y \geq 2 \\ y > 3 \end{cases}$

25. Joan and Sue have 36 books. If Joan has three times as many as Sue, how many books does each have?

26. One number is 8 less than another number. If the sum of the numbers is 12, find the numbers.

27. The width of a rectangle is 5 m less than the length. If the perimeter is 38 m, find the dimensions of the rectangle.

28. Terry invested $800 for one year, some at 5.5% simple interest and the rest at 6% simple interest. If he earned a total of $46.50 in interest, how much did he invest at 5.5%?

Let's Not Forget . . .

In Problems 29 through 36, perform the operations indicated.

29. -2^4

30. $(-2)^4$

31. $6x - (x + 3y)$

32. $\dfrac{a^2 - ab}{b - a}$

33. $(3x - 2)^2$

34. $(2x - 5)(2x + 5)$

35. $(3a + b)(2a - b)$

36. $\dfrac{x^2 + y^2}{x^2 y^3} \cdot \dfrac{xy^2}{x^4 + x^2 y^2}$

In Problems 37 through 39, simplify each expression. Watch the role of parentheses.

37. $2x + y(2x - y)$

38. $(2x + y)(x - y)$

39. $\dfrac{1}{x + 1} - \dfrac{1}{x - 1}$

In Problems 40 through 43, determine if b is a factor or a term.

40. $b + 2$

41. $b(x + 2)$

42. $\dfrac{1}{b + 2}$

43. $x(y + b)$

In Problems 44 through 46, multiply each expression by $\dfrac{x + 2}{x - 2}$.

44. $\dfrac{x - 2}{x + 2}$

45. $1 + \dfrac{x - 2}{x + 2}$

46. $\dfrac{x - 2}{x + 2}\left(1 + \dfrac{x - 2}{x + 2}\right)$

In Problems 47 through 52, identify which expressions are factored. Factor those that are not factored, if possible.

47. xy^2

48. $x^2 - 4z^2$

49. $(a + 5)^3$

50. $a^2 b - b^2 c$

51. $a^2 + 25$

52. $\dfrac{x}{2} + x^2$

In Problems 53 through 57, identify whether each problem is an expression, equation, or inequality. Solve the equations and inequalities, and perform the operation indicated with the expressions.

53. $\dfrac{2}{x + 1} - 3$

54. $\dfrac{2}{x + 1} - 3 = \dfrac{-2x}{x + 1}$

55. $-\dfrac{2}{3}x > 1$

56. $\dfrac{\dfrac{2}{xy^2}}{\dfrac{1}{xy}}$

57. $\dfrac{x^2 + 4x + 4}{x^2} \div \dfrac{x^2 + x - 2}{x^2 - x}$

► You Decide

You have a fine new home with a beautiful yard. Unfortunately, the demands of your job leave little time for upkeep. However, a very dependable local landscape company will maintain your yard. They offer three options. Option 1: $27 for mowing, $20 for fertilizing and weed control, and $10 for trimming shrubs. Option 2: $90 a month for complete service during the growing season and $25 a month for the winter months. Option 3: $750 a year for complete year-round service. Which plan do you choose and why?

► Chapter 7 Test

In Problems 1 through 5, choose the correct answer.

1. The system $\begin{cases} x - y = 0 \\ 2x - 2y = 0 \end{cases}$ is (?).

    Independent   Inconsistent
    Dependent   None of the above

2. The *x* coordinate of the solution to the system of equations $\begin{cases} 2x + 3y = 16 \\ 3x - y = 2 \end{cases}$ is (?).

    $\dfrac{18}{11}$ $\dfrac{18}{5}$
    1   2

3. The solution set for the system $\begin{cases} 2x - 4y = 8 \\ x - 2y = 8 \end{cases}$ is (?)

    $\{(16, 4)\}$ $\{(0, -2)\}$
    $\varnothing$ None of the above

4. The *y* coordinate of the solution to the system of equations $\begin{cases} y = 2 - x \\ 2x - 3y = 9 \end{cases}$ is (?).

    3   -1
    -13 5

5. The solution set for the system $\begin{cases} x + y = 0 \\ x - 2y = 0 \end{cases}$ is (?).

    $\{(0, 0)\}$ $\{(-3, 3)\}$
    $\varnothing$ None of the above

In Problems 6 through 9, solve the system of equations.

6. $\begin{cases} x = 3y \\ 3x - y = 8 \end{cases}$

7. $\begin{cases} x + 2y = 4 \\ 2x + 4y = 0 \end{cases}$

8. $\begin{cases} 3x + 2y = 0 \\ 2x - 3y = 13 \end{cases}$

9. $\begin{cases} y = x + 1 \\ 2x - y = -2 \end{cases}$

In Problems 10 through 13, graph the solution set for each system of inequalities.

10. $\begin{cases} 2x - y \leq 0 \\ x + 3y \geq 0 \end{cases}$

11. $\begin{cases} x + y \leq 3 \\ x - y > 1 \end{cases}$

12. $\begin{cases} x - 2y > 1 \\ x + y < 2 \end{cases}$

13. $\begin{cases} x \geq -2 \\ x + y < 1 \end{cases}$

14. The length of a rectangle is 7 ft more than the width. If the perimeter is 46 ft, find the dimensions of the rectangle.

15. If a head of lettuce and 2 avocados cost $2.27 and 2 heads of lettuce and 3 avocados cost $3.85, how much does 1 avocado cost?

8 Roots and Radicals

See Problem Set 8.2, Exercise 70.

Connections

The first numbers we encounter as we become educated in mathematics are the *natural numbers,* or counting numbers: 1, 2, 3, ... Soon we need zero and the negatives of the natural numbers to indicate such things as temperatures and bank balances. The set of these numbers together with the natural numbers is called the *integers.* Next, we add the fractions, or *rational numbers,* in order to measure real things such as weight and length. Rational numbers are sufficient to solve simple equations, such as the linear equations studied in Chapter 3, and to solve many day-to-day problems.

Unfortunately, the rational numbers solve only the simplest equations. In the next chapter we will study equations involving numbers such as $\sqrt{2}$, $\sqrt[3]{5}$, $2 - \sqrt{3}$, which are irrational. In preparation, this chapter continues the study of square roots and extends this idea to the *n*th root of a number. We examine the arithmetic of such numbers and when they can be simplified. The last section covers fractional exponents and their relationship to radicals.

8.1   Definitions

TAPE 13

We know that 2 and -2 are opposites and that if we square 2 or -2 the result is 4. Using an exponent of 2 indicates squaring a number. So,

$$2^2 = 4$$

$$(-2)^2 = 4$$

Thus 4 is the result of squaring 2 or -2. Integers such as 4, 9, and 16 are called perfect squares because they can be expressed as the square of an integer.

The reverse of squaring is called finding the square root. We use the symbol $\sqrt{4}$ to represent the *positive* number whose square is 4. That is,

$$\sqrt{4} = 2$$

We call $\sqrt{4}$ the **principal square root** of 4.

Using this notation, we can write $-\sqrt{4}$ to indicate the opposite of the principal square root of 4. It is the *other* number whose square is 4. Thus

$$-\sqrt{4} = -2$$

▶ EXAMPLE 1

Find the roots indicated.

(a) $\sqrt{9}$ (b) $-\sqrt{9}$ (c) $\sqrt{\dfrac{4}{9}}$

Solutions

(a) $\sqrt{9} = 3$ because 3^2 is 9 and 3 is *positive.*

(b) $-\sqrt{9} = -3$ because $\sqrt{9} = 3$ and the opposite of 3 is -3.

(c) $\sqrt{\dfrac{4}{9}} = \dfrac{2}{3}$ because $\left(\dfrac{2}{3}\right)^2 = \dfrac{4}{9}$ and $\dfrac{2}{3}$ is *positive.*   ◀

Is there a number whose square is 7? Yes, there are two such numbers, and they are opposites. We write $\sqrt{7}$ to indicate the *positive* number whose square is 7 and $-\sqrt{7}$ to indicate the negative number whose square is 7. That is,

$$(\sqrt{7})^2 = 7 \quad \text{and} \quad (-\sqrt{7})^2 = 7$$

Because 7 is not a perfect square, it is difficult to write $\sqrt{7}$ in a simpler form. However, $\sqrt{7}$ can be written as a nonrepeating, nonterminating decimal. We can *approximate* its value with a calculator or a square root table. The square root of seven is approximately 2.6, rounded to the nearest tenth. We write

$$\sqrt{7} \approx 2.6$$

Notice the *approximately equal* symbol, $\approx$.

► **EXAMPLE 2**

Approximate each square root to two decimal places.

(a) $\sqrt{11}$ (b) $-\sqrt{5}$ (c) $\sqrt{3}$

Solutions

(a) $\sqrt{11} \approx 3.32$
(b) $-\sqrt{5} \approx -2.24$
(c) $\sqrt{3} \approx 1.73$ ◄

What is $\sqrt{-4}$? Is there a real number whose square is -4? Since $(-2)^2 = 4$, there is no such real number. *Square roots of negative real numbers do not exist in our system of real numbers.*

Square Root of a Positive Number

If x is a positive real number, $\sqrt{x}$ is the *positive* number whose square is x. That is,

$$(\sqrt{x})^2 = x$$

Square Root of Zero

$$\sqrt{0} = 0$$

If we cube 2, the result is 8. An exponent of 3 indicates cubing. Thus

$$2^3 = 8$$

There is not another real number whose cube is 8, since

$$(-2)^3 = -8$$

The reverse of cubing is called finding a *cube root*. We indicate the cube root of 8 by $\sqrt[3]{8}$. So

$$\sqrt[3]{8} = 2$$

► **EXAMPLE 3**

Find $\sqrt[3]{-27}$.

Solution

$$\sqrt[3]{-27} = -3, \qquad \text{because } (-3)^3 = -27 \qquad ◄$$

Notice that the cube root of a negative number does exist and is a negative number. In fact, if x is *any real number,* $\sqrt[3]{x}$ is the number whose cube is x. That is,

$$(\sqrt[3]{x})^3 = x$$

► **EXAMPLE 4**

Find the following cube roots.

(a) $\sqrt[3]{27}$

(b) $\sqrt[3]{-8}$

(c) $-\sqrt[3]{-8}$

(d) $\sqrt[3]{-\dfrac{8}{27}}$

Solutions

(a) $\sqrt[3]{27} = 3$ because $3^3 = 27$.

(b) $\sqrt[3]{-8} = -2$ because $(-2)^3 = -8$.

(c) $-\sqrt[3]{-8} = 2$ because $-\sqrt[3]{-8} = -(-2)$.

(d) $\sqrt[3]{-\dfrac{8}{27}} = -\dfrac{2}{3}$ because $\left(-\dfrac{2}{3}\right)^3 = -\dfrac{8}{27}$. $\qquad ◄$

In addition to square roots and cube roots, we have fourth roots, fifth roots, and, in general *n*th roots.

---

*n*th **Root of a Real Number**

Suppose n is a natural number and x is a real number.

1. If n is *even* and x is not negative, $\sqrt[n]{x}$ is the *nonnegative* number whose nth power is x.

2. If n is *odd,* $\sqrt[n]{x}$ is the number whose nth power is x.

That is,

$$(\sqrt[n]{x})^n = x$$

Remember, if n is even, x must be *nonnegative,* and $\sqrt[n]{x}$ is also *nonnegative.*

Be Careful!   The expression $\sqrt[n]{x}$ is called a **radical.** We call the symbol $\sqrt[n]{}$ a **radical symbol.** Also, n is called the **index,** and x is called the **radicand.** Since square roots occur so often, we omit the index of 2 and use $\sqrt{}$ instead of $\sqrt[2]{}$.

▶ EXAMPLE 5

Find the following roots, if they exist as real numbers.

(a) $\sqrt{36}$ (b) $\sqrt[3]{-64}$ (c) $\sqrt{-16}$ (d) $\sqrt[4]{16}$ (e) $\sqrt[5]{-32}$

Solutions

(a) $\sqrt{36} = 6$ because $6^2 = 36$ and 6 is *positive.*
(b) $\sqrt[3]{-64} = -4$ because $(-4)^3 = -64.$
(c) $\sqrt{-16}$ is not a real number because -16 is *negative.*
(d) $\sqrt[4]{16} = 2$ because $2^4 = 16$ and 2 is *positive.*
(e) $\sqrt[5]{-32} = -2$ because $(-2)^5 = -32.$ ◀

Suppose we want to find $\sqrt{x^2}$. Is there a *nonnegative* number whose square is x^2? The square of x is x^2. However, we do not know if x represents a nonnegative number. To avoid this problem, we shall assume in this chapter that *all variables represent nonnegative real numbers.* So we write

$$\sqrt{x^2} = x$$

because $(x)^2 = x^2$ and we are assuming that x is *nonnegative.*

▶ EXAMPLE 6

Find the following roots. Assume that all variables represent *nonnegative* real numbers.

(a) $\sqrt{x^4}$ (b) $\sqrt{x^{10}}$ (c) $\sqrt[3]{x^3}$ (d) $\sqrt[5]{x^{10}}$ (e) $\sqrt{25x^2}$

Solutions

(a) $\sqrt{x^4} = x^2$ because $(x^2)^2 = x^4$ and x^2 is nonnegative.
(b) $\sqrt{x^{10}} = x^5$ because $(x^5)^2 = x^{10}$ and x^5 is nonnegative.
(c) $\sqrt[3]{x^3} = x$ because $(x)^3 = x^3.$
(d) $\sqrt[5]{x^{10}} = x^2$ because $(x^2)^5 = x^{10}.$
(e) $\sqrt{25x^2} = 5x$ because $(5x)^2 = 25x^2$ and $5x$ is nonnegative. ◀

In working with radicals, it is very helpful to recognize some powers of 2, 3, 4, and 5. Some of them are listed next.

n	2^n	3^n	4^n	5^n
1	2	3	4	5
2	4	9	16	25
3	8	27	64	125
4	16	81	256	625
5	32	243	1,024	3,125
6	64	729	4,096	15,625
7	128	2,187	16,384	78,125
8	256	6,561	65,536	390,625
9	512	19,683	262,144	1,953,125

Problem Set 8.1

Warm-Ups

In Problems 1 through 5, find the indicated roots if they exist. See Example 1.

1. $\sqrt{25}$

2. $\sqrt{-81}$

3. $-\sqrt{36}$

4. $\sqrt{\dfrac{4}{25}}$

5. $-\sqrt{\dfrac{4}{9}}$

In Problems 6 through 10, approximate the indicated roots to two decimal places if they exist. See Example 2.

6. $\sqrt{2}$

7. $\sqrt{-2}$

8. $-\sqrt{2}$

9. $\sqrt{8}$

10. $-\sqrt{13}$

In Problems 11 through 15, find the indicated roots if they exist. See Examples 3 and 4.

11. $\sqrt[3]{64}$

12. $\sqrt[3]{-64}$

13. $-\sqrt[3]{64}$

14. $\sqrt[3]{\dfrac{1}{64}}$

15. $-\sqrt[3]{\dfrac{1}{64}}$

In Problems 16 through 25, find the indicated roots if they exist. See Example 5.

16. $\sqrt[4]{16}$

17. $\sqrt[4]{-16}$

18. $-\sqrt[4]{16}$

19. $\sqrt[4]{\dfrac{1}{16}}$

20. $-\sqrt[4]{\dfrac{1}{16}}$

21. $\sqrt[5]{32}$

22. $\sqrt[5]{-32}$

23. $-\sqrt[5]{32}$

24. $\sqrt[5]{\dfrac{1}{32}}$

25. $-\sqrt[5]{\dfrac{1}{32}}$

In Problems 26 through 35, find the indicated roots. Assume all variables represent nonnegative real numbers. See Example 6.

26. $\sqrt{x^6}$

27. $\sqrt{y^2}$

28. $\sqrt{t^8}$

29. $\sqrt{z^{12}}$

30. $\sqrt{x^{16}}$

31. $\sqrt{4x^8}$

32. $\sqrt{25y^4}$

33. $\sqrt{9t^6}$

34. $\sqrt{100z^{18}}$

35. $\sqrt[3]{8z^3}$

Practice Exercises

In Problems 36 through 79, find the indicated roots if they exist.

36. $\sqrt{16}$

37. $\sqrt{-25}$

38. $-\sqrt{49}$

39. $-\sqrt{16}$

40. $\sqrt{\dfrac{9}{25}}$

41. $-\sqrt{\dfrac{4}{81}}$

42. $\sqrt{100}$

43. $\sqrt{-400}$

44. $\sqrt{49}$

45. $\sqrt{-100}$

46. $-\sqrt{1}$

47. $\sqrt{\dfrac{49}{1}}$

48. $\sqrt[3]{-1}$

49. $\sqrt[3]{1}$

50. $-\sqrt[3]{-1}$

51. $\sqrt[3]{-\dfrac{1}{8}}$

52. $\sqrt{0}$

53. $\sqrt{\dfrac{64}{49}}$

54. $-\sqrt{\dfrac{36}{225}}$

55. $\sqrt[3]{1000}$

56. $\sqrt[3]{-125}$

57. $-\sqrt[3]{8}$

58. $\sqrt[3]{\dfrac{1}{27}}$

59. $-\sqrt[3]{\dfrac{8}{27}}$

60. $-\sqrt[3]{-\dfrac{8}{27}}$

61. $-\sqrt[3]{1}$

62. $-\sqrt[3]{-27}$

63. $\sqrt[3]{-\dfrac{1}{125}}$

64. $-\sqrt[3]{\dfrac{27}{8}}$

65. $\sqrt[4]{81}$

66. $\sqrt[4]{-625}$

67. $-\sqrt[4]{625}$

68. $\sqrt[4]{\dfrac{1}{81}}$

69. $-\sqrt[4]{\dfrac{1}{81}}$

70. $\sqrt[5]{243}$

71. $\sqrt[5]{-243}$

72. $-\sqrt[5]{243}$

73. $\sqrt[5]{\dfrac{1}{243}}$

74. $-\sqrt[5]{\dfrac{1}{243}}$

75. $-\sqrt[3]{-27}$

76. $\sqrt[4]{256}$

77. $\sqrt[5]{0}$

78. $\sqrt{\dfrac{1}{16}}$

79. $\sqrt[4]{-1}$

In Problems 80 through 99, find each root. Assume that all variables represent nonnegative real numbers.

80. $\sqrt{x^4}$

81. $\sqrt{y^{10}}$

82. $\sqrt{t^{18}}$

83. $\sqrt{z^{20}}$

84. $\sqrt{x^{36}}$

85. $\sqrt{16x^6}$

86. $\sqrt{36y^{16}}$

87. $\sqrt{49t^2}$

88. $\sqrt{64x^8y^4}$

89. $\sqrt{25y^4z^2}$

90. $\sqrt{81s^8t^6}$

91. $\sqrt{121x^4y^2z^{18}}$

92. $\sqrt{144x^6y^4z^{12}}$

93. $\sqrt{169a^2b^2c^2}$

94. $\sqrt{121z^{14}}$

95. $\sqrt{16x^4y^2}$

96. $\sqrt{81y^6z^8}$

97. $\sqrt{144s^4t^2}$

98. $\sqrt{100x^2y^2z^2}$

99. $\sqrt{x^8y^4z^{10}}$

In Problems 100 through 103, approximate the indicated roots to two decimal places if they exist.

100. $\sqrt{3}$

101. $\sqrt{-3}$

102. $-\sqrt{3}$

103. $\sqrt{10}$

Challenge Problems

In Problems 104 through 111, estimate each root to the nearest integer.

104. $\sqrt{24}$

105. $\sqrt{15}$

106. $\sqrt{10}$

107. $\sqrt{12}$

108. $\sqrt[3]{7}$

109. $\sqrt[3]{-9}$

110. $\sqrt[3]{30}$

111. $\sqrt[3]{-26}$

In Your Own Words

112. What do we mean by the square root of a number?

113. What do we mean by the seventh root of a number?

8.2   Simplifying Radicals

TAPE 13

Let's examine the square root of a product. Consider $\sqrt{4 \cdot 16}$ and $\sqrt{4} \cdot \sqrt{16}$. Since $4 \cdot 16 = 64$,

$$\sqrt{4 \cdot 16} = \sqrt{64} = 8$$

Also,

$$\sqrt{4} \cdot \sqrt{16} = 2 \cdot 4 = 8$$

Therefore,

$$\sqrt{4 \cdot 16} = \sqrt{4} \cdot \sqrt{16}$$

This leads to an important property of radicals.

Multiplication Property of Radicals

If $\sqrt[n]{p}$ and $\sqrt[n]{q}$ exist as real numbers, then

$$\sqrt[n]{pq} = \sqrt[n]{p} \cdot \sqrt[n]{q}$$

The *n*th root of a product is the product of the *n*th roots.

We would not use this property to find $\sqrt{64}$ because 64 is a perfect square. However, it can be used to simplify radicals such as $\sqrt{75}$. Since 75 is not a perfect square, $\sqrt{75}$ is not an integer. However, 75 has a *factor* that is a perfect square.

$$\sqrt{75} = \sqrt{25 \cdot 3}$$
$$= \sqrt{25} \cdot \sqrt{3} \qquad \text{Multiplication Property}$$
$$= 5\sqrt{3}$$

So we see that we can write $\sqrt{75}$ as $5\sqrt{3}$. We say that $5\sqrt{3}$ is the *simplified form* of $\sqrt{75}$. The Multiplication Property of Radicals allows us to rewrite radical expressions with any perfect square factors of the radicand removed. We call a square root **simplified** when it is written with all perfect square factors removed from the radicand. The following examples should clarify this.

▶ **EXAMPLE 1**

Simplify each square root.

(a) $\sqrt{24}$ (b) $\sqrt{18}$ (c) $\sqrt{48}$

Solutions

(a) The largest perfect square factor of 24 is 4.

$$\sqrt{24} = \sqrt{4 \cdot 6}$$
$$= \sqrt{4} \cdot \sqrt{6} \qquad \text{Multiplication Property}$$
$$= 2\sqrt{6} \qquad\qquad \sqrt{4} = 2$$

Notice that we could have factored 24 as $3 \cdot 8$. However, neither 3 nor 8 is a perfect square, so no simplification could take place.

(b) The largest perfect square factor of 18 is 9.

$$\sqrt{18} = \sqrt{9 \cdot 2}$$
$$= \sqrt{9} \cdot \sqrt{2} \qquad \text{Multiplication Property}$$
$$= 3\sqrt{2} \qquad \sqrt{9} = 3$$

(c) The largest perfect square factor of 48 is 16.

$$\sqrt{48} = \sqrt{16 \cdot 3}$$
$$= \sqrt{16} \cdot \sqrt{3} \qquad \text{Multiplication Property}$$
$$= 4\sqrt{3} \qquad \sqrt{16} = 4$$

Notice that we could have factored 48 as $4 \cdot 12$ and removed the perfect square factor 4.

$$\sqrt{48} = \sqrt{4 \cdot 12}$$
$$= \sqrt{4} \cdot \sqrt{12}$$
$$= 2\sqrt{12}$$

However, this last expression is *not simplified*. There is a perfect square factor of 12, which is 4. We could continue on and remove it:

$$2\sqrt{12} = 2\sqrt{4 \cdot 3}$$
$$= 2\sqrt{4} \cdot \sqrt{3}$$
$$= 2 \cdot 2 \cdot \sqrt{3}$$
$$= 4\sqrt{3}$$

which is the same result as we got when we removed the *largest perfect square factor*, 16. Notice how much easier it is to remove the largest perfect square factor first.   ◄

Cube roots can also be simplified using the Multiplication Property of Radicals. Instead of looking for perfect squares as factors of the radicand, we look for perfect cube factors.

► **EXAMPLE 2**

Simplify each cube root.

(a) $\sqrt[3]{24}$ (b) $\sqrt[3]{-54}$

Solutions

(a) The largest *perfect cube* factor of 24 is 8.

$$\sqrt[3]{24} = \sqrt[3]{8 \cdot 3}$$
$$= \sqrt[3]{8} \cdot \sqrt[3]{3} \qquad \text{Multiplication Property}$$
$$= 2\sqrt[3]{3} \qquad \sqrt[3]{8} = 2$$

(b) Because the index, 3, is *odd*, the negative radicand is permitted. However, as 27 and -27 are both perfect cube factors of -54, we have a choice to make.

$$\sqrt[3]{-54} = \sqrt[3]{(27)(-2)} \quad \text{or} \quad \sqrt[3]{-54} = \sqrt[3]{(-27)(2)}$$

In this situation, we choose to remove the negative perfect cube factor.

$$\sqrt[3]{-54} = \sqrt[3]{-27 \cdot 2}$$
$$= \sqrt[3]{-27} \cdot \sqrt[3]{2} \qquad \text{Multiplication Property}$$
$$= -3\sqrt[3]{2} \qquad \sqrt[3]{-27} = -3$$

◄

► **EXAMPLE 3**

Simplify each radical. Assume that all variables represent nonnegative real numbers.

(a) $\sqrt{8x^2}$ (b) $\sqrt[3]{16y^6}$ (c) $\sqrt[3]{-16x^3}$

Solutions

(a) $\sqrt{8x^2} = \sqrt{4 \cdot 2 \cdot x^2}$
$$= \sqrt{4 \cdot x^2 \cdot 2} \qquad \text{Commutative Property}$$
$$= \sqrt{4} \cdot \sqrt{x^2} \cdot \sqrt{2} \qquad \text{Multiplication Property}$$
$$= 2x\sqrt{2}$$

(b) $\sqrt[3]{16y^6} = \sqrt[3]{8 \cdot 2y^6}$
$$= \sqrt[3]{8 \cdot y^6 \cdot 2} \qquad \text{Commutative Property}$$
$$= \sqrt[3]{8} \cdot \sqrt[3]{y^6} \cdot \sqrt[3]{2} \qquad \text{Multiplication Property}$$
$$= 2y^2\sqrt[3]{2}$$

(c) $\sqrt[3]{-16x^3} = \sqrt[3]{-8 \cdot 2x^3}$
$$= \sqrt[3]{-8 \cdot x^3 \cdot 2}$$
$$= \sqrt[3]{-8} \cdot \sqrt[3]{x^3} \cdot \sqrt[3]{2}$$
$$= -2x\sqrt[3]{2}$$

Again, to simplify a root of *odd* index and negative radicand, we remove a negative factor from the radicand.   ◄

Let's examine the quotient $\sqrt{\dfrac{36}{25}}$. Because $\left(\dfrac{6}{5}\right)^2 = \dfrac{36}{25}$, $\sqrt{\dfrac{36}{25}} = \dfrac{6}{5}$.

However, $\dfrac{6}{5}$ also equals $\dfrac{\sqrt{36}}{\sqrt{25}}$. So, we can write $\sqrt{\dfrac{36}{25}} = \dfrac{\sqrt{36}}{\sqrt{25}}$. This is the idea behind the Division Property of Radicals.

Division Property of Radicals

If $\sqrt[n]{p}$ and $\sqrt[n]{q}$ exist as real numbers and $q \neq 0$, then

$$\sqrt[n]{\dfrac{p}{q}} = \dfrac{\sqrt[n]{p}}{\sqrt[n]{q}}$$

The *n*th root of a quotient is the quotient of *n*th roots.

► **EXAMPLE 4**

Simplify each radical. Assume that all variables represent nonnegative real numbers.

(a) $\sqrt{\dfrac{7}{4}}$ (b) $\sqrt{\dfrac{8}{25y^2}}$ (c) $\sqrt[3]{\dfrac{5}{8}}$ (d) $\sqrt[3]{\dfrac{16x^3}{27}}$

Solutions

(a) $\sqrt{\dfrac{7}{4}} = \dfrac{\sqrt{7}}{\sqrt{4}}$ Division Property

       $= \dfrac{\sqrt{7}}{2}$ $\sqrt{4} = 2$

(b) $\sqrt{\dfrac{8}{25y^2}} = \dfrac{\sqrt{8}}{\sqrt{25y^2}}$ Division Property

       $= \dfrac{\sqrt{4 \cdot 2}}{5y}$ $\sqrt{25y^2} = 5y$ and $8 = 4 \cdot 2$

       $= \dfrac{\sqrt{4}\sqrt{2}}{5y}$ Multiplication Property

       $= \dfrac{2\sqrt{2}}{5y}$

       $= \dfrac{2}{5y}\sqrt{2}$

(c) $\sqrt[3]{\dfrac{5}{8}} = \dfrac{\sqrt[3]{5}}{\sqrt[3]{8}}$ Division Property

       $= \dfrac{\sqrt[3]{5}}{2}$ $\sqrt[3]{8} = 2$

(d) $\sqrt[3]{\dfrac{16x^3}{27}} = \dfrac{\sqrt[3]{16x^3}}{\sqrt[3]{27}}$ Division Property

       $= \dfrac{\sqrt[3]{8 \cdot 2 \cdot x^3}}{3}$

       $= \dfrac{\sqrt[3]{8}\sqrt[3]{2}\sqrt[3]{x^3}}{3}$ Multiplication Property

       $= \dfrac{2\sqrt[3]{2}x}{3}$

       $= \dfrac{2}{3}\sqrt[3]{2}\,x$ ◄

Procedure for Simplifying a Square Root or a Cube Root

1. If the radicand contains a fraction, use the Quotient Property to write the radical as the quotient of two radicals.

2. Remove all perfect square factors from any square roots. Remove perfect cube factors from cube roots.

3. If cube roots contain a negative radicand, be sure to remove a negative factor.

► **EXAMPLE 5**

Reduce each expression to lowest terms.

(a) $\dfrac{6\sqrt{5}}{15}$ (b) $\dfrac{\sqrt{24}}{12}$

Solutions

(a) This expression contains a radical, but the radical cannot be simplified further. However, 3 is a common factor of both the numerator and denominator of this fraction, so the fraction can be reduced.

$$\frac{6\sqrt{5}}{15} = \frac{2\sqrt{5}}{5}$$

This expression cannot be reduced further. The 5 in the denominator and the $\sqrt{5}$ in the numerator *are not* common factors!

(b) Although 12 *is* a factor of 24, it is *not* a factor of $\sqrt{24}$. However, $\sqrt{24}$ can be simplified.

$$\frac{\sqrt{24}}{12} = \frac{\sqrt{4 \cdot 6}}{12}$$

$$= \frac{\sqrt{4} \cdot \sqrt{6}}{12} \qquad \text{Multiplication Property}$$

$$= \frac{2\sqrt{6}}{12}$$

$$= \frac{\sqrt{6}}{6} \qquad \text{Common factor of 2}$$

As before, 6 and $\sqrt{6}$ *do not* have any common factors. This fraction *cannot be* reduced further.   ◄

Problem Set 8.2

Warm-Ups

In Problems 1 through 5, simplify each radical. See Example 1.

1. $\sqrt{8}$ 2. $\sqrt{12}$ 3. $\sqrt{50}$

4. $\sqrt{200}$ 5. $\sqrt{40}$

In Problems 6 through 9, simplify. See Example 2.

6. $\sqrt[3]{24}$ 7. $\sqrt[3]{54}$ 8. $\sqrt[3]{-32}$ 9. $\sqrt[3]{40}$

In Problems 10 through 13, simplify. Assume that x and y represent nonnegative real numbers. See Example 3.

10. $\sqrt{8x^4}$ 11. $\sqrt[3]{16y^3}$ 12. $\sqrt{20x^6}$ 13. $\sqrt[3]{-24x^6}$

In Problems 14 through 17, simplify. Assume that a and x represent positive numbers. See Example 4.

14. $\sqrt{\dfrac{3}{16}}$

15. $\sqrt{\dfrac{6}{49}}$

16. $\sqrt{\dfrac{12x^2}{25}}$

17. $\sqrt{\dfrac{27}{4a^2}}$

In Problems 18 through 21, reduce each expression to lowest terms. See Example 5.

18. $\dfrac{8\sqrt{3}}{24}$

19. $\dfrac{7\sqrt{5}}{35}$

20. $\dfrac{\sqrt{48}}{24}$

21. $\dfrac{\sqrt{18}}{9}$

Practice Exercises

In Problems 22 through 69, simplify each radical. Assume that all variables represent nonnegative real numbers.

22. $\sqrt{88}$

23. $\sqrt{125}$

24. $\sqrt{500}$

25. $\sqrt{600}$

26. $\sqrt{44}$

27. $\sqrt{108}$

28. $\sqrt{52}$

29. $\sqrt{68}$

30. $\sqrt{360}$

31. $\sqrt{99}$

32. $\sqrt{20}$

33. $\sqrt{112}$

34. $\sqrt{180}$

35. $\sqrt{300}$

36. $\sqrt{162}$

37. $\sqrt{32a^2b^3}$

38. $\sqrt{\dfrac{3}{16}}$

39. $\sqrt{\dfrac{3a}{16b^2}}$

40. $\sqrt{\dfrac{8b^4}{25}}$

41. $\sqrt{\dfrac{27}{49x^2}}$

42. $\sqrt[3]{-48}$

43. $\sqrt[3]{250}$

44. $\sqrt[3]{-56}$

45. $\sqrt[3]{-16}$

46. $\sqrt[3]{\dfrac{4}{125}}$

47. $\sqrt[3]{\dfrac{5}{8}}$

48. $\sqrt[3]{\dfrac{24}{125}}$

49. $\sqrt[3]{\dfrac{32}{27}}$

50. $\sqrt[3]{72}$

51. $\sqrt[3]{108}$

52. $\sqrt[3]{-96}$

53. $\sqrt[3]{56}$

54. $\sqrt[3]{-135}$

55. $\sqrt[3]{500}$

56. $\sqrt[3]{-168}$

57. $\sqrt[3]{-144}$

58. $\sqrt[3]{-250}$

59. $\sqrt[3]{160}$

60. $\sqrt[3]{-88}$

61. $\sqrt[3]{-104}$

62. $\sqrt{27x^2}$

63. $\sqrt[3]{54y^6}$

64. $\sqrt{40x^4}$

65. $\sqrt[3]{-32x^3}$

66. $\sqrt{24x^4y^2}$

67. $\sqrt[3]{72x^6y^3}$

68. $\sqrt{20x^4y^6}$

69. $\sqrt[3]{-8t^{12}}$

70. The baseball "diamond" at Candlestick Park is not a diamond at all but a square with the bases at the corners. The distance between the bases along the base path is 90 ft. The length of the hypotenuse of a right triangle is given by the formula $c = \sqrt{a^2 + b^2}$, where c is the length of the hypotenuse and a and b are the lengths of the other two sides. How far is it from home plate to second base?

71. The navigator of a ship can find the distance to the horizon in miles with the formula $d = \sqrt{1.5h}$, where h is the height of the ship's bridge in feet. How far away is the horizon if the height of the ship's bridge is 150 ft?

72. Weather forecasters estimate the duration of some storms with the formula $t = \sqrt{\dfrac{d^3}{216}}$, where t is the duration of the storm in hours and d is the diameter of the storm in miles. Find the approximate duration of a storm of this type whose diameter is 8 miles.

Challenge Problems

Simplify each radical.

73. $\sqrt[4]{32}$

74. $\sqrt[5]{64}$

75. $\sqrt[6]{128}$

76. $\sqrt{256}$

In Your Own Words

77. What do we mean by a *simplified* radical expression?

78. Describe the steps in simplifying $\dfrac{\sqrt{75}}{15}$.

8.3   Addition and Subtraction of Radicals

TAPE 13

Since radicals represent real numbers, we can add and subtract them. We know that

$$2x + 3x = 5x$$

because we can add *like* terms, but

$$2x + 3y$$

cannot be simplified because $2x$ and $3y$ are *not* like terms.

Addition of radicals is performed in the same way. For example,

$$2\sqrt{11} + 3\sqrt{11} = 5\sqrt{11}$$

because $2\sqrt{11}$ and $3\sqrt{11}$ are *like* radicals. However,

$$2\sqrt{5} + 3\sqrt{11}$$

cannot be simplified because $2\sqrt{5}$ and $3\sqrt{11}$ are *not* like radicals.

Likewise, $3\sqrt[3]{2} + 4\sqrt[3]{5}$ cannot be simplified because $3\sqrt[3]{2}$ and $4\sqrt[3]{5}$ are *not* like radicals. **Like radicals** are radicals with the same index and the same radicand.

To add or subtract radicals, combine like terms.

► EXAMPLE 1

Perform the operations indicated.

(a) $\sqrt{3} + \sqrt{3}$ (b) $3\sqrt{2} - \sqrt{2}$ (c) $8\sqrt[3]{7} + \sqrt[3]{7} - 4\sqrt[3]{7}$

Solutions

(a) $\sqrt{3} + \sqrt{3} = 2\sqrt{3}$

(b) $3\sqrt{2} - \sqrt{2} = 2\sqrt{2}$

(c) $8\sqrt[3]{7} + \sqrt[3]{7} - 4\sqrt[3]{7} = 5\sqrt[3]{7}$ ◄

Radicals that cannot be combined because they have different radicands sometimes can be simplified to a form in which they can be combined.

► **EXAMPLE 2**

Perform the operations indicated.

(a) $\sqrt{8} + \sqrt{18}$ (b) $\sqrt[3]{16} - \sqrt[3]{2}$ (c) $\sqrt{12} - \sqrt{48} + \sqrt{75}$

Solutions

(a) Because the radicands are different, these are not like radicals and cannot be combined as they are. However, they both can be simplified.

$$\sqrt{8} + \sqrt{18} = \sqrt{4 \cdot 2} + \sqrt{9 \cdot 2}$$
$$= 2\sqrt{2} + 3\sqrt{2}$$

Now we *do* have like terms.

$$= 5\sqrt{2}$$

(b) $\sqrt[3]{16} - \sqrt[3]{2} = \sqrt[3]{8 \cdot 2} - \sqrt[3]{2}$
$$= 2\sqrt[3]{2} - \sqrt[3]{2}$$
$$= \sqrt[3]{2}$$

(c) $\sqrt{12} - \sqrt{48} + \sqrt{75} = \sqrt{4 \cdot 3} - \sqrt{16 \cdot 3} + \sqrt{25 \cdot 3}$
$$= 2\sqrt{3} - 4\sqrt{3} + 5\sqrt{3}$$
$$= 3\sqrt{3}$$ ◄

► **EXAMPLE 3**

Perform the operations indicated. Assume that all variables represent nonnegative real numbers.

(a) $\sqrt{12x} + \sqrt{27x}$ (b) $\sqrt[3]{16y} - \sqrt[3]{54y}$ (c) $\sqrt{4x} - \sqrt{9x}$
(d) $\sqrt[3]{-8y} - \sqrt[3]{27y}$ (e) $\sqrt{8x^2} + \sqrt{18x^2}$

Solutions

(a) $\sqrt{12x} + \sqrt{27x} = \sqrt{4 \cdot 3x} + \sqrt{9 \cdot 3x}$
$$= 2\sqrt{3x} + 3\sqrt{3x}$$
$$= 5\sqrt{3x}$$

(b) $\sqrt[3]{16y} - \sqrt[3]{54y} = \sqrt[3]{8 \cdot 2y} - \sqrt[3]{27 \cdot 2y}$
$$= 2\sqrt[3]{2y} - 3\sqrt[3]{2y}$$
$$= -\sqrt[3]{2y}$$

(c) $\sqrt{4x} - \sqrt{9x} = 2\sqrt{x} - 3\sqrt{x}$
$$= -\sqrt{x}$$

(d) $\sqrt[3]{-8y} - \sqrt[3]{27y} = -2\sqrt[3]{y} - 3\sqrt[3]{y}$
$$= -5\sqrt[3]{y}$$

(e) $\sqrt{8x^2} + \sqrt{18x^2} = \sqrt{4x^2 \cdot 2} + \sqrt{9x^2 \cdot 2}$
$$= 2x\sqrt{2} + 3x\sqrt{2}$$
$$= 5x\sqrt{2}$$ ◄

Problem Set 8.3

Warm-Ups

In Problems 1 through 6, perform the operations indicated and simplify. Assume that all variables represent nonnegative real numbers. See Example 1.

1. $3\sqrt{5} + 6\sqrt{5}$
2. $7\sqrt{x} - 2\sqrt{x}$
3. $4\sqrt[3]{6} + 7\sqrt[3]{6}$
4. $\sqrt[3]{3} - 3\sqrt[3]{3}$
5. $2\sqrt{3} + 4\sqrt{3} - 7\sqrt{3}$
6. $5\sqrt[3]{7} - 6\sqrt[3]{7} - 3\sqrt[3]{7}$

In Problems 7 through 12, perform the operations indicated. See Example 2.

7. $\sqrt{8} + \sqrt{2}$
8. $\sqrt{24} - \sqrt{54}$
9. $\sqrt{48} - 2\sqrt{12}$
10. $2\sqrt{50} + \sqrt{8} + \sqrt{18}$
11. $3\sqrt{20} - 2\sqrt{45} + \sqrt{80}$
12. $\sqrt[3]{16} - \sqrt[3]{54} + \sqrt[3]{-128}$

In Problems 13 through 18, perform the indicated operations. Assume that x is a nonnegative number. See Example 3.

13. $\sqrt{8x} + \sqrt{2x}$
14. $\sqrt{48x^2} - 3\sqrt{12x^2}$
15. $\sqrt{4x} - 2\sqrt{16x}$
16. $5\sqrt{20x^2} - 3\sqrt{80x^2}$
17. $\sqrt[3]{-54x^3} - 3\sqrt[3]{-16x^3}$
18. $4\sqrt[3]{250x^3} - \sqrt[3]{-16x^3}$

Practice Exercises

In Problems 19 through 50, perform the operations indicated and simplify. Assume that all variables represent nonnegative real numbers.

19. $2\sqrt{3} + \sqrt{3}$
20. $5\sqrt{6} - 8\sqrt{6}$
21. $3\sqrt[3]{11} + 5\sqrt[3]{11}$
22. $\sqrt[3]{y} - 3\sqrt[3]{y}$
23. $6\sqrt{5} + \sqrt{5} - 2\sqrt{5}$
24. $8\sqrt[3]{4} - 3\sqrt[3]{4} - 2\sqrt[3]{4}$
25. $\sqrt{18} + \sqrt{2}$
26. $\sqrt{12} + \sqrt{27}$
27. $\sqrt{75} - 2\sqrt{12}$
28. $3\sqrt{72} + \sqrt{8} + \sqrt{32}$
29. $3\sqrt{45} - 3\sqrt{20} + \sqrt{5}$
30. $\sqrt{200} - 2\sqrt{8} + 3\sqrt{50}$
31. $2\sqrt{18} - \sqrt{98} - \sqrt{8}$
32. $\sqrt[3]{54} - 3\sqrt[3]{2}$
33. $\sqrt{25x} + 3\sqrt{36x}$
34. $3\sqrt{72x} - \sqrt{8x}$
35. $\sqrt[3]{16x} - 2\sqrt[3]{54x}$
36. $\sqrt[3]{-8x} + 2\sqrt[3]{27x}$
37. $\sqrt[3]{-16} + 5\sqrt[3]{-54}$
38. $4\sqrt[3]{2} - 3\sqrt[3]{-16}$
39. $\sqrt[3]{-16} - \sqrt[3]{54} - \sqrt[3]{-128}$
40. $\sqrt[3]{250} - 4\sqrt[3]{16} - 3\sqrt[3]{54}$
41. $\sqrt{20x} + \sqrt{45x}$
42. $\sqrt{24x^2} - 3\sqrt{54x^2}$
43. $\sqrt{25x} - 2\sqrt{36x}$
44. $5\sqrt{12x^2} - 3\sqrt{27x^2}$
45. $\sqrt{64x} + 3\sqrt{49x}$
46. $3\sqrt{98x} - 2\sqrt{8x}$
47. $\sqrt[3]{24x} - 2\sqrt[3]{81x}$
48. $\sqrt[3]{-27x} + 2\sqrt[3]{8x}$
49. $\sqrt[3]{-81x^3} - 3\sqrt[3]{-24x^3}$
50. $4\sqrt[3]{250x^3} + \sqrt[3]{16x^3}$

In Your Own Words

51. When can expressions containing radicals be combined?

8.4   Multiplication and Division of Radicals

TAPE 14

The Multiplication Property of Radicals that we studied in Section 8.2 tells us how to multiply radicals with the *same* index.

> ### Multiplication Property of Radicals
>
> If $\sqrt[n]{p}$ and $\sqrt[n]{q}$ exist as real numbers, then
>
> $$\sqrt[n]{p} \cdot \sqrt[n]{q} = \sqrt[n]{pq}$$

A product of nth roots is the nth root of the product of the radicands. If the index is the same, we just multiply the radicands.

► **EXAMPLE 1**

Perform the multiplication indicated and simplify.

(a) $\sqrt{2} \cdot \sqrt{3}$ (b) $\sqrt{3} \cdot \sqrt{6}$ (c) $\sqrt[3]{4} \cdot \sqrt[3]{2}$

(d) $\sqrt[3]{-9} \cdot \sqrt[3]{6}$ (e) $2\sqrt{3} \cdot 5\sqrt{7}$ (f) $(5\sqrt{6})^2$

Solutions

(a) $\sqrt{2} \cdot \sqrt{3} = \sqrt{6}$ $\sqrt{6}$ cannot be simplified further

(b) $\sqrt{3} \cdot \sqrt{6} = \sqrt{18}$ $\sqrt{18}$ can be simplified further

     $\sqrt{18} = \sqrt{9 \cdot 2}$

          $= 3\sqrt{2}$ $\sqrt{9} = 3$

(c) $\sqrt[3]{4} \cdot \sqrt[3]{2} = \sqrt[3]{8}$

          $= 2$

(d) $\sqrt[3]{-9} \cdot \sqrt[3]{6} = \sqrt[3]{-54}$

             $= \sqrt[3]{-27 \cdot 2}$

             $= -3\sqrt[3]{2}$ $\sqrt[3]{-27} =$

(e) $2\sqrt{3} \cdot 5\sqrt{7} = 2 \cdot 5 \cdot \sqrt{3} \cdot \sqrt{7}$ Co

           $= 10\sqrt{21}$

(f) $(5\sqrt{6})^2 = 5^2(\sqrt{6})^2$ Pro

         $= 25 \cdot 6$

         $= 150$ ◄

The Distributive Property and the special ⌐⌐⌐ ⌐⌐⌐ ⌐⌐⌐ ⌐⌐⌐ that we learned when we studied polynomials tell us how to perform other multiplications.

► **EXAMPLE 2**

Perform the multiplications indicated and simplify.

(a) $\sqrt{3}(\sqrt{5} + \sqrt{3})$

(b) $(2\sqrt{7} + \sqrt{3})(2\sqrt{7} - \sqrt{3})$

(c) $(2\sqrt{3} + \sqrt{5})^2$

Solutions

(a) Use the Distributive Property.

$$\sqrt{3}(\sqrt{5} + \sqrt{3}) = \sqrt{3} \cdot \sqrt{5} + \sqrt{3} \cdot \sqrt{3}$$
$$= \sqrt{15} + 3$$

(b) Use the special product *difference of two squares:*

$$(p + q)(p - q) = p^2 - q^2$$
$$(2\sqrt{7} + \sqrt{3})(2\sqrt{7} - \sqrt{3}) = (2\sqrt{7})^2 - (\sqrt{3})^2$$
$$= 2^2 \cdot (\sqrt{7})^2 - (\sqrt{3})^2$$
$$= 4 \cdot 7 - 3$$
$$= 28 - 3$$
$$= 25$$

(c) Use the special product *square of a binomial:*

$$(p + q)^2 = p^2 + 2pq + q^2$$
$$(2\sqrt{3} + \sqrt{5})^2 = (2\sqrt{3})^2 + 2 \cdot 2\sqrt{3} \cdot \sqrt{5} + (\sqrt{5})^2$$
$$= 4 \cdot 3 + 4\sqrt{15} + 5$$
$$= 12 + 4\sqrt{15} + 5$$
$$= 17 + 4\sqrt{15}$$ ◄

To divide radicals with the same index, we use the Division Property of Radicals.

Division Property of Radicals

If $\sqrt[n]{p}$ and $\sqrt[n]{q}$ exist as real numbers and $q \neq 0$, then

$$\frac{\sqrt[n]{p}}{\sqrt[n]{q}} = \sqrt[n]{\frac{p}{q}}$$

A quotient of nth roots is the nth root of the quotient of the radicands. If the index is the same, we just divide the radicands.

▶ **EXAMPLE 3**

Perform the divisions indicated and simplify. Assume that all variables represent positive real numbers.

(a) $\dfrac{\sqrt{4}}{\sqrt{2}}$ (b) $\dfrac{\sqrt{16a}}{\sqrt{2a}}$ (c) $\dfrac{\sqrt[3]{8y^4}}{\sqrt[3]{2y}}$ (d) $\dfrac{\sqrt[3]{-54}}{\sqrt[3]{2}}$

Solutions

(a) $\dfrac{\sqrt{4}}{\sqrt{2}} = \sqrt{\dfrac{4}{2}}$ Division Property

$\phantom{(a)\ \dfrac{\sqrt{4}}{\sqrt{2}}} = \sqrt{2}$

(b) $\dfrac{\sqrt{16a}}{\sqrt{2a}} = \sqrt{8}$ Division Property

$\phantom{(b)\ \dfrac{\sqrt{16a}}{\sqrt{2a}}} = \sqrt{4 \cdot 2}$

$\phantom{(b)\ \dfrac{\sqrt{16a}}{\sqrt{2a}}} = 2\sqrt{2}$

(c) $\dfrac{\sqrt[3]{8y^4}}{\sqrt[3]{2y}} = \sqrt[3]{4y^3}$ Division Property

$\phantom{(c)\ \dfrac{\sqrt[3]{8y^4}}{\sqrt[3]{2y}}} = \sqrt[3]{4} \cdot \sqrt[3]{y^3}$ Multiplication Property

$\phantom{(c)\ \dfrac{\sqrt[3]{8y^4}}{\sqrt[3]{2y}}} = \sqrt[3]{4}\, y$

(d) $\dfrac{\sqrt[3]{-54}}{\sqrt[3]{2}} = \sqrt[3]{-27}$ Division Property

$\phantom{(d)\ \dfrac{\sqrt[3]{-54}}{\sqrt[3]{2}}} = -3$ ◀

If we divide $\sqrt{5}$ by $\sqrt{3}$, we get $\dfrac{\sqrt{5}}{\sqrt{3}}$, which cannot be simplified using the procedure of Section 8.2. However, we do not consider an expression simplified if it contains a radical in the denominator. We can use the Fundamental Principle of Fractions to write the expression without a fraction in the radicand and without a radical in the denominator.

Notice what happens if we use the Fundamental Principle of Fractions and multiply the numerator and denominator by $\sqrt{3}$.

$$\frac{\sqrt{5}}{\sqrt{3}} = \frac{\sqrt{5} \cdot \sqrt{3}}{\sqrt{3} \cdot \sqrt{3}}$$

$$= \frac{\sqrt{15}}{3}$$

$$= \frac{1}{3}\sqrt{15}$$

This procedure is called **rationalizing the denominator.** Notice that the radicand is not a fraction and the denominator does not contain a radical.

If the denominator contains a square root as a factor, the denominator can be rationalized by multiplying both the numerator and denominator by the square root factor.

▶ **EXAMPLE 4**

Rationalize the denominator and simplify. Assume that all variables represent nonnegative real numbers.

(a) $\dfrac{\sqrt{3}}{\sqrt{5}}$ (b) $\dfrac{\sqrt{2t}}{\sqrt{2}}$ (c) $\dfrac{6}{5\sqrt{2}}$

Solutions

(a) $\dfrac{\sqrt{3}}{\sqrt{5}} = \dfrac{\sqrt{3} \cdot \sqrt{5}}{\sqrt{5} \cdot \sqrt{5}}$ (b) $\dfrac{\sqrt{2t}}{\sqrt{2}} = \dfrac{\sqrt{2t} \cdot \sqrt{2}}{\sqrt{2} \cdot \sqrt{2}}$ (c) $\dfrac{6}{5\sqrt{2}} = \dfrac{6 \cdot \sqrt{2}}{5\sqrt{2} \cdot \sqrt{2}}$

$\qquad\quad = \dfrac{\sqrt{15}}{5}$ $\qquad\qquad\qquad = \dfrac{\sqrt{4t}}{2}$ $\qquad\qquad\qquad = \dfrac{6\sqrt{2}}{5 \cdot 2}$

$\qquad\qquad\qquad\qquad\qquad\qquad = \dfrac{2\sqrt{t}}{2}$ $\qquad\qquad\qquad = \dfrac{3\sqrt{2}}{5}$

$\qquad\qquad\qquad\qquad\qquad\qquad = \sqrt{t}$ $\qquad\qquad\qquad$ or $\dfrac{3}{5}\sqrt{2}$

Notice in (c) that it was not necessary to multiply by $5\sqrt{2}$.   ◀

If the denominator contains a square root *term*, then we use the special product *difference of two squares*,

$$(p + q)(p - q) = p^2 - q^2$$

to rationalize the denominator.

▶ **EXAMPLE 5**

Rationalize the denominator and simplify. Assume that all variables represent nonnegative real numbers.

(a) $\dfrac{3}{\sqrt{2} - 5}$ (b) $\dfrac{\sqrt{2} - \sqrt{3}}{\sqrt{2} + \sqrt{3}}$ (c) $\dfrac{\sqrt{y}}{2\sqrt{y} - 1}$

Solutions

(a) We multiply the numerator and denominator by $\sqrt{2} + 5$ to set up the special product $(p - q)(p + q)$ in the denominator.

$$\dfrac{3}{\sqrt{2} - 5} = \dfrac{3\ (\sqrt{2} + 5)}{(\sqrt{2} - 5)\ (\sqrt{2} + 5)}$$

$$= \dfrac{3\sqrt{2} + 15}{(\sqrt{2})^2 - 5^2}$$

$$= \dfrac{3\sqrt{2} + 15}{2 - 25}$$

$$= \dfrac{3\sqrt{2} + 15}{-23} = -\dfrac{3\sqrt{2} + 15}{23}$$

(b) Here we multiply the numerator and the denominator by $\sqrt{2} - \sqrt{3}$ to set up the difference of two squares.

$$\frac{\sqrt{2} - \sqrt{3}}{\sqrt{2} + \sqrt{3}} = \frac{(\sqrt{2} - \sqrt{3})}{(\sqrt{2} + \sqrt{3})} \frac{(\sqrt{2} - \sqrt{3})}{(\sqrt{2} - \sqrt{3})}$$

$$= \frac{(\sqrt{2})^2 - 2\sqrt{2}\sqrt{3} + (\sqrt{3})^2}{(\sqrt{2})^2 - (\sqrt{3})^2}$$

$$= \frac{2 - 2\sqrt{6} + 3}{2 - 3}$$

$$= \frac{5 - 2\sqrt{6}}{-1}$$

$$= -5 + 2\sqrt{6}$$

(c) We multiply the numerator and denominator by $2\sqrt{y} + 1$.

$$\frac{\sqrt{y}}{2\sqrt{y} - 1} = \frac{\sqrt{y}\ (2\sqrt{y} + 1)}{(2\sqrt{y} - 1)\ (2\sqrt{y} + 1)}$$

$$= \frac{\sqrt{y} \cdot 2\sqrt{y} + \sqrt{y}}{(2\sqrt{y})^2 - 1^2}$$

$$= \frac{2y + \sqrt{y}}{4y - 1} \qquad \blacktriangleleft$$

Often it is convenient to rationalize a *numerator*. The procedure is the same except that the numerator is the focus of our attention.

► **EXAMPLE 6**

Rationalize the numerator in each expression.

(a) $\dfrac{3\sqrt{6}}{2a^2}$ (b) $\dfrac{2 - \sqrt{5}}{2 + \sqrt{5}}$

Solutions

(a) $\dfrac{3\sqrt{6} \cdot \sqrt{6}}{2a^2 \cdot \sqrt{6}} = \dfrac{3(\sqrt{6})^2}{2a^2\sqrt{6}}$ Multiply numerator and denominator by $\sqrt{6}$

$= \dfrac{3 \cdot 6}{2a^2\sqrt{6}}$

$= \dfrac{18}{2a^2\sqrt{6}}$

$= \dfrac{9}{a^2\sqrt{6}}$

(b) $\dfrac{2 - \sqrt{5}}{2 + \sqrt{5}} = \dfrac{(2 - \sqrt{5})\ (2 + \sqrt{5})}{(2 + \sqrt{5})\ (2 + \sqrt{5})}$ Multiply the numerator and denominator by $2 + \sqrt{5}$

$$= \frac{2^2 - (\sqrt{5})^2}{(2 + \sqrt{5})^2}$$

$$= \frac{4 - 5}{4 + 4\sqrt{5} + 5}$$

$$= \frac{-1}{9 + 4\sqrt{5}}$$ ◀

An application of radicals arises in the study of right triangles. Two special triangles are very common. The first is an isosceles right triangle. The legs are each are the same length and the angles are 45°, 45°, and 90°. In such a triangle, the length of the hypotenuse is $\sqrt{2}$ times the length of a leg. Another special right triangle occurs when the length of the hypotenuse is exactly twice the length of the shorter leg of the triangle. The angles in such a triangle are 30°, 60°, and 90°.

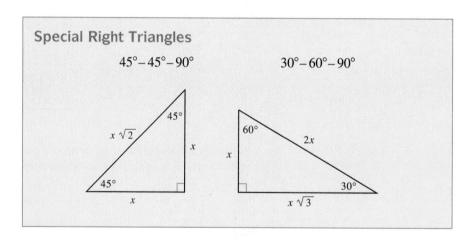

Special Right Triangles

45°–45°–90°   30°–60°–90°

► **EXAMPLE 7**

Given that the hypotenuse of a 45°–45°–90° triangle is 4 units long, what is the length of the legs?

Solution

Let the length of the legs be x units.

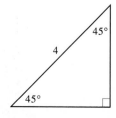

From the figure, we see that, since the hypotenuse is 4,

$$x\sqrt{2} = 4$$

$$x = \frac{4}{\sqrt{2}}$$

$$= \frac{4 \cdot \sqrt{2}}{\sqrt{2} \cdot \sqrt{2}}$$

$$= \frac{4\sqrt{2}}{2}$$

$$= 2\sqrt{2}$$

The legs are $2\sqrt{2}$ units long.   ◄

► EXAMPLE 8

If the shortest side of a 30°–60°–90° triangle is $4\sqrt{2}$ ft long, what are the lengths of the other two sides?

Solution

Let x be the length of the shorter leg.

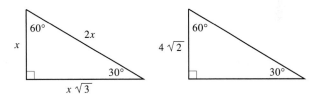

From our special 30°–60°–90° relationships, we see that

$$x = 4\sqrt{2}$$

The hypotenuse is $2x$. So

$$2x = 8\sqrt{2}$$

The other leg is $x\sqrt{3}$. Thus

$$x\sqrt{3} = 4\sqrt{2} \cdot \sqrt{3}$$

$$= 4\sqrt{6}$$

The lengths of the other sides are $8\sqrt{2}$ ft and $4\sqrt{6}$ ft.   ◄

Problem Set 8.4

Warm-Ups

In Problems 1 through 9, perform the operations indicated and simplify. Assume that all variables represent nonnegative real numbers. See Example 1.

1. $\sqrt{5} \cdot \sqrt{7}$

2. $\sqrt{3} \cdot \sqrt{11}$

3. $\sqrt[3]{3} \cdot \sqrt[3]{5}$

4. $\sqrt[3]{-2} \cdot \sqrt[3]{4}$

5. $2\sqrt{3} \cdot 4\sqrt{5}$

6. $3\sqrt{x} \cdot 2\sqrt{x}$

7. $4\sqrt{5} \cdot 3\sqrt{5}$

8. $(\sqrt{5})^2$

9. $(8\sqrt{x})^2$

In Problems 10 through 19, perform the operation indicated and simplify. See Example 2.

10. $\sqrt{3}(\sqrt{2} + \sqrt{5})$

11. $\sqrt{7}(\sqrt{6} - 1)$

12. $\sqrt{5}(\sqrt{5} + \sqrt{7})$

13. $\sqrt{t}(\sqrt{t} - \sqrt{y})$

14. $(2 + \sqrt{3})(5 - \sqrt{3})$

15. $(\sqrt{5} + 2\sqrt{2})(3\sqrt{2} - \sqrt{5})$

16. $(\sqrt{3} - 1)(\sqrt{3} + 1)$

17. $(\sqrt{5} - \sqrt{3})(\sqrt{5} + \sqrt{3})$

18. $(\sqrt{3} + 2)^2$

19. $(2\sqrt{6} + 5\sqrt{2})^2$

In Problems 20 through 27, perform the indicated divisions. Assume that a and x are positive numbers. See Example 3.

20. $\dfrac{\sqrt{6}}{\sqrt{3}}$

21. $\dfrac{\sqrt{10}}{\sqrt{2}}$

22. $\dfrac{\sqrt[3]{21}}{\sqrt[3]{7}}$

23. $\dfrac{\sqrt[3]{12}}{\sqrt[3]{2}}$

24. $\dfrac{\sqrt{32x}}{\sqrt{2x}}$

25. $\dfrac{\sqrt{45x}}{\sqrt{5x^3}}$

26. $\dfrac{\sqrt[3]{16a^4b^3}}{\sqrt[3]{2a}}$

27. $\dfrac{\sqrt[3]{81}}{\sqrt[3]{-3}}$

In Problems 28 through 33, rationalize the denominators. See Example 4.

28. $\dfrac{1}{\sqrt{3}}$

29. $\dfrac{6}{\sqrt{6}}$

30. $\dfrac{\sqrt{2} + 1}{\sqrt{2}}$

31. $\dfrac{\sqrt{3}}{4\sqrt{2}}$

32. $\dfrac{\sqrt{3} - \sqrt{2}}{2\sqrt{3}}$

33. $\dfrac{\sqrt{x} + \sqrt{y}}{\sqrt{x}}$

In Problems 34 through 38, rationalize the denominator. See Example 5.

34. $\dfrac{1}{\sqrt{2} + 3}$

35. $\dfrac{3}{\sqrt{2} - \sqrt{3}}$

36. $\dfrac{\sqrt{5}}{\sqrt{5} - 2\sqrt{3}}$

37. $\dfrac{\sqrt{3} + \sqrt{6}}{\sqrt{3} - \sqrt{6}}$

38. $\dfrac{3\sqrt{x} + 2\sqrt{y}}{3\sqrt{x} - 2\sqrt{y}}$

In Problems 39 and 40, rationalize the numerator. See Example 6.

39. $\dfrac{3\sqrt{10}}{2x}$

40. $\dfrac{3 + \sqrt{2}}{3 - \sqrt{2}}$

Practice Exercises

In Problems 41 through 91, perform the operations indicated and simplify. Assume that all variables represent nonnegative real numbers.

41. $\sqrt{3} \cdot \sqrt{5}$

42. $\sqrt{2} \cdot \sqrt{13}$

43. $\sqrt[3]{7} \cdot \sqrt[3]{4}$

44. $\sqrt[3]{9} \cdot \sqrt[3]{7}$

45. $\sqrt{3} \cdot \sqrt{6}$

46. $\sqrt{5} \cdot \sqrt{8}$

47. $\sqrt[3]{3} \cdot \sqrt[3]{9}$

48. $\sqrt[3]{-4} \cdot \sqrt[3]{4}$

49. $2\sqrt{3} \cdot 4\sqrt{2}$

50. $3\sqrt{3} \cdot 2\sqrt{2}$

51. $3\sqrt[3]{6} \cdot 2\sqrt[3]{3}$

52. $5\sqrt[3]{6} \cdot 2\sqrt[3]{2}$

53. $3\sqrt{3} \cdot 4\sqrt{8}$

54. $4\sqrt{8} \cdot 3\sqrt{2}$

55. $2\sqrt{y} \cdot 5\sqrt{y}$

56. $4\sqrt{2} \cdot 3\sqrt{2}$

57. $(\sqrt{7})^2$

58. $(\sqrt{x})^2$

59. $(3\sqrt{2})^2$

60. $(5\sqrt{2})^2$

61. $(6\sqrt{z})^2$

62. $\sqrt{2}(\sqrt{3} + \sqrt{7})$

63. $\sqrt{5}(\sqrt{3} - 1)$

64. $\sqrt{3}(\sqrt{3} + \sqrt{7})$

65. $\sqrt{x}(\sqrt{x} - \sqrt{y})$

66. $(\sqrt{3} + 2)(\sqrt{2} - 3)$

67. $(\sqrt{2} - \sqrt{5})(\sqrt{2} + 3\sqrt{5})$

68. $(3\sqrt{3} - \sqrt{5})(\sqrt{3} + \sqrt{5})$

69. $(\sqrt{x} - \sqrt{y})(\sqrt{x} - 3\sqrt{y})$

70. $(3\sqrt{6} - \sqrt{3})(\sqrt{2} + 2)$

71. $(\sqrt{2} - 1)(\sqrt{2} + 1)$

72. $(4\sqrt{7} - \sqrt{5})(2\sqrt{7} + \sqrt{5})$

73. $(\sqrt{6} - \sqrt{3})(\sqrt{6} + \sqrt{3})$

74. $(\sqrt{x} + 2\sqrt{y})(\sqrt{x} - 2\sqrt{y})$

75. $(\sqrt{2} + 3)^2$

76. $(\sqrt{5} - \sqrt{3})^2$

77. $(2\sqrt{2} + \sqrt{6})^2$

78. $(3\sqrt{3} - 4\sqrt{5})^2$

79. $(2\sqrt{10} + 3\sqrt{2})^2$

80. $\dfrac{\sqrt{15}}{\sqrt{3}}$

81. $\dfrac{\sqrt{8}}{\sqrt{2}}$

82. $\dfrac{\sqrt[3]{21}}{\sqrt[3]{3}}$

83. $\dfrac{\sqrt[3]{6}}{\sqrt[3]{2}}$

84. $\dfrac{\sqrt{80}}{\sqrt{5}}$

85. $\dfrac{\sqrt{75}}{\sqrt{3}}$

86. $\dfrac{\sqrt[3]{54}}{\sqrt[3]{-2}}$

87. $\dfrac{\sqrt[3]{-32}}{\sqrt[3]{4}}$

88. $\dfrac{\sqrt{24}}{\sqrt{2}}$

89. $\dfrac{\sqrt{54}}{\sqrt{3}}$

90. $\dfrac{\sqrt[3]{96}}{\sqrt[3]{6}}$

91. $\dfrac{\sqrt[3]{108}}{\sqrt[3]{-2}}$

In Problems 92 through 105, rationalize the denominators.

92. $\dfrac{1}{\sqrt{7}}$

93. $\dfrac{5}{\sqrt{5}}$

94. $\dfrac{\sqrt{3} + 1}{\sqrt{3}}$

95. $\dfrac{5}{2\sqrt{3}}$

96. $\dfrac{\sqrt{3}}{3\sqrt{5}}$

97. $\dfrac{\sqrt{5} - \sqrt{3}}{2\sqrt{3}}$

98. $\dfrac{w}{5\sqrt{w}}$

99. $\dfrac{\sqrt{x} - \sqrt{y}}{\sqrt{y}}$

100. $\dfrac{1}{\sqrt{3} + 2}$

101. $\dfrac{3}{\sqrt{2} + \sqrt{3}}$

102. $\dfrac{\sqrt{2}}{\sqrt{2} - 3\sqrt{3}}$

103. $\dfrac{\sqrt{2} + \sqrt{5}}{\sqrt{2} - \sqrt{5}}$

104. $\dfrac{\sqrt{7} - 2\sqrt{3}}{2\sqrt{7} + \sqrt{3}}$

105. $\dfrac{\sqrt{5} - \sqrt{6}}{\sqrt{5} + \sqrt{6}}$

In Problems 106 through 109, rationalize the numerator.

106. $\dfrac{\sqrt{15}\,p^2}{6}$

107. $\dfrac{3\sqrt{7z}}{2z}$

108. $\dfrac{\sqrt{x} - \sqrt{y}}{\sqrt{x} + \sqrt{y}}$

109. $\dfrac{\sqrt{s} + 2\sqrt{t}}{3\sqrt{s} - \sqrt{t}}$

In Problems 110 through 117, find the lengths of the sides.

110.

111.

112.

113.

114.

115.

116.

117.

In Your Own Words

115. State the Multiplication Property of Radicals.

116. State the Division Property of Radicals.

8.5   Equations Containing Square Roots

TAPE 14

In this section we solve equations containing square roots. Adding the same number to both sides of an equation or multiplying both sides of an equation by the same nonzero number will not lead us to the solution set. Somehow we must remove the square root from the equation. Recall from Section 8.1 that squaring a square root removes the radical. That is, if x is a nonnegative real number,

$$(\sqrt{x})^2 = x$$

This property of square roots along with the following result will allow us to solve equations with square roots.

> **Squaring Property**
>
> If $A = B$ has solutions, these solutions are found in the solutions of $A^2 = B^2$.

Let's solve the equation $\sqrt{x} = 2$. To remove the square root, we square both sides of the equation.

$$(\sqrt{x})^2 = 2^2$$

$$x = 4$$

The Squaring Property says that to find the solution set of the equation $\sqrt{x} = 2$ we must find the solution set of the squared equation. The solution of the squared equation is 4. This means that 4 is the only possible solution of $\sqrt{x} = 2$. We can determine if it is a solution only by checking 4 in the original equation, $\sqrt{x} = 2$.

To Check 4 in the equation $\sqrt{x} = 2$, we evaluate each side when x is 4.

$$\text{LS:} \quad \sqrt{x} = \sqrt{4} = \boxed{2}$$

$$\text{RS:} \quad \boxed{2}$$

Since the value we get on both sides is 2, we say that 4 checks. Thus 4 is in the solution set of $\sqrt{x} = 2$. There are no other possible solutions. The solution set is $\{4\}$.

> **A Procedure for Solving Equations Containing Square Roots**
>
> 1. Isolate a square root.
> 2. Square both sides.
> 3. Solve the resulting equation for all possible solutions.
> 4. Check each possible solution in the original equation.
> 5. Write the solution set.

► **EXAMPLE 1**

Solve $\sqrt{x + 1} = 5$.

Solution

STEP 1   The square root is already isolated.

$$\sqrt{x + 1} = 5$$

STEP 2   Square both sides.

$$(\sqrt{x + 1})^2 = 5^2$$

STEP 3   Solve the resulting equation.

$$x + 1 = 25 \qquad \text{Remember } (\sqrt{x + 1})^2 = x + 1$$

$$x = 24$$

STEP 4   Check 24 in the original equation.
Check 24 in $\sqrt{x + 1} = 5$.

$$\text{LS:} \quad \sqrt{x + 1} = \sqrt{24 + 1} = \sqrt{25} = \boxed{5}$$

$$\text{RS:} \quad \boxed{5}$$

24 checks and thus is in the solution set.

STEP 5   Write the solution set.

$$\{24\}$$ ◀

The next example shows that checking is necessary in solving equations with square roots.

► EXAMPLE 2

Solve $\sqrt{x - 3} + 3 = 0$.

Solution

$$\sqrt{x - 3} = -3 \qquad \text{Isolate a square root}$$
$$(\sqrt{x - 3})^2 = (-3)^2 \qquad \text{Square both sides}$$
$$x - 3 = 9$$
$$x = 12$$

Check 12 in $\sqrt{x - 3} + 3 = 0$.

$$\text{LS:} \quad \sqrt{x - 3} + 3 = \sqrt{12 - 3} + 3 = \sqrt{9} + 3 = 3 + 3 = \boxed{6}$$
$$\text{RS:} \quad \boxed{0}$$

The value on the left side is 6 and the value on the right side is 0, so 12 does *not* check. This means that 12 is *not* in the solution set. However, since 12 is the only possible solution, the solution set must be the empty set.

$$\varnothing$$ ◀

We could have noticed that the solution set would be the empty set because the square root of a real number is never negative.

► EXAMPLE 3

Solve $\sqrt{2x} - \sqrt{x + 2} = 0$.

Solution

$$\sqrt{2x} = \sqrt{x + 2} \qquad \text{Isolate a square root}$$
$$(\sqrt{2x})^2 = (\sqrt{x + 2})^2 \qquad \text{Square both sides}$$
$$2x = x + 2$$
$$x = 2$$

Check 2 in $\sqrt{2x} - \sqrt{x + 2} = 0$.

$$\text{LS:} \quad \sqrt{2x} - \sqrt{x + 2} = \sqrt{2(2)} - \sqrt{2 + 2}$$
$$= \sqrt{4} - \sqrt{4} = 2 - 2 = \boxed{0}$$

$$\text{RS:} \quad \boxed{0}$$

Thus 2 checks.

$$\{2\} \qquad \blacktriangleleft$$

The equation that results from squaring an equation with square roots may require factoring and the Property of Zero Products that we studied in Section 4.5.

► **EXAMPLE 4**

Solve $\sqrt{x + 3} + 3 = x$.

Solution

$$\sqrt{x + 3} = x - 3 \qquad \text{Isolate the radical}$$
$$(\sqrt{x + 3})^2 = (x - 3)^2 \qquad \text{Square both sides}$$

Be Careful!   The next step requires us to square a binomial. Remember that $(x - 3)^2 = x^2 - 6x + 9$. Many mistakes are made in finding the middle term. Thus we obtain

$$x + 3 = x^2 - 6x + 9$$

We make the left side 0 and then solve by factoring.

$$0 = x^2 - 7x + 6$$
$$0 = (x - 6)(x - 1) \qquad \text{Factor}$$
$$x = 6 \quad \text{or} \quad x = 1 \qquad \text{Property of Zero Products}$$

There are two possible solutions to our original equation. We *must* check *both* of them.

Check 6 in $\sqrt{x + 3} + 3 = x$.

$$\text{LS:} \quad \sqrt{x + 3} + 3 = \sqrt{6 + 3} + 3 = \sqrt{9} + 3$$
$$= 3 + 3 = \boxed{6}$$

$$\text{RS:} \quad x = \boxed{6}$$

Thus 6 checks and is in the solution set.

Check 1 in $\sqrt{x + 3} + 3 = x$.

$$\text{LS:} \quad \sqrt{x + 3} + 3 = \sqrt{1 + 3} + 3 = \sqrt{4} + 3$$
$$= 2 + 3 = \boxed{5}$$

$$\text{RS:} \quad \boxed{1}$$

Thus 1 does *not* check and is not in the solution set.

$$\{6\} \qquad \blacktriangleleft$$

Problem Set 8.5

Warm-Ups

In Problems 1 through 6, solve each equation. See Example 1.

1. $\sqrt{x + 2} = 4$

2. $\sqrt{x - 1} = 5$

3. $\sqrt{5x + 1} - 1 = 3$

4. $\sqrt{3x + 1} = 1$

5. $\sqrt{x - 4} = 0$

6. $\sqrt{2x - 3} = 3$

In Problems 7 through 10, solve each equation. See Example 2.

7. $\sqrt{1 - 2x} = -5$

8. $\sqrt{x + 3} = -4$

9. $\sqrt{4 - x} + 5 = 0$

10. $\sqrt{2x + 5} - 1 = -6$

In Problems 11 through 16, find the solution sets. See Example 3.

11. $\sqrt{3x + 1} = \sqrt{5 - x}$

12. $\sqrt{x + 9} = \sqrt{3x - 5}$

13. $\sqrt{5 - x} - \sqrt{2x + 17} = 0$

14. $\sqrt{2x + 9} + \sqrt{9 - x} = 0$

15. $\sqrt{x^2 - x - 3} = \sqrt{x^2 - 7}$

16. $\sqrt{x^2 + x + 4} - \sqrt{x^2 - 3x + 12} = 0$

In Problems 17 through 28, solve each equation. See Example 4.

17. $\sqrt{x + 3} = x + 1$

18. $\sqrt{x + 9} = x - 3$

19. $\sqrt{2 - x} = x + 4$

20. $\sqrt{2 - x} = x$

21. $\sqrt{4x + 9} = x + 1$

22. $\sqrt{2x + 20} = x + 10$

23. $\sqrt{4x + 5} = x$

24. $\sqrt{1 - 2x} = x + 1$

25. $\sqrt{2x + 6} - 3 = x$

26. $x = \sqrt{x + 4} - 2$

27. $3 = \sqrt{2x^2 + 2} - x$

28. $\sqrt{x + 5} + 7 = x$

Practice Exercises

In Problems 29 through 64, solve each equation.

29. $\sqrt{x + 7} = 3$

30. $\sqrt{x - 1} = 2$

31. $\sqrt{x + 1} = -2$

32. $\sqrt{5x + 4} = 2$

33. $\sqrt{x - 5} = 0$

34. $\sqrt{3x + 1} = 4$

35. $\sqrt{1 - 2x} = 3$

36. $\sqrt{2x + 8} - 1 = 1$

37. $\sqrt{10 - x} + 3 = 0$

38. $\sqrt{8x + 1} - 2 = 1$

39. $\sqrt{x - 3} = \sqrt{11 - x}$

40. $\sqrt{x + 8} = \sqrt{3x + 16}$

41. $\sqrt{7 - x} - \sqrt{2x + 10} = 0$

42. $\sqrt{2x + 4} - \sqrt{4 - x} = 0$

43. $\sqrt{x^2 - x + 2} = \sqrt{x^2 + 3}$

44. $\sqrt{x^2 - x - 5} - \sqrt{x^2 - 3x + 7} = 0$

45. $\sqrt{x^2 + 1} = 1 - x$

46. $\sqrt{x^2 + 2x + 1} = x + 3$

47. $\sqrt{x^2 - 3x + 2} + 2 = x$

48. $\sqrt{x^2 - 5} - 5 = x$

49. $\sqrt{x^2 - 1} = 1 + x$

50. $\sqrt{x^2 + 2x + 4} = x + 2$

51. $\sqrt{x^2 - x - 4} + 9 = x$

52. $\sqrt{x^2 + 7} - 7 = x$

53. $\sqrt{x - 1} = x - 3$

54. $\sqrt{x + 8} = x + 2$

55. $\sqrt{3 - x} = x + 3$

56. $\sqrt{3 - 2x} = x$

57. $\sqrt{3x + 4} = x - 2$

58. $\sqrt{2x + 8} = x + 4$

59. $\sqrt{6x - 9} = x$

60. $\sqrt{1 - 3x} = x - 1$

61. $\sqrt{2x + 2} - 1 = x$

62. $x = \sqrt{x + 6} + 6$

63. $2 = \sqrt{3x^2 - 2} - x$

64. $\sqrt{x + 10} + 2 = x$

65. The profit on the sale of pairs of socks can be expressed as the square root of the sum of the number of pairs produced and 5. If the profit is \$3.00, find the number of pairs produced.

66. Hasting's Lawn Service installs circular sprinkler systems using the formula $r = \sqrt{0.32A}$, where r is the radius of the circular region watered and A is the area of that region. If a radius of 5 feet is used, what will be the area of the ground watered?

Problem 66

Challenge Problems

In Problems 67 and 68, solve each equation (a and b are positive constants).

67. $\sqrt{ax} = b$

68. $\sqrt{x^2 + b} = x + a$

In Your Own Words

69. Explain how to solve an equation that contains a square root.

70. Why is checking not an optional part of the procedure when squaring both sides of an equation?

8.6   Rational Exponents

TAPE 14

In Sections 2.1 and 2.2 we studied integer exponents. In this section we examine exponents that are fractions.

   The major reason we return to exponents is to capitalize on their properties. So, in defining fractional exponents, we ensure that the five properties of exponents that we developed in Chapter 2 remain valid. In review, they are listed next.

Five Properties of Exponents
If m and n are integers and x and y are real numbers, then

1. $x^m x^n = x^{m+n}$ Product with same base

2. $(x^m)^n = x^{mn}$ Power of a power

3. $(xy)^n = x^n y^n$ Product to a power

4. $\left(\dfrac{x}{y}\right)^n = \dfrac{x^n}{y^n}; \quad y \neq 0$ Quotient to a power

5. $\dfrac{x^m}{x^n} = x^{m-n}; \quad x \neq 0$ Quotient with same base

Now suppose we define an exponent of $\frac{1}{2}$ in such a way that the properties of exponents are still valid. Then Property 2 will be true when m is $\frac{1}{2}$ and n is 2. Therefore, it must be true that

$$(x^{1/2})^2 = x^1$$
$$= x$$

So $x^{1/2}$ is a number that when squared is x. Because x is a real number squared, it must be positive. We know such a number. It is $\sqrt{x}$. Thus we define $x^{1/2}$ by

$$x^{1/2} = \sqrt{x}$$

By a similar argument,

$$(x^{1/3})^3 = x^1$$
$$= x$$

Thus $x^{1/3}$ is a number that when cubed is x. That number is $\sqrt[3]{x}$. So we define

$$x^{1/3} = \sqrt[3]{x}$$

Notice that x can be negative in this case.

These two examples suggest the following definition.

Exponent of the Form $\dfrac{1}{n}$

For a natural number n,

$$x^{1/n} = \sqrt[n]{x}, \qquad x \geq 0 \text{ if } n \text{ is even}$$

► **EXAMPLE 1**

Simplify each.

(a) $4^{1/2}$ (b) $8^{1/3}$ (c) $(-27)^{1/3}$ (d) $-9^{1/2}$ (e) $-64^{1/3}$

Solutions

(a) $4^{1/2} = \sqrt{4}$ (b) $8^{1/3} = \sqrt[3]{8}$ (c) $(-27)^{1/3} = \sqrt[3]{-27}$
$\qquad = 2$ $\qquad = 2$ $\qquad\qquad = -3$

Notice that the base is -27.

(d) $-9^{1/2} = -\sqrt{9}$
$\qquad = -3$

Notice that the base is 9 and *not* -9.

(e) $-64^{1/3} = -\sqrt[3]{64}$
$\qquad = -4$

Be Careful!

Again note that the base is 64 and *not* -64.   ◄

We can use our definition and the laws of exponents to simplify expressions such as $8^{2/3}$. Notice how this is done in the next example.

► **EXAMPLE 2**

Simplify $8^{2/3}$.

Solution

The second property of exponents allows us to rewrite this expression so that we can employ the definition of the power $1/n$.

$$8^{2/3} = (8^{1/3})^2 \qquad \text{Power of a power}$$

$$= (\sqrt[3]{8})^2 \qquad \text{Definition of exponent } \frac{1}{3}$$

$$= (2)^2$$

$$= 4 \qquad \blacktriangleleft$$

The technique employed in Example 2 leads to the following rule.

Rule for Calculating $x^{m/n}$

For m and n natural numbers,

$$x^{m/n} = (\sqrt[n]{x})^m$$

► **EXAMPLE 3**

Simplify each of the following using the rule for calculating $x^{m/n}$.

(a) $4^{3/2}$ (b) $-16^{3/2}$ (c) $(-27)^{2/3}$ (d) $-27^{2/3}$

Solutions

(a) $4^{3/2} = (\sqrt{4})^3$

$\qquad = (2)^3$

Be Careful! $\qquad = 8$

(b) The base of the exponent is 16, *not* -16.

$$-16^{3/2} = -(\sqrt{16})^3$$

$$= -4^3$$

$$= -64$$

(c) Notice that the base here is -27.

$$(-27)^{2/3} = (\sqrt[3]{-27})^2$$

$$= (-3)^2$$

$$= 9$$

Be Careful! (d) The base is 27, *not* -27.

$$-27^{2/3} = -(\sqrt[3]{27})^2$$

$$= -3^2$$

$$= -9 \qquad \text{The base is 3, } not -3. \qquad \blacktriangleleft$$

> **A Summary of Exponent Definitions**
>
> Given that n is a natural number,
>
> **1.** $x^n = x \cdot x \cdot x \cdots x$
> $\qquad\quad$ n factors of x
>
> **2.** $x^0 = 1; \quad x \neq 0$
>
> **3.** $x^{-n} = \dfrac{1}{x^n}; \quad x \neq 0$
>
> **4.** $x^{1/n} = \sqrt[n]{x}; \quad x \geq 0$ if n is even

All the properties of exponents apply when using rational numbers for exponents, just as they apply with integers.

► **EXAMPLE 4**

In each problem, perform the indicated operation. Assume that x is a nonnegative number.

(a) $x^{1/2}x^{1/3}$ $\qquad$ (b) $(5^{3/2})^2$ $\qquad$ (c) $(4x^4)^{3/2}$

(d) $\left(\dfrac{x^2}{16}\right)^{1/2}$ $\qquad$ (e) $\dfrac{5^{1/2}}{5^{1/3}}$

Solutions

(a) $x^{1/2} \cdot x^{1/3} = x^{1/2+1/3}$ $\qquad$ Product with the same base

$\qquad\qquad\quad\; = x^{5/6}$

(b) $(5^{3/2})^2 = 5^{3/2 \cdot 2}$ $\qquad$ Power to a power

$\qquad\qquad\; = 5^3$

$\qquad\qquad\; = 125$

(c) $(4x^4)^{3/2} = 4^{3/2}(x^4)^{3/2}$ $\qquad$ Product to a power

$\qquad\qquad\quad = (\sqrt{4})^3 x^6$

$\qquad\qquad\quad = 2^3 x^6 = 8x^6$

(d) $\left(\dfrac{x^2}{16}\right)^{1/2} = \dfrac{(x^2)^{1/2}}{16^{1/2}}$ $\qquad$ Quotient to a power

$\qquad\qquad\quad\;\; = \dfrac{x^1}{\sqrt{16}}$

$\qquad\qquad\quad\;\; = \dfrac{x}{4}$

(e) $\dfrac{5^{1/2}}{5^{1/3}} = 5^{1/2-1/3}$ $\qquad$ Quotient with same base

$\qquad\quad\;\; = 5^{1/6}$

◄

Calculator Box

Approximating Roots on a Calculator

With an *n*th Root Key

A few calculators have an *n*th root key. Let's approximate $\sqrt[5]{3 + \pi}$ with such a calculator.

Scientific Calculator:

Press $\boxed{(}\ \boxed{3}\ \boxed{+}\ \boxed{\pi}\ \boxed{)}\ \boxed{\sqrt[n]{}}\ \boxed{5}\ \boxed{=}$

Graphing Calculator:

Press $\boxed{5}\ \boxed{\sqrt[n]{}}\ \boxed{(}\ \boxed{3}\ \boxed{+}\ \boxed{\pi}\ \boxed{)}\ \boxed{\text{EXE}}$

In either case, we read, ▮1.43766004▮ on the display. On some calculators the *n*th root key looks like $\boxed{\sqrt[x]{}}$.

Without an *n*th Root Key

If there is no *n*th root key, we resort to fractional exponents. Using the fact that the *n*th root is the $\frac{1}{n}$-th power, we find $\sqrt[5]{3 + \pi}$ by calculating $(3 + \pi)^{1/5}$ with the exponent key.

Press

$\boxed{(}\ \boxed{3}\ \boxed{+}\ \boxed{\pi}\ \boxed{)}\ \boxed{y^x}\ \boxed{(}\ \boxed{1}\ \boxed{\div}\ \boxed{5}\ \boxed{)}\ \boxed{=}$ (or $\boxed{\text{ENTER}}$ or $\boxed{\text{EXE}}$)

These keystrokes work on both the scientific and graphing calculators.

Calculator Exercises

Find each root.

1. $\sqrt[4]{83521}$ **2.** $\sqrt[6]{117649}$ **3.** $\sqrt[7]{1.5}$ **4.** $\sqrt[5]{\sqrt{7} + 7}$

Answers

1. 17   **2.** 7   **3.** ≈ 1.0596 **4.** ≈ 1.5735

Problem Set 8.6

Warm-Ups

In Problems 1 through 12, simplify each expression.
For Problems 1 through 6, see Example 1.

1. $16^{1/2}$

2. $36^{1/2}$

3. $125^{1/3}$

4. $-36^{1/2}$

5. $-8^{1/3}$

6. $(-8)^{1/3}$

For Problems 7 through 12, see Example 3.

7. $27^{2/3}$

8. $16^{3/2}$

9. $9^{3/2}$

10. $(-1000)^{2/3}$

11. $-8^{2/3}$

12. $-4^{3/2}$

In Problems 13 through 22, perform the indicated operations. Assume that all variables represent positive real numbers. See Example 4.

13. $a^{2/3} \cdot a^{1/3}$

14. $b^{1/2} \cdot b^{1/5}$

15. $(4^{3/5})^5$

16. $(x^{1/3})^2$

17. $(-8x^6)^{1/3}$

18. $(4x^4y^6)^{1/2}$

19. $\left(\dfrac{16}{x^8}\right)^{3/2}$

20. $\left(\dfrac{27a^6}{b^9}\right)^{2/3}$

21. $\dfrac{x^{2/3}}{x}$

22. $\dfrac{t^{2/3}}{t^{1/2}}$

Practice Exercises

In Problems 23 through 46, simplify each.

23. $49^{1/2}$

24. $100^{1/2}$

25. $81^{1/2}$

26. $121^{1/2}$

27. $216^{1/3}$

28. $1000^{1/3}$

29. $343^{1/3}$

30. $(-1000)^{1/3}$

31. $(-125)^{1/3}$

32. $-81^{1/2}$

33. $-100^{1/2}$

34. $-121^{1/2}$

35. $-343^{1/3}$

36. $-1000^{1/3}$

37. $-216^{1/3}$

38. $64^{2/3}$

39. $25^{3/2}$

40. $36^{3/2}$

41. $(-8)^{2/3}$

42. $-125^{2/3}$

43. $-9^{3/2}$

44. $(-216)^{2/3}$

45. $-81^{3/2}$

46. $-216^{2/3}$

In Problems 47 through 60, perform the indicated operations. Assume that all variables represent positive real numbers. See Example 4.

47. $z^{1/3}z^{4/3}$

48. $m^{2/3} \cdot m$

49. $(x^{1/2})^3$

50. $(7^{1/3})^3$

51. $(-27y^9)^{2/3}$

52. $(4x^8)^{3/2}$

53. $\left(\dfrac{x^{12}}{-8}\right)^{1/3}$

54. $\left(\dfrac{y^{1/2}}{x}\right)^6$

55. $\dfrac{w^2}{w^{3/2}}$

56. $\dfrac{r}{r^{1/2}}$

57. $x^{1/2}(x^{1/2} - 1)$

58. $a^{2/3}(a^{1/2} + 5)$

59. $(x^{1/2} + 1)(x^{1/2} - 1)$

60. $(y^{1/3} - 2)(y^{1/3} + 2)$

Challenge Problems

A negative fraction, such as $-\frac{2}{3}$, *can always be written with a* positive *denominator and a* negative *numerator:* $-\frac{2}{3} = \frac{-2}{3}$. *Use this fact and the rule to calculate* $x^{m/n}$ *to evaluate each of the following.*

61. $8^{-2/3}$ **62.** $16^{-3/2}$ **63.** $-64^{-2/3}$

64. $(-64)^{-2/3}$ **65.** $125^{-1/3}$

In Your Own Words

66. Describe how to evaluate a number raised to a fractional power.

 # Chapter Summary

GLOSSARY

Principal square root of a nonnegative number, q: The positive number whose square is q; written $\sqrt{q}$.

Like radicals: Radicals with the same index and radicand.

The nth Root of a Real Number	If n is a natural number and x is a real number: **1.** If n is even and x is negative, $\sqrt[n]{x}$ is the nonnegative number whose n^{th} power is x. **2.** If n is odd, $\sqrt[n]{x}$ is the number whose n^{th} power is x.
Multiplication Property of Radicals	If $\sqrt[n]{p}$ and $\sqrt[n]{q}$ exist as real numbers, $$\sqrt[n]{pq} = \sqrt[n]{p} \cdot \sqrt[n]{q}$$
Division Property of Radicals	If $\sqrt[n]{p}$ and $\sqrt[n]{q}$ exist as a real numbers and $q \neq 0$, $$\sqrt[n]{\frac{p}{q}} = \frac{\sqrt[n]{p}}{\sqrt[n]{q}}$$
Solving Equations Containing Square Roots	**1.** Isolate a square root. **2.** Square both sides. **3.** Solve the resulting equation for all possible solutions. **4.** Check each possible solution in the original equation. **5.** Write the solution set.
Rational Exponents	$$x^{1/n} = \sqrt[n]{x}; \quad x \geq 0 \text{ if } n \text{ is even.}$$ $$x^{m/n} = (\sqrt[n]{x})^m; \quad x \geq 0 \text{ if } n \text{ is even.}$$

CHECK UPS

1. Find $\sqrt{\dfrac{4}{9}}$.   Section 8.1, Example 1c

2. Find $\sqrt{x^{10}}$.   Section 8.1, Example 6b

3. Simplify $\sqrt{48}$.   Section 8.2, Example 1c

4. Simplify $\sqrt[3]{\dfrac{5}{8}}$.   Section 8.2, Example 4c

5. Perform the operation indicated.   Section 8.3, Example 3b

$$\sqrt[3]{16y} - \sqrt[3]{54y}$$

6. Perform the operation indicated.   Section 8.4, Example 2b

$$(2\sqrt{7} + \sqrt{3})(2\sqrt{7} - \sqrt{3})$$

7. Rationalize the denominator.   Section 8.4, Example 5a

$$\frac{3}{\sqrt{2} - 5}$$

8. Solve $\sqrt{x + 3} + 3 = x$.   Section 8.5, Example 4

9. Simplify $(-27)^{2/3}$.   Section 8.6, Example 3c

► Review Problems

In Problems 1 through 30, perform the operations indicated and simplify.

1. $\sqrt{18} + 5\sqrt{2}$

2. $\sqrt{7} \cdot \sqrt{8}$

3. $\sqrt{2}(\sqrt{2} + \sqrt{3})$

4. $(\sqrt{2} + 5)(\sqrt{2} - 3)$

5. $(\sqrt{7} + 3)^2$

6. $(\sqrt{7} - \sqrt{11})(\sqrt{7} + \sqrt{11})$

7. $\sqrt{49x^2} + \sqrt{25x^2} - \sqrt{16x^2}$

8. $(\sqrt{7})^2$

9. $(2\sqrt{3})(6\sqrt{3})$

10. $2\sqrt{5}(3\sqrt{2} - \sqrt{5})$

11. $(\sqrt{6} - \sqrt{2})(\sqrt{6} + \sqrt{3})$

12. $\dfrac{\sqrt{18}}{\sqrt{2}}$

13. $2\sqrt[3]{5} \cdot \sqrt[3]{25}$

14. $\sqrt{48} - 2\sqrt{12} + \sqrt{27}$

15. $\dfrac{\sqrt[3]{40}}{\sqrt[3]{5}}$

16. $\sqrt[3]{27x^3} - \sqrt[3]{-8x^3}$

17. $(2\sqrt{6} - \sqrt{2})^2$

18. $(\sqrt{10} - \sqrt{2})(2\sqrt{10} + 3\sqrt{2})$

19. $\sqrt[3]{54} - \sqrt[3]{-16} + \sqrt[3]{-128}$

20. $4\sqrt{\dfrac{3}{25}} - \sqrt{\dfrac{12}{25}}$

21. $\dfrac{\sqrt{60}}{\sqrt{5}}$

22. $\sqrt[3]{54x^6y^3} - \sqrt[3]{16x^6y^3}$

23. $\dfrac{\sqrt[3]{-48}}{\sqrt[3]{-2}}$

24. $(2\sqrt{5} + 3)^2$

25. $2\sqrt{2}(\sqrt{3} - 2\sqrt{2})$

26. $(3\sqrt{11})^2$

27. $(2\sqrt{3} + 3\sqrt{2})(\sqrt{3} - \sqrt{2})$

28. $\sqrt{\dfrac{25}{4}} - \sqrt{\dfrac{16}{9}}$

29. $3\sqrt{6} - 2\sqrt{6} + \sqrt{6}$

30. $\sqrt{x}(2\sqrt{x} + \sqrt{y})$

In Problems 31 through 40, rationalize the denominator.

31. $\dfrac{4}{\sqrt{2}}$

32. $\dfrac{3}{2\sqrt{5}}$

33. $\dfrac{2}{\sqrt{x}}$

34. $\dfrac{4}{3\sqrt{2}}$

35. $\dfrac{1}{3\sqrt{5}}$

36. $\dfrac{\sqrt{3}}{\sqrt{3}+\sqrt{2}}$

37. $\dfrac{2\sqrt{2}}{\sqrt{2}-1}$

38. $\dfrac{\sqrt{5}+\sqrt{3}}{\sqrt{5}-\sqrt{3}}$

39. $\dfrac{\sqrt{3}+5\sqrt{2}}{2\sqrt{3}-\sqrt{2}}$

40. $\dfrac{x+2\sqrt{y}}{x-2\sqrt{y}}$

In Problems 41 through 50, find the solution set.

41. $\sqrt{x+4}=5$

42. $\sqrt{x+12}=x$

43. $\sqrt{x^2+7}=x+1$

44. $\sqrt{7x+15}-x=3$

45. $\sqrt{x+3}=0$

46. $\sqrt{5x+11}=x+1$

47. $\sqrt{2x+8}=x+4$

48. $\sqrt{x^2-2x+1}=\sqrt{2x-2}$

49. $\sqrt{x+2}=\sqrt{x-7}$

50. $\sqrt{x^2}=5$

In Problems 51 through 55, simplify each.

51. $-16^{1/2}$

52. $\left(\dfrac{8}{125}\right)^{1/3}$

53. $-\left(\dfrac{1}{4}\right)^{3/2}$

54. $\left(-\dfrac{1}{8}\right)^{2/3}$

55. $25^{3/2}$

In Problems 56 through 60, perform the indicated operations. Assume that all variables represent positive numbers.

56. $2^{1/2}\cdot 2^{5/2}$

57. $\dfrac{3^{3/2}}{3^{1/2}}$

58. $(9x^3)^{1/2}$

59. $\left(\dfrac{-64}{x^{12}}\right)^{2/3}$

60. $\dfrac{10^{1/2}}{10^{7/2}}$

61. The Detroit Lens Company manufactures contact lenses for wholesale distributors. The cost of manufac- turing x contact lenses is given by $C = y_0 + 97x^{2/3}$, where y_0 is the fixed start-up cost. Find the cost of manufacturing 125,000 lenses when the start-up cost is $1200.

62. Physics shows that the time in seconds for a free-falling object to fall s feet is given by $t = \dfrac{\sqrt{s}}{4}$. A stone is dropped from the top of a cliff that is 96 feet high. How many seconds will it take the stone to reach the foot of the cliff? Approximate the answer to the nearest tenth of a second.

Problem 61

Let's Not Forget . . .

In Problems 63 through 70, perform the operations indicated.

63. -3^2

64. $(-3)^2$

65. $\dfrac{-2}{-2^0}$

66. $\dfrac{(a+b)(a-b)}{b^2-a^2}$

67. $(\sqrt{5}-\sqrt{3})^2$

68. $(\sqrt{7}+\sqrt{5})(\sqrt{7}-\sqrt{5})$

69. $\dfrac{a^2-a}{b^3}\cdot\dfrac{b^3}{a^4}$

70. $(2\sqrt{6}-\sqrt{2})(\sqrt{6}+\sqrt{2})$

In Problems 71 through 73, perform the operations indicated. Watch the role of parentheses.

71. $\dfrac{x-1}{x+1} - 1$

72. $\dfrac{x-1}{x+1} - \dfrac{x-2}{x+1}$

73. $\dfrac{x-1}{x+1} - \dfrac{1}{x-1}$

In Problems 74 through 77, determine if $\sqrt{2}$ is a factor *or a* term.

74. $\sqrt{2} + 2$

75. $\sqrt{2}(\sqrt{3})$

76. $\dfrac{1}{\sqrt{2}+2}$

77. $(y - \sqrt{2})(y + \sqrt{2})$

In Problems 78 through 80, multiply each expression by $2\sqrt{3}$.

78. $2\sqrt{3}$

79. $2 + \sqrt{3}$

80. $\dfrac{1}{2\sqrt{3}}(2 + \sqrt{3})$

In Problems 81 through 85, identify which expressions are factored. Factor those that are not factored, if possible.

81. $a^2\sqrt{3} - 4\sqrt{3}$

82. x^2y^2

83. $(x - y)^3$

84. $2(b - c)$

85. $x^3 + y^3$

In Problems 86 through 90, identify whether each problem is an expression, equation, *or* inequality. *Solve the equations and inequalities and perform the operation indicated with the expressions.*

86. $\dfrac{2}{x-1} - \dfrac{1}{x+3}$

87. $\dfrac{2}{x-1} - \dfrac{1}{x+3} = \dfrac{2x}{(x-1)(x+3)}$

88. $2x - 3(x + 3) \le 0$

89. $\dfrac{\dfrac{2}{x-1}}{\dfrac{1}{x+3}}$

90. $\sqrt{x+2} = x$

► You Decide

Your grandmother has a 1973 Oldsmobile Cutlass. The gas gauge does not work, but the odometer does. Lang, her mechanic, told her that he could not replace the broken part because it was not in stock anymore. Suppose that the car gets 10 mpg in the city and 15 mpg on the highway. The gas tank holds 20 gallons. You decide on a systematic way for Grandma to figure out when to buy gas.

► Chapter 8 Test

In Problems 1 through 6, choose the correct answer.

1. $(2\sqrt{3})^2 = (?)$
   (a) 6   (b) 12
   (c) 18   (d) 36

2. $\sqrt{-4} = (?)$
   (a) 2   (b) -2
   (c) -4 (d) Is not a real number

3. $(\sqrt{5} + \sqrt{3})^2 = (?)$

   (a) 8   (b) 34

   (c) $8 + 2\sqrt{15}$ (d) $8 + \sqrt{15}$

4. $\sqrt[3]{-\dfrac{8}{27}} = (?)$

   (a) $\dfrac{3}{2}$ (b) $-\dfrac{3}{2}$

   (c) $\dfrac{2}{3}$ (d) $\dfrac{-2}{3}$

5. $\dfrac{\sqrt{54}}{\sqrt{3}} = (?)$
   (a) $2\sqrt{3}$ (b) $3\sqrt{2}$
   (c) $3\sqrt{6}$ (d) $2\sqrt{6}$

6. $3\sqrt{48} - \sqrt{12} = (?)$
   (a) $10\sqrt{3}$ (b) 18
   (c) $\sqrt{3}$ (d) $44\sqrt{3}$

In Problems 7 through 13, perform the operations indicated and simplify.

7. $(3\sqrt{7})(4\sqrt{5})$

8. $\sqrt{6}(2\sqrt{2} - \sqrt{3})$

9. $(\sqrt{3} - 2\sqrt{7})(\sqrt{3} + 2\sqrt{7})$

10. $(2\sqrt{7} + \sqrt{3})^2$

11. $(\sqrt{5} + 2\sqrt{6})(\sqrt{3} - \sqrt{5})$

12. $\dfrac{\sqrt[3]{-81}}{\sqrt[3]{3}}$

13. $\sqrt{32x^2} + \sqrt{8x^2} - 2\sqrt{18x^2}$

In Problems 14 and 15, rationalize the denominator.

14. $\dfrac{3}{2\sqrt{3}}$

15. $\dfrac{\sqrt{3} - 2}{\sqrt{2} + \sqrt{3}}$

In Problems 16 and 17, simplify each expression.

16. $-49^{1/2}$

17. $125^{2/3}$

18. Solve $\sqrt{x + 5} = \sqrt{2x + 1}$

19. Solve $\sqrt{x - 1} + 3 = x$

20. If t seconds is the time for one complete swing of a pendulum of length L feet, then $t = 2\pi\sqrt{\dfrac{L}{32}}$. A physicist's pendulum is 2 feet long. Find the time for one complete swing of this pendulum. Approximate the length to the nearest hundredth of a second.

9 Quadratic Equations

See Problem Set 9.5, Exercise 45.

> ► **Connections**

The usefulness of mathematics lies in our ability to match a mathematical concept to a physical, social, or economic problem. In Chapter 3, we saw how linear equations became mathematical models that could be used to solve applications. Hindu engineers in the ninth century were confronted with the problem of changing the size of the altars in temples. This led them to seek solutions to equations such as $x^2 - x - 6 = 0$. After electricity was discovered, scientists produced mathematical models such as $x^2 + 1 = 0$. These are examples of an important class of equations called quadratic equations.

Several methods can be used to solve such equations. Factoring is one method that we studied in Chapter 4. However, not all quadratic equations can be solved by factoring. In this chapter, we will see how to use the square root property and completing the square. We will derive the quadratic formula, which can be used to solve any quadratic equation.

9.1   Square Root Property

TAPE 14

In this section we solve second-degree equations. We saw in Section 4.5 that some second-degree equations can be solved by factoring. We now focus on the Square Root Property, which will allow us to solve second-degree equations that cannot be solved by factoring. We call second-degree equations **quadratic equations.**

> ### Quadratic Equation
>
> A **quadratic equation** in one variable is an equation that can be written in the form
>
> $$ax^2 + bx + c = 0$$
>
> where a, b, and c are real numbers with $a \neq 0$. (If $a = 0$, the equation is not quadratic.)

We refer to a as the *leading coefficient*, b as the *coefficient of x*, and c as the constant term.

Let's begin by looking at how $x^2 = 4$ can be solved by factoring.

$$x^2 = 4$$

$$x^2 - 4 = 0$$

$$(x + 2)(x - 2) = 0$$

$$\{\pm 2\} \quad \text{or} \quad \{-2, 2\}$$

Notice that the solution set is $\{\pm\sqrt{4}\}$.

How could we solve $x^2 = 5$? It looks much like $x^2 = 4$. We begin by writing the equation so that one side is 0.

$$x^2 - 5 = 0$$

We were able to factor $x^2 - 4$ as the difference of two squares. Since 5 is not a perfect square, we must extend our ideas about factoring. There is an irrational number that, when squared, is 5. Recall that

$$(\sqrt{5})^2 = 5$$

Now we can write the equation as

$$x^2 - (\sqrt{5})^2 = 0$$

This is the difference of two squares, which factors as

$$(x + \sqrt{5})(x - \sqrt{5}) = 0$$

Thus the solution set is

$$\{\pm\sqrt{5}\}$$

If the factoring steps are omitted, we can see a similarity in the two equations and in the two solution sets.

$$x^2 = 4 \qquad x^2 = 5$$
$$\{\pm\sqrt{4}\} \qquad \{\pm\sqrt{5}\}$$

Each equation is of the form $x^2 = A$, where A is a positive number. The solution set for each is $\{\pm\sqrt{A}\}$. These equations lead to the following result, which is very useful. It allows us to leave out the factoring steps.

Square Root Property

If A is a nonnegative real number, the equation

$$\mathbf{X}^2 = A$$

has the same solution set as the pair of equations

$$\mathbf{X} = \pm\sqrt{A}$$

If A is a negative number such as -3, then $\sqrt{A}$ is $\sqrt{-3}$. Since $\sqrt{-3}$ is not a real number, it is necessary to require that A be nonnegative. If A is 0, then $\sqrt{A}$ is $\sqrt{0}$, which has value 0.

► **EXAMPLE 1**

Solve the following equations using the Square Root Property.

(a) $x^2 - 9 = 0$

(b) $x^2 = 6$

Solutions

(a) We must first write the equation in the form $\mathbf{X}^2 = A$ before we can use the Square Root Property.

$$x^2 - 9 = 0$$
$$x^2 = 9$$
$$x = \pm\sqrt{9} \qquad \text{Square Root Property}$$
$$x = \pm 3$$
$$\{\pm 3\}$$

Be Careful!

(b) $x^2 = 6$
$$x = \pm\sqrt{6} \qquad \text{Square Root Property}$$
$$\{\pm\sqrt{6}\} \qquad\qquad\qquad\qquad\qquad \blacktriangleleft$$

Be careful not to omit the $\pm$ in using the Square Root Property. To omit it would mean losing a solution. The solutions can be checked in the original equation.

Be Careful!

► **EXAMPLE 2**

Solve $x^2 - 20 = 7$ by the Square Root Property.

Solution

The equation must be in the form $\mathbf{X}^2 = A$.

$$x^2 - 20 = 7$$
$$x^2 = 27 \qquad \text{Add 20 to both sides}$$
$$x = \pm\sqrt{27} \qquad \text{Square Root Property}$$
$$x = \pm 3\sqrt{3} \qquad \sqrt{27} = \sqrt{9(3)} = 3\sqrt{3}$$
$$\{\pm 3\sqrt{3}\} \qquad\qquad\qquad\qquad \blacktriangleleft$$

► **EXAMPLE 3**

Solve $3x^2 = 3$ by the Square Root Property.

Solution

The equation must be in the form $\mathbf{X}^2 = A$. We must divide by 3.

$$3x^2 = 3$$
$$x^2 = 1 \qquad \text{Divide by 3}$$
$$x = \pm\sqrt{1} \qquad \text{Square Root Property}$$
$$x = \pm 1$$
$$\{\pm 1\} \qquad\qquad\qquad\qquad \blacktriangleleft$$

The **X** in the Square Root Property need not be just a single variable; it can be an expression, such as $x - 3$.

► **EXAMPLE 4**

Solve $(x - 3)^2 = 4$ by the Square Root Property.

Solution

We treat $(x - 3)$ like the **X** in the Square Root Property. So

$$(x - 3)^2 = 4$$

$$x - 3 = \pm\sqrt{4} \qquad \text{Square Root Property}$$

$$x - 3 = \pm 2$$

$$x = 3 \pm 2 \qquad \text{Add 3 to both sides}$$

Solve the two equations.

$$x = 3 + 2 \quad \text{or} \quad x = 3 - 2$$

$$x = 5 \qquad \text{or} \quad x = 1$$

$$\{1, 5\}$$

◄

► **EXAMPLE 5**

Solve $\left(x - \dfrac{1}{2}\right)^2 = \dfrac{1}{25}$ by using the Square Root Property.

Solution

$$\left(x - \frac{1}{2}\right)^2 = \frac{1}{25}$$

$$x - \frac{1}{2} = \pm\sqrt{\frac{1}{25}} \qquad \text{Square Root Property}$$

$$x - \frac{1}{2} = \pm\frac{1}{5}$$

$$x = \frac{1}{2} \pm \frac{1}{5}$$

$$x = \frac{5}{10} \pm \frac{2}{10} \qquad \text{Common denominator}$$

$$x = \frac{5}{10} + \frac{2}{10} \quad \text{or} \quad x = \frac{5}{10} - \frac{2}{10}$$

$$x = \frac{7}{10} \qquad \text{or} \quad x = \frac{3}{10}$$

$$\left\{\frac{3}{10}, \frac{7}{10}\right\}$$

◄

► **EXAMPLE 6**

A section of the walkway between the Nursing building and the Science building is shown. To prevent students from walking on the grassy area in the isosceles right triangle, plant operations personnel plan to put up a chain around the triangle. Approximately, how much chain must they buy?

Solution

x is the length of a leg (in meters). Using the Pythagorean Theorem, we form an equation.

$$x^2 + x^2 = 12^2$$
$$2x^2 = 144$$
$$x^2 = 72 \qquad \text{Divide by 2}$$
$$x = \pm\sqrt{72} \qquad \text{Square Root Property}$$
$$x = \pm\sqrt{36 \cdot 2}$$
$$= \pm\sqrt{36}\sqrt{2}$$
$$= \pm 6\sqrt{2}$$

Since a length cannot be negative, we use $6\sqrt{2}$. The legs are $6\sqrt{2}$ m long. Check by noting that

$$(6\sqrt{2})^2 + (6\sqrt{2})^2 = 72 + 72$$
$$= 144$$

and $12^2 = 144$. They must buy $12 + 12\sqrt{2}$ meters of chain, which is approximately 29 m.   ◄

Problem Set 9.1

Warm-Ups

In Problems 1 through 10, use the Square Root Property to solve each quadratic equation. See Example 1.

1. $x^2 = 25$

2. $z^2 - 81 = 0$

3. $x^2 = 100$

4. $x^2 = 0$

5. $w^2 - 120 = 1$

6. $x^2 - 169 = 0$

7. $y^2 = 225$ **8.** $x^2 = 17$ **9.** $x^2 - 3 = 0$

10. $y^2 - 12 = 2$

In Problems 11 through 16, solve each equation using the square root property. See Example 2.

11. $z^2 = 24$ **12.** $x^2 = 48$ **13.** $x^2 - 32 = 0$

14. $y^2 - 25 = 100$ **15.** $w^2 - 54 = 0$ **16.** $x^2 - 72 = 0$

In Problems 17 through 22, solve each equation using the square root property. See Example 3.

17. $5y^2 = 5$ **18.** $2x^2 = 12$ **19.** $4x^2 - 9 = 0$

20. $\dfrac{1}{2}x^2 = 4$ **21.** $\dfrac{1}{7}x^2 = 7$ **22.** $25w^2 = 36$

In Problems 23 through 30, solve each equation using the square root property. See Example 4.

23. $(x + 3)^2 = 4$ **24.** $(w - 7)^2 = 1$ **25.** $(z + 3)^2 = 16$

26. $(x + 1)^2 = 25$ **27.** $(x + 2)^2 = 5$ **28.** $(y - 9)^2 = 7$

29. $(x + 5)^2 = 12$ **30.** $(z - 4)^2 = 27$

In Problems 31 through 36, solve each equation using the square root property. See Example 5.

31. $\left(t - \dfrac{1}{2}\right)^2 = \dfrac{1}{9}$ **32.** $\left(w - \dfrac{2}{3}\right)^2 = \dfrac{1}{4}$ **33.** $\left(x + \dfrac{2}{7}\right)^2 = \dfrac{4}{9}$

34. $\left(y + \dfrac{3}{5}\right)^2 = \dfrac{9}{16}$ **35.** $\left(x - \dfrac{1}{3}\right)^2 = \dfrac{3}{4}$ **36.** $\left(y + \dfrac{2}{5}\right)^2 = \dfrac{7}{9}$

For Problems 37 through 41, see Example 6.

37. The legs of a right triangle are 1 ft in length. Find the length of the hypotenuse.

38. Find the length of a diagonal of a square with side of 10 m.

39. Glenn leans a 20-ft ladder against his house. If the base of the ladder is 4 ft from the house, how many feet above the ground is the top of the ladder?

40. The square of 3 less than a number is 16. Find all such numbers.

41. The diagonal of a square place mat has length $7\sqrt{2}$ in. Find the length of a side of the place mat.

Practice Exercises

In Problems 42 through 75, use the Square Root Property to solve each quadratic equation.

42. $x^2 = 36$ **43.** $\dfrac{1}{2}y^2 = 8$ **44.** $x^2 - 4 = 0$

45. $z^2 - 280 = 9$ **46.** $x^2 = 144$ **47.** $x^2 = 64$

48. $2w^2 - 512 = 0$ **49.** $x^2 - 324 = 0$ **50.** $\dfrac{1}{4}y^2 = 100$

51. $x^2 = 484$ **52.** $x^2 = 21$ **53.** $4x^2 = 81$

54. $2x^2 - 4 = 0$

55. $y^2 - 23 = 0$

56. $y^2 - 250 = 0$

57. $w^2 - 63 = 0$

58. $(x + 5)^2 = 9$

59. $(w - 8)^2 = 16$

60. $(z + 1)^2 = 36$

61. $(x + 2)^2 = 64$

62. $(x + 3)^2 = 7$

63. $(y - 5)^2 = 3$

64. $(w - 2)^2 = 17$

65. $(x + 7)^2 = 11$

66. $(x + 2)^2 = 18$

67. $(z - 5)^2 = 24$

68. $(z + 4)^2 = 27$

69. $(t - 3)^2 = 125$

70. $\left(t - \dfrac{1}{2}\right)^2 = \dfrac{1}{16}$

71. $\left(w - \dfrac{2}{5}\right)^2 = \dfrac{1}{9}$

72. $\left(x + \dfrac{2}{9}\right)^2 = \dfrac{9}{16}$

73. $\left(y + \dfrac{3}{4}\right)^2 = \dfrac{81}{25}$

74. $\left(x - \dfrac{1}{5}\right)^2 = \dfrac{2}{9}$

75. $\left(y + \dfrac{2}{3}\right)^2 = \dfrac{5}{4}$

76. The legs of a right triangle are 1 ft and 3 ft in length. Find the length of the hypotenuse.

77. The length of the hypotenuse of an isosceles right triangle is 6 m. Find the length of the legs.

78. The height of a triangle is three times the base. If the area of the triangle is 24 cm², find the lengths of the base and the height.

79. The square of two more than a number is 25. Find all such numbers.

80. The diagonal of a square picture is $4\sqrt{2}$ ft. Find the length of the side of the picture.

81. Find the radius of a circular table cloth if its area is 2025π in².

82. Jackie has a 30-ft ladder leaning against her house. If the top of the ladder is 20 ft above the ground, how far away from the house is the foot of the ladder?

83. Warren Mason purchased a radio antenna tower 15 m tall. He wishes to stabilize the tower in a vertical position with three guy wires. Each guy wire is to extend from the top of the tower to the ground 6 m from the base of the tower. How much wire should Warren buy?

84. A rocket accelerates from rest for 300 ft with a constant acceleration of $\dfrac{50}{3}$ ft per second per second. How much time has elapsed? (A formula for distance under these conditions is $d = \dfrac{1}{2}at^2$, where a is acceleration and t is time.)

85. The length of the diagonal of a rectangular placemat is 18 in. If the length of one side is 12 in., what are its dimensions?

86. The cost (in dollars) of producing x radios is given by the formula $c = (x + 3)^2$. If Radio Shack budgets $1600 for producing radios, how many radios can they produce?

87. Find the radius of a circular rug of area 128π ft².

88. The length of a right triangle is five times the base. If the area of the triangle is 250 m², find the lengths of the base and the height.

89. Find the radius of a coffee can if its volume is 54π in.³ and its height is 6 in. $(V = \pi r^2 h)$

90. The distance a walnut falls from a tree to the ground is given by the formula $s = 16t^2$, where s is the distance measured in feet and t is the number of seconds that the walnut falls. How many seconds will it take a walnut to fall 81 ft to the ground?

Challenge Problems

Solve by the Square Root Property.

91. $x^2 - c = 0; \quad c > 0$

92. $(x - 2)^2 = a^2; \quad a > 0$

93. $(x - a)^2 = 4$

In Your Own Words

94. Explain when the Square Root Property can be used to solve a quadratic equation.

9.2 Completing the Square

TAPE 14

In Section 9.1 the Square Root Property allowed us to solve quadratic equations in the form $\mathbf{X}^2 = A$. In this section we examine how to write any quadratic equation in this form. Recall the following from the previous section.

Square Root Property

If A is a nonnegative real number, the equation

$$\mathbf{X}^2 = A$$

has the same solution set as the pair of equations

$$\mathbf{X} = \pm\sqrt{A}$$

► **EXAMPLE 1**

Solve $x^2 + 2x + 1 = 5$ by the Square Root Property.

Solution

We must write the equation in the form $\mathbf{X}^2 = A$ before we can apply the Square Root Property. Notice that the left side can be factored as $(x + 1)^2$. In other words, it is a perfect trinomial square.

$$x^2 + 2x + 1 = 5$$
$$(x + 1)^2 = 5 \qquad \text{Factor}$$
$$x + 1 = \pm\sqrt{5} \qquad \text{Square Root Property}$$
$$x = -1 \pm \sqrt{5} \qquad \text{Subtract 1 from both sides}$$
$$\{-1 \pm \sqrt{5}\} \quad \text{or} \quad \{-1 + \sqrt{5}, -1 - \sqrt{5}\} \qquad ◄$$

The Square Root Property can be used to solve a quadratic equation if the equation can be written in the form $\mathbf{X}^2 = A$, where A is nonnegative. Example 1 suggests that we need a perfect trinomial square as $\mathbf{X}^2$.

If we make $x^2 + 6x$ into a perfect trinomial square, we will have a binomial squared. Recall how to square a binomial:

$$(x + p)^2 = x^2 + 2px + p^2$$

We need to add a number to $x^2 + 6x$ to make the expression a perfect trinomial square. That is,

$$(x + \underline{\quad})^2 = x^2 + 6x + \underline{\quad}$$

We must find the value of p. To do this, we need to unravel the middle term, $6x$. In squaring $x + p$, the middle term is $2px$. Since $6x$ is $2 \cdot 3x$, p must be 3. Thus the number for which we are looking is 3^2. So

$$(x + 3)^2 = x^2 + 6x + 9$$

How can we look at $x^2 + 6x$ and compute the number that must be added to make a perfect trinomial square? We take the coefficient of x, which is 6. To undo the squaring, we divide 6 by 2 and then we square this result.

What number must be added to $x^2 - 2x$ to make it into a perfect trinomial square?

$$x^2 - 2x + \underline{\qquad} = (x + \underline{\qquad})^2$$

The coefficient of x is -2. We divide -2 by 2 and square the result.

$$\left(\frac{-2}{2}\right)^2 = (-1)^2 = 1$$

The number to add is 1.

$$x^2 - 2x + 1 = (x - 1)^2$$

The process of making an expression into a perfect trinomial square is called *completing the square*. We must learn how to do this before we can use the Square Root Property to solve more quadratic equations.

► **EXAMPLE 2**

What number must be added to $x^2 + 4x$ to make it a perfect trinomial square?

Solution

The coefficient of x is 4. We divide 4 by 2 and square the result.

$$\left(\frac{4}{2}\right)^2 = (2)^2 = 4$$

The number that we should add is 4. $x^2 + 4x + 4$ is a perfect trinomial square. This can be checked by noting that $(x + 2)^2 = x^2 + 4x + 4$.   ◄

► **EXAMPLE 3**

What number must be added to $y^2 - 7y$ to make it a perfect trinomial square?

Solution

The coefficient of y is -7. We divide -7 by 2 and square the result.

$$\left(\frac{-7}{2}\right)^2 = \frac{49}{4}$$

The number to add is $\frac{49}{4}$. Thus $y^2 - 7y + \frac{49}{4}$ is a perfect square.

Check by noting that $y^2 - 7y + \frac{49}{4} = \left(y - \frac{7}{2}\right)^2$.   ◄

► **EXAMPLE 4**

What number must be added to $x^2 + \frac{2}{3}x$ to make a perfect trinomial square?

Solution

The coefficient of x is $\frac{2}{3}$. We must divide $\frac{2}{3}$ by 2.

$$\frac{2}{3} \div 2 = \frac{2}{3} \cdot \frac{1}{2}$$

$$= \frac{1}{3}$$

Now we must square $\frac{1}{3}$.

$$\left(\frac{1}{3}\right)^2 = \frac{1}{9}$$

We must add $\frac{1}{9}$.

$x^2 + \frac{2}{3}x + \frac{1}{9}$ is a perfect square.

This can be checked by noting that

$$\left(x + \frac{1}{3}\right)^2 = x^2 + \frac{2}{3}x + \frac{1}{9}$$ ◄

To complete the square, we must compute the number to add to make a perfect square. This involves dividing the coefficient of x by 2. As Example 4 indicates, dividing by 2 is the same as multiplying by $\frac{1}{2}$. Sometimes it is more convenient to multiply the coefficient of x by $\frac{1}{2}$. If the coefficient of x is an integer, then dividing by 2 is the best strategy. However, if the coefficient of x is a fraction, then multiplying by $\frac{1}{2}$ is easier.

► **EXAMPLE 5**

Solve $x^2 + 6x - 7 = 0$ by completing the square.

Solution

Adding 7 to both sides, the equation becomes,

$$x^2 + 6x = 7$$

Now complete the square on the left side. First calculate the number to add. We divide the coefficient of x by 2 and square the result.

$$\left(\frac{6}{2}\right)^2 = 3^2 = 9$$

We must add 9 to *both* sides of the equation.

$$x^2 + 6x = 7$$

$$x^2 + 6x + \boxed{9} = 7 + \boxed{9} \qquad \text{Add 9 to both sides}$$

$$(x + 3)^2 = 16 \qquad \text{Factor}$$

$$x + 3 = \pm\sqrt{16} \qquad \text{Square Root Property}$$

$$x + 3 = \pm 4$$

$$x = -3 \pm 4 \qquad \text{Subtract 3 from both sides}$$

Solve the two equations.

$$x = -3 + 4 \quad \text{or} \quad x = -3 - 4$$

$$x = 1 \qquad\quad \text{or} \quad x = -7$$

$$\{-7, 1\} \qquad \blacktriangleleft$$

The coefficient of the squared term must be 1 when completing the square. If the coefficient of the squared term in the quadratic equation is not 1, make it 1 and then complete the square.

► **EXAMPLE 6**

Solve $3x^2 + 6x = 3$ by completing the square.

Solution

First we divide both sides of the equation by 3.

$$x^2 + 2x = 1$$

Now we complete the square by dividing 2 by 2 and squaring the result.

$$2 \div 2 \text{ is } 1 \quad \text{and} \quad 1^2 \text{ is } 1$$

We add 1 to both sides:

$$x^2 + 2x + 1 = 1 + 1$$

$$(x + 1)^2 = 2$$

Now we apply the Square Root Property.

$$x + 1 = \pm\sqrt{2}$$

$$x = -1 \pm \sqrt{2} \qquad \text{Subtract 1}$$

$$\{-1 \pm \sqrt{2}\} \qquad \blacktriangleleft$$

> **Solving a Quadratic Equation by Completing the Square**
> 1. Write the equation in the form $ax^2 + bx + c = 0$.
> 2. Divide both sides by the coefficient of x^2 if it is not 1.
> 3. Subtract the constant term from both sides.
> 4. Divide the coefficient of x by 2, square this result, and then add this number to both sides.
> 5. Factor the left side and simplify the right side.
> 6. Apply the Square Root Property.
> 7. Check if required.
> 8. Write the solution set.

It may not always be necessary to carry out steps 1, 2, or 3.

▶ **EXAMPLE 7**

Solve $3x^2 - 2 = 10x$ by completing the square.

Solution

STEP 1   Write the equation in the form $ax^2 + bx + c = 0$.

$$3x^2 - 10x - 2 = 0 \qquad \text{Make the right side 0}$$

STEP 2   Divide both sides by the coefficient of x^2.

$$x^2 - \frac{10}{3}x - \frac{2}{3} = 0 \qquad \text{Divide by 3}$$

STEP 3   Place the constant term on the right side.

$$x^2 - \frac{10}{3}x = \frac{2}{3} \qquad \text{Add } \frac{2}{3} \text{ to both sides}$$

STEP 4   Divide the coefficient of x by 2, square this result, and add this number to both sides.
We calculate the number to add:

$$\left(-\frac{10}{3} \cdot \frac{1}{2}\right)^2 = \left(-\frac{5}{3}\right)^2 = \frac{25}{9}$$

$$x^2 - \frac{10}{3}x + \frac{25}{9} = \frac{2}{3} + \frac{25}{9} \qquad \text{Add } \frac{25}{9} \text{ to both sides}$$

STEP 5   Factor the left side and simplify the right side.

$$\left(x - \frac{5}{3}\right)^2 = \frac{6}{9} + \frac{25}{9} \qquad \text{Factor the left side}$$

$$\left(x - \frac{5}{3}\right)^2 = \frac{31}{9} \qquad \text{Simplify the right side}$$

STEP 6   Apply the Square Root Property.

$$x - \frac{5}{3} = \pm\sqrt{\frac{31}{9}} \qquad \text{Square Root Property}$$

$$x - \frac{5}{3} = \pm\frac{\sqrt{31}}{3}$$

$$x = \frac{5}{3} \pm \frac{\sqrt{31}}{3} \qquad \text{Add } \frac{5}{3} \text{ to both sides}$$

STEP 7   Check in the original equation.

STEP 8   Write the solution set.

$$\left\{ \frac{5}{3} \pm \frac{\sqrt{31}}{3} \right\}$$

This may also be written as $\left\{ \dfrac{5 \pm \sqrt{31}}{3} \right\}$, $\left\{ \dfrac{5}{3} + \dfrac{\sqrt{31}}{3}, \dfrac{5}{3} - \dfrac{\sqrt{31}}{3} \right\}$, or

$\left\{ \dfrac{5 + \sqrt{31}}{3}, \dfrac{5 - \sqrt{31}}{3} \right\}$. ◄

Problem Set 9.2

Warm-Ups

In Problems 1 through 18, calculate the term that must be added to each binomial to make each a perfect trinomial square. For Problems 1 through 6, see Example 2.

1. $x^2 - 4x$ **2.** $x^2 + 8x$ **3.** $x^2 - 6x$

4. $y^2 - 14y$ **5.** $w^2 + 12w$ **6.** $t^2 - 22t$

For Problems 7 through 12, see Example 3.

7. $x^2 + 5x$ **8.** $z^2 + 9z$ **9.** $w^2 - w$

10. $y^2 - 3y$ **11.** $x^2 + 13x$ **12.** $x^2 - 15x$

For Problems 13 through 18, see Example 4.

13. $t^2 + \frac{1}{2}t$ **14.** $y^2 - \frac{1}{3}y$ **15.** $x^2 - \frac{2}{5}x$

16. $z^2 + \frac{2}{7}z$ **17.** $x^2 + \frac{4}{3}x$ **18.** $x^2 - \frac{6}{5}x$

In Problems 19 through 28, solve the quadratic equations by completing the square. See Example 5.

19. $x^2 + 4x - 5 = 0$ **20.** $x^2 - 2x - 8 = 0$ **21.** $y^2 + 6y = 7$

22. $z^2 - 2z = 3$ **23.** $w^2 = 2w - 1$ **24.** $x^2 = 4x - 3$

25. $y^2 + 4y + 4 = 0$ **26.** $x^2 - 6x + 9 = 0$ **27.** $x^2 - 5x = 14$

28. $x^2 = 7x - 10$

In Problems 29 through 34, solve by completing the square. See Example 6.

29. $x^2 + 6x - 2 = 0$

30. $x^2 + 4x = 1$

31. $x^2 - 8x = -11$

32. $y^2 - 10y = -22$

33. $2x^2 - 8x = -4$

34. $3x^2 - 6x - 3 = 0$

In Problems 35 through 42, solve by completing the square. See Example 7.

35. $z^2 + z - 2 = 0$

36. $y^2 - 3y = 4$

37. $2x^2 + 4x - 16 = 0$

38. $3x^2 - 3x - 36 = 0$

39. $2x^2 - 5x = 3$

40. $3z^2 = 7z - 4$

41. $2x^2 - 10x + 1 = 0$

42. $4x^2 = -8x + 1$

Practice Exercises

In Problems 43 through 72, solve each quadratic equation by completing the square.

43. $x^2 + 4x - 12 = 0$

44. $x^2 - 2x - 35 = 0$

45. $y^2 + 6y = 16$

46. $z^2 - 2z = 24$

47. $w^2 = 4w - 4$

48. $x^2 = 6x - 5$

49. $x^2 - 8x + 12 = 0$

50. $x^2 + 12x = -32$

51. $z^2 + z - 12 = 0$

52. $y^2 - 3y = 10$

53. $x^2 - 5x = -6$

54. $x^2 = 7x + 18$

55. $y^2 + 10y + 25 = 0$

56. $x^2 - 8x + 16 = 0$

57. $4y^2 + 8y + 4 = 0$

58. $5z^2 - 10z - 15 = 0$

59. $2x^2 = 3x + 2$

60. $4y^2 + y = 3$

61. $2x^2 + 10x - 28 = 0$

62. $3x^2 + 15x + 18 = 0$

63. $4y^2 + 8y - 12 = 0$

64. $5z^2 - 20z - 20 = 0$

65. $2x^2 + 3x = 1$

66. $3z^2 = 7z - 3$

67. $2x^2 - 5x = 1$

68. $4y^2 - y = 2$

69. $2z^2 - 3z + 1 = 0$

70. $3y^2 + 2y - 2 = 0$

71. $3x^2 - 2x - 1 = 0$

72. $5x^2 + 4x - 2 = 0$

73. The relationship of the speed (s) of a car in miles per hour and the length (l) of its skid marks measured in feet is $s^2 = 12l$. JoAnn was driving on a wet asphalt road and skidded into a telephone pole. The policeman measured the skid marks as 120 feet 4 inches. How fast was JoAnn driving?

74. A weather balloon is 4200 meters high and directly over a point that is 800 meters from the weather station. How far is the balloon from the station?

75. Juanita Perez invested $25,000 in a savings account at the Wachovia Bank. Interest was compounded annually, and two years later the account had $28,622.50 in it. If the bank uses the formula $A = 25,000(1 + r)^2$ to calculate the amount (A) in the account, find the interest rate (r).

76. The area of a circle is 25π square centimeters. Find the radius of the circle.

Problem 73

Challenge Problems

77. Find the height (h) of the following equilateral triangle.

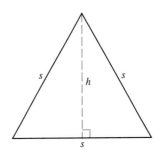

Solve each by completing the square.

78. $x^2 + 2x = c; \quad c \geq 0$

80. $x^2 + bx + c = 0$

79. $x^2 + 4bx = 4$

81. $ax^2 + bx + c = 0; \quad a > 0$

In Your Own Words

82. Explain how to solve $2x^2 - 6x + 8 = 0$ by completing the square.

9.3 The Quadratic Formula

TAPE 15

Completing the square can be used to solve any quadratic equation. However, the method is often messy and time consuming. In this section we solve the general quadratic equation $ax^2 + bx + c = 0$ by completing the square. This gives us a formula that we can use to solve all quadratic equations and avoid having to complete the square.

> ### Standard Form
>
> A quadratic equation written in the form
>
> $$ax^2 + bx + c = 0$$
>
> where a, b, and c are real numbers with $a > 0$, is said to be in **standard form.**

For example,

$$3x^2 + 5x + 8 = 0$$

is a quadratic equation in standard form with $a = 3$, $b = 5$, and $c = 8$. Likewise,

$$2x^2 + (-6)x + (-7) = 0$$

is a quadratic equation in standard form. However, we write it in the form

$$2x^2 - 6x - 7 = 0$$

Notice that $a = 2$, $b = -6$, and $c = -7$. The equation

$$x^2 - 2x = 5$$

is also a quadratic equation. To write it in standard form, we subtract 5 from both sides.

$$x^2 - 2x = 5$$

$$x^2 - 2x - 5 = 0 \qquad \text{Subtract 5 from both sides.}$$

Thus $a = 1$, $b = -2$, and $c = -5$.

As we study quadratic equations, we will see that it is often useful to write them in standard form.

► **EXAMPLE 1**

Write each quadratic equation in standard form and give the values of a, b, and c.

(a) $2x^2 + 3x = -7$ (b) $t^2 = 1$

(c) $2x = -5x^2 + 6$ (d) $\frac{1}{2}x^2 - 2x = 0$

(e) $\sqrt{2}x^2 - \sqrt{7} + x = 0$ (f) $x(x + 3) = 5$

Solutions

(a) $2x^2 + 3x = -7$

We write the equation in standard form,

$$2x^2 + 3x + 7 = 0$$

and we see that $a = 2$, $b = 3$, and $c = 7$.

(b) $t^2 = 1$

In standard form this is

$$t^2 - 1 = 0$$

Notice that having no t term means that we could write

$$t^2 + 0t - 1 = 0$$

So $a = 1$, $b = 0$, and $c = -1$.

(c) $2x = -5x^2 + 6$

We can subtract $2x$ from both sides and write

$$0 = -5x^2 - 2x + 6$$

Standard form requires that a is positive. So we multiply both sides by -1.

$$0 = 5x^2 + 2x - 6$$

Thus $a = 5$, $b = 2$, and $c = -6$.

We could add $5x^2$ to both sides of the original equation and then subtract 6 from both sides to get

$$5x^2 + 2x - 6 = 0$$

giving $a = 5$, $b = 2$, and $c = -6$.

(d) $\frac{1}{2}x^2 - 2x = 0$

This quadratic equation is in standard form with $a = \dfrac{1}{2}$, $b = -2$, and $c = 0$.

(e) $\sqrt{2}x^2 - \sqrt{7} + x = 0$

We rearrange the terms of the quadratic equation to put it in standard form.

$$\sqrt{2}x^2 + x - \sqrt{7} = 0.$$

We can see that $a = \sqrt{2}$, $b = 1$, $c = -\sqrt{7}$.

(f) $x(x + 3) = 5$

To see that this equation is a quadratic equation, we must use the distributive property.

$$x(x + 3) = 5$$
$$x^2 + 3x = 5$$

We write it in standard form,

$$x^2 + 3x - 5 = 0$$

and see that $a = 1$, $b = 3$, and $c = -5$.   ◄

Let's solve $ax^2 + bx + c = 0$ $(a > 0)$ by completing the square. We follow the steps outlined in Section 9.1.

$$ax^2 + bx + c = 0 \qquad \text{Standard form}$$

$$x^2 + \frac{b}{a}x + \frac{c}{a} = 0 \qquad \text{Divide by } a$$

$$x^2 + \frac{b}{a}x = -\frac{c}{a} \qquad \text{Subtract } \frac{c}{a}$$

To complete the square, multiply $\dfrac{b}{a}$ by $\dfrac{1}{2}$, which gives $\dfrac{b}{2a}$. The number that we must add to both sides of the equation is $\left(\dfrac{b}{2a}\right)^2$, or $\dfrac{b^2}{4a^2}$.

$$x^2 + \frac{b}{a}x + \frac{b^2}{4a^2} = -\frac{c}{a} + \frac{b^2}{4a^2} \qquad \text{Add } \frac{b^2}{4a^2}$$

$$\left(x + \frac{b}{2a}\right)^2 = \frac{b^2 - 4ac}{4a^2} \qquad \begin{array}{l}\text{Factor the left side;}\\ \quad\text{simplify the right side}\end{array}$$

$$x + \frac{b}{2a} = \pm\sqrt{\frac{b^2 - 4ac}{4a^2}} \qquad \text{Square Root Property}$$

$$x + \frac{b}{2a} = \pm\frac{\sqrt{b^2 - 4ac}}{2a} \qquad \text{Simplify}$$

$$x = -\frac{b}{2a} \pm \frac{\sqrt{b^2 - 4ac}}{2a} \qquad \text{Subtract } \frac{b}{2a}$$

$$x = \frac{-b \pm \sqrt{b^2 - 4ac}}{2a}$$

This equation is called the **Quadratic Formula.** It should be memorized. Although we derived the formula for $a > 0$, it works just as well for $a < 0$.

The Quadratic Formula

The quadratic equation $ax^2 + bx + c = 0$, with $a \neq 0$, has the same solution set as

$$x = \frac{-b \pm \sqrt{b^2 - 4ac}}{2a}$$

The Quadratic Formula can be used to solve any quadratic equation.

Using the Quadratic Formula to Solve Quadratic Equations

1. Write the equation in standard form.
2. Identify the values of a, b, and c.
3. Substitute these values into the formula and simplify.
4. Check if required.
5. Write the solution set.

▶ **EXAMPLE 2**

Solve $x^2 + 4x - 5 = 0$ by the Quadratic Formula.

Solution

We note that the equation is in standard form. So $a = 1$, $b = 4$, and $c = -5$. Substitute these values into the Quadratic Formula.

$$x = \frac{-b \pm \sqrt{b^2 - 4ac}}{2a}$$

$$= \frac{-4 \pm \sqrt{4^2 - 4(1)(-5)}}{2(1)}$$

$$= \frac{-4 \pm \sqrt{16 + 20}}{2}$$

$$= \frac{-4 \pm \sqrt{36}}{2}$$

$$= \frac{-4 \pm 6}{2}$$

$$x = \frac{-4 + 6}{2} \quad \text{or} \quad x = \frac{-4 - 6}{2}$$

$$x = 1 \quad \text{or} \quad x = -5$$

$$\{-5, 1\}$$ ◀

▶ **EXAMPLE 3**

Solve $2x^2 - 3x = 3$ by the quadratic formula.

Solution

Writing the equation in standard form, $2x^2 - 3x - 3 = 0$, we see that $a = 2$, $b = -3$, and $c = -3$. Substituting these into the quadratic formula gives

$$x = \frac{-b \pm \sqrt{b^2 - 4ac}}{2a}$$

$$= \frac{-(-3) \pm \sqrt{(-3)^2 - 4(2)(-3)}}{2(2)}$$

Notice that b has a value of -3, so $-b$ is $-(-3)$.

$$x = \frac{3 \pm \sqrt{9 + 24}}{4}$$

$$= \frac{3 \pm \sqrt{33}}{4}$$

$$\left\{ \frac{3 \pm \sqrt{33}}{4} \right\}$$

We may also write the solution set as $\left\{ \frac{3}{4} \pm \frac{\sqrt{33}}{4} \right\}$, $\left\{ \frac{3 + \sqrt{33}}{4}, \frac{3 - \sqrt{33}}{4} \right\}$,

or $\left\{ \frac{3}{4} + \frac{\sqrt{33}}{4}, \frac{3}{4} - \frac{\sqrt{33}}{4} \right\}$.   ◀

The quadratic formula often leads to expressions such as

$$\frac{-6 \pm \sqrt{24}}{8}$$

Many mistakes are made in simplifying these expressions. Notice how this can be simplified.

► EXAMPLE 4

Simplify $\dfrac{-6 \pm \sqrt{24}}{8}$.

Solution

Notice first that $\sqrt{24}$ can be simplified.

$$\sqrt{24} = \sqrt{4 \cdot 6} = \sqrt{4}\sqrt{6} = 2\sqrt{6}$$

Therefore,

$$\frac{-6 \pm \sqrt{24}}{8} = \frac{-6 \pm 2\sqrt{6}}{8}$$

Be Careful!

There are two ways to reduce these expressions. One way is to factor the numerator and divide out all common factors.

$$\frac{-6 \pm 2\sqrt{6}}{8} = \frac{2(-3 \pm \sqrt{6})}{2 \cdot 4}$$

$$= \frac{-3 \pm \sqrt{6}}{4}$$

Another way is to divide each term in the numerator by the denominator.

$$\frac{-6 \pm 2\sqrt{6}}{8} = \frac{-6}{8} \pm \frac{2\sqrt{6}}{8}$$

$$= -\frac{3}{4} \pm \frac{\sqrt{6}}{4}$$

Notice that both answers are the same, although they are written in different forms. That is,

$$\frac{-3 \pm \sqrt{6}}{4} = -\frac{3}{4} \pm \frac{\sqrt{6}}{4}$$ ◀

► EXAMPLE 5

Solve $x^2 = 4x - 2$ by the quadratic formula.

Solution

We write the equation in standard form $x^2 - 4x + 2 = 0$. From this we see that $a = 1, b = -4$, and $c = 2$. Be careful in substituting the value of b into the formula. Since b is -4, $-b$ is $-(-4)$.

*Be
Careful!*

$$x = \frac{-b \pm \sqrt{b^2 - 4ac}}{2a}$$

$$= \frac{-(-4) \pm \sqrt{(-4)^2 - 4(1)(2)}}{2(1)}$$

$$= \frac{4 \pm \sqrt{16 - 8}}{2}$$

$$= \frac{4 \pm \sqrt{8}}{2}$$

$$= \frac{4 \pm 2\sqrt{2}}{2}$$

$$= \frac{4}{2} \pm \frac{2\sqrt{2}}{2}$$

$$= 2 \pm \sqrt{2}$$

$$\{2 \pm \sqrt{2}\}$$ ◀

Each time that we use the Quadratic Formula, we must find the value of the expression $\sqrt{b^2 - 4ac}$. In Examples 2, 3, and 5, the number under the square root was a positive number. If we solve $x^2 - x + 1 = 0$ by the Quadratic Formula, $\sqrt{b^2 - 4ac}$ is $\sqrt{1^2 - 4(1)(1)}$, which is $\sqrt{-3}$. Since $\sqrt{-3}$ is not a real number, there are no real solutions to the equation $x^2 - x + 1 = 0$. This idea is examined further in Section 9.4.

Calculator Box

Storing and Recalling Numbers

Storing a Number in Memory

First, we learn how to store a number for later use. Look for a key like $\boxed{\text{STO}}$ $\boxed{\text{M}_{\text{IN}}}$ or $\boxed{\rightarrow}$. Let's put $9 + 7$ in memory.

Scientific Calculator:

Press $\boxed{9}$ $\boxed{+}$ $\boxed{7}$ $\boxed{=}$ $\boxed{\text{STO}}$ to store the displayed number.

Graphing Calculator:

Graphing calculators usually have at least 26 memories indexed by letters of the alphabet and the $\boxed{\text{ALPHA}}$ key.

Press $\boxed{9}$ $\boxed{+}$ $\boxed{7}$ $\boxed{\rightarrow}$ $\boxed{\text{ALPHA}}$ $\boxed{\text{J}}$ $\boxed{\text{EXE}}$ to store the result of $9 + 7$ in memory J. Some graphing calculators do not require the $\boxed{\text{ALPHA}}$ key when *storing* a number. With the TI-81, press $\boxed{9}$ $\boxed{+}$ $\boxed{7}$ $\boxed{\text{STO}}$ $\boxed{\text{J}}$ $\boxed{\text{ENTER}}$.

Recalling a Number from Memory

Now let's recall our stored number from memory and use it to calculate $\sqrt{9 + 7}$.

Scientific Calculator:

Clear the display and press $\boxed{\text{RCL}}$ $\boxed{\sqrt{}}$ and read ▮4▮ on the display. Some calculators use $\boxed{\text{MR}}$ instead of $\boxed{\text{RCL}}$ to recall a stored number.

Graphing Calculator:

Next, press $\boxed{\sqrt{}}$ $\boxed{\text{ALPHA}}$ $\boxed{\text{J}}$ $\boxed{\text{ENTER}}$ to compute $\sqrt{9 + 7}$.

Calculator Exercises

1. Store 266 in memory; then recall it to calculate $\sqrt{266}$.

2. Store $2 - \sqrt{2}$ in memory; then recall it to calculate $(2 - \sqrt{2})^3$.
   Store $11^2 - 4 \cdot 2 \cdot 3$ in memory for Problems 3 and 4.

3. Calculate $5 + \sqrt{\text{stored number}}$.

4. Calculate $5 - \sqrt{\text{stored number}}$.

Answers

1. ≈ 16.3095 **2.** ≈ 0.20101 **3.** ≈ 14.84886 **4.** ≈ -4.84886

Problem Set 9.3

Warm-Ups

In Problems 1 through 14, write each quadratic equation in standard form and give the values of a, b, and c for each equation. See Example 1.

1. $4x^2 - 5x + 7 = 0$

2. $x^2 = 6x$

3. $x^2 - 8x = 7$

4. $t^2 - t + 1 = 0$

5. $x^2 = 0$

6. $x^2 = 16$

7. $y = 3y^2 + 5$

8. $\frac{1}{2}x^2 = x - \frac{2}{3}$

9. $-2x^2 = 4x - 7$

10. $2w = w^2 - 3$

11. $x(x + 1) = 7$

12. $5 - 2x(x + 2) = 2x$

13. $(y + 4)(y - 5) = 0$

14. $z^2 + z(z - 4) = 0$

In Problems 15 through 34, solve each quadratic equation by using the Quadratic Formula. For Problems 15 through 26, see Example 2.

15. $x^2 + 3x + 2 = 0$

16. $x^2 + 2x - 3 = 0$

17. $x^2 - 3x = 4$

18. $x^2 - 5x = 6$

19. $2x^2 + 7x + 3 = 0$

20. $2x^2 - 5x + 2 = 0$

21. $6x^2 + x - 2 = 0$

22. $9x^2 + 3x - 2 = 0$

23. $6x^2 + 7x - 3 = 0$

24. $2x^2 = 3x + 5$

25. $2x - 1 = x^2$

26. $x^2 + 6x + 9 = 0$

For Problems 27 through 34, see Example 3.

27. $x^2 - 3x - 1 = 0$

28. $x^2 + 3x - 2 = 0$

29. $2x^2 = 7x - 4$

30. $(3x + 4)(x - 1) = -3$

31. $x^2 + 5x = -2$

32. $3x - 1 = x^2$

33. $x(x - 3) = 3$

34. $x(2x + 3) = 1$

In Problems 35 through 38, reduce each expression. See Example 4.

35. $\dfrac{4 \pm \sqrt{8}}{2}$

36. $\dfrac{-6 \pm \sqrt{12}}{4}$

37. $\dfrac{-8 \pm \sqrt{32}}{4}$

38. $\dfrac{-1 \pm \sqrt{8}}{2}$

In Problems 39 through 48, solve each quadratic equation by using the Quadratic Formula. See Example 5.

39. $x^2 + 2x - 4 = 0$

40. $x^2 - 3x - 9 = 0$

41. $2x^2 - 4x = -1$

42. $3x^2 = 2x + 1$

43. $2x^2 + 6x = -3$

44. $5x^2 - 4x = 4$

45. $3x^2 + 6x + 1 = 0$

46. $3x^2 - 8x + 4 = 0$

47. $2x^2 - 2x - 3 = 0$

48. $5x^2 + 2x - 1 = 0$

Practice Exercises

In Problems 49 through 74, solve each quadratic equation by using the Quadratic Formula.

49. $x^2 + 5x + 4 = 0$

50. $x^2 + 6x - 7 = 0$

51. $x(x - 2) = 8$

52. $x(x + 5) = 6$

53. $3x^2 + 2x - 5 = 0$

54. $14x^2 - 3x - 2 = 0$

55. $6x^2 + 11x + 4 = 0$

56. $8x^2 - 2x - 15 = 0$

57. $x^2 - 5x - 1 = 0$

58. $x^2 + 5x - 2 = 0$

59. $2x^2 = 9x - 4$

60. $3x^2 = 2 - x$

61. $x^2 + 5x = -3$

62. $5x - 5 = x^2$

63. $4x - 4 = x^2$

64. $x^2 + 8x + 16 = 0$

65. $x^2 + 2x - 1 = 0$

66. $x^2 - 4x - 2 = 0$

67. $2x^2 - 6x = -1$

68. $3x^2 = 4x + 2$

69. $2x^2 + 8x = -5$

70. $5x^2 - 2x = 2$

71. $3x^2 + 4x + 1 = 0$

72. $7x^2 - 12x - 4 = 0$

73. $2x^2 - 2x - 1 = 0$

74. $5x^2 + 6x - 2 = 0$

75. The granite monument shown is shaped like a pyramid and weighs 216,000 pounds. The formula for calculating the weight in pounds for such a pyramid is

$$W = 600s(s + 4)$$

where s is measured in feet. Approximate to the nearest foot the dimensions of the base of the pyramid.

76. The rectangular storage bin shown holds 200 bushels of grain. Approximate the dimensions of the bin to the nearest tenth of a foot. One cubic foot holds 0.8 of a bushel.

77. Happy Herman's Inc. makes and sells gourmet gift baskets. The owner has determined that the relationship between the number of employees and the number of gift baskets produced daily is

$$y = x^2 + 4x$$

where x is the number of employees and y is the number

of gift baskets made. The manager has also determined that the company can sell at least 16 gift baskets a day.
(a) How many employees must the company hire to be assured of making 16 baskets daily?
(b) How many more baskets would the company make if one more person is hired?

Problem 77

Challenge Problems

In Problems 78 through 81, write each quadratic equation in standard form and give the values of a, b, and c in each.

78. $x^2 + x + k = 0$

79. $gx^2 - 7x = -5$

80. $\pi x^2 - x = k$

81. $x^2 = \pi^2$

In Problems 82 through 84, solve by using the Quadratic Formula.

82. $x^2 + kx + 1 = 0$

83. $x^2 + 3x + k = 0$

84. $x^2 + px + q = 0$

In Your Own Words

85. Explain what the Quadratic Formula is and why it is useful.

9.4 Quadratic Equations with Complex Solutions

TAPE 15

In Section 8.1 we saw that expressions such as $\sqrt{-4}$ have no meaning as real numbers because there is no real number that equals -4 when squared. The quadratic equation $x^2 + 4 = 0$ has no solution in the set of real numbers. Any solution to this equation must be a number whose square is -4. Since there is no real number with this property, we will have to *enlarge* our set of numbers to include solutions to such equations. We call this new, larger set of numbers the **complex numbers.**

We construct the complex numbers by introducing a single new number to the set of real numbers. The new number is the number that when squared is -1. We call it i. That is, i is a number such that

$$i^2 = -1$$

The Set of Complex Numbers

The set of complex numbers is the set of all numbers that can be written as

$$a + bi$$

where a and b are real numbers and i is a number with the property

$$i^2 = -1$$

Some examples of complex numbers are

$$5 + 3i \qquad \frac{-6}{7} + 13i \qquad 1 - 2i$$

$$1 + \sqrt{22}i \qquad 6i \qquad -17$$

Note that the set of real numbers is a subset of the set of complex numbers. (The real number -17 can be written as $-17 + 0i$; thus it belongs to the set of complex numbers.)

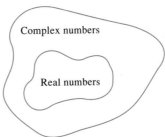

Next, let's look at $\sqrt{-4}$. This should be the number that when squared is -4. Of course, it cannot be a real number, but is it a complex number? To answer this question, notice that

$$(2i)^2 = (2)^2 \cdot i^2 = 4 \cdot (-1) = -4$$

Thus a sensible definition is $\sqrt{-4} = \sqrt{4}i$.

Negative Radicand

If k is any positive real number,

$$\sqrt{-k} = \sqrt{k}i$$

Be Careful!

Note that the i is *not* under the radical. Sometimes $\sqrt{k}i$ is written as $i\sqrt{k}$.

▶ **EXAMPLE 1**

Simplify the following complex numbers.

(a) $\sqrt{-4}$ (b) $\sqrt{-7}$ (c) $\sqrt{-12}$

Solutions

(a) $\sqrt{-4} = \sqrt{4}i = 2i$

(b) $\sqrt{-7} = \sqrt{7}i$

(c) $\sqrt{-12} = \sqrt{12}i$

$\qquad\qquad = \sqrt{4 \cdot 3}i$

$\qquad\qquad = 2\sqrt{3}i$ ◀

Be Careful!

When dealing with symbols like $\sqrt{-5}$, it is *important* to rewrite the radical as $\sqrt{5}i$ *before* doing any arithmetic.

Arithmetic with complex numbers is done like arithmetic with polynomials, remembering to replace i^2 with -1.

► **EXAMPLE 2**

Perform the operations indicated.

(a) $(3 + 5i) + (2 - 3i)$
(b) $(3 + 5i) - (2 - 3i)$
(c) $(3 + 5i)(2 - 3i)$

Solutions

(a) $(3 + 5i) + (2 - 3i) = (3 + 2) + (5i - 3i)$
$$= 5 + 2i$$

(b) $(3 + 5i) - (2 - 3i) = (3 + 5i) + (-2 + 3i)$
$$= (3 - 2) + (5i + 3i)$$
$$= 1 + 8i$$

(c) $(3 + 5i)(2 - 3i) = 6 - 9i + 10i - 15i^2$
$$= 6 + i - 15(-1) \qquad i^2 = -1$$
$$= 6 + i + 15$$
$$= 21 + i$$

Special products work with complex numbers as well.

► **EXAMPLE 3**

Perform the operations indicated.

(a) $(2 + 3i)(2 - 3i)$
(b) $(2 + 3i)^2$

Solutions

(a) The complex numbers $a + bi$ and $a - bi$ are called **complex conjugates.**
The product of conjugates is an important special product.

$$(2 + 3i)(2 - 3i) = 2^2 - (3i)^2$$
$$= 4 - 9i^2 = 4 - 9(-1)$$
$$= 4 + 9 = 13$$

(b) $(2 + 3i)^2 = 4 + 12i + 9i^2$
$$= 4 + 12i + 9(-1), \qquad i^2 = -1$$
$$= 4 + 12i - 9$$
$$= -5 + 12i$$

To do a division problem in complex numbers, we write it as a fraction and multiply the numerator and denominator by the conjugate of the denominator.

► **EXAMPLE 4**

Perform the operation indicated.

$$(2 + 3i) \div (1 + i)$$

Solution

Rewrite the problem as a fraction.

$$(2 + 3i) \div (1 + i) = \frac{2 + 3i}{1 + i}$$

Next rationalize the denominator by multiplying the numerator and denominator by the *conjugate* of the *denominator*.

$$\frac{2 + 3i}{1 + i} = \frac{(2 + 3i)\ (1 - i)}{(1 + i)\ (1 - i)}$$

$$= \frac{2 - 2i + 3i - 3i^2}{1^2 - i^2}$$

$$= \frac{2 - 2i + 3i + 3}{1 - (-1)}$$

$$= \frac{5 + i}{2} \quad \text{or} \quad \frac{5}{2} + \frac{1}{2}i$$ ◄

Now we can solve more quadratic equations.

► **EXAMPLE 5**

Solve the equations by the Square Root Property.

(a) $x^2 + 9 = 0$ (b) $(t - 1)^2 = -8$

Solution

(a) Write the equation in the form $x^2 = A$.

$$x^2 + 9 = 0$$

$$x^2 = -9$$

$$x = \pm\sqrt{-9} \qquad \text{Square Root Property}$$

$$x = \pm\sqrt{9}i$$

$$x = \pm 3i$$

$$\{\pm 3i\}$$

(b) $(t - 1)^2 = -8$

$$t - 1 = \pm\sqrt{-8} \qquad \text{Square Root Property}$$

$$t - 1 = \pm\sqrt{8}i$$

$$t - 1 = \pm 2\sqrt{2}i$$

$$t = 1 \pm 2\sqrt{2}i \qquad \text{Add 1 to both sides}$$

$$\{1 \pm 2\sqrt{2}i\}$$ ◄

▶ **EXAMPLE 6**

Solve the equations using the Quadratic Formula.

(a) $x^2 - 2x = -5$
(b) $3x^2 + 4x + 2 = 0$

Solutions

(a) Write the equation in standard form first.

$$x^2 - 2x = -5$$

$$x^2 - 2x + 5 = 0 \qquad \text{Standard form}$$

$$x = \frac{-(-2) \pm \sqrt{(-2)^2 - 4(1)(5)}}{2(1)}$$

$$= \frac{2 \pm \sqrt{4 - 20}}{2}$$

$$= \frac{2 \pm \sqrt{-16}}{2}$$

$$= \frac{2 \pm \sqrt{16}i}{2}$$

$$= \frac{2 \pm 4i}{2}$$

$$= 1 \pm 2i$$

$$\{1 \pm 2i\} \quad \text{or} \quad \{1 + 2i, 1 - 2i\}$$

(b) $3x^2 + 4x + 2 = 0$

$$x = \frac{-4 \pm \sqrt{4^2 - 4(3)(2)}}{2(3)}$$

$$= \frac{-4 \pm \sqrt{16 - 24}}{6}$$

$$= \frac{-4 \pm \sqrt{-8}}{6}$$

$$= \frac{-4 \pm \sqrt{8}i}{6}$$

$$= \frac{-4 \pm 2\sqrt{2}i}{6}$$

$$= \frac{-4}{6} \pm \frac{2\sqrt{2}}{6}i$$

$$= -\frac{2}{3} \pm \frac{\sqrt{2}}{3}i$$

$$\left\{-\frac{2}{3} \pm \frac{\sqrt{2}}{3}i\right\}$$

◀

Problem Set 9.4

Warm-Ups

In Problems 1 through 6, simplify each complex number. See Example 1.

1. $\sqrt{-9}$

2. $\sqrt{-49}$

3. $\sqrt{-5}$

4. $\sqrt{-3}$

5. $\sqrt{-8}$

6. $\sqrt{-24}$

In Problems 7 through 16, perform the operations indicated. See Example 2.

7. $(1 + i) + (2 + 3i)$

8. $(3 + 5i) - (2 + 3i)$

9. $(5 - 2i) + (3 - 5i)$

10. $(-4 - i) + (1 + 5i)$

11. $(2 + 3i) - (6 + 5i)$

12. $(5 - 6i) - (3 + 6i)$

13. $(1 + i)(2 + 3i)$

14. $(1 - i)(5 + 2i)$

15. $(1 + 2i)(3 - 2i)$

16. $(2 + 3i)(3 - 4i)$

In Problems 17 through 20, perform the multiplication indicated. See Example 3.

17. $(4 - 3i)^2$

18. $(3 + i)^2$

19. $(2 - i)(2 + i)$

20. $(5 + 2i)(5 - 2i)$

In Problems 21 through 26, perform the division indicated. See Example 4.

21. $\dfrac{1}{1 + i}$

22. $\dfrac{1}{2 - i}$

23. $\dfrac{2}{1 - 3i}$

24. $\dfrac{1 + 2i}{1 + 3i}$

25. $\dfrac{2 - i}{3 + 2i}$

26. $\dfrac{2 - i}{2 + i}$

In Problems 27 through 32, find the solution set by the Square Root Property. See Example 5.

27. $x^2 + 1 = 0$

28. $w^2 = -16$

29. $2x^2 + 14 = 0$

30. $\dfrac{1}{2}y^2 = -3$

31. $(x - 2)^2 = -12$

32. $(y + 3)^2 = -27$

In Problems 33 through 40, find the solution set by using the Quadratic Formula. See Example 6.

33. $t^2 + t + 1 = 0$

34. $x^2 + 3x = -3$

35. $5x^2 + 4x = -1$

36. $9x^2 = 3x - 1$

37. $x^2 + 2x + 5 = 0$

38. $4x^2 = 2x - 3$

39. $x^2 + x + 2 = 0$

40. $3x^2 + 2x = -1$

Practice Exercises

In Problems 41 through 60, perform the operations indicated.

41. $(1 - \sqrt{-1}) + (2 - \sqrt{-9})$

42. $(3 - \sqrt{-25}) - (2 - \sqrt{-9})$

43. $(5 + 2i) + (3 + 5i)$

44. $(-4 + i) - (1 - 5i)$

45. $(3 + \sqrt{-9}) - (5 + \sqrt{-25})$

46. $(6 - \sqrt{-49}) - (4 + \sqrt{-25})$

47. $(1 - i)(3 + 2i)$

48. $(1 + i)(2 + 4i)$

49. $(1 - 3i)(2 - 3i)$

50. $(3 + 2i)(2 - 5i)$

51. $(1 - i)^2$

52. $(2 + 3i)^2$

53. $(3 - i)(3 + i)$

54. $(4 + 3i)(4 - 3i)$

55. $\dfrac{1}{1-i}$

56. $\dfrac{1}{3+i}$

57. $\dfrac{3}{1-2i}$

58. $\dfrac{2+i}{2-3i}$

59. $\dfrac{3-i}{2+3i}$

60. $\dfrac{5-i}{5+i}$

In Problems 61 through 66, find the solution set by the Square Root Property.

61. $x^2 + 36 = 0$

62. $w^2 = -49$

63. $2x^2 + 10 = 0$

64. $\dfrac{1}{3}y^2 = -7$

65. $(x-3)^2 = -8$

66. $(y+1)^2 = -12$

In Problems 67 through 74, find the solution set by using the Quadratic Formula.

67. $t^2 - t + 3 = 0$

68. $x^2 - 3x = -3$

69. $5x^2 - 4x = -1$

70. $x^2 = 4x - 8$

71. $x^2 - 2x + 5 = 0$

72. $4x^2 = 5x - 3$

73. $2x^2 + 4x + 5 = 0$

74. $3x^2 + 6x = -4$

Challenge Problems

75. Factor $x^2 + 9$. (*Hint:* $-9 = (3i)^2$)

76. Show that $(a + bi)(a - bi) = a^2 + b^2$.

In Your Own Words

77. Why was it necessary to introduce the number i?

9.5 Summary of Solving Quadratic Equations

TAPE 15

We have learned how to solve a quadratic equation by factoring, using the Square Root Property, completing the square, and using the Quadratic Formula. In this section we examine how to choose the most appropriate method to use in solving a quadratic equation.

Any quadratic equation can be solved by using the Quadratic Formula or by completing the square. However, factoring is much easier if it can be used. The Square Root Property saves some steps if it can be used directly.

> ### Choosing a Method to Use in Solving a Quadratic Equation
>
> **1.** Determine if the equation is of the form $X^2 = A$ or can be written as $X^2 = A$ without completing the square. If so, use the Square Root Property.
> **2.** Write the equation in standard form.
> **3.** Try to factor the nonzero side. If it can be factored easily, use factoring.
> **4.** Use the Quadratic Formula if factoring cannot be used.
> **5.** Use completing the square when directed to do so.

► **EXAMPLE 1**

Solve each quadratic equation.

(a) $3x^2 = 54$ (b) $(x - 5)^2 + 8 = 0$

Solutions

(a) This equation can be written in the form $\mathbf{X}^2 = A$. The Square Root Property is appropriate.

$$3x^2 = 54$$
$$x^2 = 18 \qquad\qquad \text{Divide both sides by 3}$$
$$x = \pm\sqrt{18} \qquad\qquad \text{Square Root Property}$$
$$x = \pm\sqrt{9 \cdot 2}$$
$$x = \pm\sqrt{9}\sqrt{2}$$
$$x = \pm3\sqrt{2}$$
$$\{\pm3\sqrt{2}\}$$

(b) This equation can be written in the form $\mathbf{X}^2 = A$.

$$(x - 5)^2 = -8$$
$$x - 5 = \pm\sqrt{-8} \qquad\qquad \text{Square Root Property}$$
$$x - 5 = \pm\sqrt{8}i$$
$$x - 5 = \pm2\sqrt{2}i$$
$$x = 5 \pm 2\sqrt{2}i$$
$$\{5 \pm 2\sqrt{2}i\}$$

◄

► **EXAMPLE 2**

Solve each equation.

(a) $2x^2 - 13x + 15 = 0$ (b) $2(y^2 - y) = 24$ (c) $z^2 + z = 0$

Solutions

(a) This equation is not in the form $\mathbf{X}^2 = A$ as in Example 1. We try factoring. Notice that it is in standard form.

$$(2x - 3)(x - 5) = 0 \qquad \text{Factor}$$
$$\left\{\frac{3}{2}, 5\right\}$$

(b) This equation is not in the form $\mathbf{X}^2 = A$. We try factoring. Notice that 2 is a factor of each side of the equation. We can divide both sides by 2 before we begin.

$$\frac{2(y^2 - y)}{2} = \frac{24}{2}$$
$$y^2 - y = 12$$

Be careful to write the equation in standard form.

$$y^2 - y - 12 = 0$$

$$(y - 4)(y + 3) = 0$$

$$\{-3, 4\}$$

(c) $z^2 + z = 0$

$z(z + 1) = 0$ Factor

$\{-1, 0\}$

► **EXAMPLE 3**

Find the solution set for each equation.

(a) $(t + 1)(t + 2) = 5$
(b) $x^2 - x + 3 = 0$

Solutions

(a) Be careful to write the equation in standard form first.

$$(t + 1)(t + 2) = 5$$

$$t^2 + 3t + 2 = 5$$

$$t^2 + 3t - 3 = 0$$

Try to factor. This cannot be factored. Use the Quadratic Formula.

$$t = \frac{-3 \pm \sqrt{3^2 - 4(1)(-3)}}{2(1)}$$

$$= \frac{-3 \pm \sqrt{9 + 12}}{2}$$

$$= \frac{-3 \pm \sqrt{21}}{2}$$

$$\left\{ \frac{-3 \pm \sqrt{21}}{2} \right\}$$

(b) We use the quadratic formula.

$$x = \frac{-(-1) \pm \sqrt{(-1)^2 - 4(1)(3)}}{2(1)}$$

$$= \frac{1 \pm \sqrt{-11}}{2}$$

$$= \frac{1 \pm \sqrt{11}i}{2} = \frac{1}{2} \pm \frac{\sqrt{11}}{2}i$$

$$\left\{ \frac{1}{2} \pm \frac{\sqrt{11}}{2}i \right\}$$

◄

Calculator Box

Quadratic Formula with a Calculator

Quadratics with Real Solutions

Let's approximate the solutions for $2x^2 + 6x + 3 = 0$. Substituting in the quadratic formula yields $\dfrac{-6 \pm \sqrt{6^2 - 4 \cdot 2 \cdot 3}}{2 \cdot 2}$. It is convenient to calculate the solution using the $+$ of the $\pm$ first. We will store the square root in memory so that we won't have to calculate it twice.

Scientific Calculator:

Press $\boxed{36}$ $\boxed{-}$ $\boxed{24}$ $\boxed{=}$ $\boxed{\sqrt{}}$ $\boxed{\text{STO}}$ (store the square root) $\boxed{-}$ $\boxed{6}$ $\boxed{=}$ $\boxed{\div}$ $\boxed{4}$ $\boxed{=}$ and we read -0.633974596 on the display. (Notice that we did simple arithmetic in our head.) Now for the second solution, press $\boxed{6}$ $\boxed{+/-}$ $\boxed{-}$ $\boxed{\text{RCL}}$ $\boxed{=}$ $\boxed{\div}$ $\boxed{4}$ $\boxed{=}$ and read -2.3660254 .

Graphing Calculator:

Press $\boxed{\sqrt{}}$ $\boxed{(}$ $\boxed{36}$ $\boxed{-}$ $\boxed{24}$ $\boxed{)}$ $\boxed{\text{STO}}$ $\boxed{\text{ALPHA}}$ $\boxed{\text{H}}$ $\boxed{\text{ENTER}}$ (to store) $\boxed{-}$ $\boxed{6}$ $\boxed{\text{ENTER}}$ $\boxed{\div}$ $\boxed{4}$ $\boxed{\text{ENTER}}$ for the first solution and $\boxed{(}$ $\boxed{(-)}$ $\boxed{6}$ $\boxed{-}$ $\boxed{\text{ALPHA}}$ $\boxed{\text{H}}$ $\boxed{)}$ $\boxed{\div}$ $\boxed{4}$ $\boxed{\text{ENTER}}$ for the second.
Our two solutions are approximately -0.6340 and -2.3360.

Quadratics with Nonreal Solutions

Next, let's approximate the solutions for $3x^2 - x + 2 = 0$. Substituting in the quadratic formula gives $\dfrac{1 \pm \sqrt{1^2 - 4 \cdot 3 \cdot 2}}{3 \cdot 2}$. Here, when we calculate the expression under the square root, we get -23, a negative number. We will have two complex solutions. The value of the square root is $\sqrt{23}\,i$. Therefore, our solutions are $\dfrac{1}{6} \pm \dfrac{\sqrt{23}}{6}\,i$. Or, with either type of calculator, they are approximately $1.6667 \pm 0.7993i$.

Calculator Exercises

Approximate the solutions to each quadratic equation to the nearest thousandth.

1. $x^2 + x - 1 = 0$
2. $2x^2 - 7x + 2 = 0$

3. $3.6640x^2 + 9x - 4.4479 = 0$
4. $3x^2 - 2x + 5 = 0$

Answers

1. Approximately -1.618 and 0.618
2. Approximately 0.314 and 3.186

3. Approximately 0.422 and -2.878
4. Approximately $0.333 \pm 1.247i$

PROBLEM SET 9.5

Warm-Ups

For Problems 1 through 8, solve each equation. See Example 1.

1. $2x^2 = 10$

2. $3x^2 = -18$

3. $(x - 7)^2 = -5$

4. $(y + 1)^2 = 4$

5. $(x - 8)^2 + 12 = 0$

6. $(z - 2)(z + 2) = 4$

7. $\frac{1}{2}x^2 - 4 = 0$

8. $4t^2 - 25 = 0$

For Problems 9 through 13, solve each equation. See Example 2.

9. $x^2 + 5x = 0$

10. $t^2 = t$

11. $x^2 - 3x = 10$

12. $2z^2 = 5z + 3$

13. $(x + 4)(2x + 1) = 0$

For Problems 14 through 20, solve each equation. See Example 3.

14. $x^2 + x - 1 = 0$

15. $x^2 - x + 1 = 0$

16. $x(x - 2) = 5$

17. $x^2 + 6 = 2(x - 3)$

18. $2x^2 + 1 = -x$

19. $3x^2 - 2x + 1 = 0$

20. $4(x^2 + 4) = 12x$

Practice Exercises

In Problems 21 through 40, solve the equations.

21. $3x^2 = 15$

22. $5x^2 = -45$

23. $(x + 7)^2 = -3$

24. $(y - 8)^2 = 9$

25. $2x^2 + 12x = 0$

26. $t^2 = 4t$

27. $x^2 + 3x = 28$

28. $2z^2 = 3z + 5$

29. $x^2 - x - 1 = 0$

30. $x^2 - x + 1 = 0$

31. $9t^2 - 49 = 0$

32. $\frac{1}{3}x^2 - 9 = 0$

33. $(x - 4)(2x - 1) = 0$

34. $(z - 3)(z + 3) = 9$

35. $x(x + 2) = 5$

36. $x^2 + 4 = 2(x - 4)$

37. $(x - 1)^2 + 24 = 0$

38. $3x^2 + 2 = -x$

39. $4x^2 - 2x + 3 = 0$

40. $5(x^2 + 4) = 15x$

41. A rectangular swimming pool with dimensions 7 m by 14 m is to have a walk of uniform width around it. Find the width of the walk if the combined area of the pool and walk is 260 m².

42. The area of the following figure is x^2 square units. Find x.

43. A rectangular garden with dimensions of 8 yd by 12 yd is to have a walk of uniform width around it. Find the width of the walk if the area of the walk is equal to the area of the garden.

44. The area of the following figure is 230 square units. Find x.

45. The recreation department treats its pools chemically to control the growth of bacteria. The bacteria count (B) measured in thousands per cubic centimeter is given by $B = \frac{1}{2}d^2 - 5d + 25$, where d is the number of days after a treatment. When will the bacteria count be 37,000?

Challenge Problems

46. Solve $s = \frac{1}{2}gt^2$ for t.

47. Solve $t^2 + v_0t + s_0 = 0$ for t.

In Your Own Words

48. Explain how to choose a method to use in solving a quadratic equation.

9.6   Graphing Quadratic Functions

TAPE 15

In Section 6.3 we learned that the graph of the first-degree equation

$$y = mx + b$$

is a line. Here we will examine the graphs of second-degree equations of the form

$$y = ax^2 + bx + c, \qquad a \neq 0$$

First let's graph a simple example of such an equation, $y = x^2$. We begin by making a table of x and y coordinates and then plotting the points.

x	y
0	0
1	1
2	4
3	9
-1	1
-2	4
-3	9

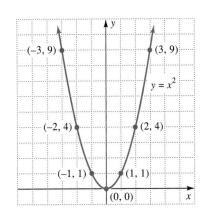

This graph is called a **parabola.** Parabolas are common in everyday life. The reflector in a flashlight is in the shape of a parabola. The cables of the Verrazano Narrows bridge hang in parabolic form, and a parabola was utilized in the design of the 10-meter Keck telescope on Mount Mauna Kea.

Notice that the parabola is drawn as a smooth curve. There are no sharp places. We make an extra effort to draw parabolas as smooth curves. The lowest point on the previous parabola (0, 0) is called the **vertex.**

► **EXAMPLE 1**

Graph the equation $y = x^2 + 1$ and identify the vertex.

Solution

Again we make a table of x–y values and sketch the graph.

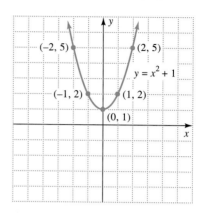

x	y
0	1
1	2
2	5
−1	2
−2	5

By inspection, we see that the vertex is the point, (0, 1).   ◄

Compare the graph in Example 1 with the graph of $y = x^2$ shown earlier in the section. Notice that the size and shape of the two curves are just the same. The graph of $y = x^2 + 1$ is exactly the graph of $y = x^2$ shifted up one unit.

► **EXAMPLE 2**

Graph $y = x^2 - 2$ and identify the vertex.

Solution

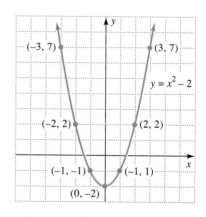

x	y
0	−2
1	−1
2	2
3	7
−1	−1
−2	2

The vertex is at (0, −2).   ◄

The graph of $y = x^2 - 2$ is exactly the graph of $y = x^2$ shifted down two units. As a matter of fact, the graph of the equation $y = x^2 + k$ is a parabola with exactly the same shape as the graph of $y = x^2$, with its vertex at (0, k).

Notice that the three graphs of parabolas that we have drawn all pass the vertical line test. Thus they are *graphs of functions*. We call such a function, $f(x) = ax^2 + bx + c, a \neq 0$, a **quadratic function.** The previous three graphs have all been graphs of quadratic functions. Note that the domain of any quadratic function is all real numbers.

Quadratic Function

A function that can be written in the form

$$f(x) = ax^2 + bx + c, \qquad a \neq 0$$

is called a **quadratic function.** Its domain is all real numbers.

► **EXAMPLE 3**

For each function, find $f(1), f(-3)$, and $f(0)$.

(a) $f(x) = 2x^2 + 3x + 4$

(b) $f(x) = x^2 - x + 2$

(c) $f(x) = -x^2 + 1$

Solutions

(a) $f(x) = 2x^2 + 3x + 4$

$f(1) = 2 \cdot 1^2 + 3 \cdot 1 + 4$

$\qquad = 2 + 3 + 4 = 9$

$f(-3) = 2(-3)^2 + 3(-3) + 4$

$\qquad = 2 \cdot 9 - 9 + 4$

$\qquad = 18 - 9 + 4 = 13$

$f(0) = 2 \cdot 0^2 + 3 \cdot 0 + 4$

$\qquad = 4$

(b) $f(x) = x^2 - x + 2$

$f(1) = 1^2 - 1 + 2$

$\qquad = 1 - 1 + 2 = 2$

$f(-3) = (-3)^2 - (-3) + 2$

$\qquad = 9 + 3 + 2 = 14$

$f(0) = 0^2 - 0 + 2 = 2$

(c) $f(x) = -x^2 + 1$

$f(1) = -(1)^2 + 1$

$\qquad = -1 + 1 = 0$

$f(-3) = -(-3)^2 + 1$

$\qquad = -9 + 1 = -8$

$f(0) = -(0)^2 + 1 = 1$ ◄

Recall that to graph the *function f,* we graph the *equation* $y = f(x)$.

► **EXAMPLE 4**

Graph $g(x) = -x^2$.

Solution

We graph the equation $y = g(x) = -x^2$. Plotting a few points gives the following graph. Notice that this is the graph of $y = x^2$ turned upside down.

x	y
0	0
1	-1
2	-4
3	-9
-1	-1
-2	-4

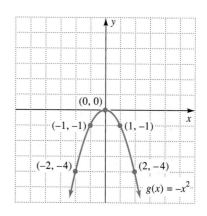

We can now make an interesting observation about graphing parabolas. The
graph of $y = px^2 + k$ is a parabola with vertex at the point $(0, k)$. It opens
downward if p is *negative,* and it opens *upward* if p is *positive.*

► **EXAMPLE 5**

Graph $h(x) = (x + 1)^2$.

Solution

We graph the equation $y = (x + 1)^2$. Plotting several points gives the following
graph.

x	y
0	1
1	4
2	9
-1	0
-2	1
-3	4

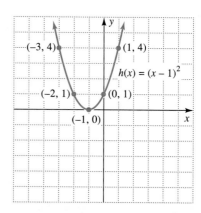

The vertex of this parabola is $(-1, 0)$. In fact, it is the graph of $y = x^2$ shifted
one unit to the left.

► **EXAMPLE 6**

Graph $f(x) = -(x - 1)^2 + 2$.

Solution

To graph f, we graph the equation, $y = f(x)$. We plot several points to see the graph.

x	y
0	1
1	2
2	1
3	-2
-1	-2
-2	-7

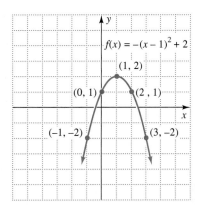

Its vertex is $(1, 2)$. It is a parabola opening *downward*.   ◄

Graphing Quadratic Functions

The graph of $y = p(x - h)^2 + k$ is a parabola with vertex at the point (h, k). The parabola opens upward if p is positive and downward if p is negative.

► **EXAMPLE 7**

Find the vertex for each parabola.

(a) $y = (x + 1)^2 + 2$ (b) $y = x^2 - 1$
(c) $y = (x + 3)^2$ (d) $y = -2(x - 4)^2 + 5$

Solutions

(a) To match the form of $y = p(x - h)^2 + k$, we see that $y = (x + 1)^2 + 2$ must be written as $y = (x - (-1))^2 + 2$. So h has a value of -1 and k has a value of 2. Thus the vertex is $(-1, 2)$.

(b) $y = x^2 - 1 = (x - 0)^2 - 1$. So the vertex is $(0, -1)$.

(c) $y = (x + 3)^2 = (x + 3)^2 + 0 = (x - (-3))^2 + 0$; therefore, the vertex is $(-3, 0)$.

(d) $y = -2(x - 4)^2 + 5$. The vertex is $(4, 5)$.   ◄

It is often necessary to complete the square in order to find the vertex and graph a parabola.

► **EXAMPLE 8**

Complete the square and find the vertex for each parabola.

(a) $y = x^2 - 8x$ (b) $y = x^2 + 2x + 5$

Solutions

(a) $y = x^2 - 8x$

We need $\left(-\dfrac{8}{2}\right)^2 = 16$ for a perfect square. We can't simply add it on, for that would change the value of the equation. However, we can add zero without harming the equation. So we add and subtract 16.

$$y = x^2 - 8x + 16 - 16$$
$$= (x - 4)^2 - 16$$

The vertex is at $(4, -16)$.

(b) $y = x^2 + 2x + 5$

Again we need to complete the square. Group the first two terms.

$$y = (x^2 + 2x) + 5$$

This time we add and subtract $\left(\dfrac{2}{2}\right)^2 = 1$.

$$y = (x^2 + 2x + 1 - 1) + 5$$
$$= (x^2 + 2x + 1) - 1 + 5$$
$$= (x + 1)^2 + 4$$

The vertex is at $(-1, 4)$.   ◄

In the next example, we use finding the vertex to aid in sketching the graph of a parabola.

► **EXAMPLE 9**

Sketch the graph of the function $f(x) = -2x^2 - 8x - 5$.

Solution

We graph the equation $y = -2x^2 - 8x - 5$. Since this is the equation of a parabola, we would like it in the form

$$y = p(x - h)^2 + k$$

In this form we can identify the vertex and draw the graph. To write the equation in the desired form, we must complete the square. First we factor -2 from the first two terms.

$$y = -2(x^2 + 4x) - 5$$

Now complete the square *inside the parentheses.*

$$y = -2(x^2 + 4x + 4 - 4) - 5$$
$$= -2(x^2 + 4x + 4) + 8 - 5 \qquad -2(-4) \text{ is } 8.$$
$$= -2(x + 2)^2 + 3$$

The vertex is $(-2, 3)$ and the parabola opens downward. We plot two more points and then sketch the graph.

 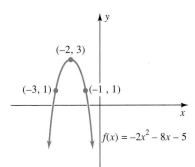

Problem Set 9.6

Warm-Ups

In Problems 1 through 4, sketch the graph of each of the given equations and identify the vertex. See Examples 1 and 2.

1. $y = x^2 + 2$ **2.** $y = x^2 - 1$ **3.** $y = x^2 - 3$ **4.** $y = x^2 + 4$

In Problems 5 through 8, find $f(2)$, $f(-1)$, and $f(0)$ for each function. See Example 3.

5. $f(x) = x^2 + x + 1$ **6.** $f(x) = x^2 - 5x + 3$
7. $f(y) = -2y^2 + 3y$ **8.** $f(t) = -3t^2 - 7t - 2$

In Problems 9 through 14, graph each function. For Problems 9 through 12, see Examples 4 and 5.

9. $f(x) = -x^2 + 1$ **10.** $g(x) = (x - 2)^2$

11. $h(x) = (x + 1)^2$ **12.** $k(x) = -x^2 - 1$

For Problems 13 and 14, see Example 6.

13. $F(x) = (x + 2)^2 - 1$

14. $G(x) = -(x - 2)^2 + 1$

In Problems 15 through 18, find the vertex for each parabola. See Example 7.

15. $y = (x - 4)^2 + 17$

16. $y = -3(x + 5)^2 - 3$

17. $y = -(x + 3)^2 + 11$

18. $y = 6(x - 2)^2 - 9$

In Problems 19 and 20, complete the square and find the vertex for each parabola. See Example 8.

19. $y = x^2 - 4x$

20. $y = x^2 + 6x - 1$

In Problems 21 and 22, sketch the graph for each quadratic function. See Example 9.

21. $f(x) = 3x^2 - 6x$

22. $g(x) = -x^2 + 6x - 7$

Practice Exercises

In Problems 23 through 30, find $f(0)$, $f(2)$, and $f(-1)$ for each function.

23. $f(x) = x^2 + 2x + 3$

24. $f(x) = x^2 + 7x + 1$

25. $f(x) = 3x^2 - x + 4$

26. $f(x) = 4x^2 + 3x - 11$

27. $f(x) = 2x^2 - 6x - 1$

28. $f(x) = 5x^2 - 4x - 9$

29. $f(x) = -x^2 + 11x - 13$

30. $f(x) = -6x^2 - 8x - 100$

In Problems 31 through 42, identify the vertex of each parabola.

31. $y = x^2 - 2$

32. $y = -x^2 + 8$

33. $y = (x - 1)^2 + 5$

34. $y = (x + 4)^2 - 7$

35. $y = -(x - 3)^2 + 2$

36. $y = -(x + 6)^2 - 12$

37. $y = 6(x + 1)^2 - 2$

38. $y = 14(x - 11)^2 + 5$

39. $y = x^2 + 12x$

40. $y = x^2 - 20x$

41. $y = 5x^2 - 20x + 13$

42. $y = -3x^2 + 12x - 10$

In Problems 43 through 56, sketch the graph of each parabola.

43. $y = x^2 - 3$

44. $y = -x^2 + 3$

45. $y = (x - 1)^2$

46. $y = (x + 4)^2$

47. $y = -(x - 2)^2$

48. $y = -(x + 1)^2$

49. $y = (x + 3)^2 + 2$

50. $y = (x - 2)^2 - 4$

51. $y = -(x + 1)^2 - 4$

52. $y = -(x - 1)^2 + 3$

53. $y = 2(x - 3)^2 + 2$

54. $y = 3(x + 4)^2 - 3$

55. $y = -3(x + 2)^2 + 3$

56. $y = -2(x - 1)^2 - 1$

In Problems 57 through 62, complete the square and find the vertex for each parabola.

57. $y = x^2 + 4x - 1$

58. $y = x^2 - 6x + 1$

59. $y = x^2 - 14x - 7$

60. $y = x^2 + 16x - 11$

61. $y = 2x^2 + 16x - 9$

62. $y = -2x^2 - 20x$

In Problems 63 through 66, sketch the graph of each function.

63. $f(x) = 2x^2 - 4x - 1$

64. $g(x) = 3x^2 + 12x + 10$

65. $f(x) = -x^2 - 6x - 7$

66. $g(x) = -2x^2 + 8x - 9$

67. Homecare Equipment Company makes wheelchairs and has determined that the relationship between the average cost of making a wheelchair and the number of wheelchairs produced per day is given by

$$y = x^2 - 14x + 140$$

where x is the number of wheelchairs produced and y is the average cost per wheelchair. What daily production of wheelchairs gives the lowest cost per wheelchair? (*Hint:* Look at the graph.)

68. The predicted path flown by one of Este's model rockets is

$$y = -2(x - 20)^2 + 800$$

where x is the horizontal distance from the launch pad in feet and y is the rocket height in feet. What is the maximum height predicted for this model?

Challenge Problems

In Problems 67 through 70, find the vertex of each parabola by completing the square.

69. $y = x^2 + x + 1$ **70.** $y = x^2 - 3x - 3$ **71.** $y = 2x^2 - x - 5$ **72.** $y = -3x^2 + 5x - 1$

73. The Great Zaborski travels in a parabolic path when shot out of a circus cannon. Her path is given by the function $F(x) = -\dfrac{1}{64}(x - 80)^2 + 100$, where x is the horizontal distance from the cannon and $F(x)$ gives her height above the circus floor. What is the maximum height she reaches and how far from the cannon should the net be placed?

Problem 73

In Your Own Words

74. Describe the procedure used to graph a function of the form $f(x) = ax^2 + bx + c$.

▶ Chapter Summary

GLOSSARY

Quadratic equation   A second-degree equation.

Complex number   A number that can be written in the form $a + bi$, where a and b are real numbers and i is a number such that $i^2 = -1$.

Square Root Property   The equation $X^2 = A$ has the same solution set as the pair of equations $X = \pm\sqrt{A}$.

Solving Quadratic Equations by Completing the Square

1. Write the equation in the form $ax^2 + bx + c = 0$.
2. Divide both sides by the coefficient of x^2 if it is not 1.
3. Subtract the constant term from both sides.
4. Divide the coefficient of x by 2, square this result, and then add this number to both sides.
5. Factor the left side and simplify the right side.
6. Apply the Square Root Property.
7. Check if required.
8. Write the solution set.

Quadratic Formula

To solve the quadratic equation $ax^2 + bx + c = 0$ $(a \neq 0)$, we use the Quadratic Formula:

$$x = \frac{-b \pm \sqrt{b^2 - 4ac}}{2a}$$

It should be used if the quadratic equation cannot be factored easily.

CHECK UPS

1. Solve $x^2 - 20 = 7$.   Section 9.1; Example 2
2. Solve $3x^2 + 6x = 3$ by completing the square.   Section 9.2; Example 6
3. Solve $x^2 = 4x - 2$ using the quadratic formula.   Section 9.3; Example 5
4. The hypotenuse of an isosceles right triangle is 12 m. Find the lengths of the legs.   Section 9.1; Example 6
5. Simplify $\sqrt{-4}$.   Section 9.4; Example 1a
6. Perform the operation indicated: $(3 + 5i)(2 - 3i)$.   Section 9.4; Example 2c
7. Solve $(t - 1)^2 = -8$ using the Square Root Property.   Section 9.4; Example 5b
8. Solve $x^2 - 2x = -5$.   Section 9.4; Example 6a
9. Solve $2(y^2 - y) = 24$ using any method.   Section 9.5; Example 2b

► Review Problems

In Problems 1 through 10, find the solution set.

1. $(x - 7)(x - 3) = 0$
2. $x(x + 4) = 0$
3. $x(x + 6) = 0$
4. $(t + 6)(t - 5) = 0$
5. $(3z + 4)(z + 7) = 0$
6. $(4x - 7)(3x + 5) = 0$
7. $x(5x - 9) = 0$
8. $(z - 5)(z + 5) = 0$
9. $(t - \sqrt{5})(t + \sqrt{5}) = 0$
10. $(z + \sqrt{15})(z - \sqrt{15}) = 0$

In Problems 11 through 20, find the solution set by factoring.

11. $x^2 + 7x + 10 = 0$
12. $t^2 - 7t + 6 = 0$
13. $z^2 + 2z = 24$

14. $x^2 - 3x - 28 = 0$ **15.** $x^2 = 2x$ **16.** $2x^2 + x = 10$

17. $3x^2 + 2x - 1 = 0$ **18.** $10z^2 = z + 3$ **19.** $w^2 = 169$

20. $4x^2 - 13x + 9 = 0$

In Problems 21 through 24, find the solution set by the Square Root Property.

21. $x^2 = -49$ **22.** $(x + 9)^2 = -4$ **23.** $(t - 2)^2 = 7$

24. $(x + 2)^2 = 32$

In Problems 25 through 30, find the solution set by completing the square.

25. $x^2 + 6x + 8 = 0$ **26.** $x^2 - 2x - 15 = 0$ **27.** $2x^2 + 4x - 16 = 0$

28. $3x^2 - 6x = 12$ **29.** $2x^2 + 6x = 6$ **30.** $3x^2 + 18x + 9 = 0$

In Problems 31 through 40, find the solution set by the Quadratic Formula.

31. $x^2 + x - 3 = 0$ **32.** $x^2 - 3x + 1 = 0$ **33.** $x^2 + 5x + 4 = 0$

34. $t^2 - 7t + 5 = 0$ **35.** $2x^2 - x + 1 = 0$ **36.** $3x^2 + 6x + 4 = 0$

37. $4x^2 = 6x - 2$ **38.** $2z^2 + 8z = -3$ **39.** $5x^2 = 7x - 1$

40. $4t + 2 = 3t^2$

In Problems 41 through 76, find the solution set by any method.

41. $(x - 3)(x + 4) = 0$ **42.** $z^2 - z = 0$ **43.** $(x - 5)^2 = 4$

44. $x^2 + 7x = 8$ **45.** $z^2 = 3z + 2$ **46.** $(x - \sqrt{3})(x + \sqrt{3}) = 0$

47. $(x + 8)^2 = -8$ **48.** $(x + 2)(x - 3) = -4$ **49.** $t^2 + t = 3$

50. $x^2 = 7x$ **51.** $x(x - 3) = 0$ **52.** $2x^2 + 9x - 5 = 0$

53. $3x^2 + 4x = 2$ **54.** $4z^2 = z - 1$ **55.** $(3x - 2)^2 = 0$

56. $(x - 4)^2 = 1$ **57.** $3x^2 + 6x + 1 = 0$ **58.** $x^2 - 12x + 36 = 0$

59. $(5x - 4)(3x + 7) = 0$ **60.** $(2x - 1)(3x + 2) = 10$ **61.** $6x^2 = 11x + 7$

62. $w^2 = 100$ **63.** $7x^2 = 8x - 1$ **64.** $10x^2 - 9x + 2 = 0$

65. $3x^2 - 5x = 0$ **66.** $(x + 7)(x - 3) = 0$ **67.** $2x^2 - 3x = 35$

68. $x^2 + 10x + 25 = 0$ **69.** $3x^2 + 7x + 3 = 0$ **70.** $3x^2 - 5x = 25$

71. $x^2 - 121 = 0$ **72.** $(x - 5)^2 = 7$ **73.** $(x + 1)^2 = 8$

74. $(x + 11)^2 = 24$ **75.** $x^2 = 144$ **76.** $x^2 + x = 0$

In Problems 77 through 84, solve each word problem.

77. Find two consecutive negative integers whose product is 306.

78. One leg in a right triangle is 3 units less than the other leg. If the length of the hypotenuse is $\sqrt{65}$, find the length of the legs.

79. The length of a rectangle is 5 units more than the width of the rectangle. If the area of the rectangle is 204 square units, find the dimensions of the rectangle.

80. One leg of a right triangle is 2 meters less than the other leg. If the area of the triangle is 24 square meters, find the lengths of both legs.

81. Jill, the cartoonist for her school newspaper, has a one-frame cartoon to complete. The frame is in the shape of a square whose diagonal measures 3.5 inches. What is the width of the cartoon frame to the nearest tenth of an inch?

82. A neighborhood baseball diamond is in the shape of a square. It is 40 feet from first base to second base. How far is it from second base to home plate?

83. West Builders and Supply Company delivered 135π cubic feet of sand to a construction site. The sandpile was in the shape of a cone with a height of 5 feet. Use the formula $V = \dfrac{1}{3}\pi r^2 h$ to find the radius of the sandpile.

84. Two 80-ft concrete walkways from the street to Jim Lebay's garage have a thickness that is $1\dfrac{1}{2}$ ft less than the width of the walkway. The cement company used 160 cubic feet of concrete to make both walkways. Find the width and thickness of the walkways to the nearest tenth of a foot.

In Problems 85 through 88, find the vertex of each parabola.

85. $y = (x + 6)^2 - 3$

86. $y = (x - 3)^2 + 5$

87. $f(x) = x^2 + 4x + 7$

88. $f(x) = x^2 - 2x - 2$

In Problems 89 through 96, sketch the graph of each function and identify the vertex of each parabola.

89. $f(x) = x^2 - 2$

90. $f(x) = -x^2 + 3$

91. $f(x) = (x + 1)^2$

92. $f(x) = -(x - 1)^2$

93. $f(x) = (x + 2)^2 - 1$

94. $f(x) = (x - 1)^2 + \dfrac{1}{2}$

95. $f(x) = x^2 + 4x - 1$

96. $f(x) = 2x^2 - 8x + 6$

In Problems 97 through 100, let g(x) $2x^2 - x + 3$ and find each value.

97. $g(2)$ **98.** $g(0)$

99. $g(-1)$ **100.** $g(-3)$

Let's Not Forget . . .

In Problems 101 through 105 perform the operations indicated.

101. -4^2

102. $(\sqrt{2} + \sqrt{3})^2$

103. $\dfrac{-2x^2}{y} \div \dfrac{x^3}{-y^3}$

104. $\dfrac{4(a - b)}{b - a}$

105. $\dfrac{0}{-7}$

In Problems 106 through 108, simplify each expression. Watch the role of the parentheses.

106. $(\sqrt{2} + \sqrt{5})(\sqrt{2} - \sqrt{5})$ **107.** $\sqrt{2} + \sqrt{5}(\sqrt{2} - \sqrt{5})$

108. $\dfrac{x - 4}{x + 3} - \dfrac{x - 2}{x + 3}$

In Problems 109 through 112, determine if $\sqrt{3}$ is a factor or a term.

109. $\sqrt{3} + 2$ **110.** $\sqrt{3}(x + 2)$ **111.** $\dfrac{1}{\sqrt{3} + 2}$

112. $x(y + \sqrt{3})$

In Problems 113 through 115, multiply each expression by $\dfrac{x}{7}$.

113. $-\dfrac{14}{x^2}$ **114.** $\dfrac{21}{x}(x - 7)$ **115.** $x^2 + 7x$

In Problems 116 through 120, label each problem as an expression, equation, or inequality. Solve the equations and inequalities and perform the indicated operation with the expressions.

116. $\dfrac{2}{x + 1} - \dfrac{1}{x + 3}$

117. $\dfrac{2}{x + 1} - \dfrac{1}{x + 3} = \dfrac{2x}{(x + 1)(x + 3)}$

118. $(x + 1)(x + 3) \le 0$

119. $\dfrac{\dfrac{2}{x + 1}}{\dfrac{1}{x + 3}}$

120. $\sqrt{x + 2} = x$

In Problems 121 through 124, identify which expressions are factored. Factor those that are not factored, if possible.

121. $y\sqrt{3} + x\sqrt{3}$ **122.** $x^3 + y^3$ **123.** $(x + y)^3$

124. $a(a + c)$

► You Decide

You have a part-time job while at college with the following salary options:

A. $120 a week for a straight 20-hour week. No over-time is expected.

B. $5 a hour plus time and a half for all hours over 20 each week. (You would be *expected* to work at least 10 hours overtime each week if you choose this option.)

Which option do you choose? State your reasons.

► Chapter 9 Test

In Problems 1 through 5, choose the letter of the correct answer.

1. The solution set for $x^2 - 2x - 8 = 0$ is (?).
  (a) $\{4, 8\}$ (b) $\{-2, 4\}$
  (c) $\{-4, 2\}$ (d) $\{2, 8\}$

2. The solution set for $2x^2 - 3x - 3 = 0$ is (?).
  (a) $\left\{\dfrac{3}{2}, -1\right\}$ (b) $\left\{\dfrac{-3 \pm \sqrt{33}}{4}\right\}$
  (c) $\left\{\dfrac{3 \pm \sqrt{33}}{4}\right\}$ (d) $\left\{\dfrac{1}{2} \pm \dfrac{\sqrt{33}}{4}\right\}$

3. $\dfrac{-3 \pm \sqrt{12}}{12} = $ (?).
  (a) $-\dfrac{1}{4} \pm \dfrac{\sqrt{3}}{6}$ (b) $\{-5, 1\}$
  (c) $\{-11, 7\}$ (d) $\dfrac{-1 \pm \sqrt{3}}{2}$

4. The solution set for $(x + 2)^2 = 9$ is (?).
  (a) $\{\pm 3\}$ (b) $\{-5, 1\}$
  (c) $\{-11, 7\}$ (d) $\{1\}$

In Problems 6 through 12 find the solution set.

5. $x^2 - 4x - 12 = 9$ **6.** $(x - 5)^2 = -7$ **7.** $x^2 + 3x = 10$

8. $2x^2 - 3x + 2 = 0$ **9.** $x^2 = 100$ **10.** $2x^2 - 2x - 1 = 0$

11. $x^2 - 6x = 0$

12. Find the solution set for $3x^2 - 6x - 3 = 0$ by completing the square.

In Problems 13 through 15, perform the operations indicated.

13. $(5 - 3i) - (2 + 7i)$ **14.** $\dfrac{2 - i}{3 + i}$

15. The hypotenuse of a right triangle is 13 m. If one leg is 7 m more than the other leg, find the length of the legs.

Photo Credits

Answers to Selected Problems

► Chapter 0

PROBLEM SET 0.1

Warm-Ups

1. $3 + 4$ **2.** $5 + 7$ **3.** $3(9)$ **4.** $11 - 6$ **5.** $21/8$ **6.** $2(14)$ **7.** $\frac{1}{2}(17)$ **8.** $1 + 5$

9. $6^2 - 4$ **10.** $2(13) - 1$ **11.** 18   **12.** 9   **13.** 16   **14.** 125   **15.** 12   **16.** 8   **17.** 14
18. 38   **19.** 11   **20.** 2   **21.** 13   **22.** 14   **23.** 9   **24.** 6   **25.** 1   **26.** 144   **27.** 9   **28.** 6
29. 20   **30.** 7   **31.** 13   **32.** 15   **33.** 5   **34.** 7   **35.** 14   **36.** 10   **37.** 15   **38.** 9

Practice Exercises

39. 12   **41.** 56   **43.** 7   **45.** 144   **47.** 127   **49.** 30   **51.** 163   **53.** 16   **55.** 1   **57.** 14
59. 14   **61.** 16   **63.** 5   **65.** 1   **67.** 7   **69.** 5   **71.** 36   **73.** 8   **75.** 1   **77.** 24   **79.** 4
81. 12   **83.** They have 14 books.   **85.** Ellen has 14 quarters.   **87.** They have 16 records.
89. Choi has 36 nickels.   **91.** 13 units   **93.** 204 units

Challenge Problems

95. $\sqrt{200} \approx 14.14$; $\sqrt{2} \approx 1.41$

PROBLEM SET 0.2

Warm-Ups

1. $2^3 \cdot 3$ **2.** $2^2 \cdot 3^2$ **3.** 5^2 **4.** 3^4 **5.** $2^4 \cdot 5$ **6.** $2^3 \cdot 5^2$.   **7.** $2^2 \cdot 3^3$ **8.** $2^2 \cdot 3^5$ **9.** $2^4 \cdot 3^2 \cdot 7$

10. $3^4 \cdot 7^2$ **11.** $\frac{2}{3}$ **12.** $\frac{1}{2}$ **13.** $\frac{5}{10}$ **14.** $\frac{9}{12}$ **15.** $\frac{24}{18}$ **16.** $\frac{99}{144}$ **17.** $\frac{1}{4}$ **18.** $\frac{1}{27}$ **19.** $\frac{4}{7}$

20. 3   **21.** $3\frac{1}{16}$ **22.** 4   **23.** $\frac{9}{4}$ **24.** $\frac{2}{7}$ **25.** $\frac{34}{33}$ **26.** $\frac{35}{36}$ **27.** 21   **28.** 16   **29.** 24

30. 120   **31.** 108   **32.** 270   **33.** $\frac{3}{5}$ **34.** $\frac{5}{12}$ **35.** $\frac{7}{24}$ **36.** $1\frac{7}{36}$ **37.** $\frac{10}{7}$ **38.** $\frac{6}{5}$ **39.** $\frac{1}{12}$

40. $\frac{1}{16}$ **41.** $\frac{2}{5}$ **42.** $\frac{5}{8}$ **43.** $\frac{53}{10}$ **44.** $\frac{22}{15}$ **45.** The result is $1\frac{2}{3}$.   **46.** The sum is $6\frac{7}{9}$.

47. Each will be $1\frac{1}{6}$ yd.   **48.** Yes, she has more than 15 lb.

Practice Exercises

49. $\frac{14}{15}$ **51.** $\frac{1}{5}$ **53.** $\frac{1}{9}$ **55.** $\frac{8}{125}$ **57.** $\frac{7}{6}$ **59.** $\frac{5}{21}$ **61.** $\frac{31}{36}$ **63.** $\frac{6}{49}$ **65.** $\frac{3}{4}$ **67.** $1\frac{1}{2}$

69. $3\frac{6}{25}$ **71.** $\frac{3}{2}$ **73.** $\frac{2}{7}$ **75.** $\frac{1}{24}$ **77.** $\frac{4}{9}$ **79.** $\frac{5}{4}$ **81.** $\frac{19}{16}$ **83.** $\frac{2}{3}$ **85.** $\frac{1}{4}$ **87.** $\frac{1}{4}$

89. It is $58\frac{7}{9}$.   **91.** Each part will be $\frac{7}{9}$ yd.   **93.** She bought 8 pencils.   **95.** It was 15 m long.

PROBLEM SET 0.3

Warm-Ups

1. 0.6   **2.** 0.375   **3.** $1.1\overline{6}$ **4.** 3.8   **5.** $2.\overline{6}$ **6.** 36.57   **7.** 111.45   **8.** 47.6035   **9.** 2
10. 18.9248   **11.** 22.565   **12.** 46.6   **13.** 91%   **14.** 77%   **15.** 9%   **16.** 80%   **17.** 14%
18. 8%   **19.** $66\frac{2}{3}\%$ **20.** 88.8%   **21.** 37.5%   **22.** $\frac{87}{100}$; 0.87   **23.** $\frac{11}{100}$; 0.11   **24.** $\frac{1}{2}$; 0.5

25. $\frac{2}{25}$; 0.08   **26.** $\frac{49}{50}$; 0.98   **27.** $\frac{9}{40}$; 0.225   **28.** $\frac{3}{1000}$; 0.003   **29.** $\frac{1}{30}$; $0.0\overline{3}$ **30.** $\frac{5}{4}$; 1.25
31. 40   **32.** 3800   **33.** 196   **34.** 36   **35.** 5227.2   **36.** 5.5   **37.** $4250   **38.** 321   **39.** 148.4

Practice Exercises

41. 2.25   **43.** $1.8\overline{3}$ **45.** $\frac{39}{50}$, 0.78   **47.** $\frac{1}{10}$, 0.1   **49.** $\frac{1}{50}$, 0.02   **51.** $\frac{11}{100}$, 0.11   **53.** $\frac{23}{200}$, 0.115

55. $\frac{1}{15}$, $0.0\overline{6}$ **57.** 97%   **59.** 3%   **61.** 22%   **63.** $22.\overline{2}\%$ **65.** 62.5%   **67.** 3400   **69.** 21
71. 4.7   **73.** 642   **75.** 8.84   **77.** 46.9945   **79.** 10.1   **81.** 193.7   **83.** 20.5   **85.** 19.32   **87.** 3
89. 4.41   **91.** 0.01   **93.** 209.1204   **95.** 9.3875   **97.** 46.033   **99.** He has $51.55.
101. No, the sum of her purchases is $9.12.   **103.** The ring is worth $5,900.   **105.** It will be $52.50.
107. He has $141.20.   **109.** It is $9.16 per pound.   **111.** No, she bought 14.97 lb.
113. The sale price was $47.20   **115.** He should study $\frac{1}{2}$ hour longer each day.

PROBLEM SET 0.4

Warm-Ups

1. $\angle a = 45°$ $\angle h = 135°$ $\angle o = 45°$ **2.** $58°$ **3.** $120°$ **4.** $95°$ **5.** $30°$
     $\angle b = 135°$ $\angle i = 90°$ $\angle p = 135°$
     $\angle c = 135°$ $\angle j = 90°$ $\angle q = 45°$
     $\angle d = 45°$ $\angle k = 90°$ $\angle r = 135°$
     $\angle e = 135°$ $\angle l = 90°$ $\angle s = 135°$
     $\angle f = 45°$ $\angle m = 135°$ $\angle t = 45°$
     $\angle g = 45°$ $\angle n = 45°$
6. $10^2 = 100$; $6^2 + 8^2 = 100$
7. (a) Two angles of one triangle are equal to two angles of the other triangle.   **(b)** Two angles of one triangle are equal to two angles of the other triangle.
8. (a) Angle–side–angle; **(b)** Angle–side–angle

Practice Exercises

9. False   **11.** False   **13.** False   **15.** False   **17.** False   **19.** False   **21.** $125°$ **23.** $20°$
25. $64°$ **27.** $13^2 = 169$; $5^2 + 12^2 = 169$
29. Corresponding sides are proportional (similar, but not congruent).   **31.** Side–angle–side (congruent)
33. False   **35.** True   **37.** True

Challenge Problems

39. Three less than the number of sides

REVIEW PROBLEMS CHAPTER 0

1. $\dfrac{4}{21}$ **3.** 15 **5.** $\dfrac{9}{2}$ **7.** 0 **9.** $6\dfrac{1}{3}$ **11.** $2(5) + 3$ **13.** $\dfrac{1}{2} \cdot 7 + 2 \cdot 4$ **15.** $15 - (4 + 6)$

17. She bought 17.22 lb of meat.   **19.** It is 150.   **21.** $\dfrac{3}{20}$; 0.15 **23.** $\dfrac{27}{200}$; 0.135 **25.** $\dfrac{5}{6}$; $0.8\overline{3}$

27. 18% **29.** $83\dfrac{1}{3}\%$ **31.** 2^4 **33.** $2^2 \cdot 5^2$ **35.** $2^4 \cdot 5^2$ **37.** 70 **39.** 80 **41.** 8 **43.** 21

45. The circumference is $13\dfrac{1}{2}\pi$ m and the area is $45\dfrac{9}{16}\pi$ sq m.

CHAPTER 0 TEST

1. (a)   **2.** (d)   **3.** (a)   **4.** (b)   **5.** (b)   **6.** 11 **7.** 10.81 **8.** $\dfrac{11}{27}$ **9.** 13 **10.** 3 **11.** 13.5

12. The area is $3\dfrac{3}{10}$ sq in.   **13.** 37.5% **14.** $3\dfrac{1}{4}$ **15.** She bought $13\dfrac{5}{8}$ yd.

► Chapter 1

PROBLEM SET 1.1

Warm-Ups

1. $\varnothing$, {1}, {4}, {1, 4}   **2.** $\varnothing$, {−2}, {0}, {2}, {−2, 0}, {−2, 2}, {0, 2}, {−2, 0, 2}   **3.** 1, 2, 4, 6, 8
4. −2, 0, 2, 4, 6, 8   **5.** 4   **6.** 2   **7.** rational   **8.** rational   **9.** irrational   **10.** rational   **11.** T
12. T   **13.** F   **14.** T   **15.** F   **16.** F
17.   **18.**   **19.**   **20.**

21. 3.5   **22.** $-\sqrt{5}$ **23.** −11/13   **24.** $-z$ **25.** 6   **26.** 35   **27.** π **28.** q **29.** 12
30. 29   **31.** 16   **32.** 18   **33.** −4   **34.** −5   **35.** −15   **36.** −3

Practice Exercises

37. 1/2, 2, 9/4, 5   **39.** $\varnothing$, {9/4}, {5}, {9/4, 5}   **41.** T   **43.** T   **45.** T   **47.** F   **49.** T
51. F   **53.** T   **55.** T   **57.** F   **59.** F   **61.** 2   **63.** 8   **65.** 5   **67.** 3.14   **69.** −8
71. −37   **73.** 48, 49, 50   **75.** 37, 39, 41

Challenge Problems

77. $\dfrac{1}{8}$ **79.** 1, 2, 3, 4, 5, 6, 7, 8, 9   **81.** 5, 10, 15, . . .   **83.** $\{x \mid x$ is an even natural number$\}$
85. $\{x \mid x$ is an integer between −3 and 3 inclusive$\}$

PROBLEM SET 1.2

Warm-Ups

1. 8   **2.** −3   **3.** 4   **4.** 6   **5.** −13   **6.** −1   **7.** −2   **8.** −14   **9.** 0   **10.** $-\dfrac{1}{12}$ **11.** −1
12. $\dfrac{15}{8}$ **13.** $\dfrac{5}{24}$ **14.** 0   **15.** $-\dfrac{5}{3}$ **16.** $-\dfrac{7}{10}$ **17.** 2.2   **18.** −12.5   **19.** 14   **20.** −3
21. 1   **22.** −23   **23.** −4   **24.** $-\dfrac{1}{8}$ **25.** −4.9   **26.** −13   **27.** −4   **28.** −9.1   **29.** $-\dfrac{7}{8}$
30. −19   **31.** $\dfrac{3}{2}$ **32.** 5; 9   **33.** 17; 23   **34.** −3   **35.** 2   **36.** 10   **37.** 4°   **38.** −30°
39. No

Practice Exercises

41. 1   **43.** −13   **45.** −2   **47.** 23   **49.** −8   **51.** 0   **53.** −6   **55.** −0.3   **57.** −4.0
59. −10   **61.** −18   **63.** $\dfrac{1}{6}$ **65.** $\dfrac{11}{4}$ **67.** 0   **69.** $-\dfrac{1}{2}$ **71.** −10.0   **73.** −0.3   **75.** −4.0
77. 1   **79.** −1   **81.** −5.9   **83.** 3   **85.** −32   **87.** $-\dfrac{19}{24}$ **89.** −1   **91.** −13   **93.** $-\dfrac{26}{15}$
95. −7.9   **97.** 17; 23   **99.** 15; 9   **101.** −1   **103.** −5°   **105.** $129   **107.** 5 yd   **109.** 1990

Challenge Problems

111. $a + b > 0$

PROBLEM SET 1.3

Warm-Ups

1. 2   **2.** 13   **3.** -10 **4.** -18 **5.** 26   **6.** -16 **7.** 3   **8.** 6   **9.** -1 **10.** 6   **11.** 0

12. 12   **13.** -25 **14.** -30 **15.** 21   **16.** 0   **17.** $\dfrac{17}{8}$ **18.** $-\dfrac{11}{24}$ **19.** $-\dfrac{5}{3}$ **20.** $\dfrac{1}{3}$ **21.** $-\dfrac{1}{2}$

22. -4.6 **23.** -7.1 **24.** -3.4 **25.** 7   **26.** -13 **27.** -4 **28.** 16   **29.** $\dfrac{7}{8}$ **30.** 7.3

31. 15   **32.** -19 **33.** $\dfrac{2}{15}$ **34.** -2.4 **35.** -7 **36.** -20 **37.** 1   **38.** 19   **39.** 12

40. -23 **41.** -17 **42.** -6 **43.** $-15°$ **44.** $-12°$

Practice Exercises

45. 3   **47.** -5 **49.** 4   **51.** 0   **53.** -4 **55.** 0   **57.** 12   **59.** 7   **61.** -5 **63.** 27   **65.** 0

67. 50   **69.** 14   **71.** -14 **73.** $\dfrac{1}{8}$ **75.** $-\dfrac{8}{5}$ **77.** $-\dfrac{1}{2}$ **79.** -5.9 **81.** 13   **83.** 6   **85.** -23

87. 20   **89.** 58   **91.** 1   **93.** 9   **95.** -4 **97.** $-18°$

Challenge Problem

99. 5

PROBLEM SET 1.4

Warm-Ups

1. 6   **2.** 0   **3.** $-\dfrac{1}{4}$ **4.** 35   **5.** -11 **6.** 3   **7.** 25   **8.** 9   **9.** 20   **10.** -28 **11.** 0

12. $\dfrac{9}{16}$ **13.** 5   **14.** -3 **15.** undefined   **16.** -2 **17.** -5 **18.** 7   **19.** 0   **20.** -3

21. -2 **22.** 1   **23.** -15 **24.** 9   **25.** -5 **26.** $-\dfrac{2}{3}$ **27.** undefined   **28.** -6 **29.** -2

30. 18   **31.** 12   **32.** -2 **33.** 1   **34.** -3 **35.** -12 **36.** 36   **37.** -4 **38.** -18

39. 6   **40.** 1.9 oz   **41.** 97 liters

Practice Exercises

43. 0   **45.** -24 **47.** 49   **49.** 36   **51.** 32   **53.** 24   **55.** d   **57.** 100   **59.** -35 **61.** $\dfrac{25}{36}$

63. -2 **65.** -4 **67.** -9 **69.** 0   **71.** 1   **73.** 8   **75.** 0   **77.** -9 **79.** -1 **81.** undefined

83. 72   **85.** -16 **87.** -5 **89.** 25   **91.** -12 **93.** 11.2 lb   **95.** \$454.02   **97.** 7630 ft

Challenge Problems

99. $14 \div (5 \cdot 11)$

PROBLEM SET 1.5

Warm-Ups

1. -3 **2.** 7   **3.** 2   **4.** 25   **5.** 3   **6.** 14   **7.** $-\dfrac{29}{3}$ **8.** -7 **9.** -2 **10.** -15 **11.** -1

12. 0   **13.** 0   **14.** -2 **15.** -1 **16.** $\dfrac{5}{2}$ **17.** 38   **18.** 5   **19.** 8   **20.** term   **21.** term

22. factor   **23.** term   **24.** factor   **25.** factor   **26.** 45   **27.** 18   **28.** 5   **29.** 4   **30.** 10

31. -3 **32.** -45 **33.** 12   **34.** -5 **35.** $\dfrac{1}{4}$ **36.** 8   **37.** -27

Practice Exercises

39. -4 **41.** 7   **43.** 14   **45.** 6   **47.** -13 **49.** 1   **51.** -2 **53.** -6 **55.** 1   **57.** 34
59. 1   **61.** term   **63.** term   **65.** term   **67.** 23, 19   **69.** 20, 20   **71.** $-4, -4$ **73.** 48, -96
75. 197.225 ft   **77.** $4\dfrac{5}{7}°$

Challenge Problems

79. equal   **81.** equal

PROBLEM SET 1.6

Warm-Ups

1. commutative   **2.** commutative   **3.** associative   **4.** commutative   **5.** distributive
6. additive identity   **7.** multiplicative inverse   **8.** additive inverse   **9.** commutative
10. multiplicative identity   **11.** multiplicative identity   **12.** distributive   **13.** additive inverse
14. associative   **15.** multiplicative inverse   **16.** 874   **17.** 48   **18.** -4 **19.** 8   **20.** $2 + 2x$
21. $3x - 12$ **22.** $-2x - 18$ **23.** $-4x + 8$ **24.** $xy + xz$ **25.** $xy - xz$ **26.** $-x, \dfrac{1}{x}$
27. $\dfrac{2}{7}, -\dfrac{7}{2}$ **28.** $-\pi, \dfrac{1}{\pi}$ **29.** $5x, -\dfrac{1}{5x}$

Practice Exercises

31. (d) Distributive   **33.** (e) Commutative   **35.** (b) Multiplicative identity
37. associative   **39.** associative   **41.** multiplicative identity   **43.** multiplicative inverse
45. additive identity   **47.** additive identity   **49.** multiplicative inverse   **51.** $\dfrac{x}{3}$ **53.** 148   **55.** 12
57. $\dfrac{3}{4}z$ **59.** $-\dfrac{5}{3}, \dfrac{3}{5}$ **61.** 0, no multiplicative inverse   **63.** $8 + 2x$ **65.** $-12 - 2y$ **67.** $xy + 3x$
69. $xy - 5x$ **71.** $3t - tx$ **73.** $3n + 2nx$ **75.** $-9c + cx$

Challenge Problems

77. no   **79.** $x^2 + 3x$ **81.** $6x^3$

PROBLEM SET 1.7

Warm-Ups

1. $41 + 6y^2$ **2.** $2x - 15$ **3.** $8xy$ **4.** $\dfrac{4}{2x}$ **5.** $10/x^2$ **6.** $3t + 19$ **7.** $2x - 2y$ **8.** $\dfrac{1}{5}(x - 5)$

9. $\dfrac{x + 4}{x - 4}$ **10.** 27 in.   **11.** $x - 4$ in. are left.   **12.** 39 lb   **13.** $x - 17$ lb   **14.** 11 quarters

15. It has $x - 70$ quarters.   **16.** $3.50; $25(x - 70)$ cents; $350 + 25(x - 70)$ cents   **17.** Don is 51.
18. Don is $x - 4$ years old.   **19.** $(x + 6)$ yrs old   **20.** $10x$ **21.** $7x^2$ **22.** $4x$ **23.** $-3abc$
24. $8x + 15$ **25.** $-3xy + 3$ **26.** $2x^2 - 3$ **27.** $13x^2y - 6xy^2$ **28.** $4x^3 - 3x^2$
29. $-x^2z + 39z^2$ **30.** $4x^2 - x - 3$ **31.** $2\sqrt{x}$ **32.** $5x - 14$ **33.** $29x - 35$ **34.** $5x + 18$
35. $-x - 7$ **36.** $60 - 25y$ **37.** $24b - 9$ **38.** $16 - 2a$ **39.** $10\sqrt{x - 1}$ **40.** 73, 75
41. The sum is 219.   **42.** $x + 2, x + 4$ **43.** The sum is $3x + 6$.   **44.** 270   **45.** The sum is $3x - 6$.

Practice Exercises

47. $6x + 23$ **49.** $\dfrac{1}{3}(2t)$ **51.** $-4x^3$ **53.** $10z$ **55.** $10x$ **57.** $-15x^2$ **59.** $24 + 11v$ **61.** $7x^2$

63. $2x^3 + 3x^2$ **65.** $-5x - 3$ **67.** $6x + 3$ **69.** $3x + 2$ **71.** $11\sqrt{2x}$ **73.** $10x - 25$
75. $17y^2 - 20$ **77.** $15x - 4$ **79.** $5x - 18$ **81.** $8 + 14a$ **83.** $-4 + 4y$ **85.** $10 - 12x$
87. $5700   **89.** $\angle B$ is 74° and $\angle C$ is 69°   **91.** 37 miles   **93.** 60 quarters
95. $102 - x$ quarters   **97.** Its measure is 124°.   **99.** Som is 25 yr old.   **101.** $x + 13$ yr old
103. They are $x + 2, x + 4,$ and $x + 6$.   **105.** The sum is 84.

Challenge Problems

107. $12\sqrt{x^2 + 3} - 5\sqrt{x^2 - 5}$ **109.** $-3\sqrt{x + 4} - 2\sqrt{2y}$

PROBLEM SET 1.8

Warm-Ups

1. no   **2.** yes   **3.** no   **4.** yes   **5.** no   **6.** yes   **7.** yes   **8.** no   **9.** $\{5\}$ **10.** $\{8\}$ **11.** $\{5\}$
12. $\{-1\}$ **13.** $\{7\}$ **14.** $\{16\}$ **15.** $\{-5\}$ **16.** $\{-25\}$ **17.** $\{0\}$ **18.** $\{-4\}$ **19.** $\{6\}$ **20.** $\{4\}$

21. $\{2\}$ **22.** $\{0\}$ **23.** $\{26\}$ **24.** $\{21\}$ **25.** $\left\{\dfrac{27}{2}\right\}$ **26.** $\left\{\dfrac{5}{6}\right\}$ **27.** $\{-5\}$ **28.** $\{7\}$ **29.** $\left\{-\dfrac{1}{2}\right\}$

30. $\left\{\dfrac{2}{5}\right\}$ **31.** $\left\{-\dfrac{5}{3}\right\}$ **32.** $\left\{\dfrac{77}{2}\right\}$ **33.** y is -18 **34.** w is -2 **35.** x is -5 **36.** t is -24

37. y is 6   **38.** s is -7 **39.** t is 75   **40.** z is -8

Practice Exercises

41. yes   **43.** no   **45.** no   **47.** no   **49.** $\{-1\}$ **51.** $\{15\}$ **53.** $\{9\}$ **55.** $\{6\}$ **57.** $\{20\}$

59. $\left\{\dfrac{5}{2}\right\}$ **61.** $\left\{\dfrac{7}{4}\right\}$ **63.** y is -2 **65.** x is -7 **67.** 23   **69.** t is -35 **71.** 7

73. $0°, 100°, -25°, -60°$ **75.** $6\dfrac{1}{4}$ ft   **77.** 2 in.

Challenge Problems

79. $2 - p$ **81.** $\dfrac{3p}{5}$ **83.** 880 ft by 1760 ft.

REVIEW PROBLEMS CHAPTER 1

1. -2 **3.** -9.2 **5.** -3 **7.** 17   **9.** -8 **11.** -1 **13.** 0   **15.** -5 **17.** -18 **19.** 29
21. -7 **23.** 30   **25.** They gained 9 yds.   **27.** $|8 + 11|$ **29.** $(2 \cdot 7)^3$ **31.** $\{10\}$ **33.** $\{15\}$
35. $\{-8\}$ **37.** T   **39.** T   **41.** F   **43.** F   **45.** T   **47.** 310 sq ft   **49.** 27 ft

Let's Not Forget

50. 1   **51.** 9   **52.** -1 **53.** 0   **54.** -5 **55.** 0   **56.** 6   **57.** -80 **58.** -53 **59.** -17
60. 7   **61.** $\pi - 2$ **62.** $2\pi - 1$ **63.** expression   **64.** equation, $\{-6\}$ **65.** equation, $\{5\}$
66. expression   **67.** expression   **68.** equation, $\{11\}$

CHAPTER 1 TEST

1. d   **2.** d   **3.** a   **4.** b   **5.** d   **6.** T   **7.** T   **8.** T   **9.** F   **10.** F   **11.** 9
12. 2   **13.** -7 **14.** 37   **15.** -1 **16.** 12   **17.** 6   **18.** 41   **19.** 4   **20.** -7

21. a. 5, b. 4, c. 3, d. 4, e. 2, f. 1   **22.** a, $-5, 0, 102$ b. $-5, -\dfrac{3}{5}, 0, \dfrac{17}{3}, 12.7, 102$ c. $\sqrt{3}$ **23.** $\{25\}$

24. $\{-4\}$ **25.** $\{6\}$

► Chapter 2

PROBLEM SET 2.1

Warm-Ups

1. 4; 16   **2.** 4; -16 **3.** -4; 16   **4.** -4; -64 **5.** 4; -64 **6.** $\dfrac{2}{3}; \dfrac{8}{27}$ **7.** x^4 **8.** $-x^9$

9. Can't be simplified.   **10.** x^9 **11.** y^{25} **12.** 6^{18} **13.** $-6z^6$ **14.** 2^{11}
15. Can't be simplified.   **16.** x^{20} **17.** y^6 **18.** 5^{16} **19.** x^7y^7 **20.** $27s^3$

21. $121k^2$ **22.** $4z^6$ **23.** $-27t^3$ **24.** $-x^5y^{10}$ **25.** $\dfrac{x^8}{y^8}$ **26.** $\dfrac{32}{t^5}$ **27.** $\dfrac{x^3}{125}$ **28.** $\dfrac{9}{x^2}$

29. $\dfrac{-125}{p^3}$ **30.** $\dfrac{1}{x^4}$ **31.** x^8 **32.** $\dfrac{1}{y^3}$ **33.** $3k^{21}$ **34.** 1   **35.** 1   **36.** $2a+1$

37. xz^{14} **38.** $32x^{20}$ **39.** $\dfrac{125}{8t^3}$ **40.** x^{11} **41.** $\dfrac{1}{49a^6}$ **42.** $\dfrac{16v^{18}}{9k^2}$ **43.** $4x^2y^2$ **44.** $\dfrac{4x^2}{y^2}$

45. Can't be simplified.   **46.** Can't be simplified.   **47.** $4x^4y^6$ **48.** Can't be simplified.

49. $4x^4y^6$ **50.** $-27x^9y^6$ **51.** $64p^9q^3$ **52.** $\dfrac{xy^6}{z^3}$ **53.** $\dfrac{36x^{14}t^2}{s^{12}}$ **54.** $\dfrac{a^3b^3c^3}{-8d^3}$

55. 1   **56.** $4yx^2$ **57.** 1   **58.** -1 **59.** -12 **60.** 1

Practice Exercises

61. 15   **63.** -13 **65.** 0   **67.** $-k^3$ **69.** 1   **71.** y^{36} **73.** p^9

75. Can't be simplified.   **77.** 2^6 **79.** s^4t^4 **81.** $169x^2$ **83.** $\dfrac{r^6}{64}$ **85.** $\dfrac{5x^2}{2}$ **87.** q **89.** 1

91. $a^{10}b^{36}$ **93.** $\dfrac{64r^3}{27}$ **95.** $256x^{12}y^{16}$ **97.** $-8x^6y^9$ **99.** $\dfrac{1}{7a}$ **101.** $\dfrac{-216x^{21}t^3}{s^{18}}$

103.

t	10 sec	20 sec	30 sec	1 min	5 min
$\left(\dfrac{1}{2t+1}\right)^2$	0.989	0.978	0.967	0.937	0.735

Challenge Problems

105. (a) $1; -1; 1; -1; 1$ (b) $(-1)^n = 1$ if n is even and $(-1)^n = -1$ if n is odd.   **107.** 1

PROBLEM SET 2.2

Warm-Ups

1. $\dfrac{1}{16}$ **2.** $\dfrac{1}{9}$ **3.** $\dfrac{1}{2}$ **4.** $\dfrac{1}{x^2}$ **5.** $\dfrac{1}{y^5}$ **6.** $\dfrac{1}{A}$ **7.** $\dfrac{1}{16}$ **8.** $-\dfrac{1}{32}$

9. $-\dfrac{1}{4}$ **10.** $\dfrac{1}{25}$ **11.** $-\dfrac{1}{125}$ **12.** $\dfrac{1}{25}$ **13.** $-\dfrac{1}{25}$ **14.** $-\dfrac{1}{125}$ **15.** $\dfrac{2}{x^3}$ **16.** $\dfrac{x}{9}$

17. $\dfrac{-2}{x^3}$ **18.** $\dfrac{-7}{v^2}$ **19.** 2   **20.** 9   **21.** $\dfrac{1}{16}$ **22.** $\dfrac{1}{16}$ **23.** $8x^4$ **24.** z^6

25. $\dfrac{-6}{x^2}$ **26.** 1   **27.** $\dfrac{1}{64}$ **28.** $\dfrac{1}{81}$ **29.** $\dfrac{5}{3}$ **30.** -2 **31.** $\dfrac{1}{x^{10}}$ **32.** z^{24}

33. $\dfrac{1}{x^3y^3}$ **34.** $\dfrac{1}{16x^4}$ **35.** $\dfrac{1}{9x^2}$ **36.** $\dfrac{1}{16s^4}$ **37.** $\dfrac{1}{2a^2}$ **38.** $\dfrac{x^2}{a^4}$ **39.** $\dfrac{1}{100}$ **40.** $\dfrac{1}{36}$

41. 125   **42.** x^7 **43.** $\dfrac{1}{r^7}$ **44.** y^8 **45.** $\dfrac{1}{k^2}$ **46.** $6x^5$ **47.** $\dfrac{32}{b^2}$ **48.** 5.28×10^3

49. 1.2345×10^{-6} **50.** 9.4×10^{-2} **51.** 7.7722×10^{13} **52.** -3.2×10^4 **53.** -2.1367×10^{-8}

54. 160,900   **55.** 0.0000543   **56.** 0.000000011   **57.** 681   **58.** $-800{,}000{,}000{,}000$

59. -0.0432101 **60.** 12,000,000,000   **61.** 200   **62.** 500   **63.** 0.00000055

Practice Exercises

65. $\dfrac{1}{27}$ **67.** $\dfrac{1}{x^{11}}$ **69.** $\dfrac{1}{T}$ **71.** $-\dfrac{1}{27}$ **73.** $-\dfrac{1}{4}$ **75.** $-\dfrac{1}{64}$ **77.** $-\dfrac{1}{64}$ **79.** $\dfrac{4}{x^2}$

81. $\dfrac{6}{x^7}$ **83.** $\dfrac{-8}{x^2}$ **85.** $\dfrac{b}{t^3}$ **87.** $\dfrac{k^2}{x^2}$ **89.** $\dfrac{3}{a^3t^2}$ **91.** 3 **93.** 4 **95.** $\dfrac{1}{81}$

97. 100 **99.** $12z$ **101.** 1 **103.** $\dfrac{1}{3^6}$ **105.** $\dfrac{x}{3}$ **107.** $\dfrac{3}{2}$ **109.** $\dfrac{1}{x^6}$ **111.** x^{36}

113. $\dfrac{1}{32x^5}$ **115.** $\dfrac{1}{64r^6}$ **117.** $-\dfrac{1}{b^4w^4}$ **119.** $\dfrac{1}{x^6}$ **121.** $\dfrac{1}{w^2}$ **123.** 3 **125.** 4 **127.** $\dfrac{11}{3}$

129. $\dfrac{27}{y^3}$ **131.** 5.5×10^{10} **133.** 9.109×10^{-31} **135.** 8.85×10^8

137. $1.0 \times 10^{-5}, 3.0 \times 10^{-3}$ **139.** $31{,}010$ **141.** 113.4 **143.** -0.00003

Challenge Problems

145. $\dfrac{15}{4}$ **147.** -1 **149.** $2, 1, \dfrac{1}{2}, \dfrac{1}{4}$

PROBLEM SET 2.3

Warm-Ups

1. $x + 6$; 1; binomial   **2.** $x^2 - 3x + 4$; 2; trinomial   **3.** $-3x^3 + 4x^2 + \dfrac{1}{2}x - 7$; 3

4. $-17y^7 + 4y^5 + \dfrac{2}{3}y + 89$; 7   **5.** $x^5 - 25$; 5; binomial   **6.** $x^3 + x^2 + 3x$; 3; trinomial

7. $y^7 - 5y^5 + 6y - 1$; 7   **8.** $-z^3 + 9$; 3; binomial   **9.** 13   **10.** -3 **11.** 6   **12.** 2
13. 13   **14.** 12   **15.** 1   **16.** 10   **17.** 7   **18.** -46 **19.** 29   **20.** -66
21. $10x$ **22.** $7x^2$ **23.** $4x$ **24.** $-11t$ **25.** $3w^3$ **26.** 0   **27.** $3a$ **28.** $3z$
29. $-21x^2$ **30.** $-4y^2$ **31.** $8x + 3$ **32.** $-5y - 5$ **33.** $7r^3 - 5$ **34.** $-3x - 2$
35. $2x^2 - 4x + 6$ **36.** $3x^2 + x - 1$ **37.** $y^2 - 5y + 1$ **38.** $y^2 + 2x^2$ **39.** $2x^2 + 3y^2$
40. $4x^3 - 3x^2$ **41.** $8x - 3$ **42.** $-6x - 3$ **43.** $-10x^3 - 1$ **44.** 8   **45.** 0

46. $2x^2 + 2$ **47.** $-10x + 2$ **48.** $x - \dfrac{1}{5}$ **49.** $3.0z - 6.3$ **50.** $5.4w - 6.3$

Practice Exercises

51. $-x + 5$; 1; binomial   **53.** $x^9 - 2x^8$; 9; binomial   **55.** $-y^7 + 5y^5 - 7y - 1$; 7   **57.** 16
59. 9   **61.** -48 **63.** -2 **65.** 21   **67.** 1   **69.** $4x$ **71.** $-x$ **73.** $-4w^3$ **75.** $2c$
77. $-6z$ **79.** $5x$ **81.** $5r^3 - 6$ **83.** $2x + 31$ **85.** $5y^2 - 9y - 5$ **87.** $6x^2 + 4y^2$
89. $x + 2$ **91.** 400; 0

Challenge Problems

93. 0 if n is odd; 2 if n is even   **95.** 160, 224, 216, 80, 0

PROBLEM SET 2.4

Warm-Ups

1. $7x + 4$ **2.** $9x + 13$ **3.** $4x^2 - 7$ **4.** $6x^3 - 3$ **5.** $8z^2 - 3$ **6.** $7x + 10$ **7.** $12x - 4$
8. $19x^2 + 4$ **9.** $-t^2 - 7$ **10.** $-5x - 12$ **11.** $3y^2 + 6y - 12$ **12.** $3z^2 - 7z + 8$
13. $5x^2 - 3x - 3$ **14.** $15x^3 + x^2 + 4x - 8$ **15.** $4x^5 + 4x^4 + x^3 - x^2 + 8x + 2$
16. $5x^6 - 4x^5 + 4x^4 + x^3 - 1$ **17.** $-3x - 1$ **18.** $-t^2 + 3$ **19.** $-6x + 9$ **20.** $z - 1$
21. $-x^2 + 8x + 7$ **22.** $x^3 - 2x^2 + x + 3$ **23.** $-7x^2 - 5$ **24.** $-x - 5$ **25.** $x - 4$
26. $-x + 4$ **27.** $x + 3$ **28.** $3x + 6$ **29.** $-2x^2 - 13$ **30.** $z^3 - 4$ **31.** $4x^4 - 3$
32. $-x + 52$ **33.** $x^3 - 2x^2 + x$ **34.** $y - 8$ **35.** $x^2 + 2x - 1$ **36.** $-2y^3 - 12y - 1$
37. $4x^3 + 6x^2 + 7x - 10$ **38.** $4y^4 - y^3 + y^2 + 6y$ **39.** $2x + 6$ **40.** $14x^2 + 8$ **41.** $-2x^2 - 12$
42. $-3x + 11$ **43.** $-6x + 11$ **44.** $-z^2 - 7$ **45.** $-10x - 2$ **46.** $7w^2 + 2w + 36$
47. $5x^2 + 3x - 2$ **48.** $s^2 + 4s - 8$ **49.** $3x^2 - 4x + 9$ **50.** $x^2 + x + 1$ **51.** $6 - 2x$
52. $6x^2 + 6x + 6$ **53.** The perimeter is $8s + 12$ ft. **54.** $2x^2 - x + 2$ feet
55. The circumference is $2\pi r - 4\pi$ inches. **56.** $3n - 3$ **57.** They have $7x - 2$ dollars.
58. $100x$ miles

Practice Exercises

59. $24x + 4$ **61.** $6y^2 + 4y - 17$ **63.** $-2x^2 - 11x + 4$ **65.** $15x^3 + 2x^2 + 4x - 8$
67. $8x^5 - 3x^4 + 13x^3 - x^2 + 7x + 12$ **69.** $6x^3 - 9x^2 + 15x - 13$ **71.** $24x^5 + 14x^4 - 4x^3 - 10x$
73. $x^2 - 15x + 10$ **75.** $z^2 - 2z - 7$ **77.** $-2x^2 - 3x + 17$ **79.** $2v^4 - 2v^3 - 4v^2 + 6v - 12$
81. $5x^2 - 4$ **83.** $6x^2 - 4x + 8$ **85.** $8t^3 - 3t^2 + 12$ **87.** $7x^2 + 5x - 5$
89. $r^2 + 4r + 2$ ft.; 119 ft **91.** The distance between them is $900x + 400$ km.
93. The sum of their ages is $3y + 7$ years.

Challenge Problems

95. $7x^n$

PROBLEM SET 2.5

Warm-Ups

1. $10x^5$ **2.** $21x^9$ **3.** $-24x^7$ **4.** $-24t^7$ **5.** z^6 **6.** $36x^{13}$ **7.** $-12s^4$ **8.** $21y^7$
9. $8z^3$ **10.** $3x^2 + 3x - 21$ **11.** $8x + 12$ **12.** $6x - 8x^2 + 2x^4$ **13.** $9x^8 - 3x^5 - 6x^4$
14. $-8t^4 - 2t^2 + 12t$ **15.** $-15x^4 + 35x^7$ **16.** $x^2 + 4x + 3$ **17.** $x^2 - x - 20$
18. $t^2 + 4t - 21$ **19.** $y^2 - 8y + 12$ **20.** $2x^2 + 21x + 27$ **21.** $5x^2 + 13x + 8$
22. $3t^2 - 2t - 5$ **23.** $5z^2 + 31z - 72$ **24.** $6x^2 - 33x + 15$ **25.** $6x^2 + 29x + 9$
26. $x^3 + 3x^2 + 3x + 2$ **27.** $x^3 - x^2 - 2x - 12$ **28.** $2x^3 + 5x^2 - 7x + 20$
29. $6t^3 - 7t^2 - 10t + 6$ **30.** $6x^3 + 8x^2 - 29x + 14$ **31.** $x^4 + 7x^3 - 4x^2 - 21x + 3$
32. $x^3 + 3x^2 + 3x + 1$ **33.** $x^3 - 9x^2 + 27x - 27$ **34.** $9t^3 - 4t$ **35.** $2y^3 + 3y^2 - 18y - 27$
36. $4x^3 + 4x^2 - 24x$ **37.** $2y^4 - y^3 - 3y^2$ **38.** $x^2 - 2x + 1$ square feet
39. The area is $x^2 + 3x$ sq ft. **40.** The area is $3x^2 + 14x + 8$ square meters
41. The area is $3h^2 - 2h$ sq yd. **42.** The area is $\pi r^2 + 4\pi r + 4\pi$ cm^2
43. The product is $x^2 + 2x$. **44.** $n^3 + 6n^2 + 8n$ **45.** The product is $y^2 - y$.

Practice Exercises

47. $18r^6$ **49.** $-24w^5$ **51.** $21x^{10}$ **53.** $-2x^2$ **55.** $4x - 20$ **57.** $6y - 18y^2 + 2y^3$
59. $-21t^3 + 24t^2 - 6t$ **61.** $x^2 + 7x + 10$ **63.** $t^2 - 9$ **65.** $3t^2 + 29t + 18$ **67.** $2x^2 - x - 10$
69. $12x^2 - x - 6$ **71.** $6r^2 + 11r - 35$ **73.** $18 - 2x^2$ **75.** $x^2 - 16$ **77.** $y^2 + 10y + 25$
79. $5t^2 - 38t + 21$ **81.** $15x^2 - 19x - 10$ **83.** $6s^2 - s - 35$ **85.** $6 + 13x - 8x^2$ **87.** $x^3 - 1$
89. $x^3 + 64$ **91.** $6x^3 - 7x^2 - 11x + 12$ **93.** $s^3 - 3s^2 + 3s - 1$ **95.** $3t^3 - 75t$
97. The area is $2x^2 - x$ sq cm. **99.** The product is $z^2 + 6z + 8$.
101. The area is $2x^2 - x$ sq yd; 28 sq yd

Challenge Problems

103. $\frac{1}{4}x^2 - 1$ **105.** $\frac{2}{9}z^2 + z - 2$ **107.** $p^3 + q^3$

PROBLEM SET 2.6

Warm-Ups

1. $x^2 + 2x + 1$ **2.** $x^2 - 14x + 49$ **3.** $x^2 - 2x + 1$ **4.** $x^2 + 4x + 4$ **5.** $16s^2 + 24s + 9$
6. $25t^2 - 40t + 16$ **7.** $4x^2 - 4x + 1$ **8.** $9x^2 - 12x + 4$
9. $81 + 18y + y^2$ **10.** $64 - 16t + t^2$ **11.** $x^2 - 1$ **12.** $x^2 - 36$ **13.** $4x^2 - 9$ **14.** $9x^2 - 25$
15. $36x^2 - 81$ **16.** $49 - 121w^2$ **17.** $9x^2 - 49$ **18.** $100 - 81t^2$ **19.** $81 - 25t^2$ **20.** $1 - t^2$
21. $-2x - 1$ **22.** $-6 - x$ **23.** $-12x^3 + 40x^2 + 30x$ **24.** $2t - 3t^2$ **25.** $2x^2 + 3x + 9$
26. $4x - 2$ **27.** $2y^2 - y - 16$ **28.** $11x^2 - 16x - 49$
29. The area is $9y^2 + 12y + 4$ sq ft. **30.** The area is $\pi r^2 - 2\pi r + \pi$ square meters
31. The area is $x^2 - 9$ sq ft. **32.** It is $n^2 + 2n + 1$ **33.** It is $q^2 - 2q + 1$.

Practice Exercises

35. $x^2 + 6x + 9$ **37.** $y^2 - 25$ **39.** $16t^2 + 40t + 25$ **41.** $4t^2 - 49$ **43.** $4x^2 - 12x + 9$
45. $4z^2 + 4z + 1$ **47.** $121q^2 - 25$ **49.** $9w^2 - 24w + 16$ **51.** $25 - 144k^2$
53. $81g^2 + 36g + 4$ **55.** $-6x^3 + 39x^2 + 17x$ **57.** $-4z^2 + 4z - 1$ **59.** $-8w - 23$
61. $-3t - 18$ **63.** $-t^3 - t^2 + 7t$ **65.** The area is $\pi r^2 + 14\pi r + 49\pi$ sq in.
67. The area is $x^2 - 4$ sq m. **69.** $36t^2 - 60t + 25$ trees; 4489 trees

Challenge Problems

71. $\frac{4}{9} - \frac{1}{4}x^2$

PROBLEM SET 2.7

Warm-Ups

1. $a^2 - a + 1$ **2.** $-4 + 2t^3$ **3.** $1 - \frac{1}{x} + \frac{1}{x^2}$ **4.** $\frac{z^4}{5} - \frac{z^2}{3} - \frac{1}{15}$ **5.** $-x + 1$ **6.** $\frac{1}{y^2} - \frac{2}{5y} + 1$

7. $3x^3 - x$ **8.** $z^6 - \dfrac{2}{z^2} + \dfrac{4}{z^3}$ **9.** $-3x + 1 - \dfrac{1}{x^2}$ **10.** $\dfrac{1}{2x} + 1$ **11.** $x + \dfrac{3x^2}{2}$ **12.** $t^2 - 7$

13. $1 + \dfrac{2x}{3}$ **14.** $x - y$ **15.** $\dfrac{1}{3} - \dfrac{x^2}{2}$ **16.** $x + 2$ **17.** $x + 4$ **18.** $x - 5$ **19.** $x + 3 + \dfrac{1}{x + 1}$

20. $y + 4 + \dfrac{3}{y + 2}$ **21.** $t - 2 + \dfrac{2}{t + 3}$ **22.** $2x - 1$ **23.** $3x + 2$ **24.** $2x - 7$

25. $s + 2 + \dfrac{1}{s - 1}$ **26.** $2x - 1 + \dfrac{3}{x + 4}$ **27.** $3z - 2 + \dfrac{-4}{z - 3}$ **28.** $x^2 + x + 1$

29. $x^2 + 2x - 7 + \dfrac{4}{x + 2}$ **30.** $t^2 - t - 1$ **31.** $x^2 + x - 1 + \dfrac{1}{x + 2}$ **32.** $x^2 - 2x - 3 + \dfrac{-4}{x - 1}$

33. $x^2 + 2x - 2 + \dfrac{4}{x + 2}$ **34.** $x^2 + 3x + 2$ **35.** $2y^2 - y + 1 + \dfrac{-3}{2y + 3}$

36. $x^2 + 5x - 1 + \dfrac{2}{x - 5}$ **37.** $3v^2 - 2v - 1 + \dfrac{-5}{2v - 1}$ **38.** $x^2 - 3x + 2 + \dfrac{1}{3x + 2}$

39. $z^2 - 2z + 2 + \dfrac{-8}{4 + 2z}$

Practice Exercises

41. $\dfrac{3y^4}{2} - 1$ **43.** $x^3 + 4x^4$ **45.** $t^2 - 6$ **47.** $\dfrac{1}{4} - \dfrac{x^3}{3}$ **49.** $x + 3$ **51.** $x - 2$

53. $y + 4 + \dfrac{-1}{y + 3}$ **55.** $x + 2$ **57.** $3x - 2 + \dfrac{-12}{x - 3}$ **59.** $3x + 2 + \dfrac{-3}{x + 2}$

61. $x^2 + x + 1$ **63.** $t^2 - t + 2$ **65.** $x^2 + x + 2 + \dfrac{-5}{x - 2}$ **67.** $x^2 + x - 1$

69. $2x^2 + 3x + 5 + \dfrac{4}{2x - 3}$ **71.** $4x^2 + 3x + 2 + \dfrac{6}{4x - 3}$ **73.** $2x + 1$ hills; 13 hills

Challenge Problems

75. $2x^3 - 2x^2 - 3x + 5 + \dfrac{-4}{2x - 1}$ **77.** $x^2 + 3x - 4$ **79.** $x^2 + 2x + 6 + \dfrac{-1}{x^2 + 1}$

REVIEW PROBLEMS CHAPTER 2

1. x^7 **3.** $\dfrac{1}{4z^8}$ **5.** $-\dfrac{1}{49}$ **7.** t^{10} **9.** 1 **11.** (a) 0.000000163 (b) 8,503,000 **13.** -25

15. $5x^5 - 2x^4 - 3x^3 - 12x + 5$ **17.** $-5x^2 + 25x^7$ **19.** $x^3 - 2x^2 + 2x - 1$ **21.** $3x - 5$
23. $x^3 - 6x^2 + 12x - 8$ **25.** $7x - 3$ **27.** $-3x^4y^4$ **29.** $x^2 + 14x + 49$ **31.** $4t^2 - 12t + 9$
33. The length is $3x + 5$ ft. **35.** The area is $y^2 - 9$ sq in. **37.** $1.50, \$1.75, \$1.85, \$2.50, \$3.75, \$6.25
39. $2x + 1$ trees

Let's Not Forget . . .

40. -16 **41.** $9x - 12$ **42.** $90 - x$ **43.** $3x^2 - 3x + 5$ **44.** $2x^3$ **45.** -16
46. $-\dfrac{1}{16}$ **47.** $\dfrac{1}{16}$ **48.** 16 **49.** -1 **50.** 0 **51.** undefined **52.** $\dfrac{-2}{x}$ **53.** $x - 1$

54. $x^2 - 9$ **55.** $-2x - 9$ **56.** $x^2 - 3x + 3$ **57.** factor   **58.** term   **59.** factor   **60.** x
61. $x - 1$ **62.** $x - 3$ **63.** Equation, $\{-9\}$ **64.** Expression, $2x + 6$ **65.** Expression, $8x$
66. Equation, $\{5\}$ **67.** Expression, -5

CHAPTER 2 TEST

1. (a)   **2.** (d)   **3.** (c)   **4.** (b)   **5.** (b)   **6.** (a) 2.33×10^{-3} (b) 5.763×10^{2}
7. (a) 9400   (b) 0.00004321   **8.** $6 - 4x$ **9.** $2x^2 - x - 3$ **10.** $2x^3 - 3x^2 - 4x$

11. $9x^2 - 12x + 4$ **12.** $x^3 + 6x^2 + 12x + 8$ **13.** $6x^2 - 8x + 9 + \dfrac{-11}{x + 1}$

14. $4x^3 - x^2 + 2x + 11$ **15.** 4   **16.** $4 - y^2$ **17.** $2x^7 - 3x^6 + 3x^3$
18. $x^2 - x - 16$ **19.** The area is $2y^2 + 3y - 2$ sq in.   **20.** The length is $3s + 22$ yd.

► **Chapter 3**

PROBLEM SET 3.1

Warm-Ups

1. $\{-9\}$ **2.** $\{-18\}$ **3.** $\{0\}$ **4.** $\{-5\}$ **5.** $\left\{-\dfrac{1}{2}\right\}$ **6.** $\{7\}$ **7.** $\{12\}$ **8.** $\{-6\}$

9. $\{15\}$ **10.** $\{4\}$ **11.** $\{4\}$ **12.** $\{5\}$ **13.** $\left\{\dfrac{1}{2}\right\}$ **14.** $\{-5\}$ **15.** $\{2\}$ **16.** $\{0\}$

17. $\{0\}$ **18.** $\{1\}$ **19.** $\{-3\}$ **20.** $\left\{\dfrac{4}{3}\right\}$ **21.** $\{5\}$ **22.** $\{3\}$ **23.** $\{-1\}$ **24.** $\{-2\}$

25. $\{1\}$ **26.** $\{-3\}$ **27.** $\{1\}$ **28.** $\{-2\}$ **29.** $\{-3\}$ **30.** $\{5\}$ **31.** $\{1\}$ **32.** $\{-5\}$

33. $\{-1\}$ **34.** $\{-2\}$ **35.** $\{3\}$ **36.** $\{12\}$ **37.** $\{7\}$ **38.** $\left\{-\dfrac{1}{2}\right\}$ **39.** $\{3\}$ **40.** $\{-2\}$

41. $\{5\}$ **42.** $\{1\}$ **43.** 5   **44.** -9 **45.** 9   **46.** 3   **47.** -1 **48.** 2   **49.** -1 **50.** 35 bolts
51. 16 in.   **52.** 5 mi   **53.** 11 nickels and 16 dimes   **54.** 4.5 ft   **55.** 16 cm   **56.** 2 meters

Practice Exercises

57. $\{-1\}$ **59.** $\{7\}$ **61.** $\{-3\}$ **63.** $\{5\}$ **65.** $\{-4\}$ **67.** $\{5/2\}$ **69.** $\{-14\}$ **71.** $\{11\}$

73. $\{-3\}$ **75.** $\{1\}$ **77.** $\{3\}$ **79.** $\left\{-\dfrac{16}{7}\right\}$ **81.** $\left\{\dfrac{2}{3}\right\}$ **83.** $\{-7\}$ **85.** $\{-1\}$ **87.** -3
89. 8   **91.** 1   **93.** 7 cm   **95.** 8 pairs

Challenge Problems

97. $\{2q\}$

PROBLEM SET 3.2

Warm-Ups

1. $\left\{\dfrac{11}{4}\right\}$ **2.** $\left\{-\dfrac{4}{5}\right\}$ **3.** $\{0\}$ **4.** $\left\{\dfrac{4}{7}\right\}$ **5.** $\left\{\dfrac{3}{4}\right\}$ **6.** $\{-1\}$ **7.** $\left\{-\dfrac{1}{6}\right\}$ **8.** $\{7\}$

9. $\{-18\}$ **10.** $\left\{\dfrac{3}{2}\right\}$ **11.** $\{11\}$ **12.** $\{-9\}$ **13.** $\{-15\}$ **14.** $\{-2\}$ **15.** $\left\{\dfrac{7}{3}\right\}$ **16.** $\{-2\}$

17. $\{18\}$ **18.** $\{-7\}$ **19.** $\{4\}$ **20.** $\left\{\dfrac{1}{2}\right\}$ **21.** $\{2\}$ **22.** $\{3\}$ **23.** $\{-1\}$ **24.** $\{1\}$

25. $\left\{\dfrac{1}{3}\right\}$ **26.** $\{1\}$ **27.** $\{4\}$ **28.** $\{-8\}$ **29.** $\{-5\}$ **30.** $\{-6\}$ **31.** $\{1\}$ **32.** $\{6\}$

33. $\{3\}$ **34.** $\{5\}$ **35.** $\{-15\}$ **36.** $\{-2\}$ **37.** $\varnothing$ **38.** $\{x \mid x \text{ is a real number}\}$

39. He worked 27 hrs.   **40.** 17   **41.** She has 20 dimes.   **42.** 15, 17, 19   **43.** They are $-1, 0, 1, 2$.

44. 20, 22, 24   **45.** She deposited \$570.   **46.** 55 mph   **47.** He rode 8 hrs.

48. 5 meters   **49.** It is 500.   **50.** 144   **51.** It contains 20 gal.   **52.** 14 L

Practice Exercises

53. $\{2\}$ **55.** $\{2\}$ **57.** $\{3\}$ **59.** $\{-1\}$ **61.** $\{3\}$ **63.** $\{-2\}$ **65.** $\{-1\}$ **67.** $\left\{-\dfrac{1}{9}\right\}$

69. $\varnothing$ **71.** $\{10\}$ **73.** $\{2\}$ **75.** $\{x \mid x \text{ is a real number}\}$ **77.** $\{1\}$ **79.** $\left\{\dfrac{1}{4}\right\}$ **81.** $\{1\}$

83. $\{-0.4\}$ **85.** $\{-1\}$ **87.** $\{-1\}$ **89.** $\{-9\}$ **91.** $\{-6\}$ **93.** $\left\{\dfrac{3}{2}\right\}$ **95.** $\{x \mid x \text{ is a real number}\}$

97. $\left\{-\dfrac{2}{3}\right\}$ **99.** $\{x \mid x \text{ is a real number}\}$ **101.** It is 480.   **103.** He borrowed \$750.

105. He has 8.   **107.** It is 15 ft high.   **109.** He bought 8 tapes.

Challenge Problems

111. $\left\{\dfrac{4}{3}q\right\}$

PROBLEM SET 3.3

Warm-Ups

1. $w = \dfrac{A}{l}$ **2.** $b = \dfrac{2A}{h}$ **3.** $r = \dfrac{C}{2\pi}$ **4.** $t = \dfrac{I}{pr}$ **5.** $h = \dfrac{3V}{\pi r^2}$ **6.** $m = \dfrac{E}{c^2}$ **7.** $C = \dfrac{5F - 160}{9}$

8. $x = 1 - y$ **9.** $y = 1 - x$ **10.** $x = a - y$ **11.** $x = a + y$ **12.** $x = \dfrac{q}{2}$

13. $x = -\dfrac{p}{3}$ **14.** $y = \dfrac{k - ax}{2}$ **15.** $y = -2k$ **16.** $y = \dfrac{z - ax}{b}$ **17.** $y = -\dfrac{z}{b}$

18. $x = \dfrac{bD - bC}{2}$ **19.** $y = \dfrac{4f + 4L}{a}$ **20.** $z = \dfrac{14C + Cs}{b}$ **21.** $w = -\dfrac{2kq}{a}$ **22.** $x = \dfrac{5 - 2a}{2}$

23. $y = \dfrac{1 + 3q}{3}$ **24.** $w = ug - 3$ **25.** $x = 6p - a$ **26.** $y = \dfrac{b - 2x}{2}$ **27.** $w = \dfrac{2bL - ak}{a}$

Practice Exercises

29. $m = \dfrac{rF}{v^2}$ **31.** $F = \dfrac{9C + 160}{5}$ **33.** $z = 5 - 2x$ **35.** $x = k + 3z$ **37.** $x = -\dfrac{q}{2}$

39. $x = -\dfrac{7}{p}$ **41.** $x = -\dfrac{a}{k}$ **43.** $y = h - 5x$ **45.** $y = \dfrac{k - ax}{4}$ **47.** $y = \dfrac{ax - z}{b}$

49. $x = 6 + k$ **51.** $y = -3w$ **53** $z = pT + pq$ **55.** $y = \dfrac{-2f - 2L}{q}$ **57.** $w = -\dfrac{k^2 t}{a}$

59. $y = \dfrac{1 + 2p}{2}$ **61.** $x = \dfrac{kp - 2}{p}$ **63.** $x = \dfrac{C + AB}{3A}$ **65.** $s = -A$ **67.** $x = \dfrac{3r - 2d}{2}$

69. $x = -2r$ **71.** $w = \dfrac{5dL - ck}{c}$ **73.** $S = \dfrac{100P - DP}{100}$

Challenge Problems

75. $P_2 = \dfrac{P_1 V_1 T_2}{T_1 V_2}$

PROBLEM SET 3.4

Warm-Ups

1. It is 34.   **2.** 62, 64   **3.** They are 6 ft and 10 ft.   **4.** 14 in., 28 in.   **5.** 9 m by 11 m
6. 22 cm by 27 cm   **7.** 17 yd by 51 yd   **8.** 14 m by 49 m   **9.** Homer is 64 and Pete is 32.
10. Sam is 23 and John is 26.   **11.** Ellen is 30 and Sarah is 34.   **12.** Traci is 20 and Bill is 50.
13. 10 nickels and 12 dimes   **14.** 20 fives and 10 tens.   **15.** 24 nickels and 30 pennies.
16. 381   **17.** She should add 3 liters.   **18.** 7 gal   **19.** 630 barrels of $16 crude   **20.** 90 pounds
21. It will take $2\frac{1}{2}$ hr.   **22.** 2 hr   **23.** It will take 2 hr.   **24.** 2 hr

Practice Exercises

25. They are 69 and 71.   **27.** 24 m by 27 m   **29.** Kitty is 16 and Charlie is 32.
31. 11 nickels and 22 dimes   **33.** They are 34 and 35.   **35.** The angles are 36° and 144°.
37. Kimberly is 6 and Tom is 24.   **39.** 30 nickels, 20 dimes, 125 pennies   **41.** 10 lb of $3.50 pecans
43. It will take $3\frac{1}{3}$ hr.   **45.** She should add 6 liters.   **47.** It will take 3 hr.   **49.** They are 24° and 66°.
51. 30 yd by 120 yd   **53.** J.R. is 9 and Kim is 15.   **55.** 5 nickels and 15 pennies
57. 25 gallons of the $35 oil   **59.** It will take $2\frac{1}{2}$ hr.   **61.** It is 43.   **63.** It is 19 mm.
65. Daughter is 14 and Louise is 42.   **67.** 300 nickels, 600 dimes, and 1500 quarters
69. 108 lb are required.   **71.** It will take $1\frac{1}{2}$ hr.

Challenge Problems

73. $54\frac{6}{11}$ miles

PROBLEM SET 3.5

Warm-Ups

1. $-8, -2, 0, 2$ **2.** 13, 40   **3.** 40, 400   **4.** 0, 2, 10   **5.** 1, 2, 3, 4, 5, 6   **6.** (graph)

7. (graph) -1 **8.** (graph) 5 **9.** (graph) -3 **10.** $\{x \mid x > 13\}$ (graph) 13

11. $\{x \mid x < 18\}$ (graph) 18 **12.** $\{x \mid x \le -1\}$ (graph) -1 **13.** $\{x \mid x \ge 20\}$ (graph) 20 **14.** $\{x \mid x < 9\}$ (graph) 9

15. $\{x \mid x \ge 6\}$ (graph) 6 **16.** $\{x \mid x \le -4\}$ (graph) -4 **17.** $\{x \mid x \le 3\}$ (graph) 3 **18.** $\{y \mid y > 10\}$ (graph) 10

19. $\{t \mid t \le -\frac{8}{3}\}$

20. $\{s \mid s \ge -14\}$

21. $\{x \mid x > -4\}$

22. $\{x \mid x < 12\}$

23. $\{w \mid w \ge -25\}$

24. $\{x \mid x \le -10\}$

25. $\{x \mid x \le -10\}$

26. $\{x \mid x \ge 2\}$

27. $\{x \mid x > -1\}$

28. $\{x \mid x < -2\}$

29. $\{t \mid t \ge -2\}$

30. $\{s \mid s \le \frac{2}{7}\}$

31. $\{k \mid k > \frac{1}{3}\}$

32. $\{x \mid x < \frac{1}{2}\}$

33. $\{y \mid y \ge -2\}$

34. $\{x \mid x \ge 3\}$

35. $\{x \mid x < -5\}$

36. $\{y \mid y > -1\}$

37. $\{w \mid w < 1\}$

38. $\{x \mid x \ge -1\}$

39. $\{z \mid z > -3\}$

40. $\{x \mid x < -2\}$

41. $\{x \mid x > 3\}$

42. $\{y \mid y \le -5\}$

43. $\{z \mid z \ge -1\}$

44. $\{x \mid x \le \frac{2}{3}\}$

45. $\{x \mid x > \frac{9}{10}\}$

46. $\{x \mid x > -\frac{4}{3}\}$

47. $\{x \mid x > \frac{16}{3}\}$

48. $\{x \mid x < -\frac{3}{2}\}$

49. $\{x \mid x > 8\}$

50. $\{x \mid x \le -1\}$

51. $\{x \mid x \ge \frac{1}{2}\}$

52.

53.

54.

55.

Practice Exercises

57. $\{x \mid x < 30\}$

59. $\{x \mid x \ge -26\}$

61. $\{x \mid x > 8\}$

63. $\{z \mid z > -8\}$

65. $\{x \mid x > -\frac{7}{2}\}$

67. $\{s \mid s \ge -49\}$

69. $\{x \mid x < 0\}$

71. $\{s \mid s \le -5\}$

73. $\{x \mid x \text{ is a real number}\}$

75. $\{t \mid t \ge \frac{5}{2}\}$

77. $\{k \mid k > -4\}$

79. $\{y \mid y \ge 2\}$

s81. $\{v \mid v \le \frac{17}{18}\}$

83. $\{x \mid x < 10\}$

85. $\emptyset$

87.

89.

91. $\{g \mid -\frac{7}{3} \le g \le 2\}$

93. $\{x \mid 0 \le x < 3\}$

Challenge Problems

95. $\left\{ x \mid x > \dfrac{p}{2} \right\}$ **97.** $\left\{ x \mid x < \dfrac{2}{p} \right\}$

PROBLEM SET 3.6

Warm-Ups

1. $y < -2$ **2.** $t + 5 \le 15$ **3.** $s + (s + 1) > 0$ **4.** $|y| \ge 5$ **5.** $-3z \le 30$ **6.** $m^2 > 0$
7. $2x + 3 > 5$ **8.** $|-k| > 0$ **9.** greater than or equal to $10\frac{1}{2}$ ft   **10.** 8 ft   **11.** no longer than 12 in.
12. at least 86   **13.** at least 6.6   **14.** 1 and 2, or 2 and 3

Practice Exercises

15. $xy \ge -3$ **17.** $\frac{1}{3}w + 5 \ge 9$ **19.** $-x < 0$ **21.** $\frac{1}{2}p \le -4$ **23.** $\sqrt{p} > 7$ **25.** $\frac{5}{3} < x$
27. $|k| \ge 4$ **29.** at least 9.05   **31.** all numbers less than 6   **33.** at least 134   **35.** 11 or less
37. at least 45   **39.** all numbers less than 5   **41.** The sides could be at most 16 in. each.

Challenge Problems

43. He would need 118!

REVIEW PROBLEMS CHAPTER 3

1. $\{20\}$ **3.** $\{-9\}$ **5.** $\{2\}$ **7.** $\{x \mid x < 20\}$ **9.** $\{4\}$ **11.** $\{y \mid y > 5\}$ **13.** $\{1\}$
15. $\{x \mid x \ge 6\}$ **17.** $\left\{ -\dfrac{1}{2} \right\}$ **19.** $x = t - s$ **21.** $x = \dfrac{b + c}{a}$ **23.** $r = \dfrac{K + 2at}{2}$

25. $x = \dfrac{3bh}{a}$ **27.** 43 and 83   **29.** Carol is 60 and Russ is 64.

31. 5 grams of 50% gold should be added.   **33.** all numbers less than or equal to $-\dfrac{6}{5}$

35. 15 in. by 15 in.   **37.** The Miami airport served 26.6 million passengers in 1991.
39. The Mesozoic Era lasted 160 million years.
41. She would weigh 108 pounds on Earth.

Let's Not Forget . . .

42. $-6x^2$ **43.** $-6x - 2x^2$ **44.** undefined   **45.** $x^2 - 18x + 81$ **46.** -36 **47.** $\dfrac{3}{z}$
48. $2 + x^2 - x$ **49.** $x^2 + x - 2$ **50.** $2x + x^2 - 1$ **51.** term   **52.** factor   **53.** factor
54. term   **55.** $-3x$ **56.** $3(x + 7)$ **57.** $3x - 2$ **58.** expression; $5x - 19$
59. inequality; $\{x \mid x > 5\}$ **60.** equation; $\{-1\}$ **61.** expression; $-2x - 7$
62. inequality; $\left\{ x \mid x \ge -\dfrac{1}{2} \right\}$ **63.** equation; $\{-1\}$

CHAPTER 3 TEST

1. (d)　　**2.** (c)　　**3.** (a)　　**4.** (a)　　**5.** (b)　　**6.** (b)　　**7.** $\{1\}$　　**8.** $\{x \mid x \le 0\}$　　**9.** $\{2\}$

10. $\left\{x \mid x < -\dfrac{2}{3}\right\}$　　**11.** $x = -\dfrac{2t}{a}$　　**12.** $h = \dfrac{3V}{\pi r^2}$　　**13.** 17 and 33.

14. Billy is 16 and his sister is 11.　　**15.** 13 nickels and 7 dimes

---

► **Chapter 4**

PROBLEM SET 4.1

Warm-Ups

1. $3(2x + 1)$　　**2.** $2(x - 2)$　　**3.** $x(y + z)$　　**4.** $m(n - p)$　　**5.** $5(x - 3y)$　　**6.** $9(2a + b)$
7. $3a(3b - 2c)$　　**8.** $7r(s + 2)$　　**9.** $6(x + 1)$　　**10.** $3(z - 1)$　　**11.** 6　　**12.** 9　　**13.** 9　　**14.** x^2
15. xy　　**16.** ab^2c　　**17.** $3x$　　**18.** $9q^2$　　**19.** $4xy$　　**20.** $5rst$　　**21.** $5x(x - 3)$
22. $9ab(2a + 3b)$　　**23.** $25x^2yz(z - 3)$　　**24.** $15z^2(2z + 3)$　　**25.** $3rst(2rst + 1)$　　**26.** $7y^5(3 - 2y^2)$
27. $13x^2(2x^4 - x^2 + 3)$　　**28.** $4y^5(8y^3 + 9y^2 - 12)$　　**29.** $7m^2(n - 3mn^2 + 1)$
30. $a^2b^3c^5(bc^2 - ab^2c^3 - a^3)$　　**31.** $8rst(7r^3s + 8rst - 3)$　　**32.** $-5(y + 2)$
33. $-4(x - 1)$　　**34.** $-2(z + 1)$　　**35.** $-5y^2(3y^3 - 2y^2 + 4)$　　**36.** $-3ab(4a + 5b - c)$
37. $(s + t)(r + 2)$　　**38.** $(a - b)(z + w)$　　**39.** $(y - z)(x^2 + y^2)$　　**40.** $(c - d)(x - y)$
41. $(p + r)(q - 1)$　　**42.** $(x + y^2)(m^2 + 1)$　　**43.** $(s^2 - 2)(3r^3 - 2w)$　　**44.** $(x^2 + y^2)(x^4 - z^7)$
45. $(s + t)(m + n)$　　**46.** $(x - y)(z + w)$　　**47.** $(y - z)(r^2 + s^2)$　　**48.** $(p - r)(a + b)$
49. $(px + qz)(r - 3)$　　**50.** $(s^2 + t^2)(x^2 - 7)$　　**51.** $(z^5 + x)(2 - y)$　　**52.** $(x^2 + 2)(2 - w)$
53. $(y^2 - 2)(3x - 2w)$　　**54.** $(m^3 - y^2)(m^3 - z)$　　**55.** $(a + b)(x^2 + 1)$　　**56.** $(s - t)(a - 1)$

Practice Exercises

57. $5(x + 2)$　　**59.** $r(w + z)$　　**61.** $3(2x - 5y)$　　**63.** $-3a(3b - 2c)$　　**65.** $2x(x^2 + 2)$
67. $3xy(x - 5)$　　**69.** $15x^2yz^2(x - 2)$　　**71.** $-3r^3st(2t^2 - 1)$　　**73.** prime　　**75.** $24r^3s^2(3r + 2)$
77. $7x^3(2x^3 - 3x + 4)$　　**79.** $7ab(3a - 2b + 1)$　　**81.** $12rst(9r^3s + 8rst - 7)$　　**83.** $(a - b)(r + 2t)$
85. $(x^2 + 1)(x - y)$　　**87.** $(a^2 - b)(2r^2 - 1)$　　**89.** $(a - b)(s + 2t)$　　**91.** $(y^2 + 1)(y - x)$
93. $(q + r)(p + 3)$　　**95.** $(a^2 - b)(2 - z)$　　**97.** $(a + 2)(3 + y)$　　**99.** $(x + 3)(x^2 + 1)$

Challenge Problems

101. $a(s + t)(m + n)$　　**103.** $6(c + d)(b - d)$

PROBLEM SET 4.2

Warm-Ups

1. $(x - 3)(x + 3)$　　**2.** $(y - 9)(y + 9)$　　**3.** $(2x - 3)(2x + 3)$　　**4.** $(x - 7)(x + 7)$
5. $(4x - 5)(4x + 5)$　　**6.** $(x - y)(x + y)$　　**7.** $(a - 2b)(a + 2b)$　　**8.** $(2b - 1)(2b + 1)$
9. $(ab - 6)(ab + 6)$　　**10.** $(2t - xy)(2t + xy)$　　**11.** $(r - st)(r + st)$　　**12.** prime
13. $(2z - 7)(2z + 7)$　　**14.** $(3x - m)(3x + m)$　　**15.** $(10c - 3d)(10c + 3d)$　　**16.** $2(x - 5)(x + 5)$

17. $x(x - 1)(x + 1)$ **18.** $4(x - 2)(x + 2)$ **19.** $2(2x - 3)(2x + 3)$
20. $a(a - b)(a + b)$ **21.** $3(2b - 1)(2b + 1)$ **22.** $5(z^2 + 4)$ **23.** $2a(a - 2)(a + 2)$
24. $3a(b - 3)(b + 3)$ **25.** $4c(c - 2)(c + 2)$ **26.** $x^2(1 + x^2)$ **27.** $(y - x)(y^2 + yx + x^2)$
28. $(c + d)(c^2 - cd + d^2)$ **29.** $(x + 5)(x^2 - 5x + 25)$ **30.** $(x + 4)(x^2 - 4x + 16)$
31. $(z - 1)(z^2 + z + 1)$ **32.** $(d - 2c)(d^2 + 2cd + 4c^2)$ **33.** $(2x + 3)(4x^2 - 6x + 9)$
34. $(3k + 4)(9k^2 - 12k + 16)$ **35.** $(2p - 1)(4p^2 + 2p + 1)$ **36.** $(2x - y)(4x^2 + 2xy + y^2)$
37. $(3t + 2s)(9t^2 - 6st + 4s^2)$ **38.** $(5z - 3w)(25z^2 + 15zw + 9w^2)$ **39.** $2(y + x)(y^2 - yx + x^2)$
40. $3(z - 2)(z^2 + 2z + 4)$ **41.** $q(2p - 1)(4p^2 + 2p + 1)$ **42.** $2(x + 4)(x^2 - 4x + 16)$
43. $8(z - 1)(z^2 + z + 1)$ **44.** $b(2a - 3)(4a^2 + 6a + 9)$ **45.** $8(2x - y)(4x^2 + 2xy + y^2)$
46. $2(2p + q)(4p^2 - 2pq + q^2)$ **47.** $-5(z + 1)(z^2 - z + 1)$

Practice Exercises

49. $(y - 8)(y + 8)$ **51.** $(x - 11)(x + 11)$ **53.** $(2x - y)(2x + y)$ **55.** $(2b - 9)(2b + 9)$
57. $(t - 2xy)(t + 2xy)$ **59.** $4(x^2 + 1)$ **61.** $9(x - m)(x + m)$ **63.** $2(x - 3)(x + 3)$
65. $16(2x - 1)(2x + 1)$ **67.** $s(s - t)(s + t)$ **69.** $8(z^2 + 9)$ **71.** $2x(5y - z)(5y + z)$
73. $(c - d)(c^2 + cd + d^2)$ **75.** $(x - 4)(x^2 + 4x + 16)$ **77.** $(d + 2c)(d^2 - 2cd + 4c^2)$
79. $(3k - 4)(9k^2 + 12k + 16)$ **81.** $(2x + y)(4x^2 - 2xy + y^2)$ **83.** $(5z + 3w)(25z^2 - 15wz + 9w^2)$
85. $3(z + 2)(z^2 - 2z + 4)$ **87.** $2(x - 4)(x^2 + 4x + 16)$ **89.** $b(2a + 3)(4a^2 - 6a + 9)$
91. $2(2p - q)(4p^2 + 2pq + q^2)$

Challenge Problems

93. $(x^3 - 2)(x^3 + 2)$ **95.** $(x^2 + 4)(x + 2)(x - 2)$ **97.** $(x - 1)(x^2 + x + 1)(x^6 + x^3 + 1)$
99. $[(a + b) + 2][(a + b)^2 - 2(a + b) + 4]$ **101.** $(4 - a - b)(4 + a + b)$

PROBLEM SET 4.3

Warm-Ups

1. $(x + 2)(x + 1)$ **2.** $(x + 3)(x + 1)$ **3.** $(x + 4)(x + 1)$ **4.** $(x + 5y)(x + y)$
5. $(x + 6)(x + 1)$ **6.** $(x + 2)^2$ **7.** $(x + 2)(x + 3)$ **8.** $(x + 4)(x + 2)$
9. $(x - 3)(x - 5)$ **10.** $(x - 3)(x - 4)$ **11.** $(x - 2y)(x - y)$ **12.** $(x - 3)(x - 1)$
13. $(x - 4)^2$ **14.** $(x - 5)(x - 1)$ **15.** $(x - 6)(x - 1)$ **16.** $(x - 2)(x + 1)$
17. $(a - 3b)(a + b)$ **18.** $(z + 4)(z - 1)$ **19.** $(x + 5)(x - 1)$ **20.** $(y - 6)(y + 1)$
21. $(x + 2)(x - 3)$ **22.** $(x - 4)(x + 2)$ **23.** $(x - 5)(x + 2)$ **24.** $(t + 4)(t - 3)$
25. $(x + 4)(x - 5)$ **26.** $(2x + 1)(x + 1)$ **27.** $(3b + 1)(b + 1)$ **28.** $(2x - 1)(x - 2)$
29. $(3x - z)(x - 2z)$ **30.** $(5w + 2)(w + 1)$ **31.** $(3x + 1)(x + 3)$ **32.** $(3x - 2)^2$
33. $(5x - 1)(x - 3)$ **34.** $(2x + 1)(x - 1)$ **35.** $(3c + 1)(c - 1)$ **36.** $(2x - 1)(x + 2)$
37. $(3x - 2)(x - 1)$ **38.** $(3x + 1)(2x + 1)$ **39.** $(2x + 3)(x + 2)$ **40.** $(5x - 1)(2x - 1)$
41. $(3a - 2)(a - 4)$ **42.** $(3x + y)(2x - 3y)$ **43.** $(7x - 2)(x + 5)$ **44.** $(3t + 4)(2t + 5)$
45. $(4x - 7)(2x - 3)$ **46.** $(4p + 5)(3p - 5)$ **47.** $2(x + 2)(x - 3)$ **48.** $3(x - 4)(x - 1)$
49. $2(2x + 1)(x - 4)$ **50.** $a(2x + 3)(x + 3)$

Practice Exercises

51. $(x + 1)(x + 7)$ **53.** $(x + 1)(x + 9)$ **55.** $(x + 2)(x + 6)$ **57.** $(x - 1)^2$ **59.** $(x - 3)(x - 6)$
61. $(y + 2)(y - 1)$ **63.** prime   **65.** $(x + 6)(x - 1)$ **67.** $(a + 2)(a - 4)$ **69.** $(x + 3)(x - 4)$
71. $(5x + 1)(x + 1)$ **73.** $(5a - b)(a - b)$ **75.** $(3x + 1)(x + 2)$ **77.** $(7b - 2)^2$
79. $(2a - b)(a + b)$ **81.** $(3x - 1)(x + 2)$ **83.** prime   **85.** $(2c - 1)(3c - 1)$

87. $(3x - 2)(2x + 3)$ **89.** $(2x - 5)(3x - 4)$ **91.** $(2a + b)(8a - 9b)$ **93.** $(3x - 2)(8x - 3)$
95. $(6x + 5)(x - 5)$ **97.** $4a(x + 2)^2$ **99.** $a^2(a - b)(a + 2b)$

Challenge Problems

101. $(x^2 - 2)(x^2 - 3)$ **103.** $(x - 1)(x + 1)(x - 2)(x + 2)$

PROBLEM SET 4.4

Warm-Ups

1. $25(x - 1)(x + 1)$ **2.** $(2w - 3)(4w^2 + 6w + 9)$ **3.** $(3b - x)(3b + x)$ **4.** $45a^2b^2c(5a^3 - 2)$
5. $(s + 3)(s^2 - 3s + 9)$ **6.** $x(x - 1)$ **7.** $4(x^2 + 1)$ **8.** $(7 - 2s)(7 + 2s)$ **9.** $(z - 4)(y + 1)$
10. prime   **11.** $2(2 - b)(2 + b)$ **12.** prime   **13.** $(4 - 5x)(4 + 5x)$
14. $(w + 5x)(w^2 - 5wx + 25x^2)$ **15.** $(x - y)(ab + c)$ **16.** $(z + 1)(z^2 - z + 1)$
17. $16(5 - y)(25 + 5y + y^2)$ **18.** $(1 - y)(z - 2)(z + 2)$ **19.** $(x + 2)(x - 1)$ **20.** prime
21. $3(x - 9)(x - 1)$ **22.** $(3r + 1)(3r - 8)$ **23.** $(x + 3)^2$ **24.** $9(2a^3 + 4b^3c^3 + 3b^2c^2)$
25. $(7x - 1)(3x - 2)$ **26.** $(y - 7)(y + 1)$ **27.** $(x - 6)(x + 1)$ **28.** prime
29. $(z + x)(a - 1)(a + 1)$ **30.** $x^3(x^2 + 1)(x + 1)$ **31.** $(z + y)(y + a)$ **32.** $(b + x)(2a + y)$

Practice Exercises

33. $(x + 7)(x + 2)$ **35.** $45r^2t^4(3rt + 5)$ **37.** $(1 - 3a)(1 + 3a + 9a^2)$ **39.** $(x + y)(a + 1)$
41. Factored completely   **43.** $2(4 - x)(4 + x)$ **45.** $(t - x)(r - 1)(r^2 + r + 1)$
47. $(z - xy)(z + xy)$ **49.** $(a + b)(ab + 1)$ **51.** $(x + 2)(x + 1)$ **53.** $(2x + y)(4x^2 - 2xy + y^2)$
55. $(x + 6)^2$ **57.** prime   **59.** $(2y + z)(4y^2 - 2yz + z^2)$ **61.** $(x + 5)(x - 4)$ **63.** $(9 - x)^2$
65. $(a - 5)(a^2 + 5a + 25)$ **67.** $a^5b^5c^3(a^2c - abc^3 + b)$

Challenge Problems

69. $(x^2 - 3)(x^2 + 2)$ **71.** $5(x^2 + 5)(x - 2)(x + 2)$ **73.** $(x^3 - 3)(x^2 + 3)$

PROBLEM SET 4.5

Warm-Ups

1. $\{-7, 9\}$ **2.** $\{0, 6\}$ **3.** $\{-7, 1\}$ **4.** $\{-6, 6\}$ **5.** $\{0, 8\}$ **6.** $\{-8\}$ **7.** $\left\{-3, \frac{5}{2}\right\}$ **8.** $\left\{-\frac{7}{3}, \frac{5}{2}\right\}$
9. $\left\{-\frac{5}{6}, \frac{5}{6}\right\}$ **10.** $\left\{-\frac{9}{2}, -\frac{3}{5}\right\}$ **11.** $\left\{0, -4, \frac{8}{7}\right\}$ **12.** $\left\{-\frac{9}{5}, \frac{1}{2}\right\}$ **13.** $\{9\}$ **14.** $\{0\}$
15. $\left\{-8, -\frac{9}{2}\right\}$ **16.** $\left\{-2, \frac{7}{5}, 5\right\}$ **17.** $\{-1, 1\}$ **18.** $\left\{-\frac{3}{2}\right\}$ **19.** $\left\{-\frac{1}{4}, \frac{5}{3}\right\}$ **20.** $\{-\sqrt{2}, \sqrt{2}\}$
21. $\{-\sqrt{5}, \sqrt{5}\}$ **22.** $\{-\sqrt{7}, \sqrt{7}\}$ **23.** $\{\pm 1\}$ **24.** $\{\pm 2\}$ **25.** $\{\pm 8\}$ **26.** $\left\{\pm\frac{5}{2}\right\}$ **27.** $\left\{\pm\frac{2}{3}\right\}$
28. $\left\{\pm\frac{9}{10}\right\}$ **29.** $\{0, 5\}$ **30.** $\{-1, 0\}$ **31.** $\{-2, -1\}$ **32.** $\{-5, -1\}$ **33.** $\{-1, 6\}$ **34.** $\{3, 5\}$
35. $\{-4\}$ **36.** $\{-1, 9\}$ **37.** $\left\{\frac{3}{2}, 2\right\}$ **38.** $\left\{-2, \frac{5}{3}\right\}$ **39.** $\{-4, 1\}$ **40.** $\{1\}$ **41.** $\{3, 5\}$
42. $\{-3, 2\}$ **43.** $\{-2, 0, 2\}$ **44.** $\{-2, 0, 3\}$ **45.** $\{-3, 1, 8\}$ **46.** $\{0, \pm 3\}$

Practice Exercises

47. $\{\pm 4\}$ **49.** $\{0, \pm 9\}$ **51.** $\left\{\pm\dfrac{5}{4}\right\}$ **53.** $\{0, 4\}$ **55.** $\{0, 3\}$ **57.** $\left\{-3, -\dfrac{1}{2}, 1\right\}$ **59.** $\{-4, -2\}$

61. $\{-6, 1\}$ **63.** $\{3\}$ **65.** $\{-3, 2\}$ **67.** $\left\{-\dfrac{5}{2}, 3\right\}$ **69.** $\left\{-\dfrac{2}{3}, \dfrac{1}{2}\right\}$ **71.** $\{3, 4\}$ **73.** $\{-3, 6\}$

Challenge Problems

75. $\{\pm 2\}$ **77.** $\{3\}$

PROBLEM SET 4.6

Warm-Ups

1. The integers are 8 and 9.   **2.** They are 5 and 9.   **3.** The temperatures were $-9°$ and $-7°$.
4. He has 8.   **5.** The integers are 3 and 4.   **6.** They are 6 in. and 8 in.   **7.** The distance is 1300 ft.
8. The height is 15 feet.   **9.** The sides are 5, 12 and 13 units.   **10.** 15 ft up the wall.
11. The sides are 10, 24, and 26 units.   **12.** The dimensions are 9 m by 12 m.
13. The dimensions are $\dfrac{1}{2}$ yd by 2 yd.   **14.** The dimensions are 15 cm by 25 cm.
15. The dimensions are 12 in. by 4 in.   **16.** Base is 10 ft; height is 5 ft.
17. The base is 4 inches and the height is 16 inches.   **18.** 4 in. and 8 in.
19. The base is 14 meters and the height is 6 meters.   **20.** x is 10 units.

Practice Exercises

21. They are 7 and 8.   **23.** The noon temperature was $-8°$.   **25.** The integers are 2 and 3.
27. She needs 34 in. of gold string.   **29.** The sides are 9, 12, 15 units.
31. The sides are 16, 30, and 34 units.   **33.** The dimensions are 8 mi by 14 mi.
35. The dimensions are 15 cm by 13 cm.   **37.** The base is 40 feet and the height is 10 feet.
39. It is 6 units.   **41.** They are 10 and 15, but we can't determine who is which age.

REVIEW PROBLEMS CHAPTER 4

1. $(z + 4t)(z - 4t)$ **3.** prime   **5.** $(x + t)(t + 2)$ **7.** $(2u - v)(4u^2 + 2uv + v^2)$
9. $a^2 b^2 c^3 (bc - a + bc^2)$ **11.** $4(3d + 1)$ **13.** $(a + b)(z^2 - 5)$ **15.** $2(x - 3)(x + 3)$
17. $(x + 3)^2$ **19.** $(xy - 4)(x^2 y^2 + 4xy + 16)$ **21.** Factored completely   **23.** $(x + 7)(x + 2)$
25. $(k - 5j)(k^2 + 5jk + 25j^2)$ **27.** $(1 - 2n)(k^2 + 4)$ **29.** $b(2b + 1)(b - 3)$ **31.** $\{3, 7\}$
33. $\{\pm\sqrt{15}\}$ **35.** $\{0, 3\}$ **37.** $\{-2, 5\}$ **39.** $\{\pm 4\}$ **41.** The integers are -22 and -23.
43. The dimensions are 11 by 16 units.   **45.** They are 14 in., 25 in., and 25 in.
47. There are 25 seats in each of 20 rows.   **49.** The other side is 13 ft.
51. The Polish athletes won 16 medals.

Let's Not Forget. . . .

52. x^2 **53.** 0   **54.** $6t + 1$ **55.** $y^2 - 2y + 1$ **56.** $x - 5$ **57.** -9 **58.** $3x - 2$
59. $x^2 + x - 2$ **60.** $x^2 + 2x - 1$ **61.** term   **62.** factor   **63.** term   **64.** x **65.** $4x + 28$
66. $6y + 8$ **67.** expression; $x^2 + x - 6$ **68.** equation; $\{-3, 2\}$ **69.** inequality; $\{x \mid x < -4\}$
70. expression; $5x + 13$ **71.** $y(a + 1)$ **72.** $(x + 1)(x^2 - x + 1)$ **73.** factored   **74.** factored

CHAPTER 4 TEST

1. b   **2.** c   **3.** b   **4.** d   **5.** d   **6.** $(2a - 3)(4a^2 + 6a + 9)$ **7.** $(2x - 1)(x + 5)$
8. $(a - b)(x + 2)$ **9.** $(p - q)(x - y)$ **10.** $(a - 4)(a + 2)$ **11.** $(x + 7)^2$ **12.** $-3cd(2c + d)$
13. $(2y + 5)(2y - 5)$ **14.** $9a^2b^2(3b - 4a + 5a^2b^2)$ **15.** prime   **16.** $\{\pm 9\}$ **17.** $\{-2, 6\}$
18. $\{-3, 0, 1\}$ **19.** The dimensions are 13 yd by 18 yd.   **20.** The sides are 5 m, 12 m, and 13 m.

► Chapter 5

PROBLEM SET 5.1

Warm-Ups

1. -1 **2.** $-\dfrac{1}{2}$ **3.** $\dfrac{1}{3}$ **4.** $-\dfrac{1}{4}$ **5.** 0   **6.** No values of x.   **7.** 0   **8.** 1   **9.** 3   **10.** -5

11. -8 **12.** 0   **13.** 2, 5   **14.** $-9, -1$ **15.** 4, 11   **16.** $-1, 1$ **17.** $-3, 3$ **18.** $-5, 5$

19. $\dfrac{5}{-x}; \dfrac{-5}{x}$ **20.** $-\dfrac{8}{x}; \dfrac{-8}{x}$ **21.** $\dfrac{-17}{x^2}; \dfrac{17}{-x^2}$ **22.** $-\dfrac{x + 4}{x}; \dfrac{x + 4}{-x}$ **23.** $\dfrac{-11x^3}{2x - 3}; -\dfrac{11x^3}{2x - 3}$

24. $\dfrac{-(x + 7)}{7 - x}; \dfrac{x + 7}{x - 7}$ **25.** $-\dfrac{3x}{x + 4}; \dfrac{3x}{-(x + 4)}$ **26.** $-\dfrac{6 - 5x}{x^3}; \dfrac{5x - 6}{x^3}$ **27.** $\dfrac{-(x + 6)}{x - 3}; \dfrac{x + 6}{3 - x}$

28. (a) 40 mph; (b) $\dfrac{360}{t - 1}$; (c) 45 mph   **29.** (a) $\dfrac{1}{600}$ hr or 6 sec; (b) $\dfrac{1}{4(s + 100)}$; (c) $\dfrac{1}{480}$ hr or 7.5 sec

Practice Exercises

31. 0   **33.** $-\dfrac{2}{3}$ **35.** $\dfrac{7}{4}$ **37.** no values of x **39.** 2   **41.** -5 **43.** -2 **45.** 2   **47.** 16

49. $-4, -2$ **51.** $-2, 2$ **53.** $-7, 7$ **55.** $-\dfrac{9}{2x}; \dfrac{-9}{2x}$ **57.** $-\dfrac{x - 1}{x}; \dfrac{x - 1}{-x}$ **59.** $\dfrac{3 - x}{3 + x}; \dfrac{x - 3}{-(3 + x)}$

61. $-\dfrac{9 + 8x}{x}; \dfrac{-(9 + 8x)}{x}$ **63.** (a) \$.05; (b) \$.0525; (c) 32 oz jar is a better buy. (d) $\dfrac{x}{k}$ dollars per oz.

65. (a) 10 min; (b) 7.5 min; (c) $\dfrac{2500}{15,000 + r}$ hr or $\dfrac{2500 \cdot 60}{15,000 + r}$ min.

Challenge Problems

67. $\dfrac{1}{2}$ **69.** $-\dfrac{1}{2}$ **71.** $\dfrac{4}{3}$ **73.** no real value of x **75.** $-\sqrt{5}, \sqrt{5}$

PROBLEM SET 5.2

Warm-Ups

1. $\dfrac{2x}{3y}$ **2.** $\dfrac{2}{3x}$ **3.** $2x$ **4.** $\dfrac{2y^2}{3}$ **5.** $\dfrac{x + 2}{2y}$ **6.** $\dfrac{-2x}{y}$ **7.** $\dfrac{4x}{y}$ **8.** $\dfrac{n}{3 + n}$

9. $\dfrac{1}{x + 1}$ **10.** $y - 5$ **11.** Cannot be reduced   **12.** $\dfrac{2x + 3y}{3y}$ **13.** y **14.** Cannot be reduced.

15. $\dfrac{x+y}{x}$ **16.** $\dfrac{x-1}{x+1}$ **17.** $\dfrac{4}{3}$ **18.** $\dfrac{1}{2}$ **19.** $\dfrac{-x}{2(a+b)^3}$ **20.** $\dfrac{x^2}{4}$ **21.** $\dfrac{x}{2}$ **22.** $\dfrac{3y}{xz}$

23. $\dfrac{2}{x-4}$ **24.** $\dfrac{1+x}{2}$ **25.** $\dfrac{x^2(2z-1)}{4}$ **26.** $\dfrac{xyz}{a-t}$ **27.** $\dfrac{x+3}{x-3}$ **28.** $\dfrac{x+1}{3}$

29. $\dfrac{2x}{x+1}$ **30.** $\dfrac{x-2}{x-1}$ **31.** $\dfrac{3(3x-y)}{3x+y}$ **32.** $\dfrac{x+2}{x-2}$ **33.** $\dfrac{x-3}{x-1}$ **34.** -1 **35.** -1 **36.** -2

37. -2 **38.** $\dfrac{-1}{2x+3}$ **39.** $\dfrac{-1}{x-2}$ or $\dfrac{1}{2-x}$ **40.** $\dfrac{1}{x^2-x+1}$ **41.** $\dfrac{1}{9+3y+y^2}$ **42.** $\dfrac{x+y}{5}$

43. $\dfrac{(x-y)^2}{x^2+xy+y^2}$ **44.** $\dfrac{1}{x+1}$ **45.** $\dfrac{1+2x+4x^2}{1-2y}$ **46.** $\dfrac{3x}{3x^2}$ **47.** $\dfrac{21x^2}{6x^3}$ **48.** $\dfrac{8ay}{12a^2x}$

49. $\dfrac{10}{5(x-1)}$ **50.** $\dfrac{-6x-54}{x^2-81}$ **51.** $\dfrac{x^2-x-2}{x^2-4x+4}$ **52.** $\dfrac{-48x-112}{6x^2-7x-49}$

Practice Exercises

53. $\dfrac{4x}{3y}$ **55.** $\dfrac{2x^2}{t}$ **57.** $\dfrac{x-3}{3y}$ **59.** $\dfrac{-2}{x}$ **61.** $\dfrac{1}{(s+5)^2}$ **63.** Cannot be reduced **65.** $\dfrac{x+7}{x-7}$

67. $\dfrac{1}{2}$ **69.** $\dfrac{t^2}{3}$ **71.** $\dfrac{2}{s}$ **73.** $\dfrac{1-x}{3}$ **75.** $\dfrac{klm}{a+t}$ **77.** $\dfrac{x-1}{3}$ **79.** $\dfrac{x+2}{x+1}$ **81.** $\dfrac{x+1}{x-1}$

83. $\dfrac{x+3}{x+1}$ **85.** $\dfrac{-1}{t-1}$ or $\dfrac{1}{1-t}$ **87.** -1 **89.** $\dfrac{1}{s^2+s+1}$ **91.** $\dfrac{x-y}{7}$ **93.** $\dfrac{5x+y}{x+1}$

95. $\dfrac{10y}{5y^2}$ **97.** $\dfrac{14abx}{21a^2x^2}$ **99.** $\dfrac{-x^2-7x}{x^2-49}$ **101.** $\dfrac{x^2+x-2}{x^2+4x+4}$ **103.** $\dfrac{-10x+30}{6x^2-13x-15}$

Challenge Problems

105. $(x-2)(x^2+4)$ **107.** $-(x+1)(x^2+1)$ **109.** $\dfrac{2y^2-5}{y(2y^2-1)}$ **111.** $\dfrac{r-s}{r+s}$

PROBLEM SET 5.3

Warm-Ups

1. $\dfrac{8x^2}{21y^2}$ **2.** $\dfrac{9ax}{50b^4}$ **3.** $\dfrac{22m^4}{91n^3}$ **4.** $\dfrac{3xz}{20}$ **5.** $\dfrac{-3nx}{my}$ **6.** $90x$ **7.** $-3x$ **8.** $\dfrac{-x^2}{3}$

9. $\dfrac{x+5}{x+6}$ **10.** $\dfrac{2(x-1)}{x+1}$ **11.** $\dfrac{5kx}{6}$ **12.** $x+1$ **13.** $\dfrac{2x-3}{2}$ **14.** $\dfrac{5}{9}$

15. $\dfrac{(2x+1)(x+1)}{(x-1)^2}$ **16.** $\dfrac{(x+2)(2x-1)}{2(x-2)}$ **17.** $\dfrac{s+3}{(s+2)(s-3)}$ **18.** $(t+1)(t-1)$

19. $\dfrac{(x+1)(x+4)}{(x-1)(x-4)}$ **20.** $\dfrac{-x}{y}$ **21.** -2 **22.** $\dfrac{4x^2}{9y^2}$ **23.** $\dfrac{21a^3}{20b^2x}$ **24.** $\dfrac{-6m^3}{n^3}$

25. $\dfrac{14}{15}$ **26.** $\dfrac{11ny}{12x}$ **27.** $\dfrac{-x}{40y^2}$ **28.** $\dfrac{-3x^2}{2y^2}$ **29.** $\dfrac{2y^2}{7}$ **30.** $\dfrac{15}{8x^2}$ **31.** $-(1+2x)$

32. $\dfrac{x}{4y(y-2)}$ **33.** $x-1$ **34.** $\dfrac{x+1}{2(x-1)}$ **35.** $\dfrac{9x}{4}$ **36.** $\dfrac{(x-2)^2}{2x-1}$ **37.** $\dfrac{(x-1)(2x+1)}{(x+1)^2}$

38. $\dfrac{2(x-3)}{3(x+3)}$ **39.** $\dfrac{y+2}{y+5}$ **40.** $\dfrac{(2x+1)(x-2)}{(x-1)(x+1)}$ **41.** $-2(x+1)$ **42.** $\dfrac{-2(q-3)}{q-5}$

Practice Exercises

43. $\dfrac{15x^2}{28y^2}$ **45.** $\dfrac{27m^3}{55n^4}$ **47.** $\dfrac{4n^2y}{m^2x}$ **49.** $\dfrac{-9x^2}{4y^2}$ **51.** $\dfrac{3x}{5y^3}$ **53.** $\dfrac{-13x}{10y}$ **55.** $2y$ **57.** $\dfrac{x+5}{x-6}$

59. $\dfrac{-7kx^2}{12}$ **61.** $\dfrac{y^2}{4}$ **63.** $\dfrac{2x+1}{3x-2}$ **65.** $\dfrac{12w^2}{35}$ **67.** $\dfrac{-8xz^2}{5}$ **69.** $x-2$ **71.** $\dfrac{x}{5y(y-1)}$

73. $\dfrac{2-r}{2(r+2)}$ **75.** $x-3$ **77.** $\dfrac{(x-1)^2}{2x+1}$ **79.** $\dfrac{(x-1)(2x+1)}{(x+1)^2}$ **81.** $\dfrac{(x+1)(2x-1)}{(x-1)^2}$

83. $\dfrac{y+2}{(y+3)(y-2)}$ **85.** $\dfrac{3(x-2)}{4-x}$ **87.** $\dfrac{x+1}{x-1}$ **89.** $\dfrac{(2x-1)(x+2)}{(x+1)(x-1)}$

Challenge Problems

91. $-\dfrac{x+2}{x-2}$ or $\dfrac{x+2}{2-x}$ **93.** $\dfrac{(x-3)(x-7)}{(x+2)(4x+1)}$

PROBLEM SET 5.4

Warm-Ups

1. $\dfrac{13}{x}$ **2.** $\dfrac{2x+3}{x+2}$ **3.** 2 **4.** $\dfrac{4t-1}{2-t}$ **5.** $\dfrac{-1}{x-1}$ **6.** $\dfrac{2t+1}{2-t}$ **7.** $x+2$ **8.** $\dfrac{x+2}{x-1}$

9. $x-2$ **10.** $\dfrac{x+1}{x-2}$ **11.** 35 **12.** $x(x+2)$ **13.** 27 **14.** x^3 **15.** 180 **16.** $18x$

17. $x^2(x-2)$ **18.** x^7y^5z **19.** $5x(x+5)$ **20.** $(x+2)^2(x-2)$ **21.** $3(x+2)(x-3)$

22. $54x^3y^4$ **23.** $\dfrac{15y+2x}{3xy}$ **24.** $\dfrac{2x^2+21}{14x}$ **25.** $\dfrac{10-3x}{2x}$ **26.** $\dfrac{9-2x^2}{6x}$ **27.** $\dfrac{2x-7}{x(x+7)}$

28. $\dfrac{4x-1}{x(x-1)}$ **29.** $\dfrac{r+6}{(r+2)(r-2)}$ **30.** $\dfrac{5}{(3z-2)(z+1)}$ **31.** $\dfrac{2t^2}{(t+1)(t-1)}$ **32.** $\dfrac{4x-1}{(x+1)(2x-3)}$

33. $\dfrac{2+x}{2x^2}$ **34.** $\dfrac{x+6}{3x^3}$ **35.** $\dfrac{3-10a}{4a^2}$ **36.** $\dfrac{14-x}{12xy}$ **37.** $\dfrac{8y-9x}{30x^2y^2}$ **38.** $\dfrac{2y+15x}{6x^2y}$

39. $\dfrac{15x}{4(1-5x)}$ **40.** $\dfrac{-9}{z+9}$ **41.** $\dfrac{k+9}{8(k-6)}$ **42.** $\dfrac{x+7}{(x-7)^2}$ **43.** $\dfrac{y-4}{(y+4)^2}$ **44.** $\dfrac{7}{2(k-7)}$

45. $\dfrac{9d-10}{15(d+2)}$ **46.** $\dfrac{2x-1}{(x-1)(x+1)}$ **47.** $\dfrac{2x+1}{(x-1)^2}$ **48.** $\dfrac{-x-10}{(x+2)^2(x-2)}$

49. $\dfrac{2c^2+5c}{(c+4)(c-3)(c+1)}$ **50.** $\dfrac{2t^2+7t-5}{(t-5)(t+3)(t-1)}$ **51.** $\dfrac{x+7}{7}$ **52.** $\dfrac{3x+1}{3}$

53. $\dfrac{y^2-2}{y}$ **54.** $\dfrac{3x^3-1}{x}$ **55.** $\dfrac{x+5}{x+3}$ **56.** $\dfrac{z+3}{z+2}$ **57.** $\dfrac{-3x-2}{1+2x}$ **58.** $\dfrac{8-3t}{t-1}$ **59.** $\dfrac{2x}{x-1}$

60. $\dfrac{x+3}{x+1}$ **61.** $\dfrac{x-1}{2x-1}$ **62.** $\dfrac{6}{t-14}$ **63.** $\dfrac{a-1}{b-1}$ **64.** 1

Practice Exercises

65. 24 **67.** 8 **69.** 252 **71.** $x^2(x+3)$ **73.** $(2x+1)(x-3)(x+3)$ **75.** 1 **77.** $\dfrac{3y-1}{5-y}$

79. $\dfrac{8t}{3(2-3t)}$ **81.** $\dfrac{4y+3x^2}{6x^3y}$ **83.** $\dfrac{2x+4}{x(x-4)}$ **85.** $\dfrac{r^2+4r-4}{(r-2)(r+2)}$ **87.** $\dfrac{5a}{(a-3)(a+2)}$

89. $\dfrac{2t-9}{12t^3}$ **91.** $\dfrac{2y+5}{(y+5)^2}$ **93.** $\dfrac{1}{12(d+3)}$ **95.** $\dfrac{-3x-5}{(x+1)^2}$ **97.** $\dfrac{3c}{(c-4)(c+3)(c-1)}$ **99.** $\dfrac{x}{x-2}$

101. $\dfrac{3-4x}{x}$ **103.** $\dfrac{1}{s-11}$

105. (a) $\dfrac{500}{r+100}$ hr (b) $\dfrac{250}{r}$ hr (c) $\dfrac{500}{r+100}+\dfrac{250}{r}$ hr (d) $\dfrac{750r+25,000}{r^2+100r}$ hr (e) Approximately 4.22 hr

107. (a) $\dfrac{26}{t}$ km/min (b) $\dfrac{26}{60t}$ km/hr (c) $\dfrac{26}{t+4}$ km/min (d) $\dfrac{26}{60(t+4)}$ km/hr

Challenge Problems

109. $\dfrac{x+1}{x^2+5x+25}$

PROBLEM SET 5.5

Warm-Ups

1. $\dfrac{1}{3}$ **2.** $\dfrac{3}{5}$ **3.** $\dfrac{3}{2}$ **4.** x **5.** $\dfrac{1}{3y}$ **6.** $2x$ **7.** $\dfrac{2}{9}$ **8.** $\dfrac{y}{x}$ **9.** $\dfrac{3}{4w}$ **10.** $\dfrac{x+1}{x-1}$

11. $\dfrac{x}{4}$ **12.** $\dfrac{3(2-3x)}{2(3-4x)}$ **13.** $\dfrac{5x}{14}$ **14.** $\dfrac{3}{2z(z+1)}$ **15.** $\dfrac{2(x+1)}{x-1}$ **16.** $\dfrac{3}{5}$

17. $\dfrac{1}{11}$ **18.** $-\dfrac{14}{11}$ **19.** $\dfrac{1+x}{1-x}$ **20.** $\dfrac{2-3y}{4+5y}$ **21.** $\dfrac{2z+3}{4z-5}$ **22.** $\dfrac{z-2}{2}$ **23.** $\dfrac{2}{2s+1}$

24. $\dfrac{3-x}{6}$ **25.** $\dfrac{y+xy}{x+xy}$ **26.** $\dfrac{y+x}{y-x}$ **27.** $\dfrac{x+2}{5x-3}$

Practice Exercises

29. $\dfrac{5}{4}$ **31.** $\dfrac{5}{11}$ **33.** $\dfrac{-2}{17}$ **35.** $\dfrac{3y}{4}$ **37.** $\dfrac{2}{9}$ **39.** $\dfrac{2w}{5}$ **41.** $\dfrac{x^2}{3}$ **43.** $\dfrac{7x^2}{10}$ **45.** $\dfrac{x+2}{x-2}$

47. $\dfrac{3-4y}{5+6y}$ **49.** $6-z$ **51.** $-\dfrac{2k+5}{2}$ **53.** $\dfrac{1}{y-1}$

Challenge Problems

55. no

PROBLEM SET 5.6

Warm-Ups

1. $\{-1\}$ **2.** $\{1\}$ **3.** $\{-1\}$ **4.** $\left\{\dfrac{3}{2}\right\}$ **5.** $\{10\}$ **6.** $\{3\}$ **7.** $\{3\}$ **8.** $\{5\}$

9. $\{7\}$ **10.** $\{-9\}$ **11.** $\left\{\dfrac{14}{3}\right\}$ **12.** $\{2\}$ **13.** $\{2\}$ **14.** $\{3\}$ **15.** $\{1\}$ **16.** $\{10\}$

17. $\left\{\dfrac{1}{2}\right\}$ **18.** $\{3\}$ **19.** $\varnothing$ **20.** $\varnothing$ **21.** $\varnothing$ **22.** $\varnothing$ **23.** $\varnothing$ **24.** $\varnothing$

25. $\dfrac{x}{3}$ **26.** $\dfrac{5}{2y}$ **27.** $\dfrac{1-x}{x+1}$ **28.** $\dfrac{w+5}{2(w+3)}$ **29.** $\{-3,-1\}$ **30.** $\{-1,4\}$

31. $\{-3,-2\}$ **32.** $\{-3,4\}$

Practice Exercises

33. $\{10\}$ **35.** $\left\{\dfrac{1}{2}\right\}$ **37.** $\{2\}$ **39.** $\{-1\}$ **41.** $\{3\}$ **43.** $\{15\}$ **45.** $\{8\}$ **47.** $\{2\}$ **49.** $\dfrac{6-x}{6x}$

51. $\{0\}$ **53.** $\{4\}$ **55.** $\{-3,1\}$ **57.** $\varnothing$ **59.** $\{0\}$ **61.** $\varnothing$

Challenge Problems

63. $\{2\}$ **65.** $\{Q-R\}$ **67.** $\{R\}$ **69.** $\{10\}$

PROBLEM SET 5.7

Warm-Ups

1. (a) 16/12 or 4/3 (b) 12/16 or 3/4 (c) 8/7 (d) 16/8 or 2/1 (e) 12/28 or 3/7   **2.** 15

3. 56   **4.** 8   **5.** 25   **6.** 90 ft   **7.** There are 200 in.   **8.** 100 lb

9. Five tomato plants can be planted.   **10.** 350 drops

11. She used 9500 gal.   **12.** 6   **13.** 3   **14.** 6   **15.** The pieces are 12 m and 18 m.   **16.** $\dfrac{1}{3}$

17. The numbers are 2 and 8.   **18.** 45 min   **19.** Frank ran 10 km/hr.   **20.** 20 mph

Practice Exercises

21. 8 ft   **23.** There are 15 lb in the lighter and 25 lb in the heavier.

25. The investment contained 130 shares.   **27.** The number is -4.

29. The speeds are 40 mph and 50 mph.   **31.** 49 ft

Challenge Problems

33. There are no such *even* integers.

REVIEW PROBLEMS CHAPTER 5

1. 0   **3.** $-2, 3$ **5.** $\dfrac{84x}{4x^3}$ **7.** $\dfrac{x^2-2x+1}{x^2+x-2}$ **9.** $\dfrac{3t}{5}$ **11.** $\dfrac{x}{x+2}$ **13.** $\dfrac{x}{x+3}$ **15.** $\dfrac{-3x^2}{2y}$

17. $\dfrac{3x+1}{2(3x-2)}$ **19.** $\dfrac{1}{6(2x+3)}$ **21.** $\dfrac{-1}{t^2(4t+1)}$ **23.** $\dfrac{2x^2-3}{2x}$ **25.** $\dfrac{1}{w-2}$ **27.** -1 **29.** $\dfrac{1}{5}$

31. $\dfrac{y+1}{y-1}$ **33.** $\{3\}$ **35.** $\{8\}$ **37.** $\varnothing$ **39.** $\{-1,3\}$ **41.** There are 40 wolves in the area.

43. The numbers are -1 and 1.   **45.** The cost of the bookbag was $61.25.

47. The height of poster will be 40 in.

Let's Not Forget . . .

49. -25 **50.** $x^2 - 10x + 25$ **51.** 25 **52.** undefined   **53.** x **54.** $\dfrac{7}{x-1}$ **55.** $-5x + 14$

56. $3x^2 - 6x - 7$ **57.** $3x^2 - 13x + 14$ **58.** $-5x - 7$ **59.** $\dfrac{2}{y}$ **60.** $\dfrac{2}{y(x-1)}$

61. term   **62.** term   **63.** factor   **64.** factor   **65.** 1 **66.** $2x^2 + 14x$ **67.** $4x + 6$

68. expression; $-(x+3)$ **69.** equation; $\{1\}$ **70.** inequality; $\left\{ x \mid x < \dfrac{1}{5} \right\}$ **71.** expression; $\dfrac{11}{4x}$

72. factored **73.** $(x-3)(x+3)$ **74.** $(x-3)^2$ **75.** $(x+3)(x^2-3x+9)$ **76.** factored

77. $3(x-3)(x^2+3x+9)$

CHAPTER 5 TEST

1. c   **2.** b   **3.** c   **4.** d   **5.** a   **6.** d   **7.** b   **8.** a   **9.** $-\dfrac{5}{2}$ **10.** $\dfrac{1+x^2}{x(1-x)}$

11. $\dfrac{t-4}{(t+4)(t+2)}$ **12.** $x+3$ **13.** $\dfrac{x+5}{x+3}$ **14.** $\dfrac{s+4}{s-4}$ **15.** $\{1\}$ **16.** $\varnothing$ **17.** $\{-1, 7\}$

18. (a) 3/2; (b) 2/5   **19.** There are 187.5 miles between the cities.   **20.** The numbers are 3 and 12.

► Chapter 6

PROBLEM SET 6.1

Warm-Ups

Answers to Problems 13 to 19 may vary.

1.   **2.**   **3.**   **4.**

$(2, 3)$ $(1, -3)$ $(-2, 2)$ $(-2, -3)$

5. I   **6.** II   **7.** IV   **8.** III   **9.** $(2, 0)$ **10.** $(0, 2)$ **11.** $(0, 0), (1, -1)$ **12.** $(2, 1), (-2, -1)$

13. $(0, 4), (4, 0), (2, 2)$ **14.** $(0, 0), (3, 1), (6, 2)$ **15.** $(0, 0), (2, 1), (4, 2)$ **16.** $(0, -6), (3, 0), (4, 2)$

17. $(0, 4), (1, 4), (2, 4)$ **18.** $(3, 0), (0, -2), (6, 2)$ **19.** $\left(\dfrac{7}{3}, 0\right), \left(\dfrac{7}{3}, 1\right), \left(\dfrac{7}{3}, 2\right)$

20. $(0, 5), \left(\dfrac{10}{3}, 0\right), (2, 2)$

21.   **22.**   **23.**   **24.**

25.

26.

27.

28.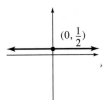

Practice Exercises

Answers to Problems 51 to 57 may vary.

29.

31.

33.

35.

37.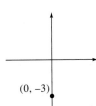

39. IV   **41.** II   **43.** $(0, -4)$ **45.** $(0, 0), (1, 1)$ **47.** $(0, -3), (-1, -8)$

49. $(7, 7), (7, 3)$ **51.** $(3, 0), (0, 3), (1, 2)$ **53.** $(0, 0), (4, 1), (-4, -1)$ **55.** $(0, -3), (1, -3), (2, -3)$
57. $(8, 0), (8, 1), (8, 2)$

59.

61.

63.

65.

67.

69.

71.

73.

75.

77.

79.

81.

Challenge Problems

83. (a)   (b) 100; 100 gal (c) −4; 4 min   **85.** $x - y = 0$ **87.** $2x - y = 0$

PROBLEM SET 6.2

Warm-Ups

1. 1   **2.** −2   **3.** $\dfrac{1}{4}$ **4.** $\dfrac{3}{2}$ **5.** 1   **6.** −1   **7.** −7   **8.** $\dfrac{1}{2}$ **9.** $-\dfrac{3}{5}$ **10.** $\dfrac{1}{5}$ **11.** −2   **12.** $\dfrac{1}{4}$

13. 0   **14.** no defined slope   **15.** no defined slope   **16.** $-\dfrac{1}{2}$ **17.** 3   **18.** $\dfrac{1}{4}$ **19.** $-\dfrac{2}{3}$

20. no defined slope   **21.** 0   **22.** $\dfrac{1}{3}$; −3   **23.** -4; $\dfrac{1}{4}$ **24.** no defined slope; 0

25. $-\dfrac{5}{3}$; $\dfrac{3}{5}$ **26.** −1; 1   **27.** 1; −1   **28.** 2; $-\dfrac{1}{2}$

Practice Exercises

29. 1   **31.** $\dfrac{1}{5}$ **33.** $-\dfrac{3}{5}$ **35.** 1   **37.** −1   **39.** 0   **41.** no defined slope   **43.** −5   **45.** 1

47. $-\dfrac{1}{2}$ **49.** $\dfrac{1}{3}$ **51.** $-\dfrac{1}{2}$ **53.** $\dfrac{1}{3}$; −3   **55.** $\dfrac{1}{2}$; −2   **57.** 0; no defined slope   **59.** −1; 1

61. 2; $-\dfrac{1}{2}$ **63.** $\dfrac{3}{5}$; $-\dfrac{5}{3}$ **65.** The slope is $-\dfrac{1}{16}$.   **67.** Its slope is $-\dfrac{11}{80}$.   **69.** $\dfrac{7}{18}$

Challenge Problems

71. no slope   **73.** 0

PROBLEM SET 6.3

Warm-Ups

1.   **2.**   **3.**   **4.**

5.

$y = 3x - 1$
$m = 3$
$b = -1$

6.

$y = -\frac{1}{2}x + 1$
$m = -\frac{1}{2}$
$b = 1$

7. $x - 2y = 8$ **8.** $2x + 3y = 10$
9. $x - y = 0$ **10.** $2x - 3y = -7$
11. $3x - 4y = 29$ **12.** $3x - y = 0$
13. $3x - 2y = -6$ **14.** $y = x + 2$
15. $y = \frac{2}{3}x - 1$ **16.** $x = 2$
17. $x = 1$ **18.** $y = 2$ **19.** $y = -2$ **20.** $x = 1$

Practice Exercises

21.

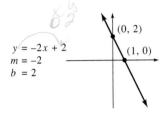

$y = -2x + 2$
$m = -2$
$b = 2$

23.

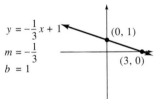

$y = x - 3$
$m = 1$
$b = -3$

25.

$y = \frac{2}{3}x + 2$
$m = \frac{2}{3}$
$b = 2$

27.

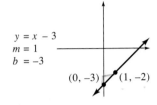

$y = -\frac{1}{3}x + 1$
$m = -\frac{1}{3}$
$b = 1$

29.

$y = \frac{3}{4}x - 1$
$m = \frac{3}{4}$
$b = -1$

31. $x + y = -2$ **33.** $3x + 2y = 3$

35. $y = 2x - 2$ **37.** $x = -1$ **39.** $x = 2$ **41.** $x + 7y = 26$ **43.** $x - 3y = 22$ **45.** $x = 1$
47. $3x - 2y = 6$

Challenge Problems

49. $x = 0$ **51.** $y = 2x$ **53.** $x - 2y = -5$ **55.** $y = 2$ **57.** $c - 15n = 400$

PROBLEM SET 6.4

Warm-Ups

1.

2.

3.

4.

5.

Practice Exercises

7.

9.

11.

13.

15.

17.

19. (a) $.95c + 1.39n \leq 10$

   (b) At most, 3 notebooks.

21.

23.

25.

27.

29.

31.

PROBLEM SET 6.5

Warm-Ups

1. function   **2.** function   **3.** not a function   **4.** function

5. $f = \{(-1, 3), (0, 2), (1, 3)\}; \{2, 3\}$ **6.** $g = \{(1, 1), (2, 0), (3, 1), (4, 0)\}; \{0, 4\}$

7. $f(-1) = 2; f(0) = 0; f(1) = 0; f(7) = 42$ **8.** $g(0) = 1; g(1) = 1; g(-1) = 0; g(7) = \dfrac{4}{25}$

9. $h(0) = 1; h(3) = 1; h(-1) = 5; h(-3) = 19$ **10.** {all real numbers}   **11.** $\{x \mid x \neq 2; x \neq 3\}$

12. {all real numbers}   **13.** $\{x \mid x \geq 5\}$ **14.**   **15.**

16. function   **17.** not a function   **18.** domain: $\{x \mid -2 \leq x \leq 4\}$ **19.** domain: $\{x \mid -\pi < x < \pi\}$

                                                              range: $\{y \mid -1 \leq y \leq 4\}$ range: $\{y \mid -1 \leq y \leq 1\}$

20. domain: $\{x \mid x < 0\}$ **21.** domain: {all real numbers}
   range: $\{y \mid y > 0\}$ range: {all real numbers}

Practice Exercises

23. $g = \left\{(1, 12), (2, 6), (3, 4), (4, 3), \left(5, \dfrac{12}{5}\right)\right\}$; $\left\{\dfrac{12}{5}, 3, 4, 6, 12\right\}$

25. $G = \left\{\left(-1, \dfrac{1}{2}\right), (0, 0), \left(1, \dfrac{1}{2}\right)\right\}$; $\left\{0, \dfrac{1}{2}\right\}$ **27.** $f(-2) = 8; f(-1) = 3; f(0) = 0; f(1) = -1; f(2) = 0$

29. $h(-3) = 4; h(0) = 1; h(1) = 2$ **31.** {all real numbers}   **33.** $\{t \mid t \neq -4\}$

35.   **37.**   **39.**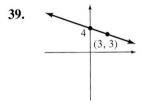

41. function   **43.** not a function   **45.** not a function
47. domain: $\{x \mid -2 \leq x \leq 2\}$; range: $\{y \mid -1 \leq y \leq 1\}$
49. domain: $\{x \mid -4 \leq x \leq 4\}$; range: $\{y \mid -3 \leq y \leq 3\}$ **51.** domain: $\{x \mid x > 0\}$; range: $\{y \mid y > 0\}$
53. domain: {all real numbers}; range: {all real numbers}

Challenge Problems

55. (a) yes   (b) domain: $\{t \mid 0 \leq t \leq 24\}$; range: $\{T \mid -10 \leq T \leq 20\}$ (c) 0°F; 20°F

REVIEW PROBLEMS CHAPTER 6

1. I   **3.** III   **5.** yes   **7.** yes   **9.** yes   **11.** $(-2, 0), (0, -2)$ **13.** $(1, 0), (0, -3)$ **15.** $(0, -4)$
17.   **19.**   **21.** 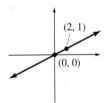   **23.**

25. $-\dfrac{1}{3}$ **27.** no defined slope   **29.** 0   **31.** $y = -2x + 4; m = -2; b = 4$

33. $y = -\dfrac{2}{3}x + 2; m = -\dfrac{2}{3}; b = 2$ **35.** $y = \dfrac{2}{3}x; m = \dfrac{2}{3}; b = 0$ **37.** $4x - y = 5$

39. $2x + y = 3$ **41.** $y = -3$ **43.** $x = -1$
45.   **47.**   **49.**

51. $f(2) = 2; f(-1) =$

Let's Not Forget . . .

54. -4 **55.** 9   **56.** $-4x - 3$ **57.** $-(a + b)$ **58.** 0   **59.** $\dfrac{x^2}{x + 2}$ **60.** $x^2 + 4xy + 4y^2$

61. $a^2 - 16$ **62.** $x^2 + 2x - 2$ **63.** $x^2 - 4$ **64.** $\dfrac{x}{x + 1}$ **65.** term   **66.** term   **67.** factor

68. factor   **69.** 1   **70.** $\dfrac{2}{x + 1} + 5$ **71.** $x^2 - 1$ **72.** $y(4 + y)$

73. $(x + y)(x - y)$ **74.** factored   **75.** factored   **76.** $(x + y)^2$ **77.** $(x - 2)(x^2 + 2x + 4)$

78. expression; $\dfrac{-4x - 15}{x + 4}$ **79.** equation; $\{-3\}$ **80.** inequality; $\{x \mid x \geq -15\}$

81. expression; $\dfrac{2 + x^2}{1 + 3x}$ **82.** expression; $3x - 7$

CHAPTER 6 TEST

1. (c)   **2.** (b)   **3.** (a)   **4.** (a)   **5.** (d)   **6.** (b)   **7.** (a)

8.

9.

10.

11.

12.

13.

14. $3x - y = 5$ **15.** $y = -3$ **16.** 27   **17.** a, b, and e

---

▶ Chapter 7

PROBLEM SET 7.1

Warm-Ups

1.

2.

3.

4.

5.

6.

7.

8.

9.
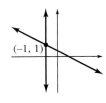

10. one, independent   **11.** no solution, inconsistent   **12.** one, independent   **13.** one, independent
14. Infinite number; dependent   **15.** one, independent   **16.** No solutions; inconsistent
17. infinite number, dependent   **18.** one, independent

Practice Exercises

19.

21.

23.

25.

27. inconsistent   **29.**   **31.**   **33.** inconsistent

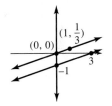

35. one, independent   **37.** one, independent   **39.** one, independent   **41.** infinite number, dependent

Challenge Problems

43. $\{(a, b)\}$ **45.** $\{(a, b)\}$

PROBLEM SET 7.2

Warm-Ups

1. $\{(1, 1)\}$ **2.** $\{(-2, -6)\}$ **3.** $\{(0, 0)\}$ **4.** $\{(1, 3)\}$ **5.** $\{(-1, 5)\}$ **6.** $\{(-2, 2)\}$
7. $\{(3, -1)\}$ **8.** $\{(2, -1)\}$ **9.** $\{(1, -1)\}$ **10.** $\{(4, -3)\}$ **11.** $\{2, -5)\}$
12. $\{(-1, -1)\}$ **13.** $\{(1, -2)\}$ **14.** $\{(2, 1)\}$ **15.** $\{(-6, -1)\}$ **16.** $\varnothing$ **17.** dependent   **18.** $\varnothing$

Practice Exercises

19. $\{(1, 3)\}$ **21.** dependent   **23.** $\{(-1, 7)\}$ **25.** $\{(5, 2)\}$ **27.** $\varnothing$ **29.** dependent
31. $\left\{\left(\frac{2}{3}, 1\right)\right\}$ **33.** $\{(5, 3)\}$ **35.** $\{(-4, -7)\}$ **37.** $\{(2, -2)\}$ **39.** $\{(1, 0)\}$ **41.** $\left\{\left(-\frac{9}{2}, 3\right)\right\}$

Challenge Problems

43. $\{(4 - k, k)\}$ **45.** $\{(1, 3), (-1, -3)\}$

PROBLEM SET 7.3

Warm-Ups

1. $\{(4, 0)\}$ **2.** $\{(3, 3)\}$ **3.** $\{(2, 2)\}$ **4.** $\{(3, 1)\}$ **5.** $\{(2, -1)\}$ **6.** $\{(-1, 1)\}$
7. $\{(3, -2)\}$ **8.** $\{(2, 5)\}$ **9.** $\{(3, -1)\}$ **10.** $\varnothing$ **11.** dependent   **12.** $\varnothing$

Practice Exercises

13. $\{(3, 0)\}$ **15.** $\{(3, 3)\}$ **17.** $\{(3, -5)\}$ **19.** $\varnothing$ **21.** $\{(8, 2)\}$ **23.** $\varnothing$
25. $\{(3, -7)\}$ **27.** $\{(-5, 4)\}$ **29.** dependent   **31.** $\{(-2, -3)\}$ **33.** $\left\{\left(-\frac{1}{2}, \frac{1}{2}\right)\right\}$

Challenge Problems

35. $\{(a, 0)\}$ **37.** $\left\{\left(\frac{2a + b}{3}, \frac{a - b}{3}\right)\right\}$

PROBLEM SET 7.4

Warm-Ups

1. 16 and 32   **2.** They are 4 and 6.   **3.** 3270   **4.** 26 red and 52 blue   **5.** 15 m by 21 m
6. 13 ft and 25 ft   **7.** 25 yd by 75 yd   **8.** 85 fives; 20 tens
9. 1250 attended Friday and 1600 Saturday   **10.** 16 29¢ stamps and 20 19¢ stamps   **11.** $2.19
12. $20,000 at 8% and $30,000 at 11%   **13.** He borrowed $9000 at 10% and $6000 at 7%
14. $3500 at 10% and $2500 at 8%   **15.** Airspeed is 175 mph and wind speed is 25 mph.   **16.** 85 mph

Practice Exercises

17. -6 and 12   **19.** They are $15 and $20.   **21.** 1200   **23.** $1.99   **25.** 99¢
27. 37 classic and 47 caffeine free   **29.** 7 cm by 20 cm   **31.** 40 white and 80 black   **33.** 12 ft and 8 ft
35. 1 lb of peppermints and 3 lb of chocolates   **37.** Pencils are 29¢.
39. Wool is $7.98 and cotton $2.99.   **41.** Ms. Gould borrowed $200 at 9% and $600 at 7%.
43. The crew rate in still water is $5\frac{1}{2}$ mph and the rate of the current is $1\frac{1}{2}$ mph.
45. The current's speed is $\frac{1}{2}$ mph and the boat's speed in still water is $6\frac{1}{2}$ mph.

Challenge Problems

46. $4770

PROBLEM SET 7.5

Warm-Ups

1.

2.

3.

4.

5.

6.

7.

8.

9.

10.

11.

12.

Practice Exercises

13.

15.

17.

19.

21.

23.

25.

27.

29.

31.

33.

35.

Challenge Problems

37.   **39.**

REVIEW PROBLEMS CHAPTER 7

1. independent   **3.** dependent   **5.** indepenent   **7.** $\left\{\left(\frac{3}{2}, \frac{3}{2}\right)\right\}$ **9.** $\{(-1, -2)\}$ **11.** $\{(-1, 1)\}$

13. $\{(-2, 1)\}$ **15.** $\{(1, 1)\}$ **17.** $\{(0, 0)\}$

19.   **21.**   **23.**

25. Joan has 27 and Sue has 9.   **27.** 7 m by 12 m

Let's Not Forget . . .

29. -16 **30.** 16 **31.** $5x - 3y$ **32.** $-a$ **33.** $9x^2 - 12x + 4$ **34.** $4x^2 - 25$

35. $6a^2 - ab - b^2$ **36.** $\dfrac{1}{x^3 y}$ **37.** $2x + 2xy - y^2$ **38.** $2x^2 - xy - y^2$ **39.** $\dfrac{-2}{(x + 1)(x - 1)}$

40. term   **41.** factor   **42.** term   **43.** term   **44.** 1   **45.** $\dfrac{2x}{x - 2}$ **46.** $1 + \dfrac{x - 2}{x + 2}$ **47.** factored

48. $(x + 2z)(x - 2z)$ **49.** factored   **50.** $b(a^2 - bc)$ **51.** prime   **52.** $x\left(\dfrac{1}{2} + x\right)$

53. expression; $\dfrac{-3x - 1}{x + 1}$ **54.** equation; $\varnothing$ **55.** inequality; $\left\{x \mid x < -\dfrac{3}{2}\right\}$ **56.** expression; $\dfrac{2}{y}$

57. expression; $\dfrac{x + 2}{x}$

CHAPTER 7 TEST

1. (c)   **2.** (d)   **3.** (c)   **4.** (b)   **5.** (a)   **6.** $\{(3, 1)\}$ **7.** $\varnothing$ **8.** $\{(2, -3)\}$ **9.** $\{(-1, 0)\}$

10.   **11.**   **12.**   **13.**

14. 8 ft by 15 ft   **15.** 69¢

► Chapter 8

PROBLEM SET 8.1

Warm-Ups

1. 5　　**2.** not a real number　　**3.** −6　　**4.** $\frac{2}{5}$　　**5.** $-\frac{2}{3}$　　**6.** 1.41　　**7.** not a real number　　**8.** −1.41

9. 2.83　　**10.** −3.61　　**11.** 4　　**12.** −4　　**13.** −4　　**14.** $\frac{1}{4}$　　**15.** $-\frac{1}{4}$　　**16.** 2

17. not a real number　　**18.** −2　　**19.** $\frac{1}{2}$　　**20.** $-\frac{1}{2}$　　**21.** 2　　**22.** −2　　**23.** −2　　**24.** $\frac{1}{2}$

25. $-\frac{1}{2}$　　**26.** x^3　　**27.** y　　**28.** t^4　　**29.** z^6　　**30.** x^8　　**31.** $2x^4$　　**32.** $5y^2$

33. $3t^3$　　**34.** $10z^9$　　**35.** $2z$

Practice Exercises

37. not a real number　　**39.** −4　　**41.** $-\frac{2}{9}$　　**43.** not a real number　　**45.** not a real number　　**47.** 7

49. 1　　**51.** $-\frac{1}{2}$　　**53.** $\frac{8}{7}$　　**55.** 10　　**57.** −2　　**59.** $-\frac{2}{3}$　　**61.** −1　　**63.** $-\frac{1}{5}$　　**65.** 3　　**67.** −5

69. $-\frac{1}{3}$　　**71.** −3　　**73.** $\frac{1}{3}$　　**75.** 3　　**77.** 0　　**79.** not a real number　　**81.** y^5　　**83.** z^{10}　　**85.** $4x^3$

87. $7t$　　**89.** $5y^2z$　　**91.** $11x^2yz^9$　　**93.** $13abc$　　**95.** $4x^2y$　　**97.** $12s^2t$　　**99.** $x^4y^2z^5$

101. not a real number　　**103.** 3.16

Challenge Problems

105. 4　　**107.** 3　　**109.** −2　　**111.** −3

PROBLEM SET 8.2

Warm-Ups

1. $2\sqrt{2}$　　**2.** $2\sqrt{3}$　　**3.** $5\sqrt{2}$　　**4.** $10\sqrt{2}$　　**5.** $2\sqrt{10}$　　**6.** $2\sqrt[3]{3}$　　**7.** $3\sqrt[3]{2}$　　**8.** $-2\sqrt[3]{4}$

9. $2\sqrt[3]{5}$　　**10.** $2x^2\sqrt{2}$　　**11.** $2y\sqrt[3]{2}$　　**12.** $2x^3\sqrt{5}$　　**13.** $-2x^2\sqrt[3]{3}$　　**14.** $\frac{\sqrt{3}}{4}$　　**15.** $\frac{\sqrt{6}}{7}$　　**16.** $\frac{2x\sqrt{3}}{5}$

17. $\frac{3\sqrt{3}}{2a}$　　**18.** $\frac{\sqrt{3}}{3}$　　**19.** $\frac{\sqrt{5}}{5}$　　**20.** $\frac{\sqrt{3}}{6}$　　**21.** $\frac{\sqrt{2}}{3}$

Practice Exercises

23. $5\sqrt{5}$　　**25.** $10\sqrt{6}$　　**27.** $6\sqrt{3}$　　**29.** $2\sqrt{17}$　　**31.** $3\sqrt{11}$　　**33.** $4\sqrt{7}$　　**35.** $10\sqrt{3}$　　**37.** $4ab\sqrt{2b}$

39. $\frac{\sqrt{3a}}{4b}$　　**41.** $\frac{3\sqrt{3}}{7x}$　　**43.** $5\sqrt[3]{2}$　　**45.** $-2\sqrt[3]{2}$　　**47.** $\frac{\sqrt[3]{5}}{2}$　　**49.** $\frac{2\sqrt[3]{4}}{3}$　　**51.** $3\sqrt[3]{4}$　　**53.** $2\sqrt[3]{7}$

55. $5\sqrt[3]{4}$　　**57.** $-2\sqrt[3]{18}$　　**59.** $2\sqrt[3]{20}$　　**61.** $-2\sqrt[3]{13}$　　**63.** $3y^2\sqrt[3]{2}$　　**65.** $-2x\sqrt[3]{4}$　　**67.** $2x^2y\sqrt[3]{9}$

69. not a real number

Challenge Problems

71. The horizon is 15 mi away.   **73.** $2\sqrt[4]{2}$ **75.** $2\sqrt[6]{2}$

PROBLEM SET 8.3

Warm-Ups

1. $9\sqrt{5}$ **2.** $5\sqrt{x}$ **3.** $11\sqrt[3]{6}$ **4.** $-2\sqrt[3]{3}$ **5.** $-\sqrt{3}$ **6.** $-4\sqrt[3]{7}$ **7.** $3\sqrt{2}$ **8.** $-\sqrt{6}$
9. 0 **10.** $15\sqrt{2}$ **11.** $4\sqrt{5}$ **12.** $-5\sqrt[3]{2}$ **13.** $3\sqrt{2x}$ **14.** $-2n\sqrt{3}$ **15.** $-6\sqrt{x}$
16. $-2x\sqrt{5}$ **17.** $3x\sqrt[3]{2}$ **18.** $22x\sqrt[3]{2}$

Practice Exercises

19. $3\sqrt{3}$ **21.** $8\sqrt[3]{11}$ **23.** $5\sqrt{5}$ **25.** $4\sqrt{2}$ **27.** $\sqrt{3}$ **29.** $4\sqrt{5}$ **31.** $-3\sqrt{2}$
33. $23\sqrt{x}$ **35.** $-4\sqrt[3]{2x}$ **37.** $-17\sqrt[3]{2}$ **39.** $-\sqrt[3]{2}$ **41.** $5\sqrt{5x}$ **43.** $-7\sqrt{x}$
45. $29\sqrt{x}$ **47.** $-4\sqrt[3]{3x}$ **49.** $3x\sqrt[3]{3}$

PROBLEM SET 8.4

Warm-Ups

1. $\sqrt{35}$ **2.** $\sqrt{33}$ **3.** $\sqrt[3]{15}$ **4.** -2 **5.** $8\sqrt{15}$ **6.** $6x$ **7.** 60 **8.** 5 **9.** $64x$
10. $\sqrt{6} + \sqrt{15}$ **11.** $\sqrt{42} - \sqrt{7}$ **12.** $5 + \sqrt{35}$ **13.** $t - \sqrt{ty}$ **14.** $7 + 3\sqrt{3}$ **15.** $7 + \sqrt{10}$
16. 2 **17.** 2 **18.** $7 + 4\sqrt{3}$ **19.** $74 + 40\sqrt{3}$ **20.** $\sqrt{2}$ **21.** $\sqrt{5}$ **22.** $\sqrt[3]{3}$ **23.** $\sqrt[3]{6}$
24. 4 **25.** $\dfrac{3}{x}$ **26.** $2ab$ **27.** -3 **28.** $\dfrac{\sqrt{3}}{3}$ **29.** $\sqrt{6}$ **30.** $\dfrac{2 + \sqrt{2}}{2}$ **31.** $\dfrac{\sqrt{6}}{8}$ **32.** $\dfrac{3 - \sqrt{6}}{6}$
33. $\dfrac{x + \sqrt{xy}}{x}$ **34.** $-\dfrac{\sqrt{2} - 3}{7}$ **35.** $-3(\sqrt{2} + \sqrt{3})$ **36.** $-\dfrac{5 + 2\sqrt{15}}{7}$ **37.** $-3 - 2\sqrt{2}$
38. $\dfrac{9x + 12\sqrt{xy} + 4x}{9x - 4y}$ **39.** $\dfrac{15}{x\sqrt{10}}$ **40.** $\dfrac{7}{11 - 6\sqrt{2}}$

Practice Exercises

41. $\sqrt{15}$ **43.** $\sqrt[3]{28}$ **45.** $3\sqrt{2}$ **47.** 3 **49.** $8\sqrt{6}$ **51.** $6\sqrt[3]{18}$ **53.** $24\sqrt{6}$ **55.** $10y$
57. 7 **59.** 18 **61.** $36z$ **63.** $\sqrt{15} - \sqrt{5}$ **65.** $x - \sqrt{xy}$ **67.** $2\sqrt{10} - 13$
69. $x - 4\sqrt{xy} + 3y$ **71.** 1 **73.** 3 **75.** $11 + 6\sqrt{2}$ **77.** $14 + 8\sqrt{3}$ **79.** $58 + 24\sqrt{5}$
81. 2 **83.** $\sqrt[3]{3}$ **85.** 5 **87.** -2 **89.** $3\sqrt{2}$ **91.** $-3\sqrt[3]{2}$ **93.** $\sqrt{5}$ **95.** $\dfrac{5\sqrt{3}}{6}$ **97.** $\dfrac{\sqrt{15} - 3}{6}$
99. $\dfrac{\sqrt{xy} - y}{y}$ **101.** $-3(\sqrt{2} - \sqrt{3})$ **103.** $-\dfrac{7 + 2\sqrt{10}}{3}$ **105.** $2\sqrt{30} - 11$ **107.** $\dfrac{21}{2\sqrt{7z}}$
109. $\dfrac{s - 4t}{3s - 7\sqrt{st} + 2t}$ **111.** $5, \dfrac{5\sqrt{3}}{3}, \dfrac{10\sqrt{3}}{3}$ **113.** $\sqrt{2}, \sqrt{2}, 2$ **115.** $\dfrac{1}{2}, \dfrac{\sqrt{2}}{2}, \dfrac{\sqrt{2}}{2}$
117. $\sqrt{6}, \sqrt{3}, \sqrt{3}$

PROBLEM SET 8.5

Warm-Ups

1. $\{14\}$ **2.** $\{26\}$ **3.** $\{3\}$ **4.** $\{0\}$ **5.** $\{4\}$ **6.** $\{6\}$ **7.** $\varnothing$ **8.** $\varnothing$ **9.** $\varnothing$ **10.** $\varnothing$ **11.** $\{1\}$
12. $\{7\}$ **13.** $\{-4\}$ **14.** $\varnothing$ **15.** $\{4\}$ **16.** $\{2\}$ **17.** $\{1\}$ **18.** $\{7\}$ **19.** $\{-2\}$ **20.** $\{1\}$
21. $\{4\}$ **22.** $\{-10, -8\}$ **23.** $\{5\}$ **24.** $\{0\}$ **25.** $\{-3, -1\}$ **26.** $\{0\}$ **27.** $\{-1, 7\}$ **28.** $\{11\}$

Practice Exercises

29. $\{2\}$ **31.** $\varnothing$ **33.** $\{5\}$ **35.** $\{-4\}$ **37.** $\varnothing$ **39.** $\{7\}$ **41.** $\{-1\}$ **43.** $\{-1\}$ **45.** $\{0\}$
47. $\{2\}$ **49.** $\{-1\}$ **51.** $\varnothing$ **53.** $\{5\}$ **55.** $\{-1\}$ **57.** $\{7\}$ **59.** $\{3\}$ **61.** $\{\pm 1\}$
63. $\{-1, 3\}$ **65.** Four pairs of socks were produced.

Challenge Problems

67. $\left\{\dfrac{b^2}{a}\right\}$

PROBLEM SET 8.6

Warm-Ups

1. 4   **2.** 6   **3.** 5   **4.** -6 **5.** -2 **6.** -2 **7.** 9   **8.** 64   **9.** 27   **10.** 100   **11.** -4
12. -8 **13.** a **14.** $\varnothing$ **15.** 64   **16.** $\{z\}$ **17.** $-2x^2$ **18.** $\{7\}$ **19.** $64/x^{12}$ **20.** $\{1\}$
21. $1/x^{1/3}$ **22.** $t^{1/6}$

Practice Exercises

23. 7   **25.** 9   **27.** 6   **29.** 7   **31.** -5 **33.** -10 **35.** -7 **37.** -6
39. 125   **41.** 4   **43.** -27 **45.** -729 **47.** $t^{5/3}$ **49.** $x^{3/2}$ **51.** $9y^6$
53. $-\dfrac{x^4}{2}$ **55.** $w^{1/2}$ **57.** $x - x^{1/2}$ **59.** $x - 1$

Challenge Problems

61. $\dfrac{1}{4}$ **63.** $-\dfrac{1}{16}$ **65.** $\dfrac{1}{5}$

REVIEW PROBLEMS CHAPTER 8

1. $8\sqrt{2}$ **3.** $2 + \sqrt{6}$ **5.** $16 + 6\sqrt{7}$ **7.** $8x$ **9.** 36   **11.** $6 + 3\sqrt{2} - 2\sqrt{3} - \sqrt{6}$
13. 10   **15.** 2   **17.** $26 - 8\sqrt{3}$ **19.** $\sqrt[3]{2}$ **21.** $2\sqrt{3}$ **23.** $2\sqrt[3]{3}$ **25.** $2\sqrt{6} - 8$
27. $\sqrt{6}$ **29.** $2\sqrt{6}$ **31.** $2\sqrt{2}$ **33.** $\dfrac{2\sqrt{x}}{x}$ **35.** $\dfrac{\sqrt{5}}{15}$ **37.** $4 + 2\sqrt{2}$ **39.** $\dfrac{16 + 11\sqrt{6}}{10}$
41. $\{21\}$ **43.** $\{3\}$ **45.** $\varnothing$ **47.** $\{-4, -2\}$ **49.** $\varnothing$ **51.** -4 **53.** $-\dfrac{1}{8}$ **55.** 125
57. 3   **59.** $\dfrac{16}{x^8}$ **61.** The cost is \$243,700.

Let's Not Forget . . .

63. -9 **64.** 9 **65.** 2 **66.** -1 **67.** $8-2\sqrt{15}$ **68.** 2 **69.** $\dfrac{a-1}{a^3}$ **70.** $10 + 2\sqrt{3}$

71. $\dfrac{-2}{x+1}$ **72.** $\dfrac{1}{x+1}$ **73.** $\dfrac{x^2-3x}{(x+1)(x-1)}$ **74.** term   **75.** factor   **76.** term   **77.** term

78. 12 **79.** $4\sqrt{3}+6$ **80.** $2+\sqrt{3}$ **81.** $\sqrt{3}(a+2)(a-2)$ **82.** factored

83. factored   **84.** factored   **85.** $(x+y)(x^2-xy+y^2)$ **86.** expression; $\dfrac{x+7}{(x-1)(x+3)}$

87. equation; $\{7\}$ **88.** inequality; $\{x \mid x \geq -9\}$ **89.** expression; $\dfrac{2(x+3)}{x-1}$ **90.** equation; $\{2\}$

CHAPTER 8 TEST

1. b   **2.** d   **3.** c   **4.** d   **5.** b   **6.** a   **7.** $12\sqrt{35}$ **8.** $4\sqrt{3}-3\sqrt{2}$ **9.** -25

10. $31+4\sqrt{21}$ **11.** $\sqrt{15}-5+6\sqrt{2}-2\sqrt{30}$ **12.** -3 **13.** 0 **14.** $\dfrac{\sqrt{3}}{2}$

15. $-\sqrt{6}+3+2\sqrt{2}-2\sqrt{3}$ **16.** -7 **17.** 25 **18.** $\{4\}$ **19.** $\{5\}$

20. The time is $\pi/2$ seconds or ≈ 1.57 seconds.

► Chapter 9

PROBLEM SET 9.1

Warm-Ups

1. $\{\pm 5\}$ **2.** $\{\pm 9\}$ **3.** $\{\pm 10\}$ **4.** $\{0\}$ **5.** $\{\pm 11\}$ **6.** $\{\pm 13\}$ **7.** $\{\pm 15\}$ **8.** $\{\pm\sqrt{17}\}$

9. $\{\pm\sqrt{3}\}$ **10.** $\{\pm\sqrt{14}\}$ **11.** $\{\pm 2\sqrt{6}\}$ **12.** $\{\pm 4\sqrt{3}\}$ **13.** $\{\pm 4\sqrt{2}\}$ **14.** $\{\pm 5\sqrt{5}\}$

15. $\{\pm 3\sqrt{6}\}$ **16.** $\{\pm 6\sqrt{2}\}$ **17.** $\{\pm 1\}$ **18.** $\{\pm\sqrt{6}\}$ **19.** $\left\{\pm\dfrac{3}{2}\right\}$ **20.** $\{\pm 2\sqrt{2}\}$ **21.** $\{\pm 7\}$

22. $\left\{\pm\dfrac{6}{5}\right\}$ **23.** $\{-5, -1\}$ **24.** $\{6, 8\}$ **25.** $\{-7, 1\}$ **26.** $\{-6, 4\}$ **27.** $\{-2 \pm \sqrt{5}\}$

28. $\{9 \pm \sqrt{7}\}$ **29.** $\{-5 \pm 2\sqrt{3}\}$ **30.** $\{4 \pm 3\sqrt{3}\}$ **31.** $\left\{\dfrac{1}{6}, \dfrac{5}{6}\right\}$ **32.** $\left\{\dfrac{1}{6}, \dfrac{7}{6}\right\}$ **33.** $\left\{\dfrac{-20}{21}, \dfrac{8}{21}\right\}$

34. $\left\{-\dfrac{27}{20}, \dfrac{3}{20}\right\}$ **35.** $\left\{\dfrac{1}{3} \pm \dfrac{\sqrt{3}}{2}\right\}$ **36.** $\left\{-\dfrac{2}{5}, \pm\dfrac{\sqrt{7}}{3}\right\}$ **37.** The hypotenuse is $\sqrt{2}$ ft.   **38.** $10\sqrt{2}$ m

39. The ladder is $8\sqrt{6}$ ft above the ground.   **40.** -1 and 7 **41.** The side is 7 in.

Practice Exercises

43. $\{\pm 4\}$ **45.** $\pm 17\}$ **47.** $\{\pm 8\}$ **49.** $\{\pm 18\}$ **51.** $\{\pm 22\}$ **53.** $\left\{\pm\dfrac{9}{2}\right\}$ **55.** $\{\pm\sqrt{23}\}$

57. $\{\pm 3\sqrt{7}\}$ **59.** $\{4, 12\}$ **61.** $\{-10, 6\}$ **63.** $\{5 \pm \sqrt{3}\}$ **65.** $\{-7 \pm \sqrt{11}\}$ **67.** $\{5 \pm 2\sqrt{6}\}$

69. $\{3 \pm 5\sqrt{5}\}$ **71.** $\left\{\dfrac{1}{15}, \dfrac{11}{15}\right\}$ **73.** $\left\{-\dfrac{51}{20}, \dfrac{21}{20}\right\}$ **75.** $\left\{-\dfrac{2}{3} \pm \dfrac{\sqrt{5}}{2}\right\}$ **77.** The legs are $3\sqrt{2}$ m.

79. There are two such numbers, -7 and 3.   **81.** The radius is 45 in.   **83.** Warren should buy $9\sqrt{29}$ m.

85. The dimensions of the place mat are $6\sqrt{5}$ in. by 12 in.   **87.** The radius is $8\sqrt{2}$ ft.   **89.** The radius is 3 in.

Challenge Problems

91. $\{\pm\sqrt{c}\}$ **93.** $\{a \pm 2\}$

PROBLEM SET 9.2

Warm-Ups

1. 4   **2.** 16   **3.** 9   **4.** 49   **5.** 36   **6.** 121   **7.** $\frac{25}{4}$ **8.** $\frac{81}{4}$ **9.** $\frac{1}{4}$ **10.** $\frac{9}{4}$ **11.** $\frac{169}{4}$

12. $\frac{225}{4}$ **13.** $\frac{1}{16}$ **14.** $\frac{1}{36}$ **15.** $\frac{1}{25}$ **16.** $\frac{1}{49}$ **17.** $\frac{4}{9}$ **18.** $\frac{9}{25}$ **19.** $\{-5, 1\}$ **20.** $\{-2, 4\}$

21. $\{-7, 1\}$ **22.** $\{-1, 3\}$ **23.** $\{1\}$ **24.** $\{1, 3\}$ **25.** $\{-2\}$ **26.** $\{3\}$ **27.** $\{-2, 7\}$ **28.** $\{2, 5\}$

29. $\{-3 \pm \sqrt{11}\}$ **30.** $\{-2 \pm \sqrt{5}\}$ **31.** $\{4 \pm \sqrt{5}\}$ **32.** $\{5 \pm \sqrt{3}\}$ **33.** $\{2 \pm \sqrt{2}\}$

34. $\{1 \pm \sqrt{2}\}$ **35.** $\{-2, 1\}$ **36.** $\{-1, 4\}$ **37.** $\{-4, 2\}$ **38.** $\{-3, 4\}$ **39.** $\left\{-\frac{1}{2}, 3\right\}$

40. $\left\{1, \frac{4}{3}\right\}$ **41.** $\left\{\frac{5}{2} \pm \frac{\sqrt{23}}{2}\right\}$ **42.** $\left\{-1, \pm \frac{\sqrt{5}}{2}\right\}$

Practice Exercises

43. $\{-6, 2\}$ **45.** $\{-8, 2\}$ **47.** $\{2\}$ **49.** $\{2, 6\}$ **51.** $\{-4, 3\}$ **53.** $\{2, 3\}$ **55.** $\{-5\}$ **57.** $\{-1\}$

59. $\left\{-\frac{1}{2}, 2\right\}$ **61.** $\{-7, 2\}$ **63.** $\{-3, 1\}$ **65.** $\left\{\frac{-3}{4} \pm \frac{\sqrt{17}}{4}\right\}$ **67.** $\left\{\frac{5}{4} \pm \frac{\sqrt{33}}{4}\right\}$ **69.** $\left\{\frac{1}{2}, 1\right\}$

71. $\left\{\frac{-1}{3}, 1\right\}$ **73.** JoAnn was driving at 38 mph.   **75.** The interest rate is 7%.

Challenge Problems

77. $h = \frac{\sqrt{3}}{2}s$ **79.** $\{-2b \pm 2\sqrt{1 + b^2}\}$ **81.** $\left\{\frac{-b}{2a} \pm \frac{\sqrt{b^2 - 4ac}}{2a}\right\}$

PROBLEM SET 9.3

Warm-Ups

1. $a = 4; b = -5; c = 7$ **2.** $x^2 - 6x = 0; a = 1, b = -6, c = 0$

3. $x^2 - 8x - 7 = 0; a = 1; b = -8; c = -7$ **4.** $a = 1, b = -1, c = 1$

5. $a = 1; b = 0; c = 0$ **6.** $x^2 - 16 = 0; a = 1, b = 0, c = -16$

7. $3y^2 - y + 5 = 0; a = 3; b = -1; c = 5$ **8.** $\frac{1}{2}x^2 - x + \frac{2}{3} = 0; a = \frac{1}{2}, b = -1, c = \frac{2}{3}$

9. $2x^2 + 4x - 7 = 0; a = 2; b = 4; c = -7$ **10.** $w^2 - 2w - 3 = 0; a = 1, b = -2, c = -3$

11. $x^2 + x - 7 = 0; a = 1; b = 1; c = -7$ **12.** $2x^2 + 6x - 5 = 0; a = 2, b = 6, c = -5$

13. $y^2 - y - 20 = 0; a = 1; b = -1; c = -20$ **14.** $2z^2 - 4z = 0; a = 2, b = -4, c = 0$

15. $\{-2, -1\}$ **16.** $\{-3, 1\}$ **17.** $\{-1, 4\}$ **18.** $\{-1, 6\}$ **19.** $\left\{-3, -\frac{1}{2}\right\}$ **20.** $\left\{\frac{1}{2}, 2\right\}$

21. $\left\{-\frac{2}{3}, \frac{1}{2}\right\}$ **22.** $\left\{-\frac{2}{3}, \frac{1}{3}\right\}$ **23.** $\left\{-\frac{3}{2}, \frac{1}{3}\right\}$ **24.** $\left\{-1, \frac{5}{2}\right\}$ **25.** $\{1\}$ **26.** $\{-3\}$ **27.** $\left\{\frac{3 \pm \sqrt{13}}{2}\right\}$

28. $\left\{\frac{-3 \pm \sqrt{17}}{2}\right\}$ **29.** $\left\{\frac{7 \pm \sqrt{17}}{4}\right\}$ **30.** $\left\{\frac{-1 \pm \sqrt{13}}{6}\right\}$ **31.** $\left\{\frac{-5 \pm \sqrt{17}}{2}\right\}$ **32.** $\left\{\frac{3 \pm \sqrt{5}}{2}\right\}$

33. $\left\{\dfrac{3 \pm \sqrt{21}}{2}\right\}$ **34.** $\left\{\dfrac{-3 \pm \sqrt{17}}{4}\right\}$ **35.** $\{2 \pm \sqrt{2}\}$ **36.** $\left\{-\dfrac{3}{2} \pm \dfrac{\sqrt{3}}{2}\right\}$ **37.** $\{-2 \pm \sqrt{2}\}$

38. $\left\{-\dfrac{1}{2} \pm \sqrt{2}\right\}$ **39.** $\{-1 \pm \sqrt{5}\}$ **40.** $\left\{\dfrac{3 \pm 3\sqrt{5}}{2}\right\}$ **41.** $\left\{1 \pm \dfrac{\sqrt{2}}{2}\right\}$ **42.** $\left\{-\dfrac{1}{3}, 1\right\}$

43. $\left\{-\dfrac{3}{2} \pm \dfrac{\sqrt{3}}{2}\right\}$ **44.** $\left\{\dfrac{2}{5} \pm \dfrac{2\sqrt{6}}{5}\right\}$ **45.** $\left\{-1 \pm \dfrac{\sqrt{6}}{3}\right\}$ **46.** $\left\{\dfrac{2}{3}, 2\right\}$ **47.** $\left\{\dfrac{1}{2} \pm \dfrac{\sqrt{7}}{2}\right\}$

48. $\left\{-\dfrac{1}{5} \pm \dfrac{\sqrt{6}}{5}\right\}$

Practice Exercises

49. $\{-4, -1\}$ **51.** $\{-2, 4\}$ **53.** $\left\{-\dfrac{5}{3}, 1\right\}$ **55.** $\left\{-\dfrac{4}{3}, -\dfrac{1}{2}\right\}$ **57.** $\left\{\dfrac{5 \pm \sqrt{29}}{2}\right\}$ **59.** $\left\{\dfrac{1}{2}, 4\right\}$

61. $\left\{\dfrac{-5 \pm \sqrt{13}}{2}\right\}$ **63.** $\{2\}$ **65.** $\{-1 \pm \sqrt{2}\}$ **67.** $\left\{\dfrac{3}{2} \pm \dfrac{\sqrt{7}}{2}\right\}$ **69.** $\left\{-2 \pm \dfrac{\sqrt{6}}{2}\right\}$ **71.** $\left\{-1, -\dfrac{1}{3}\right\}$

73. $\left\{\dfrac{1}{2} \pm \dfrac{\sqrt{3}}{2}\right\}$ **75.** The base is about 17 ft by 21 ft.

77. (a) Three employees must be hired. (b) 11 more baskets could be made

Challenge Problems

79. $gx^2 - 7x + 5 = 0; a = g, b = -7, c = 5$ **81.** $x^2 - \pi^2 = 0; a = 1, b = 0, c = -\pi^2$

83. $\left\{\dfrac{-3 \pm \sqrt{9 - 4k}}{2}\right\}$

PROBLEM SET 9.4

Warm-Ups

1. $3i$ **2.** $7i$ **3.** $\sqrt{5}\,i$ **4.** $\sqrt{3}\,i$ **5.** $2\sqrt{2}\,i$ **6.** $2\sqrt{6}\,i$ **7.** $3 + 4i$ **8.** $1 + 2i$

9. $8 - 7i$ **10.** $-3 + 4i$ **11.** $-4 - 2i$ **12.** $2 - 12i$ **13.** $-1 + 5i$ **14.** $7 - 3i$ **15.** $7 + 4i$

16. $18 + i$ **17.** $7 - 24i$ **18.** $8 + 6i$ **19.** 5 **20.** 29 **21.** $\dfrac{1}{2} - \dfrac{1}{2}i$ **22.** $\dfrac{2}{5} + \dfrac{1}{5}i$ **23.** $\dfrac{1}{5} + \dfrac{3}{5}i$

24. $\dfrac{7}{10} - \dfrac{1}{10}i$ **25.** $\dfrac{4}{13} - \dfrac{7}{13}i$ **26.** $\dfrac{3}{5} - \dfrac{4}{5}i$ **27.** $\{\pm i\}$ **28.** $\{\pm 4i\}$ **29.** $\{\pm \sqrt{7}i\}$ **30.** $\{\pm \sqrt{6}\,i\}$

31. $\{2 \pm 2\sqrt{3}i\}$ **32.** $\{-3 \pm 3\sqrt{3}\,i\}$ **33.** $\left\{-\dfrac{1}{2} \pm \dfrac{\sqrt{3}}{2}i\right\}$ **34.** $\left\{-\dfrac{3}{2} \pm \dfrac{\sqrt{3}}{2}i\right\}$ **35.** $\left\{-\dfrac{2}{5} \pm \dfrac{1}{5}i\right\}$

36. $\left\{\dfrac{1}{6} \pm \dfrac{\sqrt{3}}{6}i\right\}$ **37.** $\{-1 \pm 2i\}$ **38.** $\left\{\dfrac{1}{4} \pm \dfrac{\sqrt{11}}{4}i\right\}$ **39.** $\left\{-\dfrac{1}{2} \pm \dfrac{\sqrt{7}}{2}i\right\}$ **40.** $\left\{-\dfrac{1}{3} \pm \dfrac{\sqrt{2}}{3}i\right\}$

Practice Exercises

41. $3 - 4i$ **43.** $8 + 7i$ **45.** $-2 - 2i$ **47.** $5 - i$ **49.** $-7 - 9i$ **51.** $-2i$ **53.** 10

55. $\dfrac{1}{2} + \dfrac{1}{2}i$ **57.** $\dfrac{3}{5} + \dfrac{6}{5}i$ **59.** $\dfrac{3}{13} - \dfrac{11}{13}i$ **61.** $\{\pm 6i\}$ **63.** $\{\pm \sqrt{5}i\}$ **65.** $\{3 \pm 2\sqrt{2}i\}$

67. $\left\{\dfrac{1}{2} \pm \dfrac{\sqrt{11}}{2}i\right\}$ **69.** $\left\{\dfrac{2}{5} \pm \dfrac{1}{5}i\right\}$ **71.** $\{1 \pm 2i\}$ **73.** $\left\{-1 \pm \dfrac{\sqrt{6}}{2}i\right\}$

Challenge Problems

75. $(x - 3i)(x + 3i)$

PROBLEM SET 9.5

Warm-Ups

1. $\{\pm\sqrt{5}\}$ **2.** $\{\pm\sqrt{6}\,i\}$ **3.** $\{7 \pm \sqrt{5}i\}$ **4.** $\{-3, 1\}$ **5.** $\{8 \pm 2\sqrt{3}i\}$ **6.** $\{\pm2\sqrt{2}\}$

7. $\{\pm2\sqrt{2}\}$ **8.** $\left\{\pm\dfrac{5}{2}\right\}$ **9.** $\{-5, 0\}$ **10.** $\{0, 1\}$ **11.** $\{-2, 5\}$ **12.** $\left\{-\dfrac{1}{2}, 3\right\}$

13. $\left\{-4, -\dfrac{1}{2}\right\}$ **14.** $\left\{\dfrac{-1 \pm \sqrt{5}}{2}\right\}$ **15.** $\left\{\dfrac{1}{2} \pm \dfrac{\sqrt{3}}{2}i\right\}$ **16.** $\{1 \pm \sqrt{6}\}$ **17.** $\{1 \pm \sqrt{11}i\}$

18. $\left\{-\dfrac{1}{4} \pm \dfrac{\sqrt{7}}{4}i\right\}$ **19.** $\left\{\dfrac{1}{3} \pm \dfrac{\sqrt{2}}{3}i\right\}$ **20.** $\left\{\dfrac{3}{2} \pm \dfrac{\sqrt{7}}{2}i\right\}$

Practice Exercises

21. $\{\pm\sqrt{5}\}$ **23.** $\{-7 \pm \sqrt{3}i\}$ **25.** $\{-6, 0\}$ **27.** $\{-7, 4\}$ **29.** $\left\{\dfrac{1}{2} \pm \dfrac{\sqrt{5}}{2}\right\}$ **31.** $\left\{\pm\dfrac{7}{3}\right\}$

33. $\left\{\dfrac{1}{2}, 4\right\}$ **35.** $\{-1 \pm \sqrt{6}\}$ **37.** $\{1 \pm 2\sqrt{6}i\}$ **39.** $\left\{\dfrac{1}{4} \pm \dfrac{\sqrt{11}}{4}i\right\}$ **41.** The walk is 3 m wide.

43. The walk is 2 yd wide. **45.** The bacteria count will be 37,000 after 12 days.

Challenge Problems

47. $t = \dfrac{-v_0 \pm \sqrt{v_0^2 - 4s_0}}{2}$

PROBLEM SET 9.6

Warm-Ups

1.

2.

3.

4.

5. $f(2) = 7; f(-1) = 1; f(0) = 1$ **6.** $f(2) = -3; f(-1) = 9; f(0) = 3$ **7.** $f(2) = -2; f(-1) = -5;$
$f(0) = 0$ **8.** $f(2) = -28; f(-1) = 2; f(0) = -2$

9.

10.

11.

12.

13.

14.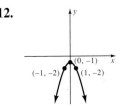

15. (4, 17)   **16.** (−5, −3)   **17.** (−3, 11)   **18.** (2, −9)   **19.** (2,−4)   **20.** (−3, −10)

21.

22.

Practice Exercises

23. $f(0) = 3; f(2) = 11; f(−1) = 2$ **25.** $f(0) = 4; f(2) = 14; f(−1) = 8$
27. $f(0) = −1; f(2) = −5; f(−1) = 7$ **29.** $f(0) = −13; f(2) = 5; f(−1) = −25$
31. (0, −2)   **33.** (1, 5)   **35.** (3, 2)   **37.** (−1, −2)   **39.** (−6, −36)   **41.** (2, −7)

43.

45.

47.

49.

51.

53.

55.

57. (−2, −5)   **59.** (7, −56)   **61.** (−4, −41)

63.

65.

67. 7 wheelchairs

Challenge Problems

69. $\left(-\dfrac{1}{2}, \dfrac{3}{4}\right)$ **71.** $\left(\dfrac{1}{4}, -\dfrac{41}{8}\right)$ **73.** Maximum height, 100 ft; net should be placed 160 ft from the cannon.

REVIEW PROBLEMS CHAPTER 9

1. {3, 7}   **3.** {−6, 0}   **5.** $\left\{-7, -\dfrac{4}{3}\right\}$ **7.** $\left\{0, \dfrac{9}{5}\right\}$ **9.** {±√5}   **11.** {−5, −2}

13. $\{-6, 4\}$ **15.** $\{0, 2\}$ **17.** $\left\{-1, \dfrac{1}{3}\right\}$ **19.** $\{\pm 13\}$ **21.** $\{\pm 7i\}$ **23.** $\{2 \pm \sqrt{7}\}$ **25.** $\{-4, -2\}$

27. $\{-4, 2\}$ **29.** $\left\{-\dfrac{3}{2} \pm \dfrac{\sqrt{21}}{2}\right\}$ **31.** $\left\{-\dfrac{1}{2} \pm \dfrac{\sqrt{13}}{2}\right\}$ **33.** $\{-4, -1\}$ **35.** $\left\{\dfrac{1}{4} \pm \dfrac{\sqrt{7}}{4}i\right\}$

37. $\left\{\dfrac{1}{2}, 1\right\}$ **39.** $\left\{\dfrac{7}{10} \pm \dfrac{\sqrt{29}}{10}\right\}$ **41.** $\{-4, 3\}$ **43.** $\{3, 7\}$ **45.** $\left\{\dfrac{3}{2} \pm \dfrac{\sqrt{17}}{2}\right\}$

47. $\{-8, \pm 2\sqrt{2}i\}$ **49.** $\left\{-\dfrac{1}{2} \pm \dfrac{\sqrt{13}}{2}\right\}$ **51.** $\{0, 3\}$ **53.** $\left\{-\dfrac{2}{3} \pm \dfrac{\sqrt{10}}{3}\right\}$ **55.** $\left\{\dfrac{2}{3}\right\}$

57. $\left\{-1 \pm \dfrac{\sqrt{6}}{3}\right\}$ **59.** $\left\{-\dfrac{7}{3}, \dfrac{4}{5}\right\}$ **61.** $\left\{-\dfrac{1}{2}, \dfrac{7}{3}\right\}$ **63.** $\left\{\dfrac{1}{7}, 1\right\}$ **65.** $\left\{0, \dfrac{5}{3}\right\}$

67. $\left\{-\dfrac{7}{2}, 5\right\}$ **69.** $\left\{-\dfrac{7}{6} \pm \dfrac{\sqrt{13}}{6}\right\}$ **71.** $\{\pm 11\}$ **73.** $\{-1 \pm 2\sqrt{2}\}$ **75.** $\{\pm 12\}$

77. The integers are -18 and -17.   **79.** The dimensions are 12 units by 17 units.   **81.** 2.5 in.
83. 9 ft   **85.** $(-6, -3)$ **87.** $(-2, 3)$ **89.**   **91.**

93.   **95.**   **97.** 9   **99.** 6

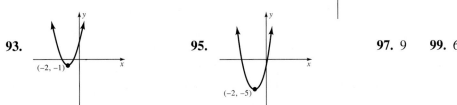

Let's Not Forget . . .

101. -16 **102.** $5 + 2\sqrt{6}$ **103.** $\dfrac{2y^2}{x}$ **104.** -4 **105.** 0   **106.** -3

107. $\sqrt{2} + \sqrt{10} - 5$ **108.** $-\dfrac{2}{x + 3}$

109. Term   **110.** Factor   **111.** Term   **112.** Term   **113.** $-\dfrac{2}{x}$ **114.** $3x - 21$ **115.** $\dfrac{x^3}{7} + x^2$

116. Expression; $\dfrac{x + 5}{(x + 1)(x + 3)}$ **117.** Equation; $\{5\}$ **118.** Inequality; $\{x \mid -3 \le x \le -1\}$

119. Expression; $\dfrac{2x + 6}{x + 1}$ **120.** Equation; $\{2\}$ **121.** $\sqrt{3}\,(y + x)$ **122.** $(x + y)(x^2 - xy + y^2)$
123. Factored   **124.** Factored

CHAPTER 9 TEST

1. (b)   **2.** (c)   **3.** (a)   **4.** (b)   **5.** $\{-3, 7\}$ **6.** $\{5 \pm \sqrt{7}i\}$ **7.** $\{-5, 2\}$ **8.** $\left\{\dfrac{3}{4} \pm \dfrac{\sqrt{7}}{4}i\right\}$

9. $\{\pm 10\}$ **10.** $\left\{\dfrac{1}{2} \pm \dfrac{\sqrt{3}}{2}\right\}$ **11.** $\{0, 6\}$ **12.** $\{1 \pm \sqrt{2}\}$ **13.** $3 - 10i$ **14.** $\dfrac{1}{2} - \dfrac{1}{2}i$
15. The legs are 5 m and 12 m long.

Index

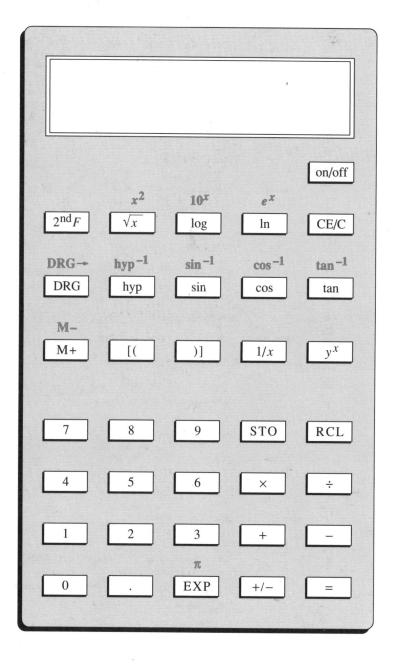